An Introduction to
Partial Differential Equations with MATLAB®

Second Edition

Published Titles

CHAPMAN & HALL/CRC APPLIED MATHEMATICS
AND NONLINEAR SCIENCE SERIES

An Introduction to
Partial Differential
Equations with
MATLAB®
Second Edition

Matthew P. Coleman

Fairfield University
Connecticut, USA

CRC Press
Taylor & Francis Group
Boca Raton London New York

CRC Press is an imprint of the
Taylor & Francis Group, an **informa** business

A CHAPMAN & HALL BOOK

CRC Press
Taylor & Francis Group
6000 Broken Sound Parkway NW, Suite 300
Boca Raton, FL 33487-2742

© 2013 by Taylor & Francis Group, LLC
CRC Press is an imprint of Taylor & Francis Group, an Informa business

No claim to original U.S. Government works

Printed on acid-free paper
Version Date: 20130327

International Standard Book Number-13: 978-1-4398-9846-8 (Hardback)

Library of Congress Cataloging-in-Publication Data

Coleman, Matthew P.
An introduction to partial differential equations with MATLAB / Matthew P. Coleman. -- Second edition.
pages cm -- (Chapman & Hall/CRC applied mathematics & nonlinear science)
Includes bibliographical references and index.
ISBN 978-1-4398-9846-8 (hardback)
1. Differential equations, Partial--Computer-assisted instruction. 2. MATLAB. I. Title.

QA371.35.C66 2013
515'.353028553--dc23   2012050932

Visit the Taylor & Francis Web site at
http://www.taylorandfrancis.com

and the CRC Press Web site at
http://www.crcpress.com

To Mom and Dad

Contents

Preface   xi

Prelude to Chapter 1   1

1   Introduction   3
    1.1   What are Partial Differential Equations? 3
    1.2   PDEs We Can Already Solve 6
    1.3   Initial and Boundary Conditions 10
    1.4   Linear PDEs—Definitions 12
    1.5   Linear PDEs—The Principle of Superposition 16
    1.6   Separation of Variables for Linear, Homogeneous PDEs 19
    1.7   Eigenvalue Problems 25

Prelude to Chapter 2   41

2   The Big Three PDEs   43
    2.1   Second-Order, Linear, Homogeneous PDEs with Constant Co-
        efficients 43
    2.2   The Heat Equation and Diffusion 44
    2.3   The Wave Equation and the Vibrating String 54
    2.4   Initial and Boundary Conditions for the Heat and Wave Equa-
        tions 59
    2.5   Laplace's Equation—The Potential Equation 66
    2.6   Using Separation of Variables to Solve the Big Three PDEs . . 71

Prelude to Chapter 3   77

3   Fourier Series   79
    3.1   Introduction 79
    3.2   Properties of Sine and Cosine 80
    3.3   The Fourier Series 89
    3.4   The Fourier Series, Continued 95
    3.5   The Fourier Series—Proof of Pointwise Convergence 104
    3.6   Fourier Sine and Cosine Series 117
    3.7   Completeness 124

Prelude to Chapter 4   **127**

4 Solving the Big Three PDEs on Finite Domains   **129**
     4.1 Solving the Homogeneous Heat Equation for a Finite Rod . . . 129
     4.2 Solving the Homogeneous Wave Equation for a Finite String . 138
     4.3 Solving the Homogeneous Laplace's Equation on a Rectangular
         Domain 147
     4.4 Nonhomogeneous Problems 153

Prelude to Chapter 5   **161**

5 Characteristics   **163**
     5.1 First-Order PDEs with Constant Coefficients 163
     5.2 First-Order PDEs with Variable Coefficients 174
     5.3 The Infinite String 180
     5.4 Characteristics for Semi-Infinite and Finite String Problems . 192
     5.5 General Second-Order Linear PDEs and Characteristics 201

Prelude to Chapter 6   **211**

6 Integral Transforms   **213**
     6.1 The Laplace Transform for PDEs 213
     6.2 Fourier Sine and Cosine Transforms 220
     6.3 The Fourier Transform 230
     6.4 The Infinite and Semi-Infinite Heat Equations 242
     6.5 Distributions, the Dirac Delta Function and Generalized Fourier
         Transforms 254
     6.6 Proof of the Fourier Integral Formula 266

Prelude to Chapter 7   **275**

7 Special Functions and Orthogonal Polynomials   **277**
     7.1 The Special Functions and Their Differential Equations 277
     7.2 Ordinary Points and Power Series Solutions; Chebyshev, Her-
         mite and Legendre Polynomials 285
     7.3 The Method of Frobenius; Laguerre Polynomials 292
     7.4 Interlude: The Gamma Function 300
     7.5 Bessel Functions 305
     7.6 Recap: A List of Properties of Bessel Functions and Orthogonal
         Polynomials 317

Prelude to Chapter 8   **327**

8 Sturm–Liouville Theory and Generalized Fourier Series   **329**
     8.1 Sturm–Liouville Problems 329
     8.2 Regular and Periodic Sturm–Liouville Problems 337

8.3 Singular Sturm–Liouville Problems; Self-Adjoint Problems . . 345

8.4 The Mean-Square or L^2 Norm and Convergence in the Mean . 354

8.5 Generalized Fourier Series; Parseval's Equality and Completeness 361

Prelude to Chapter 9 373

9 PDEs in Higher Dimensions 375

9.1 PDEs in Higher Dimensions: Examples and Derivations 375

9.2 The Heat and Wave Equations on a Rectangle; Multiple Fourier Series 386

9.3 Laplace's Equation in Polar Coordinates: Poisson's Integral Formula 402

9.4 The Wave and Heat Equations in Polar Coordinates 414

9.5 Problems in Spherical Coordinates 425

9.6 The Infinite Wave Equation and Multiple Fourier Transforms . 439

9.7 Postlude: Eigenvalues and Eigenfunctions of the Laplace Operator; Green's Identities for the Laplacian 456

Prelude to Chapter 10 463

10 Nonhomogeneous Problems and Green's Functions 465

10.1 Green's Functions for ODEs 465

10.2 Green's Function and the Dirac Delta Function 484

10.3 Green's Functions for Elliptic PDEs (I): Poisson's Equation in Two Dimensions 500

10.4 Green's Functions for Elliptic PDEs (II): Poisson's Equation in Three Dimensions; the Helmholtz Equation 516

10.5 Green's Functions for Equations of Evolution 525

Prelude to Chapter 11 537

11 Numerical Methods 539

11.1 Finite Difference Approximations for ODEs 539

11.2 Finite Difference Approximations for PDEs 551

11.3 Spectral Methods and the Finite Element Method 565

A Uniform Convergence; Differentiation and Integration of Fourier Series 579

B Other Important Theorems 585

C Existence and Uniqueness Theorems 591

D A Menagerie of PDEs 601

E MATLAB Code for Figures and Exercises 613

F   Answers to Selected Exercises   **627**

References   **647**

Index   **655**

Preface

Many problems in the physical world can be modeled by partial differential equations, from applications as diverse as the flow of heat, the vibration of a ball, the propagation of sound waves, the diffusion of ink in a glass of water, electric and magnetic fields, the spread of algae along the ocean's surface, the fluctuation in the price of a stock option, and the quantum mechanical behavior of a hydrogen atom. However, as with any area of applied mathematics, the field of PDEs is interesting not only because of its applications, but because it has taken on a mathematical life of its own. The author has written this book with both ideas in mind, in the hope that the student will appreciate the usefulness of the subject and, at the same time, get a glimpse into the beauty of some of the underlying mathematics.

This text is suitable for a two-semester introduction to partial differential equations and Fourier series for students who have had basic courses in multivariable calculus (through Stokes's and the Divergence Theorems) and ordinary differential equations. Over the years, the author has taught much of the material to undergraduate mathematics, physics and engineering students at Penn State and Fairfield Universities, as well as to engineering graduate students at Penn State and mathematics and engineering graduate students at Fairfield. It is assumed that the student has *not* had a course in real analysis. Thus, we treat *pointwise* convergence of Fourier series and do not talk about *mean-square* convergence until Chapter 8 (and, there, in terms of the Riemann, and not the Lebesgue, integral). Further, we feel that it is not appropriate to introduce so subtle an idea as *uniform* convergence in this setting, so we discuss it only in the Appendices.

Approach and Suggestions for Instructor

One may approach the teaching of PDEs in one of two ways: either based on type of equation, or based on method of solution. While appreciating the importance of the former idea, we have chosen the latter approach, as it

1. allows us to treat problems in one spatial dimension before dealing with those in higher dimensions, and

2. allows the text to be used for a one-semester course without the need to jump around.

A typical one-semester course would cover the core Chapters 1–6. Beyond that, one might consider doing Chapter 7, or the beginnings of Chapters 9 and 10, or Chapter 11. Alternatively, if the students already are familiar with special functions, one may wish to cover Chapter 8 or most of Chapter 9.

Motivation

The author believes that it is essential to provide the students with motivation (other than grade) for each of the various topics. We have tried, as far as possible, to provide such motivation, both physical and mathematical (so, for example, the Fourier series is introduced only after the need for it, through solving the heat equation via separation of variables, has been established). Further, we begin by considering PDEs on bounded domains before looking at unbounded domains, because

1. This approach allows us to get to Fourier series early on.

2. Problems on bounded domains are more natural than those on unbounded domains, at least in one dimension.

Further, and in this same vein, we have provided a Prelude to each chapter, the purpose of which is to describe the topics to be covered in the chapter, so as to tell the student what is coming and why it is coming, and to put the material into its historical setting, as well.

Exercises

Of course, mathematics is not a spectator sport, and can only be learned by doing. Thus, it goes without saying that the exercises are a key part of the text. Basically, they are of four types:

1. "solve-the-problem" exercises,

2. proofs,

3. "extend-the-material" exercises,

4. graphical exercises.

Types (1) and (2) are self-explanatory. As for type (3), there are some topics that we choose to present as exercises. In some cases, these will be problems that are similar enough to those already solved in the text. In others, they may involve material which we feel is important, but which is not necessary in later parts of the text. As some of these may be quite difficult, we make sure to lead the student through them when necessary. Alternatively, the instructor may choose to present the material herself in class.

Lastly, as PDEs is such a visual subject, we've provided a number of graphical exercises. Some of these can be done by hand, but the majority are to be performed using MATLAB® (and these are labeled **MATLAB**).

MATLAB

This text has been written so that it can be used without access to software. That said, it makes little sense to write a book on such a visual and intuitive subject as PDEs without taking advantage of one of the multitude of mathematical software packages available these days. We have chosen MATLAB because it is, by far, the most user-friendly of the packages we've tried, because of its excellent graphics capabilities, and because it seems to be the software-of-choice among the engineering community (while making strong inroads in math and physics, as well). While we have used the latest version of MATLAB, 8.0, most, if not all, of the code will run in many of the earlier versions.

This text does not pretend to be an introduction to MATLAB. There are a number of good books available for that purpose (for example, that by Davis listed in the Bibliography). *For those wishing to use the MATLAB exercises, we assume that the student is familiar with the rudiments of the package—how to get it up-and-running, how M-files work, etc.* What we have done is to use MATLAB to generate the tables and the more "mathematical" figures in the book, for which we've supplied the MATLAB code in Appendix E, and also on the author's website at

www.faculty.fairfield.edu/mcoleman

The exercises labeled **MATLAB**, then, are to be done using this code, with slight variations provided by the student (for example, changing the input function, the viewing window, the number of steps in a *for loop* and the like). Little actual programming is required of the student.

Acknowledgments

It has been a pleasure working again with Bob Stern, my editor at Chapman and Hall/CRC Press, and I thank him for his continued support, encouragement and patience. Thanks also to Robert Ross and to my ever-patient project editor, Michele Dimont.

I am grateful to a number of people for their suggestions for this second edition, including L. Benjauthrit, U.E. George, A. Madzvamuse, R. Medina, K.A. Nguyen, C. Wdowski and G. Zahedi. I must give special thanks to my colleague Mark Demers, whose many thoughtful comments and suggestions have, to my view, led to a much improved second edition, and to Michelle Ghrist and her PDE class at the U.S. Air Force Academy, for their very thorough list of suggested changes. In addition, I must not forget all of the students I have taught using the first edition of this book at Fairfield U.

I must thank Robin Campbell yet again for her excellent work in typing the manuscript and providing many of the figures, and Artem Aleksenko who, building on the first edition work of Mike Capriotti, updated the MATLAB code for this new edition.

Finally, I would like to give special thanks to Yuet-Ying Yu.

Of course, any errors are the sole responsibility of the author.

Matthew P. Coleman
mcoleman@fairfield.edu
March 1, 2013

MATLAB and Simulink are registered trademarks of The MathWorks, Inc. For production information, please contact:

The MathWorks, Inc.
3 Apple Hill Drive
Natick, MA 01760-2098 USA
Tel: 508-647-7000
Fax: 508-647-7001
Email: info@mathworks.com
Web: www.mathworks.com

Prelude to Chapter 1

We have seen how physical problems often give rise to ordinary differential equations (henceforth, ODEs). These same and similar physical problems, when involving more than one independent variable, lead us to, instead, *partial* differential equations (PDEs). A PDE, therefore, will look very much like an ODE, except that the unknown function will be a function of *several* variables - and, of course, any derivatives that appear must be *partial* derivatives. Although we shall find a number of PDEs which are solved in the same way that we solved ODEs, this happy state of affairs will be short-lived. Indeed, two- and higher-dimensional mathematical objects exhibit a wealth of behavior which we do not see in one-dimensional objects. Similarly, PDEs, as a rule, will exhibit much more complicated behavior and, therefore, be much harder to solve than ODEs.

In this first chapter, we introduce PDEs, and we point out those which can be solved like ODEs. Historically, many of these simpler PDEs were overlooked by earlier mathematicians, simply because they weren't interesting (they already knew how to solve ODEs) or important (the really interesting physical problems led to PDEs which could not be solved in this manner). So, in the 18th century, we see famous mathematicians jumping right into the more difficult equations, those which we will begin to discuss in Chapter 2.

We also treat in this chapter the PDE analogs of other ideas that were studied in ODEs: initial and boundary conditions, and the important concept of a linear PDE. We then introduce one of the most important tools for solving linear PDEs, the method of separation of variables. The so-called product solutions which are derived via separation of variables were studied as early as the first half of the 18th century, first by Daniel Bernoulli (1700–1782, son of John Bernoulli) in his study of the vibrating chain, then by the great Leonhard Euler (1707–1783) and, most notably, by Jean Le Rond d'Alembert (1717–1783), in their work on the wave equation (which models the vibrations of a string).

Finally, we turn back to ODEs and look at the so-called eigenvalue problems which arise when we apply separation of variables to PDEs. At this point, we consider only simpler special cases of this type of problem, reserving a more complete study for Chapter 8.

1

Introduction

1.1 What are Partial Differential Equations?

Roughly speaking, a *partial differential equation (PDE)* is similar to an *ordinary differential equation (ODE)*, except that the dependent variable is a function of not just one, but of several independent variables. Let's be more precise. Given a function $u = u(x_1, x_2, \ldots, x_n)$, a **partial differential equation (PDE)** in u is an equation which relates any of the partial derivatives of u to each other and/or to any of the variables $x_1, x_2, \ldots, x_n$ and u.

Before doing some examples, we introduce a bit of notation: Instead of the somewhat unwieldy $\frac{\partial u}{\partial x}, \frac{\partial^3 u}{\partial x^2 \partial y}$ and the like, we will use subscripts whenever possible. We write

$$u_x = \frac{\partial u}{\partial x}.$$

For higher order derivatives, we read the subscripts from left to right. So, for example,

$$u_{xy} = \frac{\partial}{\partial y}\left(\frac{\partial u}{\partial x}\right) = \frac{\partial^2 u}{\partial y \partial x}.$$

However, for all practical purposes, the order of differentiation will not matter to us. So, for example, we'll have

$$u_{xzyx} = u_{zxxy} = u_{yxzx}, \text{ etc.}$$

Examples

1. $u_x + u = 0$ is a PDE in $u = u(x, y)$. However, it also could be a PDE in $u = u(x, y, z)$. In general, we only know the number of independent variables from the context.

2. $2u_x + 3u_z = 0$ is a PDE in $u(x, z)$, although, more likely, it is a PDE in $u(x, y, z)$.

3. $u_x u_{yy} - xy^3 u = e^u$ is a PDE in $u(x, y)$.

4. $z^2 u_{xxy} - x \cos y \, u_{yy} + u_y - e^{3z} u = \tan y^2 z$ is a PDE in $u(x, y, z)$.

5. $u_{xxz}^5 + u_{xxyz} = u_{zzz}$ is a PDE in $u(x, y, z)$.

Very important in the categorization of PDEs, as it is for ODEs, is a PDE's *order*. We define the **order** of a PDE to be the order of the highest derivative which appears in the equation. So, for example, the orders of the PDEs in Examples 1–5 are, respectively, 1, 1, 2, 3 and 4.

As with ODEs, the dependent variable in a PDE generally is *unknown* and we wish to *solve* for it. A **solution** of a PDE, then, is any function u which satisfies the PDE identically, that is, for all possible values of the independent variables.

Examples

6. $u(x, y) = e^{-x}, u(x, y) = 5e^{-x}, u(x, y) = ye^{-x}$ and $u(x, y) = y^3 \cos y \ e^{-x}$ all are solutions of $u_x + u = 0$, since, in each case, $u_x = -u$.

7. $u(x, y, z) = y^2$ and $u(x, y, z) = y^5 \cos y$ are both solutions of $2u_x + 3u_z = 0$. In fact, so is $u(x, y, z) = f(y)$ for any (well *almost* any—see below) function f. Also, $u(x, y, z) = e^{3x-2z}$ and $u(x, y, z) = ye^{3x-2z}$ are solutions.

8. $u(x, y) = cx + d$ is a solution of

$$u_{xxz}^5 + u_{xxyz} = u_{zzz}$$

for any choice of the constants c and d.

Frequently, we will seek functions which are solutions of a given PDE in some restricted region. Also, in order to ensure that there is never a problem with the order of differentiation, *we will require any solution u of an nth-order PDE to have the property that all of the nth partial derivatives of u exist and are continuous.*

We conclude this section with a list of important PDEs which arise from physical problems (most of which we will study in some detail):

$u_t + cu_x = 0$	**convection** (or **advection** or **transport**) equation
$u_t + uu_x = 0$	**Burger's equation** (from the study of the dynamics of gases)
$u_x^2 + u_y^2 = 1$	**eikonal equation** (from optics)
$u_t = \alpha^2 u_{xx}$	**heat equation** (in one space variable)
$u_{tt} = c^2(u_{xx} + u_{yy})$	**wave equation** (in two space variables)
$u_{xx} + u_{yy} + u_{zz}$ $+ [E - V(x, y, z)]u = 0$	**Schrödinger's equation** (time independent, in three space variables; from quantum mechanics)
$u_{rr} + \frac{1}{r}u_r + \frac{1}{r^2}u_{\theta\theta} = 0$	**Laplace's equation** (in polar coordinates)
$u_{tt} + \alpha^4 u_{xxxx} = 0$	**Euler–Bernoulli beam equation**

In Appendix D we provide a complete list of the PDEs covered in this text, along with many other important PDEs, organized by application.

Exercises 1.1

1. Find the order of each PDE:

    a) The *convection* or *advection equation*, $u_t + cu_x = 0$

    b) The *wave equation*, $u_{tt} = c^2 u_{xx}$

    c) The *eikonal equation*, $u_x^2 + u_y^2 = 1$

    d) The *Euler–Bernoulli beam equation*, $u_{tt} + a^4 u_{xxxx} = 0$

    e) $u_{xxyyz} - u^8 + u_{xx}^6 = 0$

2. Show that each given function is a solution of the corresponding PDE:

    a) $u = x^2 y,\ xu_x - 2yu_y = 0$

    b) $u = x \sin y,\ u_{xx} - u_{yy} = u$

    c) $u = yf(x),\ u_{yy} = 0$ (where f is any function with continuous second derivative)

    d) $u = e^{x+2y} + e^{x-2y},\ 4u_{xx} - u_{yy} = 0$

    e) $u = e^x \cos y + ax + by$, Laplace's equation in two dimensions in rectangular coordinates $u_{xx} + u_{yy} = 0$ (where a and b are any constants)

    f) $u = xyz,\ 2xu_x - yu_y - zu_z = 0$

    g) $u = x^2 y^3 z^2 - xz^3,\ 3x^2 u_{xx} + 2yu_y + 2xy^3 z^2 u_{zzz} = 0$

3. Consider the *convection equation* $u_t + cu_x = 0$, where c is a constant.

    a) Show that $u = \sin(x - ct)$, $u = \cos(x - ct)$ and $u = 5(x - ct)^2$ are solutions.

    b) Show that $u = 7\sin(x - ct)$, $u = 3\cos(x - ct)$ and $u = 7\sin(x - ct) - 3\cos(x - ct)$ also are solutions.

    c) Show that $u = f(x - ct)$ is a solution for "any" function f.

    d) Why is "any" in quotation marks in part (c)?

4. Consider the *one-dimensional wave equation* $u_{tt} = c^2 u_{xx}$, where c is a constant.

    a) Show that all of the functions in Exercise 3 are solutions of this equation, as well.

    b) Show that $u = g(x + ct)$ also satisfies the wave equation for "any" function g.

5. Consider the *eikonal equation* $u_x^2 + u_y^2 = 1$.

    a) Show that $u = x$ and $u = y$ are solutions.

    b) Are $u = 3x$ and $u = -4y$ solutions?

   c) Is $u = x + y$ a solution?

   d) Find all solutions of the form $u = ax + by$, where a and b are constants.

6. Consider the simple first-order PDE $u_x = 0$, where $u = u(x, y)$.

   a) Find all solutions. (Compare this problem with that of finding all solutions of the *ODE* $\frac{dy}{dx} = 0$.) Describe them (compare to Exercise 3).

   b) Describe the set of solutions which satisfy the additional requirement that $u(0, 0) = 0$. How many are there?

   c) Do the same, but for the requirement $u(0, y) = y^2 - \cos y$.

   d) Do the same, but for $u(x, 0) = x^3$.

1.2 PDEs We Can Already Solve

Let's go back and look at Exercise 6 of the previous section. However, first, remember how we would solve the ODE

$$\frac{dy}{dx} = 0.$$

We integrated both sides to get $y = $ constant (after having proved in calculus that $y = $ constant is the only function whose derivative is identically zero).

We can do the same with PDEs—except that we must remember that the derivatives are *partial* derivatives, so any antiderivatives we take will be, in a sense, "partial antiderivatives" or "partial integrals." That is, we "anti-differentiate" with respect to one variable while treating the other variables as constants.

So for the PDE $u_x = 0$, any function which is independent of x will be a solution. (Further, similarly to above, these will be the *only* solutions.) To be more precise, in order to find all functions $u = u(x, y)$ which solve

$$u_x = 0, \tag{1.1}$$

we get

$$u = \int 0 \, dx = f(y) \tag{1.2}$$

where f is any* arbitrary function of y (and where $\int \ldots dx$ is, as we've mentioned, any antiderivative with respect to x while treating y as a constant).

---

*Again, from our definition of solution, f' must be continuous.

Since

$$u = f(y) \tag{1.3}$$

represents all possible solutions of (1.1), we call (1.2) the **general solution** of (1.1). So, where the general solution of an ODE involves arbitrary constants, the general solution of a PDE involves **arbitrary functions**.

With these ideas in mind, there already are *plenty* of PDEs we can solve.

Example 1 Find all solutions $u = (x, y)$ of $u_x = x^2 + y^2$. We have

$$u = \int (x^2 + y^2) dx = \frac{x^3}{3} + xy^2 + f(y),$$

where $f(y)$ is an arbitrary function of y.

Example 2 Find the general solution of $u_y = xz + yz$. We have

$$u = \int (xz + yz) dy = xyz + \frac{y^2 z}{2} + f(x, z),$$

where f is an arbitrary function *of x and z*.

We need not restrict ourselves to equations of the first order.

Example 3 Find the general solution of $u_{xx} = 12xy$. We integrate twice, of course:

$$u_x = \int 12xy \, dx = 6x^2 y + f(y),$$

where f is an arbitrary function of y, then

$$u = \int (6x^2 y + f(y)) dx = 2x^3 y + xf(y) + g(y),$$

where g is an arbitrary function of y.

Example 4 Do the same for $u_{xy} = \cos x$. First, we have

$$u_x = \int \cos x \, dy = y \cos x + f(x).$$

Then,

$$u = \int (y \cos x + f(x)) dx.$$

Now, what is $\int f(x) dx$? If we antidifferentiate $f(x)$ with respect to x, we just get another function of x. However, we also get an "arbitrary constant," that is, in this case, an arbitrary function of y. So

$$\int f(x) dx = f_1(x) + g(y)$$

and our general solution is

$$u = y \sin x + f_1(x) + g(y).$$

Finally, since f, f_1 and g are arbitrary, we drop the subscript:

$$u = y \sin x + f(x) + g(y),$$

where f and g are arbitrary functions.

Example 5 Find the general solution of $u_x + 2u = y$. Remember that we would solve the *ODE* $\frac{dy}{dx} + 2y = 5$ by using the integrating factor e^{2x}. We may do the same here:

$$u_x + 2u = y$$

$$\Rightarrow \quad \frac{\partial}{\partial x}(e^{2x}u) = ye^{2x}$$

$$\Rightarrow \quad e^{2x}u = \frac{1}{2}ye^{2x} + f(y)$$

or

$$u = \frac{1}{2}y + e^{-2x}f(y)$$

for arbitrary f.

Example 6 Find the general solution of $u_{xxz} = x + y - z$. We have

$$u_{xx} = xz + yz - \frac{z^2}{2} + f(x, y).$$

Then,

$$u_x = \frac{x^2 z}{2} + xyz - \frac{xz^2}{2} + f_1(x, y) + g(y, z)$$

and

$$u = \frac{x^3 z}{6} + \frac{x^2 yz}{2} - \frac{x^2 z^2}{4} + f_2(x, y) + xg(y, z) + h(y, z).$$

Example 7 There are *many* other types of PDEs that we may solve at this point. For example, the PDE $u_{yy} + u = 0$ looks like the ODE $y'' + y = 0$. Since the latter has solution $y = c_1 \cos x + c_2 \sin x$, the former will have general solution

$$u = f(x) \cos y + g(x) \sin y$$

for arbitrary f and g. Similarly, the PDE $u^2 u_x = x$ will behave like the *separable ODE* $y^2 \frac{dy}{dx} = x$, which has solution $y = \sqrt[3]{\frac{3}{2}x^2 + c}$. Therefore, the PDE's solution is

$$u = \sqrt[3]{\frac{3}{2}x^2 + f(y)}$$

for arbitrary f. (However, we must remember that, while we may multiply and divide by the differentials dx and dy in the ODE, this generally is *not* true of the "numerator" and "denominator" in the *partial* derivative $\frac{\partial u}{\partial x}$.)

Exercises 1.2

Find the general solution of each PDE. The solution u is a function of the variables which appear, unless otherwise stated.

1. $u_y = 2x$

2. $u_x = \sin x + \cos y$

3. $u_x = \sin x + \cos y$, $u = u(x, y, z)$

4. $u_{yy} = x^2 y$

5. $u_{xy} = x - y$

6. $u_{xxy} = 0$

7. $u_{xxyy} = \sin 2x$

8. $u_{xzz} = x - yz + y^3$

9. $u_{xyzz} = 0$

10. $u_{xyyzz} = xyz$

11. $u_y - 4u = 0$, $u = u(x, y)$

12. $u_x + 3u = e^x$, $u = u(x, y)$

13. $u_x - y^2 u = 0$

14. $u_x + 3u = xy^2 + y$

15. $u_y + xu = 2$

16. $u_x - zu = y - z$

17. $u_{xx} + u_x - 2u = 0$, $u = u(x, y)$

18. Find all solutions of the PDE $u_y = 2x$ which also satisfy the additional requirement that

    a) $u(x, 0) = \sin x$
    b) $u(x, 3) = \sin x$
    c) $u(0, y) = 3y$

19. Find all solutions $u(x, y)$ of the PDE $u_x - 2u = 0$ which also satisfy the additional requirement that

    a) $u(0, y) = y^2$
    b) $u(1, y) = y^2$
    c) $u(x, 1) = x^2$

1.3   Initial and Boundary Conditions

In some of the exercises in the previous sections, we were asked to solve a PDE and then to find the subset of those solutions which also satisfied an additional requirement. These **side conditions** are part and parcel of the study of PDEs. As these conditions arise naturally in physical settings, let's introduce them by way of a specific physical problem.

In Section 1.1 we mentioned the **heat equation**

$$u_t = \alpha^2 u_{xx}. \tag{1.4}$$

Here, α is a constant and $u = u(x,t)$ represents the temperature at any point x along a narrow piece of material, at any time t. (See Figure 1.1—we will have *much* more to say about this equation in Chapter 2 and beyond.) We are asked to find the temperature function, that is, to solve the PDE. Now, as we would like to predict the temperature of a particular piece of material, we would like to find the *one* solution of the PDE that does so. Certainly, there must be some additional requirements at our disposal to narrow down the general solution to one, unique solution.

$x{-}axis$

FIGURE 1.1
Metal rod; $u(x,t)$ = temperature at point x, at time t.

First, it seems fairly clear that we cannot know the temperature at later times if we don't know the temperature *now* or, at least, at some definite point in time. So we should hope that we are given, or can measure, the so-called *initial* temperature of the material at each point x, at some specified time $t = t_0$. That is, we would like to be given the function f for which

$$u(x,t_0) = f(x), \qquad 0 \le x \le L. \tag{1.5}$$

We call this an **initial condition**. In practice, the initial time generally is taken, or arranged, to be $t_0 = 0$.

What additional requirements will we need? Well, it will turn out that PDE (1.4) is derived under the assumption that the whole piece of material is insulated except, possibly, at its ends, and that the heat "flows" only in the x-direction. Therefore, it seems that we will need to know what is going on at the endpoints. In fact, the endpoints generally are under the control of the experimenters—so, for example, the left end may be held at a constant temperature of u_0 degrees, that is,

$$u(0,t) = u_0, \qquad t > 0. \tag{1.6}$$

Alternatively, the right end may be insulated. We will see that, mathematically, this means that

$$u_x(L,t) = 0, \qquad t > 0. \tag{1.7}$$

Equations (1.6) and (1.7) are called **boundary conditions**, and a system like the one consisting of PDE (1.4), subject to conditions (1.5), (1.6) and (1.7), is called an **initial-boundary-value problem**.

As for simpler equations, in Exercises 18 and 19 in the previous section we were asked to solve a first-order PDE subject to only one side condition. In practice, one of the variables often will represent time, so the side condition will be an initial condition, and the problem will be an **initial-value problem**. (In fact, when treating first-order PDEs in Chapter 5, we will *always* refer to the side condition as an initial condition and the system as an initial-value problem.)

Now, it turns out that the initial-boundary-value problem (1.4), (1.5), (1.6), (1.7) has a unique solution. We call such a problem a **well-posed** problem. Similarly, the problems in Exercises 6c of 1.1, and Exercises 18a, 18b, 19a and 19b of 1.2 all are well-posed. Those in Exercises 6b and 6d of 1.1 and 18c and 19c of 1.2 are *not* well-posed.

To be precise, an initial-value or initial-boundary-value problem is well-posed if

1) A solution to it *exists*.

2) There is only *one* such solution (i.e., the solution is *unique*).

3) The problem is *stable*.[†]

Property (3), the *stability condition*, need not concern us. (Most, but not all, of the problems considered in this book will be stable.)

By the way, remember from ODEs that, if an equation is of order n, we generally need n initial conditions to determine a unique solution. For PDEs, the situation is much more complicated. However, notice that our heat equation example has one time derivative and one initial condition, while it has two x-derivatives and two x-boundary conditions. This often is the case. So, for example, in order that the *finite* vibrating string problem be well-posed, we will require two initial and two boundary conditions.

Exercises 1.3

The idea of well-posedness applies to ODEs, as well. Again, remember that an nth-order linear ODE, with n conditions assigned at the same x-value— *initial conditions*—"usually" has a unique solution. This may *not* be the case,

---

[†]Basically a problem is stable if, whenever we change the initial or boundary conditions by a "little bit," the solution also changes by only a little bit—where we can, of course, quantify what we mean by a *little bit*.

however, for the ODE problems in Exercises 1–4. Each of these problems is called a **boundary-value problem**, and we will study these problems in detail in Section 1.7. For now, decide whether each of these problems is well-posed, in terms of existence and uniqueness of solutions.

1. $y'' + y = 0, y(0) = y(2) = 0, 0 \leq x \leq 2$

2. $y'' + y = 0, y(0) = y(\pi) = 0, 0 \leq x \leq \pi$

3. $y'' + y' - 2y = 0, y(0) = 0, y'(1) = 0, 0 \leq x \leq 1$

4. $y'' + 25y = 0, y(0) = 1, y(\pi) = -1, 0 \leq x \leq \pi$

5. For which values of the constant L is the following boundary-value problem well-posed?

$$y'' + 4y = 8x, y(0) = A, y(L) = B, \qquad 0 \leq x \leq L$$

Explain why each of the following problems is not well-posed.

6. $u_{xx} = 0, u(0, y) = y^2, u(1, y) = 3y, u(x, 0) = x + 2, x \geq 0, 0 \leq y \leq 1$

7. $u_{xx} + u_{yy} = 0, u_x(0, y) = u_x(1, y) = u_y(x, 0) = u_y(x, 2) = 0, 0 \leq x \leq 1, 0 \leq y \leq 2$

1.4   Linear PDEs—Definitions

Almost every PDE which we have met so far is what is called a **linear** PDE, which is defined in exactly the same manner as a linear ODE. Remember that the latter was any ODE which could be written in the form

$$a_0(x)y^{(n)} + a_1(x)y^{(n-1)} + \cdots + a_{n-1}(x)y' + a_n(x)y = f(x),$$

where $y = y(x)$ and $y^{(k)} = \frac{d^k y}{dx^k}$. However, a more fruitful way of looking at it is to define the so-called **operator,**[‡] L, by

$$L[y] = a_0(x)y^{(n)} + a_1(x)y^{(n-1)} + \cdots + a_{n-1}(x)y' + a_n(x)y.$$

It is then easy to show that, if c is any constant and y any function in the domain of L, then

$$L[cy] = cL[y],$$

[‡]An operator is a "function of functions," as it were. That is, it is a function which has the property that its domain and range each consists of a certain class of functions.

and that, if y_1 and y_2 are any functions in the domain of L, then

$$L[y_1 + y_2] = L[y_1] + L[y_2].$$

We use the idea of an operator to define linear PDEs. First, given a PDE in $u = u(x_1, x_2, \ldots, x_n)$, we write the equation in the form

$$L[u] = f(x_1, x_2, \ldots, x_n),$$

where f is a given function.

Example 1 The heat equation, $u_t = \alpha^2 u_{xx}$, can be written as

$$L[u] = 0,$$

where L is the operator defined by

$$L[u] = u_t - \alpha^2 u_{xx}.$$

Example 2 The PDE $u_x + yu_y - xy^2 + \sin y = 0$ can be written as

$$L[u] = xy^2 - \sin y,$$

where L is defined by

$$L[u] = u_x + yu_y.$$

Then, we define a linear PDE as follows:

Definition 1.1 *The PDE*

$$L[u] = f$$

is a **linear PDE** *if*

1) $L[cu] = cL[u],$ $\hspace{4cm}$ (1.8)
    for all constants c and all functions u in the domain of L, and

2) $L[u_1 + u_2] = L[u_1] + L[u_2],$ $\hspace{3cm}$ (1.9)
    for all functions u_1 and u_2 in the domain of L.

Also, if an operator satisfies both (1.8) and (1.9), we say that it is a **linear operator***. If an operator or PDE fails to be linear, we call it a* **nonlinear operator** *or PDE (and we do not call the operator L—for "linear"—if we know that it is nonlinear).*

We will prove, in Exercise 8, that L is linear if and only if

$$L[c_1 u_1 + c_2 u_2] = c_1 L[u_1] + c_2 L[u_2], \tag{1.10}$$

$$\text{for all constants } c_1 \text{ and } c_2, \text{ and all functions}$$

$$u_1 \text{ and } u_2 \text{ in the domain of } L.$$

We note that any discussion of linearity of PDEs is based upon the theorems from calculus that tell us that first partial derivatives are linear (that is, that $\frac{\partial}{\partial x}(cu) = cu_x$ and $\frac{\partial}{\partial x}(u_1 + u_2) = u_{1_x} + u_{2_x}$), from which it also follows that all higher-order partial derivatives are linear (see Exercise 9).

In Examples 3–5, determine if the given PDE is linear or nonlinear.

Example 3 $u_x + 5u = x^2 y$. The operator is $L[u] = u_x + 5u$ and we have

$$\begin{aligned}
L[c_1 u_1 + c_2 u_2] &= (c_1 u_1 + c_2 u_2)_x + 5(c_1 u_1 + c_2 u_2) \\
&= c_1 u_{1_x} + c_2 u_{2_x} + 5c_1 u_1 + 5c_2 u_2 \\
&= c_1(u_{1_x} + 5u_1) + c_2(u_{2_x} + 5u_2) \\
&= c_1 L[u_1] + c_2 L[u_2]
\end{aligned}$$

and L is linear, so the PDE is linear, as well.

Example 4 The eikonal equation, $u_x^2 + u_y^2 = 1$. We have $L[u] = u_x^2 + u_y^2$. Consider, then, $L[cu]$:

$$\begin{aligned}
L[cu] &= (cu)_x^2 + (cu_y)^2 \\
&= c^2 u_x^2 + c^2 u_y^2.
\end{aligned}$$

The question is, do we have $L[cu] = cL[u]$ for *all* constants c and *all* functions u (in the domain of L)? That is, is

$$c^2 u_x^2 + c^2 u_y^2 = cu_x^2 + cu_y^2?$$

Certainly, the answer is *no*. To be more precise, the equation *may* be true for certain constants c and/or functions u, but we need only find *one* counterexample, that is, *one* case involving a particular c and a particular u for which the equality doesn't hold (e.g., try $c = 2$ and $u = x$). Therefore, the PDE is nonlinear.

Example 5 $y^2 u_{xx} + u_{yy} = 1$. Here, $L[u] = y^2 u_{xx} + u_{yy}$ and

$$\begin{aligned}
L[c_1 u_1 + c_2 u_2] &= y^2(c_1 u_1 + c_2 u_2)_{xx} + (c_1 u_1 + c_2 u_2)_{yy} \\
&= c_1 y^2 u_{1_{xx}} + c_2 y^2 u_{2_{xx}} + c_1 u_{1_{yy}} + c_2 u_{2_{yy}} \\
&= c_1(y^2 u_{1_{xx}} + u_{1_{yy}}) + c_2(y^2 u_{2_{xx}} + u_{2_{yy}}) \\
&= c_1 L[u_1] + c_2 L[u_2],
\end{aligned}$$

so this PDE is linear.

In practice, the things that make ODEs nonlinear also make PDEs nonlinear, for example, powers of u and its derivatives $(\sqrt{u}, \frac{1}{u_x}, u_{yy}^3, \ldots)$, products involving u and its derivatives $(u_x u_y, u u_{xxy}, \ldots)$, various functions of u and its derivatives $(e^u, \cos u_x, \ldots)$ and the like.

As with linear ODEs, we distinguish between homogeneous and nonhomogeneous equations.

Definition 1.2 *Given the* linear *PDE* $L[u] = f$, *if* $f \equiv 0$ *on some region (that is,* f *is the* zero-function *on some region), we say that the PDE is* **homogeneous** *on that region. Otherwise, the PDE is* **nonhomogeneous**.

Example 6 The PDE $x u_{xx} - 5 u_{xy} + y^2 u_x = 0$ is homogeneous (on the x-y plane).

Example 7 The PDE $u_x + 5u = x^2 y$ is nonhomogeneous (on the x-y plane).

Example 8 $u_x = \begin{cases} 1, \text{ if } x<0 \text{ or } y<0 \\ 0, \text{ otherwise} \end{cases}$ is nonhomogeneous on the x-y plane, but it is *homogeneous* on the first quadrant.

Example 9 $u_x^2 + u_y^2 = 0$ cannot be said to be homogeneous or nonhomogeneous, because it is not a *linear* PDE to start with.

Exercises 1.4

In Exercises 1–7, determine whether the PDE is linear or nonlinear, and prove your result. If it is linear, decide if it is homogeneous or nonhomogeneous. If it is nonlinear, point out the term or terms which make it nonlinear.

1. Burger's equation, $u_t + u u_x = 0$

2. $u_{xxy} - (\sin x) u_{yy} + x - y = 0$

3. $2 u_y - 5 u^3 = x$

4. $u_{xx} = \sin u$

5. The *three-dimensional heat equation*, $u_t = \alpha^2 (u_{xx} + u_{yy} + u_{zz})$, where α^2 is a constant.

6. Poisson's equation is two dimensions (in polar coordinates),

$$u_{rr} + \frac{1}{r} u_r + \frac{1}{r^2} u_{\theta\theta} = f(r, \theta).$$

7. $\sqrt{1 + x^2 y^2} u_{xyy} - \cos(xy^3) u_{xxy} + e^{-y^3} u_x - (5x^2 - 2xy + 3y^2) u = 0$

8. Prove that the operator $L[u]$ is linear if and only if it satisfies property (1.10), that is, prove that $L[u]$ satisfies properties (1.8) and (1.9) if and only if it satisfies property (1.10).

9. We know from calculus (and from Exercise 8) that $\frac{\partial}{\partial x}(c_1 u_1 + c_2 u_2) = c_1 u_{1_x} + c_2 u_{2_x}$, for all constants c_1 and c_2 and all differentiable functions u (and that the same is true not only for x but, of course, for any independent variable).

   a) Use this fact to prove that the following higher-order derivatives are linear operators, as well.

      i) $L[u] = u_{yy}$

      ii) $L[u] = u_{xxy}$

   b) Use mathematical induction to prove that the operator $L[u] = \frac{\partial^n u}{\partial x^n} = u_{\underbrace{xx\cdots x}_{n\ \text{times}}}$ is linear.

10. Prove that, if u_1 and u_2 are solutions of the homogeneous PDE $L[u] = 0$, then so is the function $c_1 u_1 + c_2 u_2$, for any choice of the constants c_1 and c_2. Is this true for nonhomogeneous PDEs, as well?

11. If u_1 and u_2 are solutions of the nonhomogeneous equation $L[u] = f$, what can we say about the function $u_1 - u_2$?

12. Use mathematical induction to prove that, if L is linear,

$$L[c_1 u_1 + c_2 u_2 + \cdots + c_n u_n] = c_1 L[u_1] + c_2 L[u_2] + \cdots + c_n L[u_n]$$

for all constants $c_1, c_2, \ldots, c_n$ and all functions $u_1, u_2, \ldots, u_n$ in the domain of L.

1.5   Linear PDEs—The Principle of Superposition

Here, again, we take our cue from the theory of linear ODEs.

Definition 1.3 *Given functions $u_1, u_2, \ldots, u_n$, any function of the form*

$$c_1 u_1 + c_2 u_2 + \cdots + c_n u_n,$$

where $c_1, c_2, \ldots, c_n$ are constants, is called a **linear combination** *of $u_1, u_2, \ldots, u_n$.*

The following theorem follows immediately from the result of Exercise 12 of the previous section.

Theorem 1.1 *If $u_1, u_2, \ldots, u_n$ are solutions of the linear, homogeneous PDE $L[u] = 0$, then so is any linear combination of $u_1, u_2, \ldots, u_n$.* (*This is the* **principle of superposition of solutions** *for linear PDEs.*)

PROOF The fact that $u_1, u_2, \ldots, u_n$ are solutions gives us

$$L[u_1] = L[u_2] = \cdots = L[u_n] = 0.$$

Then, for any linear combination $c_1 u_1 + c_2 u_2 + \cdots + c_n u_n$,

$$L[c_1 u_1 + c_2 u_2 + \cdots + c_n u_n] = c_1 L[u_1] + c_2 L[u_2] + \cdots + c_n L[u_n]$$
$$= c_1 \cdot 0 + c_2 \cdot 0 + \cdots + c_n \cdot 0 = 0.$$

∎

Now, in the theory of ODEs, for an nth-order linear, homogeneous equation, we need only find n linearly independent solutions. Then, the general solution consists of all possible (finite) linear combinations of these solutions. However, life is much more complicated in the realm of PDEs. Often, we will need to find *infinitely* many solutions, $u_1, u_2, \ldots$, of a linear, homogeneous PDE before we are in a position to construct a general solution

$$u = c_1 u_1 + c_2 u_2 + \cdots = \sum_{n=1}^{\infty} c_n u_n. \tag{1.11}$$

And since this infinite linear combination actually is an infinite series, questions of convergence come to the forefront. Indeed, for any given choice of the coefficients, expression (1.10) may diverge for all values of x, or it may converge for some values of x but not for others.

Suffice it to say that, throughout this book, we will assume that, whenever (1.11) converges, it satisfies the linearity condition

$$L\left[\sum_{n=1}^{\infty} c_n u_n \right] = \sum_{n=1}^{\infty} c_n L[u_n] \tag{1.12}$$

and, therefore, that if each u_n is a solution of $L[u] = 0$, then so is the linear combination, (1.11), of these solutions. When (1.12) holds, we say that we may *differentiate the series term-by-term.*

Exercises 1.5

In Exercises 1–4, verify directly that the principle of superposition holds for any two solutions, u_1 and u_2, of the given PDE.

1. $y u_x - x^2 u_y + 2u = 0$

2. The heat equation in two space variables, $u_t = \alpha^2(u_{xx} + u_{yy})$

3. Laplace's equation in three space variables, $u_{xx} + u_{yy} + u_{zz} = 0$

4. The wave equation in three space variables,

$$u_{tt} + c^2(u_{xx} + u_{yy} + u_{zz}) = 0$$

5. Use (1.12) to show that the function $u(x,t) = \sum\limits_{n=1}^{\infty} c_n e^{-n^2 t} \sin nx$ is a solution of the heat equation $u_t = u_{xx}$ (whenever the series converges, of course).

6. Show directly that the principle of superposition does *not* hold for the PDE $u_x + u^2 = 0$, $u = u(x,y)$, by finding two different solutions, then finding a linear combination of them that is *not* a solution.

7. We may also prove theorems for solutions of *non*homogeneous PDEs that are analogous to those for ODEs. Prove that the general solution of the nonhomogeneous PDE $L[u] = f$ is $u = u_h + u_p$, where u_p is any one particular solution of $L[u] = f$, and u_h is the general solution of the **associated homogeneous PDE** $L[u] = 0$, as follows:

    a) First, prove that $u_h + u_p$ always is a solution of $L[u] = f$.

    b) Next, prove that, if u is any particular solution of $L[u] = f$, then we can always write

$$u = u_{h'} + u_p,$$

where $u_{h'}$ is a particular case of the solution u_h.

Illustrate the theorem that we proved in Exercise 7 for the nonhomogeneous PDEs in Exercises 8–12. You may refer to the corresponding exercises in Section 1.2.

8. $u_y = 2x$

9. $u_x = \sin x + \cos y$

10. $u_{xxy} = 12x$

11. $u_{zz} = x + y$

12. $u_{xx} + u_x - 2u = 6$, $u = u(x,y)$.

13.   a) If v is a solution of the PDE $L[u] = f$, and w is a solution of $L[u] = g$, find a solution of the PDE $L[u] = \alpha f + \beta g$, where α and β are any two constants.

    b) Use what you did in part (a) to find a solution of the PDE $u_{xx} + u_{yy} = 3x - 5y$.

1.6 Separation of Variables for Linear, Homogeneous PDEs

In the mid-1700s, Daniel Bernoulli and, later, Jean le Rond d'Alembert experimented with a new technique for producing solutions of linear, homogeneous PDEs. This method, called **separation of variables**,[§] entails the reduction of a PDE to an ODE (or, more commonly, to a number of ODEs, each corresponding to a different independent variable), a recurrent theme in the study of PDEs.

Definition 1.4 *Given a PDE in* $u = u(x,y)$, *we say that* u *is a* **product solution** *if*

$$u(x,y) = f(x)g(y)$$

for functions f *and* g. *More generally,* $u = u(x_1, x_2, \ldots, x_n)$ *is a product solution of a PDE in the* n *variables* $x_1, x_2, \ldots, x_n$ *if*

$$u(x_1, x_2, \ldots, x_n) = f_1(x_1)f_2(x_2)\ldots f_n(x_n)$$

for functions $f_1, f_2, \ldots, f_n$. *(See Exercise 23.)*

In practice, it is more common to write $u(x,y) = X(x)Y(y)$, $u(x,y,z) = X(x)Y(y)Z(z)$, etc.

How does the method work? Let's look at some examples.

Example 1 Find all product solutions of the first-order, linear, homogeneous PDE $u_x + u_y = 0$.

We search for all solutions of the form $u(x,y) = X(x)Y(y)$. Using the facts that

$$u_x = X'Y \quad \text{and} \quad u_y = XY',$$

we substitute into the PDE and get

$$X'Y + XY' = 0. \tag{1.13}$$

How does this help us? Well, a little algebra (specifically, dividing both sides by XY[¶]) gives us

$$\frac{X'}{X} = -\frac{Y'}{Y}, \tag{1.14}$$

[§]When studying a linear, homogeneous PDE, the first question that a mathematician usually asks is, "Is the equation separable?"

[¶]Of course, if either X or Y is the zero-function, then we may not divide by XY. However, in this case, u is the zero-function, which is already known to be a solution to *any* linear, homogeneous PDE. "Officially," we may use this method only on two-dimensional regions where $X(x)Y(y) \neq 0$ although, in practice, this turns out not to be an issue.

that is, we have managed to **separate** the variable x from the variable y. We say that the equation is **separable** and that we have **separated the variables**. Now, we have a situation where a function of x equals a function of y, that is, where

$$f(x) = g(y)$$

for *all* values of x and y in the domain of the problem. So choose *any* such x-value, $x = x_0$. We then have

$$f(x_0) = g(y)$$

for all values of y, that is, that $g(y)$ is a *constant* function! Then, it follows that $f(x)$ is a constant function, as well!

So, at this point, we have

$$u(x,y) = X(x)Y(y) \text{ is a solution } \Rightarrow \frac{X'}{X} = -\frac{Y'}{Y} = \lambda \qquad (1.15)$$

for some real constant λ. Conversely, given any real constant λ, if (1.15) is satisfied, then $u = XY$ is a solution of the PDE (why?).

Equation (1.15) actually is two equations:

$$\frac{X'}{X} = \lambda \quad \text{and} \quad \frac{Y'}{Y} = -\lambda.$$

Therefore, we conclude that $u = XY$ is a solution of the PDE if and only if X and Y satisfy the ODEs

$$X' - \lambda X = 0 \quad \text{and} \quad Y' + \lambda Y = 0$$

for the same λ. The product solutions, thus, are

$$X(x) = e^{\lambda x} \quad \text{and} \quad Y(y) = e^{-\lambda y}$$

or

$$u(x,y) = e^{\lambda(x-y)},$$

for any real constant, λ. Further, any linear combination of these solutions is, again, a solution.

As we shall see in the following chapter, although it looks as though we have found only solutions which are linear combinations of product solutions, in many cases that will be enough to solve any well-posed problem involving the given PDE. In the process of solving these initial-boundary-value problems, we shall find that only certain values of λ will lead to *nontrivial* solutions of the problem, that is, to solutions other than the zero-function.

Example 2 Find all product solutions of the heat equation, $u_t = u_{xx}$.

We let $u(x,t) = X(x)T(t)$ and substitute:

$$X(x)T'(t) = X''(x)T(t).$$

Again, dividing both sides by u gives us

$$\frac{T'}{T} = \frac{X''}{X} = \text{constant.}$$

For the sake of convenience (we'll see why later), we call the constant $-\lambda$:

$$\frac{T'}{T} = \frac{X''}{X} = -\lambda$$

or

$$X'' + \lambda X = 0 \quad \text{and} \quad T' + \lambda T = 0.$$

So, for each real number λ, we must solve these two ODEs. Now, the form of the solution of the first PDE will depend on the sign of λ (this did *not* happen in the previous example), so we must consider three cases.

Case 1: $\lambda > 0$

$$X = c \cos \sqrt{\lambda}\, x + d \sin \sqrt{\lambda}\, x, T = e^{-\lambda t}$$

and

$$u = e^{-\lambda t}[c \cos \sqrt{\lambda}\, x + d \sin \sqrt{\lambda}\, x].$$

Case 2: $\lambda = 0$

$$X = cx + d, T = 1$$

and

$$u = cx + d.$$

Case 3: $\lambda < 0$

$$X = c e^{\sqrt{-\lambda} x} + d e^{-\sqrt{-\lambda} x}, T = e^{-\lambda t}$$

and

$$u = e^{-\lambda t}[c e^{\sqrt{-\lambda} x} + d e^{-\sqrt{-\lambda} x}].$$

In each case, c and d are arbitrary constants. Again, any linear combination of solutions is a solution.

Example 3 Separate the PDE $3u_{yy} - 5u_{xxxy} + 7u_{xxy} = 0$.
   Again, let $u = XY$:

$$3XY'' - 5X'''Y' + 7X''Y' = 0.$$

Then, dividing by XY doesn't help us, but dividing by XY' gives us

$$\frac{3Y''}{Y'} = \frac{5X''' - 7X''}{X} = -\lambda$$

or

$$5X''' - 7X'' + \lambda X = 0 \quad \text{and} \quad 3Y'' + \lambda Y' = 0.$$

Example 4 Separate the PDE (in $u(x, y, z)$),

$$u_x - 2u_{yy} + 3u_z = 0.$$

We let $u(x, y, z) = X(x)Y(y)Z(z)$ and, substituting, get

$$X'YZ - 2XY''Z + 3XYZ' = 0.$$

Let's divide by $u = XYZ$ and see what happens:

$$\frac{X'}{X} - \frac{2Y''}{Y} + \frac{3Z'}{Z} = 0.$$

At the very least, we can separate any one of the variables from the other two. For example, we can write

$$\frac{X'}{X} = \frac{2Y''}{Y} - \frac{3Z'}{Z} = -\lambda_1,$$

where we have concluded, as before, that each side of the separated equation must be constant. Now, we immediately get the ODE

$$X' + \lambda_1 X = 0.$$

As for the second half, we can rewrite it as

$$\frac{2Y''}{Y} = \frac{3Z'}{Z} - \lambda_1,$$

and we have separated the variables y and z. Hence, we conclude that

$$\frac{2Y''}{Y} = \frac{3Z'}{Z} - \lambda_1 = -\lambda_2$$

for any real λ_2, or

$$2Y'' + \lambda_2 Y = 0 \quad \text{and} \quad 3Z' + (\lambda_2 - \lambda_1)Z = 0.$$

Therefore, $u = XYZ$ is a solution if and only if there exist constants λ_1 and λ_2 such that X, Y and Z satisfy the three ODEs above (with, of course, the same λ_1 and the same λ_2 in each).

We do *not* want to give the impression that all linear, homogeneous PDEs are separable—in fact, "most" are not separable. However, many of the equations which are important in applications *are* separable (rather, many of the simplifications which are made in deriving PDEs are made *so that* the resulting PDEs are linear and, often, separable). It is very easy to prove that a PDE is separable—by separating it! However, it is more difficult to *prove* that a PDE is *not* separable.

Exercises 1.6

In Exercises 1–21, separate the PDE into a system of ODEs.

1. $3u_x - 2u_y = 0$

2. $5u_x + 4u_y - 2u = 0$

3. $y^2 u_x + x^2 u_y = 0$

4. $u_{xx} - u_y + u = 0$

5. The wave equation, $u_{tt} - u_{xx} = 0$

6. Laplace's equation, $u_{xx} + u_{yy} = 0$

7. $u_{xx} + 2u_{yy} - u_x + 3u_y = 0$

8. $u_{xx} - xu_y + xu = 0$

9. $-iu_t = u_{xx} - x^2 u$ (This is the one-dimensional Schrödinger's equation for a harmonic oscillator. Here, i is the imaginary constant with the property $i^2 = -1$.)

10. $x^2 u_{xx} + 2u_x - 3u_y - yu = 0$

11. Laplace's equation in polar coordinates, $u_{rr} + \frac{1}{r} u_r + \frac{1}{r^2} u_{\theta\theta} = 0$

12. $ru_{rr} + u_r - ru_t = 0$ (this equation gives the intensity of the magnetic field inside a solenoid)

13. The Euler–Bernoulli beam equation, $u_{tt} + u_{xxxx} = 0$

14. $u_x + u_y - u_z = 0$

15. $u_x + u_y + u_z + u = 0$

16. $u_{xx} - u_y + u_z = 0$

17. $x^2 u_x - y^3 u_y - 4zu_z = 0$

18. The two-dimensional heat equation, $u_t = u_{xx} + u_{yy}$

19. The two-dimensional wave equation, $u_{tt} = u_{xx} + u_{yy}$

20. The three-dimensional Laplace equation, $u_{xx} + u_{yy} + u_{zz} = 0$

21. Schrödinger's equation (with zero potential), $u_{xx} + u_{yy} + u_{zz} + u = 0$

In Exercises 22–35, find all product solutions of the PDE (each PDE already was separated in Exercises 1–21).

22. $3u_x - 2u_y = 0$

23. $5u_x + 4u_y - 2u = 0$

24. $y^2 u_x + x^2 u_y = 0$

25. $u_{xx} - u_y + u = 0$

26. The wave equation, $u_{tt} - u_{xx} = 0$

27. Laplace's equation, $u_{xx} + u_{yy} = 0$

28. $u_{xx} + 2u_{yy} - u_x + 3u_y = 0$

29. Laplace's equation in polar coordinates, $u_{rr} + \frac{1}{r}u_r + \frac{1}{r^2}u_{\theta\theta} = 0$

30. $u_x + u_y - u_z = 0$

31. $u_x + u_y + u_z + u = 0$

32. $u_{xx} + u_y + u_z = 0$

33. The two-dimensional heat equation, $u_t = u_{xx} + u_{yy}$

34. The two-dimensional wave equation, $u_{tt} = u_{xx} + u_{yy}$

35. The three-dimensional Laplace equation, $u_{xx} + u_{yy} + u_{zz} = 0$

36. Prove that if $f(x) = g(y, z)$ for all x, y and z, then f and g both are constant functions.

37. One also may try to separate variables in other ways.

    a) Find all solutions of the PDE $u_x + u_y = 0$ of the form $u(x, y) = X(x) + Y(y)$.

    b) Do the same for the eikonal equation, $u_x^2 + u_y^2 = 1$.

38. In Section 1.3, we saw that we often are interested in solving a PDE subject to certain auxiliary conditions, namely, initial and boundary conditions. In fact, when we solve an initial-boundary-value problem using separation of variables, we will find it much easier to solve if we also *separate the boundary conditions*. For each of the boundary conditions given below, separate the variables, that is, decide what each of them tells you about product solutions $u(x, t) = X(x)T(t)$. (In each case, a is a constant.)

    a) $u(a, t) = 0$ (the so-called **Dirichlet boundary condition**)

    b) $u_x(a, t) = 0$ (the **Neumann condition**)

    c) $\alpha u_x(a, t) + \beta u(a, t) = 0$, where α and β are constants (the **Robin condition**)

    d) $u_{xx}(a, t) = 0$

    e) $u_{xxx}(a, t) = 0$

(The last two boundary conditions are encountered in connection with the Euler–Bernoulli beam PDE, for example.)

39. Decide whether the given function is a *product* function, that is, if it can be written in the form $f(x)g(y)$ for functions f and g. If it is not, justify your answer.

    a) $u(x,y) = e^{xy}$

    b) $u(x,y) = e^{2x-3y}$

    c) $u(x,y) = y^2 - xy + 1$

    d) $u(x,y) = \sin(x+y)$

1.7 Eigenvalue Problems

In Section 1.3, we discussed the heat equation, subject to initial and boundary conditions. Suppose, for instance, we're solving the heat equation $u_t = u_{xx}$ on the interval $0 \le x \le 1$. Suppose, further, that the equation is subject to the boundary conditions

$$u(0,t) = u(1,t) = 0$$

for $t > 0$. We first separate the PDE, resulting in the ODEs

$$X'' + \lambda X = 0 \quad \text{and} \quad T' + \lambda T = 0.$$

Then, as in Exercise 38 in the previous section, we separate the boundary conditions, as follows:

$u(0,t) = X(0)T(t) = 0$ for all $t > 0 \Rightarrow X(0) = 0$, or $T(t) = 0$ for all $t > 0$.

So we have two types of product solutions of the PDE which satisfy the left boundary condition: those which satisfy $X(0) = 0$ and those for which $T(t) \equiv 0$, that is, those for which $T(t)$ is the zero-function. But the latter gives us the zero-solution (which we already know is a solution). So the only *nontrivial* product solutions which satisfy the left boundary condition are those which satisfy $X(0) = 0$.

Similarly, the only nontrivial product solutions which satisfy the right boundary condition will satisfy $X(1) = 0$. So, we actually need to solve the system

$$X'' + \lambda X = 0 \qquad T' + \lambda T = 0$$
$$X(0) = X(1) = 0.$$

The X-system looks like the problems in Section 1.3, Exercises 1–5. Essentially, then, it is an *ODE boundary-value problem*. However, it differs from the latter in that it includes the *parameter* λ. Remember that we solve the X-ODE *for each real number* λ. For each λ, the ODE has infinitely many solutions. Now, though, we need to find which of these solutions "survive" the boundary conditions—that is, we shall see that, for "most" real numbers λ, the only solution that also satisfies the boundary conditions is the zero-solution, $X(x) \equiv 0$. Thus, we need to identify those values of λ for which the X-system has *nontrivial* solutions (and, of course, find those solutions). Then we will solve the T-equation, but only for these values of λ, and form the nontrivial product solutions of the PDE and boundary conditions.

These values of λ are called **eigenvalues**[||] of the X-system, and the corresponding nontrivial solutions are the **eigenfunctions** associated with λ. The system itself is an example of an ODE **eigenvalue problem**.

Let's calculate some eigenthings.

Example 1 Find all eigenvalues and eigenfunctions of the eigenvalue problem

$$y'' + \lambda y = 0, \quad y = y(x), \qquad 0 < x < 1$$
$$y(0) = y(1) = 0.$$

As with ODE initial-value problems, and the boundary-value problems from the exercises in Section 1.3, we first find the general solution of the ODE, then apply the boundary conditions. To do this, we set $y = e^{rx}$ and find the characteristic equation $r^2 + \lambda = 0$. It now becomes apparent that we need to treat the cases $\lambda > 0$, $\lambda = 0$ and $\lambda < 0$ separately.

Case 1: $\lambda < 0$

If $\lambda < 0$, then we can write $\lambda = -k^2$ for some real number k with $k > 0$. Then, the characteristic equation $r^2 - k^2 = 0$ leads to the two independent solutions e^{kx} and e^{-kx}. However, it turns out that life is much easier if we use, instead, the functions

$$y_1 = \frac{e^{kx} + e^{-kx}}{2} = \cosh(kx)$$

and

$$y_2 = \frac{e^{kx} - e^{-kx}}{2} = \sinh(kx)$$

(see Exercise 22). Then, the general solution is

$$y = c_1 \cosh(kx) + c_2 \sinh(kx).$$

[||]These eigenvalues are similar to those which we see in Linear Algebra, where the eigenvalues of a matrix are those real numbers for which the matrix equation $A\vec{v} = \lambda\vec{v}$ has a nontrivial solution. Here, we are looking for real numbers for which the *functional* equation $Ly = \lambda y$ has a nontrivial solution, where $Ly = -y''$.

Applying the left end boundary condition, we have

$$y(0) = 0 = c_1 \cosh 0 + c_2 \sinh 0$$
$$= c_1.$$

So the only solutions that survive this boundary condition are those of the form

$$y = c_2 \sinh(kx).$$

Then, applying the right end boundary condition gives us

$$y(1) = 0 = c_2 \sinh k$$

and, since $k > 0$, we have $\sinh k \neq 0$ and, therefore, $c_2 = 0$. Therefore, for each negative number λ, the only solution which survives the boundary conditions is the zero-function

$$y \equiv 0.$$

Therefore, *there are no negative eigenvalues.*

Case 2: $\lambda = 0$
In this case, the ODE is just $y'' = 0$, with general solution

$$y = c_1 x + c_2.$$

Then,

$$y(0) = 0 = c_2 \quad \text{and} \quad y(1) = 0 = c_1 + c_2,$$

so $c_1 = c_2 = 0$, and $\lambda = 0$ *is not a eigenvalue.*

Case 3: $\lambda > 0$
If $\lambda > 0$, we can write $\lambda = k^2$ for some real number k with $k > 0$. Then, the characteristic equation $r^2 + k^2 = 0$ leads to the two linearly independent solutions $\cos(kx)$ and $\sin(kx)$ and, therefore, to the general solution

$$y = c_1 \cos(kx) + c_2 \sin(kx).$$

Applying the left end boundary condition, we have

$$y(0) = 0 = c_1 \cos 0 + c_2 \sin 0$$
$$= c_1.$$

So the only solutions which survive this boundary condition are

$$y = c_2 \sin(kx).$$

Then, the other boundary condition gives us

$$y(1) = 0 = c_2 \sin k.$$

As in Case 1, this forces $c_2 = 0$ *except in those cases where k is a number with the property* $\sin k = 0$. For these latter values of k, we need *not* have $c_2 = 0$; in fact, there is no restriction on c_2, so the term $c_2 \sin(kx)$ *survives both boundary conditions*. In other words, these values of k give us the eigenvalues $\lambda = k^2$ of the problem; for each such k, the functions

$$y = c \sin(kx)$$

are the associated eigenfunctions. In practice, we say that *the* eigenfunction is $y = \sin(kx)$, realizing that any constant multiple of an eigenfunction is an eigenfunction (why?).

So the eigenvalues are those numbers $\lambda = k^2$ where $\sin k = 0$. Therefore, we have

$$k = \pi, 2\pi, 3\pi, \ldots = n\pi, \quad n = 1, 2, 3, \ldots \text{ (remember: } k > 0)$$

and

$$\lambda = \pi^2, 4\pi^2, 9\pi^2, \ldots = n^2\pi^2, \quad n = 1, 2, 3, \ldots .$$

We write the eigenvalues as

$$\lambda_n = n^2\pi^2, \quad n = 1, 2, 3, \ldots$$

and the corresponding eigenfunctions as

$$y_n = \sin(n\pi x), \quad n = 1, 2, 3, \ldots .$$

Example 2 Do the same for

$$y'' + \lambda y = 0,$$
$$y'(0) = y'(3) = 0.$$

Case 1: $\lambda < 0, \lambda = -k^2, k > 0$
    We have
$$y = c_1 \cosh(kx) + c_2 \sinh(kx),$$

so that
$$y' = c_1 k \sinh(kx) + c_2 k \cosh(kx).$$

(See Exercise 22.) Then,

$$y'(0) = 0 = c_2 k \Rightarrow c_2 = 0$$

and

$$y'(3) = 0 = c_1 k \sinh 3k \Rightarrow c_1 = 0,$$

so there are no negative eigenvalues.

Case 2: $\lambda = 0$

The general solution is

$$y = c_1 x + c_2$$

so that

$$y' = c_1.$$

Then,

$$y'(0) = y'(3) = 0 = c_1.$$

Therefore, the function $y = c_2$ survives both boundary conditions, so $\lambda = 0$ is an eigenvalue. We write

$$\lambda_0 = 0$$

with eigenfunction

$$y_0 = 1.$$

Case 3: $\lambda > 0, \lambda = k^2, k > 0$

We have the general solution

$$y = c_1 \cos(kx) + c_2 \sin(kx),$$

so that

$$y' = -c_1 k \sin(kx) + c_2 k \cos(kx).$$

Then,

$$y'(0) = 0 = c_2 k \Rightarrow c_2 = 0;$$
$$y'(3) = 0 = -c_1 k \sin(3k) \Rightarrow c_1 = 0$$

unless

$$\sin(3k) = 0, \text{ that is, } 3k = \pi, 2\pi, 3\pi, \ldots$$

or

$$k = \frac{n\pi}{3}, \qquad n = 1, 2, 3, \ldots .$$

Therefore, we have eigenvalues

$$\lambda_n = \frac{n^2 \pi^2}{9}, \qquad n = 1, 2, 3, \ldots$$

with associated eigenfunctions

$$y_n = \cos \frac{n\pi x}{3}, \qquad n = 1, 2, 3, \ldots .$$

Example 3 Do the same for

$$x^2 y'' + xy' - \lambda y = 0$$
$$y(1) = y(e) = 0.$$

First, note that this is a Cauchy–Euler equation, for $x > 0$. We let $y = x^r$ and determine the values of r that give us solutions. So

$$y = x^r \Rightarrow x^2 r(r-1)x^{r-2} + x \cdot rx^{r-1} - \lambda x^r = 0$$
$$\Rightarrow r(r-1) + r - \lambda = 0$$
$$\Rightarrow r^2 - \lambda = 0.$$

Again, we must consider three cases.

Case 1: $\lambda > 0, \lambda = k^2, k > 0$
We have $r = \pm k$, so our two linearly independent solutions are x^k and x^{-k}, giving us the general solution

$$y = c_1 x^k + c_2 x^{-k}.$$

Then,

$$y(1) = 0 = c_1 + c_2$$
$$y(e) = 0 = c_1 e^k + c_2 e^{-k}$$

which imply

$$c_2 = -c_1$$
$$c_1(e^k - e^{-k}) = 0.$$

Since $k > 0$, the latter implies that $c_1 = 0$, so $c_2 = 0$ as well, and we have no positive eigenvalues.

Case 2: $\lambda = 0$
In this case, we have the repeated root $r = 0$, giving us the linearly independent solutions x^0 and $x^0 \ln x$. So the general solution is

$$y = c_1 + c_2 \ln x.$$

Then,

$$y(1) = 0 = c_1$$
$$y(e) = 0 = c_1 + c_2$$

so, again, $c_1 = c_2 = 0$, and $\lambda = 0$ is not an eigenvalue.

Case 3: $\lambda < 0, \lambda = -k^2, k > 0$
Here we have the roots $r = \pm ik$ and corresponding linearly independent solutions $\cos(k \ln x)$ and $\sin(k \ln x)$. The general solution is

$$y = c_1 \cos(k \ln x) + c_2 \sin(k \ln x).$$

Applying the boundary conditions, we have

$$y(1) = 0 = c_1 \Rightarrow c_1 = 0$$
$$y(e) = 0 = c_2 \sin k$$

and the latter equation forces $c_2 = 0$, *except for those values of k satisfying* $\sin k = 0$; that is, c_2 is arbitrary when

$$k = \pi, 2\pi, 3\pi, \ldots$$

or

$$\lambda = -\pi^2, -4\pi^2, -9\pi^2, \ldots .$$

Therefore, the eigenvalues are

$$\lambda_n = -n^2\pi^2, \qquad n = 1, 2, 3, \ldots,$$

and the associated eigenfunctions are

$$y_n = \sin(n\pi \ln x), \qquad n = 1, 2, 3, \ldots .$$

Example 4 Do the same for

$$y'' + \lambda y = 0$$
$$y(0) = y(1) + y'(1) = 0.$$

Case 1: $\lambda < 0, \lambda = -k^2, k > 0$

We have

$$y = c_1 \cosh kx + c_2 \sinh kx.$$

Then,

$$y' = c_1 k \sinh kx + c_2 k \cosh kx$$

and, applying the boundary conditions, we have

$$y(0) = 0 = c_1$$
$$y(1) = 0 = c_2(\sinh k + k \cosh k).$$

So we must have $c_2 = 0$, except for those values of $k > 0$ satisfying

$$\sinh k + k \cosh k = 0.$$

Essentially, then, we wish to find all positive roots of the function

$$f(x) = \sinh x + x \cosh x.$$

Now, $f(0) = 0$. For $x > 0$, let's consider f':

$$f'(x) = 2 \cosh x + x \sinh x > 0 \text{ for } x > 0.$$

Therefore (see Exercise 23), $f(x)$ has no roots when $x > 0$ and the problem has no negative eigenvalues.

Case 2: $\lambda = 0$

The general solution here is

$$y = c_1 x + c_2$$

and, since $y' = c_1$, the boundary conditions give

$$y(0) = 0 = c_2$$
$$y(1) + y'(1) = 0 = 2c_1 + c_2.$$

Therefore, $c_1 = c_2 = 0$, so $\lambda_0 = 0$ is not an eigenvalue.

Case 3: $\lambda > 0, \lambda = k^2, k > 0.$

Here, as usual,

$$y = c_1 \cos kx + c_2 \sin kx,$$

so

$$y' = -c_1 k \sin kx + c_2 k \cos x.$$

Then,

$$y(0) = 0 = c_1$$
$$y(1) = 0 = c_2(\sin k + k \cos k).$$

This system has only the solution $c_1 = c_2 = 0$ *unless* k is such that

$$\sin k + k \cos k = 0.$$

Therefore, the eigenvalues correspond to those values of k satisfying

$$-k = \tan k.$$

How do we solve for k? We don't—because we can't! However, we can show that there are infinitely many such values of k, by looking at the graphs of $y = -k$ and $y = \tan k$ for $k > 0$, which we have plotted in Figure 1.2. In fact, it looks as though $y = -k$ intersects each branch of $y = \tan k$ exactly once. Therefore, our eigenvalues correspond to values of k_n, $n = 1, 2, 3, \ldots$, satisfying

$$\frac{\pi}{2} < k_1 < \frac{3\pi}{2}$$
$$\frac{3\pi}{2} < k_2 < \frac{5\pi}{2}$$
$$\vdots$$
$$\frac{(2n-1)\pi}{2} < k_n < \frac{(2n+1)\pi}{2} \ .$$
$$\vdots$$

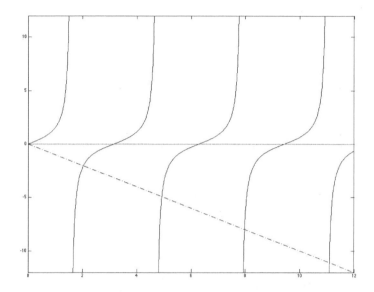

FIGURE 1.2
MATLAB graph of the intersection of the functions $y = -k$ and $y = \tan k$ for $k > 0$.

Therefore, the eigenvalues are those $\lambda_n > 0$ satisfying $-\sqrt{\lambda_n} = \tan \sqrt{\lambda_n}$, with associated eigenfunctions

$$y_n = \sin \sqrt{\lambda_n} \; x.$$

We have also solved this same problem using the MATLAB routine BVP4C. The first five eigenvalues are given in Table 1.1, and the first five eigenfunctions (*normalized* by requiring that $y'(0) = 1$) are plotted in Figure 1.3. Note that the solutions *do* seem to satisfy the condition $y'(1) = -y(1)$.

n	λ_n
1.	4.116
2.	24.142
3.	63.664
4.	122.897
5.	201.863

TABLE 1.1
First five eigenvalues of the problem in Example 4, computed using the MATLAB routine BVP4C.

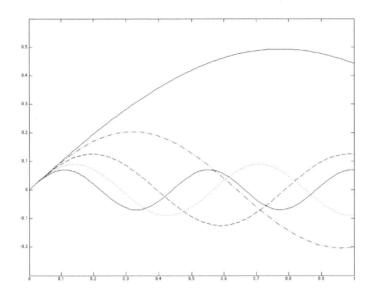

FIGURE 1.3
MATLAB graph of the first five eigenfunctions in Example 4, using the routine BVP4C (highest to lowest, respectively, at $x = 0.2$).

It may seem odd to include such an example, but this problem illustrates the fact that many eigenvalue problems *cannot be solved explicitly.*

Further, this type of eigenvalue problem often shows up in applications. See, for example, Exercises 5e and 5f in Section 4.1.

Example 5 Here, we briefly introduce a more general technique for solving these eigenvalue problems. Suppose we wish to find the positive eigenvalues of

$$y'' + \lambda y = 0,$$
$$y(0) + y'(0) = y(1) = 0.$$

Proceeding as before, we have

$$y = c_1 \cos kx + c_2 \sin kx, \qquad \lambda = k^2,$$

and we must find those values of k for which the system

$$c_1 + c_2 k = 0$$
$$c_1 \cos k + c_2 \sin k = 0.$$

Now, if we write these equations in matrix form,

$$\begin{bmatrix} 1 & k \\ \cos k & \sin k \end{bmatrix} \begin{bmatrix} c_1 \\ c_2 \end{bmatrix} = \begin{bmatrix} 0 \\ 0 \end{bmatrix},$$

we see that $\lambda = k^2$ is an eigenvalue if and only if

$$\begin{vmatrix} 1 & k \\ \cos k & \sin k \end{vmatrix} = 0.$$

This leads to the equation

$$k = \tan k,$$

with solution similar to that in the previous example (with one additional concern—see Exercise 24).

Exercises 1.7

In Exercises 1–14, solve the eigenvalue problem, that is, find all eigenvalues and associated eigenfunctions.

1. $y'' + \lambda y = 0, y(0) = y(5) = 0$

2. $y'' + \lambda y = 0, y'(0) = y'\left(\frac{1}{2}\right) = 0$

3. $y'' + \lambda y = 0, y'(0) = y(\pi) = 0$

4. $y'' + \lambda y = 0, y(0) = y'(4) = 0$

5. $y'' + \lambda y = 0, y(0) - y'(0) = y(1) - y'(1) = 0$

6. $y'' + \lambda y = 0, y(0) + y'(0) = y(2) + y'(2) = 0$

7. $x^2 y'' + 3xy' + \lambda y = 0, y(1) = y(e^2) = 0$

8. $y'' + 2y' + (\lambda + 1)y = 0, y(0) = y(\pi) = 0$

9. $y'' + \lambda y = 0, y(-1) = y(1) = 0$

10. $y'' + 2y' + \lambda y = 0, y(-2) = y(2) = 0$

11. $y^{(4)} + \lambda y = 0, y(0) = y''(0) = y(1) = y''(1) = 0$ (Hint: Let $\lambda = k^4, 0, -k^4$.)

12. $y^{(4)} + \lambda y = 0, y'(0) = y'''(0) = y'(\pi) = y'''(\pi) = 0$

13. $y'' + \lambda y = 0, y(0) = y(2) - y'(2) = 0$

14. $y^{(4)} + \lambda y''(0) = 0, y(0) = y''(0) = y(1) = y''(1) = 0$

15. Solve each eigenvalue/eigenfunction problem two ways:

    i) by hand

    ii) **MATLAB:** Using the MATLAB routine BVP4C

a) $y'' + \lambda y = 0, y(0) + y'(0) = y(1) = 0$
b) $y'' + \lambda y = 0, 2y(0) - y'(0) = y(2) = 0$
c) Exercise 13
d) $y^{(4)} + \lambda y'' = 0, y(0) = y''(0) = y(1) = y'(1) = 0$

16. Find all eigenvalues and eigenfunctions of the ODE $y'' + \lambda y = 0$ subject to the boundary conditions (where $L > 0$)

    a) $y(0) = y(L) = 0$

    b) $y(0) = y'(L) = 0$

    c) $y'(0) = y(L) = 0$

    d) $y'(0) = y'(L) = 0$

17. Find all product solutions of the heat equation $u_t = u_{xx}$ which also satisfy the boundary conditions $u(0,t) = u(5,t) = 0$. (Refer to Exercise 1, above, and to Example 2 of the previous section.)

18. Find all product solutions of the wave equation $u_{tt} = u_{xx}$ which also satisfy the boundary conditions $u(0,t) = u(1,t) = 0$. (Refer to Example 1, and to Exercises 5 and 26 of the previous section.)

19.   a) Show that the ODE $y'' + \lambda y = 0$ has nontrivial periodic solutions of period L, that is, which satisfy

$$y(x + L) = y(x) \quad \text{for all} \quad x$$

    if $\lambda = \lambda_n = \left(\frac{2\pi n}{L}\right)^2, n = 0, 1, 2, \dots$.

    b) Show that the statement actually is an if *and only if.* (Hint: Write $c_1 \cos kx + c_2 \sin kx = c_3 \cos(kx - c_4)$.)

Hence, show that the only solutions of $y'' + \lambda y = 0$ of period 2π are the functions

$$y_0 = 1, \quad y_n = c_1 \cos nx + c_2 \sin nx, \quad n = 1, 2, \dots .$$

    c) Solve, instead, the eigenvalue problem

$$y'' + \lambda y = 0$$
$$y(-\pi) = y(\pi)$$
$$y'(-\pi) = y'(\pi)$$

and show that we get the same eigenvalues and eigenfunction as above.

20. **Sturm Comparison Theorem:** The *Sturm Comparison Theorem* (due to Jacques Charles François Sturm, whom we'll meet in Chapter 8) says, as a special case, that if y_1 and y_2 are nonzero solutions of $y'' + \lambda_1 y = 0$ and $y'' + \lambda_2 y = 0$, respectively, with $0 < \lambda_1 < \lambda_2$, then, between any two consecutive roots of y_1, there is a root of y_2.

    a) Show that the theorem is true for the eigenvalues and eigenfunctions of each problem in Exercise 16.

    b) **MATLAB:** Exhibit this theorem graphically for the first 10 eigenfunctions from Example 4.

21. Suppose we are asked to solve the eigenvalue problem

$$y'' + \lambda y = 0, \qquad a \le x \le b$$
$$y(a) = y(b) = 0.$$

    a) Show that the change of variable $z = \frac{L}{b-a}(x - a)$ transforms the problem to what, essentially, is the problem in Exercise 16a.

    b) Use this transformation and the solution of Exercise 16a to solve this problem.

    c) Use this method to solve Exercise 9.

22.   a) Explain why we may say that $y = c_1 \cosh kx + c_2 \sinh kx$ is the general solution of $y'' - k^2 y = 0$.

    b) Show that $\frac{d}{dx}(\cosh x) = \sinh x$ and $\frac{d}{dx}(\sinh x) = \cosh x$.

23. Use Rolle's Theorem to prove that if $f(0) = 0$ and $f'(x) > 0$ for $x > 0$, then f has no positive roots.

24.   a) **MATLAB:** Graph $y = x$ and $y = \tan x$, $x \ge 0$, on the same set of axes.

    b) Prove that $f(x) = x$ and $g(x) = \tan x$ do not intersect on the interval $0 < x < \frac{\pi}{2}$.

25. In this exercise we prove that if y_1 and y_2 are eigenfunctions, corresponding to different eigenvalues, of the problem $y'' + \lambda y = 0$ subject to either $y(0) = 0$ or $y'(0) = 0$ at the left end and either $y(L) = 0$ or $y'(L) = 0$ at the right end, then

$$\int_0^L y_1 y_2 \, dx = 0.$$

(In this case, we say that y_1 and y_2 are **orthogonal** on $0 \le x \le L$.) To this end, suppose that y_1 and y_2 are eigenfunctions corresponding to the eigenvalues λ_1 and λ_2, respectively, with $\lambda_1 \ne \lambda_2$. Then,

a) Show that $(\lambda_1 - \lambda_2) \int_0^L y_1 y_2 \, dx = \int_0^L (y_1 y_2'' - y_1'' y_2) dx.$

b) Use integration by parts to prove **Green's first identity**

$$\int_0^L y_1 y_2'' \, dx = y_1 y_2' \Big|_{x=0}^{x=L} - \int_0^L y_1' y_2' \, dx.$$

c) Now prove **Green's second identity**

$$\int_0^L (y_1 y_2'' - y_1'' y_2) dx = (y_1 y_2' - y_1' y_2) \Big|_{x=0}^{x=L}.$$

d) Conclude that

$$\int_0^L y_1 y_2 \, dx = 0.$$

e) Show that we can*not* do anything similar for

$$\int_a^b (y_1 y_2' - y_2 y_1') dx.$$

How about for

$$\int_a^b (y_1 y_2' + y_2 y_1') dx?$$

(Note that the second integral results from replacing y_1' by $-y_1'$ in the first. We'll see the significance of this in Chapter 8.)

26. **Rayleigh quotient:** Suppose that λ_n is an eigenvalue, with eigenfunction y_n, of any of the four eigenvalue problems

$$y'' + \lambda y = 0, \qquad 0 < x < L,$$
$$y(0) = 0 \quad \text{or} \quad y'(0) = 0,$$
$$y(L) = 0 \quad \text{or} \quad y'(L) = 0.$$

a) Use Green's first identity, from the previous exercise, to show that

$$\lambda_n = \frac{\int_0^L [y_n'(x)]^2 dx}{\int_0^L [y_n(x)]^2 dx}.$$

This is the **Rayleigh**** **quotient** for λ_n, in terms of y_n.

b) Conclude that we must have $\lambda_n \geq 0$, that is, that the problems have no negative eigenvalues. In which case(s) is 0 an eigenvalue?

**After the great British scientist John William Strutt, Lord Rayleigh (1842–1919).

c) Show that the eigenvalue problem

$$y'' + \lambda y = 0, \qquad 0 < x < L,$$
$$y(0) - ay'(0) = y(L) + by'(L) = 0$$

has no negative eigenvalues if $a > 0$ and $b > 0$. (These boundary conditions will show up in our discussion of the heat equation.)

27. Consider the *non*homogeneous boundary-value problem

$$y'' + \alpha y = f(t), \alpha = \text{ constant}$$
$$y(0) = y(L) = 0$$

(where $f(t)$ is continuous on $0 \le x \le L$).

a) Show that if α is *not* an eigenvalue of the associated homogeneous problem $y'' + \lambda y = 0$, $y(0) = y(L) = 0$, then the nonhomogeneous problem has a unique solution.

b) Show that if α *is* an eigenvalue of the homogeneous problem, then the nonhomogeneous problem may or may not have a solution. In this case, for which functions $f(t)$ will the problem have a solution? Is the solution unique?

(Compare this problem to the nonhomogeneous problem from Linear Algebra,

$$(A - \lambda I)\vec{v} = \vec{b}.)$$

Prelude to Chapter 2

In this chapter, we provide physical derivations for the three most important PDEs, the heat equation, the wave equation and Laplace's equation, each in two independent variables. We also derive the appropriate initial and boundary conditions in order that these problems be well-posed on finite domains. Finally, we'll solve special cases of these initial-boundary-value problems, and we'll see that we are only one step away from solving them in general—with that last step to be filled in Chapter 3.

Although many mathematicians in the late 18th and early 19th centuries investigated the problem of heat conduction, the name of Joseph Fourier (1768–1830) has become synonymous with this particular problem. Fourier played an important role in the French Revolution and, when Napoleon came to power, Fourier was appointed Chair of Mathematics at the newly formed École Normale. Fourier became so successful that Napolean decided to take him along on his ill-fated invasion of Egypt in 1798. The French had a successful landing and met almost no resistance, but the British destroyed the French fleet in Alexandria harbor, stranding Napoleon's army, most of whom—including Fourier – were stuck in Egypt for more than two years!

During this time and after, motivated by the problem of better designing cannons so that they would cool quickly after firing, Fourier continued thinking about the conduction of heat. He soon was able to derive the heat equation and to solve it using trigonometric series about which we'll say *much* more in Chapter 3.

The study of the vibrating string seems to have been prompted by the writings of French composer and music theorist Jean-Philippe Rameau (1683–1764) and, in particular, by the appearance in 1722 of his famous textbook on harmony. In 1727, John Bernoulli (1667–1748, brother of James and father of Daniel) approximated a continuous string by a massless string loaded with a finite number of discrete masses. Although Bernoulli seems to have "taken the limit," the wave equation as we know it did not appear until the 1760s, in the works of Euler and d'Alembert.

Laplace's equation, or the potential equation, actually appeared first in 1752 in a paper by Euler. The paper dealt with the motion of fluids and was influenced by Daniel Bernoulli's seminal work *Hydrodynamics*, which appeared in 1738 and in which he coined the term "potential function." Pierre-Simon de Laplace (1749–1827) got his name attached to the equation through his rederivation and use of it in connection with the problem of gravitational attraction. He wrote a number of important papers on the topic in the 1770s and 1780s, but his greatest contribution was his landmark five-volume work, *Traité de mécanique céleste* (Treatise on celestial mechanics), in which Laplace compiled *all* of the important work, since Newton, on Newtonian gravitation and its role in the solar system. In particular, Laplace's main goal was to prove that the solar system is *stable*, a problem that has returned to the forefront with recent advances in dynamical systems and the study of *chaos*.

Of course, these are but a few of the highlights of the rich and varied history of the Big Three PDEs.

2

The Big Three PDEs

2.1 Second-Order, Linear, Homogeneous PDEs with Constant Coefficients

In this chapter we begin to look at the "Big Three PDEs"—the *heat equation* (or *diffusion equation*), the *wave equation* and *Laplace's equation* (or the *potential equation*)—each in two independent variables. Each is a second-order, linear, homogeneous PDE with constant coefficients. The general such equation is

$$au_{xx} + bu_{xy} + cu_{yy} + du_x + fu_y + gu = 0, \tag{2.1}$$

where, again, $u = u(x, y)$ and, of course, a, b, c, d, f and g are constants.

We study equation (2.1) in detail in Section 5.4. In particular, there we'll classify these equations as in the following definition and give reasons for such a classification.

Definition 2.1 *Equation (2.1) is said to be:*
   Hyperbolic, *if $b^2 - 4ac > 0$*
   Parabolic, *if $b^2 - 4ac = 0$*
   Elliptic, *if $b^2 - 4ac < 0$*

We mention this classification now because, as we'll see in the exercises, the heat equation is parabolic; the wave equation is hyperbolic; and Laplace's equation is elliptic. In fact, each of these equations is, in some sense, the standard example or "prototypical equation" of its type. As a result, the mathematical importance of these three PDEs goes far beyond their connection with physical problems.

Exercises 2.1

1. a) Show that the heat equation in one space variable, $u_t = \alpha^2 u_{xx}$, where α is a constant, is parabolic.

    b) Show that the wave equation in one space variable, $u_{tt} = c^2 u_{xx}$, where c is a constant, is hyperbolic.

    c) Show that Laplace's equation in two space variables, $u_{xx} + u_{yy} = 0$, is elliptic.

d) Show that, in each of the above, if we interchange the independent variables, the classification remains the same.

2.2   The Heat Equation and Diffusion

Although we know that *heat* is a form of energy which results from the motion of molecules, at a macroscopic level it appears to flow from warmer to cooler regions. We would like to use this idea of heat flow in order to study its conduction throughout a long, thin piece of material—a *rod*. We will derive a PDE which must be satisfied by the *temperature function* of the rod.

First, a word on the derivations found in this book. We will be providing the simplest, "barebones" derivations, the purpose being to give the student who is approaching these ideas for the first time an intuitive feel for what is involved. As such, we will make some approximations which may seem ad hoc or based on hindsight (we already *know* what the heat equation is!). Be assured that our assumptions are reasonable and that they can be made rigorous; we'll provide more rigorous derivations in the exercises, or we'll point the student to an appropriate reference.

We begin with a *very* brief derivation of the heat equation, filling in the gaps afterwards. Along the way, we'll introduce certain simplifying assumptions, as needed.

We have, then, a *rod* of length L, placed along the x-axis (as in Figure 1.1). We wish to determine the **temperature function**

$$u(x,t) = \text{ temperature at point } x, \text{ at time } t.$$

See Figure 2.1. As with most PDE derivations, we start by looking at an arbitrary small piece of the rod, from x to $x + \Delta x$, as in Figure 2.2 (this piece often is called a **differential element** of the rod). We will measure, in two different ways, the rate at which heat enters the element.

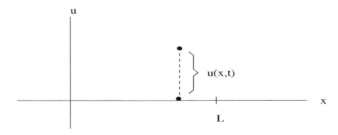

FIGURE 2.1
Temperature function for a rod of length L.

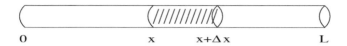

FIGURE 2.2
Differential element of length Δx.

First, heat content will be defined so that the amount of heat contained in the element (at any time t) is proportional to its temperature and its length. Then, the rate at which heat is entering the element is its time derivative, that is,

$$\text{rate} \sim \frac{\partial}{\partial t}(u \Delta x) = u_t \Delta x.$$

As for the second way to calculate this rate, *we assume that the rod is insulated* except, possibly, at its ends. Therefore, heat enters or leaves the element only at its endpoints. **Fourier's Law** will tell us that the rate at which heat flows across a given cross section is proportional to u_x at that point. Therefore, the above rate also is proportional to the

rate at which heat enters the right end

+ rate at which heat enters the left end

$$\sim u_x(x + \Delta x, t) - u_x(x, t).$$

Therefore, we have

$$u_t \Delta x \sim u_x(x + \Delta x, t) - u_x(x, t)$$

or

$$u_t \sim \frac{u_x(x + \Delta x, t) - u_x(x, t)}{\Delta x},$$

and, letting $\Delta x \to 0$, we have

$$u_t \sim u_{xx}$$

or

$$u_t = \text{constant} \cdot u_{xx},$$

which is the heat equation!

Okay, let's go back and clean things up a bit. (What is the constant and, in particular, what is its sign?) First, suppose we have a homogeneous piece of material of mass m, at constant temperature T. Then, we define the heat content of the material to be

$$\sigma m T,$$

where the proportionality constant σ is called the **specific heat** of the given substance. Applying this to the element in Figure 2.2, and *supposing that the element's cross sectional area A, mass density* (mass per unit volume) ρ *and specific heat are constant*, we arrive at a heat content of approximately

$$\sigma \rho A \Delta x \; u \left(x + \frac{\Delta x}{2}, t \right), \tag{2.2}$$

where $\rho A \Delta x$ is the mass of the element, and where we have approximated the variable temperature using the temperature at the element's midpoint (but see Exercise 4). The time rate of change of this heat content, then, is

$$\sigma \rho A \Delta x \; u_t \left(x + \frac{\Delta x}{2}, t \right). \tag{2.3}$$

Now, for *Fourier's Law*: First, in defining the temperature function as we have done, we are assuming that the rod is sufficiently thin so that the temperature is essentially constant throughout any cross section. This, in turn, coupled with the assumption that the rod is insulated (except, possibly, at the ends), allows us to assume that heat flows only in the x-direction.

Fourier's Law then states, for the heat problem, that the rate of left-to-right flow of heat per unit area, i.e., the **flux** of heat, through any cross section, is

$$\Phi(x, t) = -k u_x(x, t), \tag{2.4}$$

where the ratio k is called the material's **thermal conductivity** and $-u_x$ is called the **temperature gradient** (Why gradient? We'll see, in Chapter 9). See Figure 2.3.

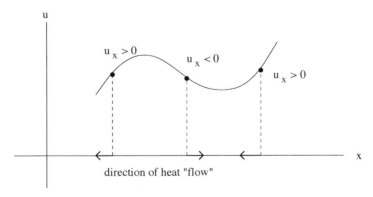

FIGURE 2.3
Fourier's Law and the direction of heat flow.

Then, assuming that k is constant along the rod (and remembering the assumption that A is constant), the rate of change of the element's heat content is the

rate at which heat enters the right end

+ rate at which heat enters the left end

$$= -A\Phi(x + \Delta x, t) + A\Phi(x, t)$$

$$= kA[u_x(x + \Delta x, t) - u_x(x, t)].^* \qquad (2.5)$$

Finally, we equate (2.3) and (2.5):

$$\sigma \rho A \Delta x u_t \left(x + \frac{\Delta x}{2}, t \right) = kA[u_x(x + \Delta x, t) - u_x(x, t)],$$

and, dividing by Δx and letting $\Delta x \to 0$, we have the heat equation

$$u_t = \alpha^2 u_{xx}, \qquad (2.6)$$

where the constant $\alpha^2 = \frac{k}{\sigma \rho}$ is called the **thermal diffusivity** of the material. Heat flow is not the only application of this PDE.

OTHER APPLICATIONS OF THE HEAT EQUATION

Diffusion

Heat conduction is a specific example of the process of **diffusion**—we say that heat diffuses through the rod, just as a drop of ink diffuses throughout a container of water. In general, let $u(x, t)$ represent the concentration (mass per unit volume), at point x at time t, of whatever it is that's diffusing. In this case, equation (2.4) is known as **Fick's Law** (actually, Fick's First Law of Diffusion), and the constant $k^\dagger$ is called the **coefficient of diffusion**. The resulting PDE is

$$u_t = ku_{xx}.$$

(So, for example, temperature can be thought of as the concentration of heat.)

---

*But see Exercise 3.

†*Not* the k from the heat equation derivation.

Electric current in a long, insulated cable

If $i(x,t)$ and $E(x,t)$ represent the current and voltage in a long, insulated cable, it can be shown that both i and E satisfy the **telegraph equation**

$$u_{xx} = LCu_{tt} + (RC + LG)u_t + RGu.$$

The constants are defined in Exercise 10. If we may neglect L and G, we see that i and E satisfy

$$u_{xx} = RCu_t,$$

the heat/diffusion equation. Again, see Exercise 10.

Financial mathematics—the Black–Scholes equation

In the study of options pricing, a very important—and relatively new (1973)—model is the so-called **Black–Scholes equation**[‡]

$$\frac{\sigma^2 s^2}{2}V_{ss} + rsV_s - rV + V_t = 0.$$

Here, $V = V(s,t)$, t is time, s is the market value of the asset being optioned, σ is the constant *volatility* of the asset and r is the constant interest rate. Although the equation is separable, it's possible via a change of variable to turn it into the heat equation

$$u_\tau = \frac{\sigma^2}{2}u_{xx}.$$

See Exercise 11.

It's nice to interpret the heat equation graphically. If we plot u in terms of x, for fixed t, then u_{xx} is just the concavity of the graph. Since $u_t = \alpha^2 u_{xx}$, we have the following possibilities:

$$u \text{ concave down } \Rightarrow u_t < 0 \Rightarrow u \text{ decreasing}$$
$$u \text{ concave up } \Rightarrow u_t > 0 \Rightarrow u \text{ increasing}.$$

Thus, heat flows in such a way that it *smooths out* the temperature function. See Figure 2.4.

[‡]For a derivation see, e.g., *Financial Calculus* by Baxter and Rennie.

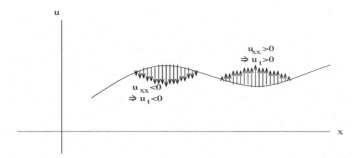

FIGURE 2.4
Relationship between temperature change and concavity of temperature graph.

We'll discuss initial and boundary conditions for the heat equation in Section 2.4.

EQUATIONS OF CONTINUITY AND CONSERVATION LAWS

When deriving the heat equation, if there is no source, we arrive at the statement

$$\sigma\rho A u_t \Delta x = -A[\Phi(x + \Delta x, t) - \Phi(x, t)]$$

(see (2.3) and (2.5)), or, letting $\Delta x \to 0$,

$$\sigma\rho u_t + \Phi_x = 0.$$

If σ and ρ were *not* constant, we would have

$$(\sigma\rho u)_t + \Phi_x = 0.$$

This is the one-dimensional version of what is called the **equation of continuity** for heat flow and, since it really is a statement of the conservation of heat energy, we refer to it as a **conservation law**. In general, the one-dimensional equation of continuity/conservation law in any similar situation is

$$\rho_t + \Phi_x = 0, {}^{\S}$$

where ρ is the concentration and Φ is the flux of the "substance" involved. (Convince yourself that $\sigma\rho u$ is, in fact, the "heat concentration.")

Examples abound—the equation of continuity shows up whenever we have something which is diffusing or flowing.

---

§In higher dimensions we have $\rho_t + \nabla \cdot \Phi = \rho_t + \operatorname{div} \Phi = 0.$

Fluid flow

Suppose we have a liquid in one-dimensional flow through a pipe with constant cross sectional area A. If $\rho(x,t)$ is the mass density (mass per unit volume) of the liquid, and if $v(x,t)$ is the velocity at point x, time t, then ρ and v satisfy the continuity equation

$$\rho_t + (\rho v)_x = 0.$$

See Exercise 13.

Electric current

If electricity flows along a very thin wire, with charge density $\rho(x,t)$ (charge per unit length) and current $i(x,t)$, then ρ and i satisfy the continuity equation

$$\rho_t + i_x = 0.$$

See Exercise 12.

One final note: The heat equation is homogeneous, of course. However, if we suppose there is an additional heat source/sink along or within the rod, given by

$$f(x,t) = \text{time rate at which heat is added/removed,}$$
$$\text{per unit volume, at point } x \text{ at time } t,$$

then the result is the *nonhomogeneous heat/diffusion equation*

$$u_t = \alpha^2 u_{xx} + F(x,t), \tag{2.7}$$

where

$$F(x,t) = \frac{1}{\sigma\rho} f(x,t)$$

(see Exercise 8). Of course, any source terms will appear in the equation of continuity, as well. So, for example, with heat source/sink f, the equation becomes

$$(\sigma\rho u)_t + \Phi_x = f.$$

Exercises 2.2

1. a) What are the dimensions of α^2 in the heat equation? (Use *calories* for heat content; you can use "time" and "length" for the remaining quantities.)

    b) What are the dimensions of the specific heat, σ?

    c) What are the dimensions of the thermal conductivity, k?

    d) What are the dimensions of the source term $F(x,t)$ in the nonhomogeneous heat equation?

e) What are the dimensions of k in the *diffusion* equation?

2. Write down the heat equation (homogeneous) which corresponds to the given data. (Throughout, heat is measured in calories, temperature is measured in °C and the other basic units are measured in centimeters, grams and seconds.)

   a) Thermal diffusivity $= .72$ cm^2/sec

   b) Specific heat $= .215$ cal/g-°C
      Density $= 2.7$ g/cm^3
      Thermal conductivity $= .63$ cal/cm-sec-°C

   c) Specific heat $= .09$ cal/g-°C
      Density $= 8.9$ g/cm^3
      Thermal conductivity $= .92$ cal/cm-sec-°C

3. Use Taylor series to show that we're justified in writing

   $$f(x + \Delta x) - f(x) = f'(x)\Delta x$$

   in these physical derivations. Thus, in (2.5), we can immediately write

   $$u_x(x + \Delta x, t) - u_x(x, t) = u_{xx}(x, t)\Delta x.$$

4. Show that if u is a solution of the heat equation $u_t = \alpha^2 u_{xx} + F(x, t)$, then so is $u + c_1 x + c_2$ for *any* choice of the constants c_1 and c_2.

5. Show that if u satisfies the heat equation $u_t = \alpha^2 u_{xx}$, and if we make the change of variable $\tau = t - t_0$, where t_0 is any constant, then the new function of x and τ still satisfies the same PDE. (This will mean that it doesn't matter what we call the *initial time* in our heat equation problems.)

6. Give the details in the derivation of the diffusion equation (2.7).

7. When deriving expression (2.2) for the heat content of the rod element, we approximated the temperature of the element using the temperature at its midpoint. However, we did not (and *could not*) use a similar approximation for u_x in expression (2.5). Provide a more rigorous derivation of the heat equation as follows:

   a) Write down an integral which represents the *exact* heat content of the element at time t.

   b) Replace expression (2.2) with this integral, and arrive at the heat equation, using the equation

   $$\frac{\partial}{\partial t} \int_a^b g(x, t)dx = \int_a^b g_t(x, t)dt,$$

   as well as the *Mean Value Theorem for Integrals*.

8.   a) Derive the nonhomogeneous heat equation (2.7). You will need to write down an approximate expression similar to (2.2) in order to deal with the source term $f(x, t)$.

     b) Do the same as in part (a) but, instead, proceeding as in Exercise 7 and using an integral to represent the effect of the source term.

9. In order to generalize a PDE like the heat equation, so that it is applicable to a greater variety of problems, it is necessary to *relax* the simplifying assumptions.

   a) Suppose that σ and ρ are not constant, but are functions of x, $\sigma = \sigma(x)$ and $\rho = \rho(x)$. Show that the heat equation still takes the same form

   $$u_t = \alpha^2 u_{xx},$$

   where $\alpha^2 = \alpha^2(x) = \frac{k}{\sigma(x)\rho(x)}$.

   b) Suppose, instead, that σ and ρ are constant, but that the thermal conductivity k depends on x, $k = k(x)$. Show that, in this case, the heat equation becomes

   $$u_t = \frac{1}{\sigma\rho} \frac{\partial}{\partial x}(k(x)u_x).$$

   (Hint: Remember the proof of the product rule.) Note: One easily can imagine more complicated situations where these basic quantities depend on t and even on the temperature, u. In the latter case, the heat equation will be nonlinear.

10. **Electric current in a long, insulated cable:** Suppose we have an insulated wire with current $i = i(x, t)$ and voltage $E = E(x, t)$. Let R be the resistance, L be the inductance, C be the capacitance and G be the conductance (or leakage), all per unit length and all constant, of the wire. Then, if we look at a differential element of the wire from x to $x + \Delta x$, the potential drop along this element gives us

$$-\Delta E = i(x, t)R\Delta x + L\frac{\partial i}{\partial t}(x, t)\Delta x.$$

Also, the capacitance and inductance lead to

$$-\Delta i = GE\Delta x + C\frac{\partial E}{\partial t}(x, t)\Delta x.$$

   a) Show that letting $\Delta x \to 0$ gives us the two first-order PDEs:

$$E_x + Ri + Li_t = 0,$$
$$i_x + GE + CE_t = 0.$$

b) Differentiate the first equation by x and the second equation by t, then, along with the second equation above, eliminate i_x and i_{xt} to arrive at

$$E_{xx} = LCE_{tt} + (RC + LG)E_t + RGE.$$

c) Instead, differentiate the first equation in part (a) by t and the second equation by x, then, along with the first equation in (a), eliminate E_t and E_{xt} to arrive at

$$i_{xx} = LCi_{tt} + (RC + LG)i_t + RGi.$$

Thus, E and i both satisfy the *telegraph equation*.

d) If the inductance and leakage are very small and can be neglected, show that E and i both satisfy the heat/diffusion equation.

11. a) Use separation of variables to turn the Black–Scholes equation

$$\frac{\sigma^2 s^2}{2}V_{ss} + rsV_s - rV + V_t = 0$$

into two ODEs.

b) Instead of separating variables, it is possible via change-of-variables to turn the Black–Scholes equation into the heat equation, as follows:

(i) Show that the change of variables $x = \ln(c_1 s)$, $\tau = c_2 - t$, for any choice of the constants c_1 and c_2, turns the Black–Scholes equation into the PDE

$$V_\tau = \frac{\sigma^2}{2}V_{xx} + \left(r - \frac{\sigma^2}{2}\right)V_x - rV.$$

(ii) Now show that the V_x and V terms can be eliminated by choosing appropriate constants α and β for which the substitution

$$V(x,\tau) = U(x,\tau)e^{\alpha x + \beta \tau}$$

turns this PDE into the heat equation $U_\tau = \frac{\sigma^2}{2}U_{xx}$.

12. Derive the equation of continuity for electric current.

13. Derive the equation of continuity for fluid flow. (Hint: The volume of liquid flowing through a cross section at x during time interval Δt is, approximately, length $\cdot$ cross sectional area $\cdot$ density $= [v(x,t)\Delta t] \cdot A \cdot \rho(x,t)$.)

2.3　The Wave Equation and the Vibrating String[¶]

Suppose we have a perfectly flexible string which, for the time being, is nailed down at both ends. If we pluck the string, or pull it and let it go, it will begin to *vibrate*. We would like to be able to determine the shapes of these vibrations. To that end, we derive a PDE—the **wave equation**—which must be satisfied by the *position function* of the string.

To be precise, suppose our string has length L and that it is nailed down at the endpoints $(0,0)$ and $(L,0)$, as in Figure 2.5. The string's motion is described by its position function

$$u(x,t) = \text{height of string at point } x, \text{ at time } t.$$

(Here we are tacitly assuming that each point of the string moves only in the u-direction and that the motion is restricted to the x-u plane. Thus, each point on the string occupies the same x-coordinate at all times; therefore, the statement "string at point x" is unambiguous.)

Now, we proceed as we did for the heat equation, that is, we look at an arbitrary *differential element* of the string, of length Δx, as in Figure 2.6. We intend to apply *Newton's 2nd Law, $F = ma$*, to the vertical motion of the string.

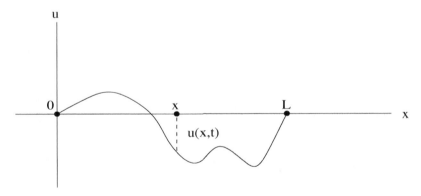

FIGURE 2.5
Displacement function for a string of length L.

[¶]For a *very* rigorous derivation of the wave equation, see *A First Course in Partial Differential Equations* by H. F. Weinberger.

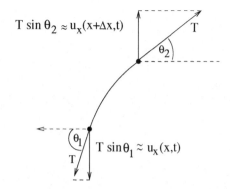

$T \sin \theta_2 \approx u_x(x+\Delta x, t)$

T

θ_2

θ_1

$T \sin \theta_1 \approx u_x(x, t)$

T

FIGURE 2.6
Forces acting on a differential element of length Δx.

We see that the only forces acting upon the isolated element are due to the pull of the rest of the string at either end—that is, the **tension** at each end. Assuming that the tension T is constant along the string (an approximation),* and that it is tangent to the string at each point (a consequence of the perfect flexibility of the string), we see that the vertical force components are $T \sin \theta_1$ at the left end and $T \sin \theta_2$ at the right end, as in Figure 2.6. Now, u_x is the slope of the string, so we have

$$\tan \theta_1 = u_x(x, t)$$

and

$$\tan \theta_2 = u_x(x + \Delta x, t).$$

This gives us a nonlinear relationship between the angles θ_1 and θ_2 and the function u. In order to get a *linear* relationship, we make the following assumption: u_x *is small.* In this case, θ_1 and θ_2 are small, as well, and for small values of θ, we have

$$\tan \theta \approx \theta \approx \sin \theta.$$

(Remember the Taylor series for these functions: $\tan \theta = \theta + \frac{\theta^3}{3} + \cdots$, $\sin \theta = \theta - \frac{\theta^3}{3!} + \cdots$.) Therefore, $T \sin \theta_1 \approx T u_x(x, t)$ and $T \sin \theta_2 \approx T u_x(x + \Delta x, t)$, and the sum of the vertical forces acting on the element is

$$T[u_x(x + \Delta x, t) - u_x(x, t)].$$

The mass of the element is $\rho \Delta x$, where ρ is the constant linear mass density (linear because, here, it represents mass per unit *length* of the string)

---

*This actually follows from *Hooke's Law*, along with the assumption, which we make below, that u_x is small.

and, since u_{tt} is the acceleration at each point, we approximate mass times acceleration by

$$\rho \Delta x \, u_{tt} \left(x + \frac{\Delta x}{2}, t \right).$$

Newton's 2nd Law applied to the element gives

$$\rho \Delta x \, u_{tt} \left(x + \frac{\Delta x}{2}, t \right) = T[u_x(x + \Delta x, t) - u_x(x, t)]$$

and, dividing by Δx and letting $\Delta x \to 0$, we have the **wave equation**

$$u_{tt} = c^2 u_{xx},$$

where $c = \sqrt{\frac{T}{\rho}}$ is called the **wave speed** (we'll see why in Section 5.3).

Notice that we have neglected, among other possible effects, the force of gravity. Suppose that we do wish to include such an external force, referred to as a **load**, given by

$$f(x, t) = \text{vertical force per unit length at point } x, \text{ at time } t.$$

Then, the result will be the *nonhomogeneous wave equation*

$$u_{tt} = c^2 u_{xx} + F(x, t), \tag{2.8}$$

where $F(x, t) = \frac{1}{\rho} f(x, t)$.

OTHER APPLICATIONS OF THE WAVE EQUATION

Longitudinal vibrations of a rod

Take a rod like the one we used in our derivation but, instead, attach one end to a wall and hit the other end with a hammer (horizontally, as in Figure 2.7). Then, each cross section will vibrate horizontally. If we let

$$u(x, t) = \text{left-right displacement of the cross section which}$$
$$\text{originally was at } x, \text{ at time } t,$$

then u satisfies

$$u_{tt} = k u_{xx},$$

the wave equation! Here, the constant k is a measure of how elastic the rod is, and is called **Young's modulus.**[†]

[†]After the scientist Thomas Young (1773–1829), who also made major contributions to the study of light.

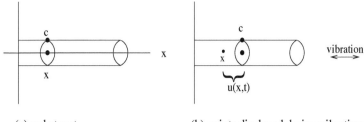

(a) rod at rest   (b) points displaced during vibration

FIGURE 2.7
Longitudinal vibration of a rod.

Torsional vibrations of a rod

Take the *same* rod (in this case, assuming its cross sections are circular) and, instead, *twist* it and release. The rod then undergoes *torsional vibration* (see Figure 2.8). If we let

$\theta(x,t) = angle\ of\ twist$ of the cross section which is at x, at time t,

then θ satisfies ... surprise ...

$$\theta_{tt} = c^2\theta_{xx},$$

where $c^2 = \frac{G}{\rho}$, ρ is the density and G is called the *shear modulus* of the material.

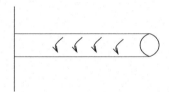

FIGURE 2.8
Torsional vibration of a rod.

Electric current in a long, insulated cable

In the previous section, we showed that the current i and voltage E both satisfy the telegraph equation

$$u_{xx} = LCu_{tt} + (RC + LG)u_t + RGu.$$

If, instead of neglecting L and G, we are able to neglect R and G (which turns out to be the case when dealing with high frequencies), then i and E both satisfy

$$u_{xx} = LCu_{tt}.$$

Again, it's good to look at a graphical integration of the wave equation. Fixing t, we again have that u_{xx} is the concavity, and it is proportional to the *acceleration* u_{tt}. Hence,

$$u \text{ concave down } \Rightarrow u_{tt} < 0 \Rightarrow \text{ downward force/acceleration}$$
$$u \text{ concave up } \Rightarrow u_{tt} > 0 \Rightarrow \text{ upward force/acceleration.}$$

See Figure 2.9.

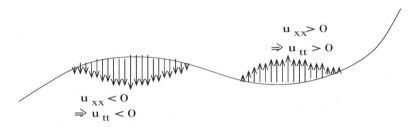

FIGURE 2.9
Relationship between acceleration and concavity of string.

Exercises 2.3

1. Write down the homogeneous wave equation for the following data:
   tension = 6 dynes
   density = 2 g/cm

2. Show that, if u is a solution of the wave equation $u_{tt} = c^2 u_{xx} + F(x,t)$, then so is $u + c_1 x + c_2 t + c_3$ for *any* choice of the constants c_1, c_2 and c_3.

3. As in Exercise 9 of the previous section, consider what happens with the homogeneous wave equation when the assumptions are relaxed.

   a) Suppose that the tension remains constant, but the density is a function of x, $\rho = \rho(x)$. Show that the form of the wave equation remains the same, that is, that the resulting PDE is

   $$u_{tt} = \frac{T}{\rho(x)} u_{xx}.$$

   b) Suppose instead that the density is constant, but that $T = T(x)$. Show that, in this case, the PDE becomes

   $$u_{tt} = \frac{1}{\rho} \frac{\partial}{\partial x}(T(x) u_x).$$

4. In deriving the homogeneous wave equation, suppose that the string is vibrating in a medium which offers resistance. The simplest model for such frictional resistance is a force per unit length which is proportional to the velocity. Show that the inclusion of such a *damping* term leads to the **damped** or **dissipative wave equation**

$$u_{tt} = c^2 u_{xx} - \beta u_t.$$

(You should assume that the medium is homogeneous so that the damping coefficient is constant.)

2.4   Initial and Boundary Conditions for the Heat and Wave Equations

As we suggested in Section 1.3, in order for the heat equation for a finite rod to be well-posed,* we must be supplied with an initial condition and two boundary conditions, one at each end. The initial condition generally is of the form

$$u(x, t_0) = f(x), \qquad 0 \le x \le L$$

and specifies the *initial temperature* at each point of the rod.

It turns out that the wave equation for a finite string requires *two* initial conditions, as well as two boundary conditions, in order to be well-posed. These initial conditions generally are of the form

$$u(x, t_0) = f(x), \qquad 0 \le x \le L, \tag{2.9}$$

$$u_t(x, t_0) = g(x), \qquad 0 \le x \le L, \tag{2.10}$$

and specify, respectively, the *initial shape* of the string and the *initial velocity* at each point.

BOUNDARY CONDITIONS FOR THE HEAT EQUATION

Temperature/concentration specified at an end (Dirichlet condition[†])

---

*Most of the problems we consider will be well-posed. For a detailed discussion of the issues involved, you may want to look at some of the higher level books mentioned in the references. In particular, the classic text by Churchill and Brown, *Fourier Series and Boundary Value Problems*, contains an excellent treatment of the uniqueness of solutions of the Big Three PDEs. Also, see Appendix C.

[†]After the Prussian mathematician Peter Gustav Lejeune-Dirichlet (1805–1859).

Suppose the left or right end is held at $0°$ throughout the duration of the experiment. We then have the *homogeneous* **Dirichlet boundary condition**

$$u(0,t) = 0, \qquad t \geq t_0$$

or

$$u(L,t) = 0, \qquad t \geq t_0. \tag{2.11}$$

More generally, an end may be held at any temperature. In fact, the temperature need not be constant, but can be a given function of t, in which case we have the *nonhomogeneous* Dirichlet condition

$$u(0,t) = g_1(t), \qquad t \geq t_0$$

or

$$u(L,t) = g_2(t), \qquad t \geq t_0. \tag{2.12}$$

Flux specified at an end (Neumann condition[‡])

If an end of the rod is insulated, so that no heat enters or leaves the rod at that end, then the flux is zero there. From (2.4), we then have the *homogeneous* **Neumann boundary condition**

$$u_x(0,t) = 0, \qquad t > t_0,\,^{§}$$

or

$$u_x(L,t) = 0, \qquad t > t_0. \tag{2.13}$$

As with the Dirichlet condition, the flux need not be zero, but may be any specified function of t, in which case (2.4) gives us the *nonhomogeneous* Neumann condition

$$k u_x(0,t) = g_1(t), \qquad t > t_0,$$

or

$$-k u_x(L,t) = g_2(t), \qquad t > t_0. \tag{2.14}$$

[‡]After another Prussian/German, Carl Gottfried Neumann (1832–1925).

[§]You may notice that the boundary conditions are specified in some cases for $t \geq t_0$ and in others for $t > t_0$. Although this need not concern us, it is a consequence of the various existence-uniqueness-of-solution theorems connected with these problems.

End in contact with material held at constant temperature/concentration (Robin condition)

Suppose, instead, that an end of the rod is immersed in a large container of water which is held at $0°$. (We suppose the container is of sufficient size, and the water is in motion, so that the water in contact with the end of the rod remains at $0°$.) To deal with this type of condition, we assume that **Newton's Law of Cooling** applies, namely, that the outward flux of heat at such a boundary is proportional to the temperature difference between the two media. In this particular case, then, we have the *homogeneous* **Robin boundary condition**

$$ku_x(0,t) = h[u(0,t) - 0], \qquad t > t_0,$$

or

$$-ku_x(L,t) = h[u(L,t) - 0], \qquad t > t_0,$$

where h is called the **heat-exchange coefficient**.

Again, more generally, the temperature of the water/medium may be specified as a function of t, in which case we get the *nonhomogeneous* Robin condition

$$ku_x(0,t) = h[u(0,t) - g_1(t)], \qquad t > t_0,$$

or

$$-ku_x(L,t) = h[u(L,t) - g_2(t)], \qquad t > t_0.$$

Figure 2.10 illustrates a typical set of boundary conditions.

Note that each of these three types of boundary conditions is linear. In the following examples, we have some typical heat equation initial-boundary-value problems for a finite rod.

Example 1

$$u_t = u_{xx}, \qquad 0 < x < 5, t > 0 \P$$
$$u(x,0) = x(5-x), \qquad 0 \le x \le 5$$
$$u(0,t) = u(5,t) = 0, \qquad t > 0.$$

¶From now on we will not include the domain $a < x < b$, $t > t_0$, unless it is not obvious.

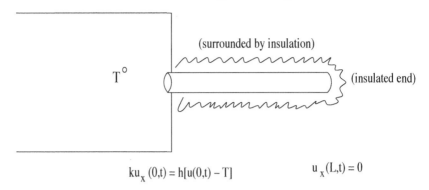

$$ku_x(0,t) = h[u(0,t) - T] \qquad u_x(L,t) = 0$$

FIGURE 2.10
A typical setup for the heat problem.

Example 2

$$\dot{u}_t = 1.17u_{xx}$$
$$u(x,0) = \sin x - 3\cos 4x$$
$$u_x(0,t) = 3, u_x(9,t) - 2u(9,t) = e^{-t}.$$

BOUNDARY CONDITIONS FOR THE WAVE EQUATION

It turns out that the Dirichlet, Neumann and Robin boundary conditions are applicable to the wave equation, as well. The most common conditions found in connection with the vibrating string are the homogeneous Dirichlet condition

$$u(0,t) = 0 \quad \text{or} \quad u(L,t) = 0, \qquad t > t_0, \tag{2.15}$$

and the homogeneous Neumann condition

$$u_x(0,t) = 0 \quad \text{or} \quad u_x(L,t) = 0, \qquad t > t_0. \tag{2.16}$$

The physical meaning of each is fairly obvious: (2.15) just means that the end is nailed down at height zero, exactly as in Figure 2.11, while (2.15) means that the slope of the end is held at zero. Alternatively, since Tu_x is the vertical component of the tension, (2.16) may be interpreted as saying that there is no such force exerted upon the end of the string and, thus, nothing is "pulling" the slope away from zero.

It is not difficult to imagine generalizing these boundary conditions to (2.10) or (2.12). Similar ideas are used to derive the Robin condition for the wave equation, as well.‖

‖For a more comprehensive, physical look at the boundary conditions, see Stanley Farlow's excellent physical/intuitive book, *Partial Differential Equations for Scientists and Engineers.*

The following, then, are typical examples of the wave equation initial-boundary-value problem for a finite string.

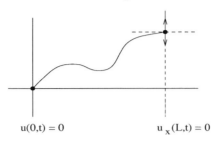

FIGURE 2.11
Boundary conditions for the vibrating string.

Example 3

$$u_{tt} = u_{xx}$$
$$u(x,0) = x(1-x)$$
$$u_t(x,0) = \sin x$$
$$u(0,t) = 2, u(1,t) = -3.$$

Example 4

$$u_{tt} = 5u_{xx}$$
$$u(x,0) = 0$$
$$u_t(x,0) = x^2$$
$$u(0,t) = te^{-t}, u_x(3,t) = 0.$$

Exercises 2.4

1. Set up the heat/diffusion initial-boundary-value problem for the given data:

    a) Length of rod = 5 cm
    Thermal diffusivity = 1.2 cm^2/sec
    Left end: held at 20°
    Right end: insulated
    Initial temperature: 50°

    b) Length = 3.7 cm
    Specific heat = .215 cal/g-°C
    Density = 2.78/cm^3
    Thermal conductivity = .63 cal/cm-sec-°C
    Left end: in container of water at 0° (h = .8 cal/cm^2-sec-°C)
    Right end: held at 0°
    Initial temperature: $f(x) = x(3.7 - x)$

   c) Length = 22 cm
      Specific heat = .09 cal/g-°C
      Density = 8.9 g/cm^3
      Thermal conductivity = .92 cal/cm-sec-°C
      Left end: insulated
      Right end: in container of water at 15° (h = .65 cal/cm^2-sec-°C)
      Initial temperature: linear function, 0° at left end, 15° at right
        end

2. Two rods, each 7 cm long, are identical, except that one is at a constant temperature of 40° and the other is at a constant temperature of $-20°$. Our experiment starts when these two rods are placed in perfect contact with one another, end to end, so that they form, in effect, one rod. If the other ends of the rods are insulated, write down the initial and boundary conditions for this problem.

3. **Steady state** solutions of the heat equation (or any other equation involving time, for that matter) are solutions which are time independent, that is, of the form $u = u(x)$. Find all steady state solutions of the heat equation for a rod of length L, if

   a) The left end is held at constant temperature T_1, and the right end is held at T_2.

   b) The left end is insulated, and the right end is held at constant temperature T.

   c) The left end is held at constant temperature T, and the right end satisfies the Robin condition $ku_x(L,t) + hu(L,t) = 0$.

4.   a) Set up a wave initial-boundary-value problem for the following **plucked string**. Suppose we have a string of length = 8 cm, with wave speed = 1 cm/sec, with each end held at the same height. The string is set into motion by the act of *plucking* it, that is, by holding it at its midpoint, pulling up a distance of 2 cm and releasing it. (This act of plucking is essentially what goes on when one plays a guitar, mandolin, harp or similar instrument.)

   b) Do the same for a string of length = 4 cm, with tension = 3 dynes and density = 1 g/cm. Suppose that the left end is held in position 1 cm below the x-axis, while the right end is held 2 cm above the x-axis. The motion is started by taking the taut string at rest and hitting it with a long, flat object which has a downward velocity of 5 cm/sec and which hits, simultaneously, each point on the string (excepting the endpoints).

   c) Do the same for the string in part (b), except suppose that the object with which the string is struck is only 2 cm long and that it hits the string symmetrically with respect to its midpoint. (A piano string is sounded this way, by a hammer striking a segment of the string. Of course, in this case, the string is *much* longer than 4 cm, and the hammer hits a very small fraction of the total length.)

5. Suppose we include the effect of gravity on a string of length L, with wave speed c and constant density ρ.

   a) Show that the string satisfies the PDE $u_{tt} = c^2 u_{xx} - g$, where g is the constant acceleration due to gravity at the earth's surface.

   b) If the ends of the string are nailed down at the same height, what shape does the string take if it just hangs and doesn't vibrate?

6. Apply the change of variable $\tau = t - t_0$ to the problem

$$u_t = \alpha^2 u_{xx}$$
$$u(x, t_0) = f(x)$$
$$u(0, t) = u(L, t) = 0.$$

Thus, *without loss of generality* (mathematicians say *WLOG*), we need only consider such heat/diffusion problems with initial time $t_0 = 0$. (You should convince yourself that this is true for *any* of the initial-boundary-value problems we have discussed.)

7. a) Given the heat problem with *nonhomogeneous* boundary conditions

$$u_t = u_{xx}$$
$$u(x, 0) = f(x)$$
$$u(0, t) = 10, u(5, t) = 30,$$

   find a function $v(x) = c_1 x + c_2$ so that the new unknown

$$w(x, t) = u(x, t) - v(x)$$

   satisfies the heat problem with *homogeneous* boundary conditions. What is the new initial condition for w?

   b) Do the same for the general problem

$$u_t = \alpha^2 u_{xx}$$
$$u(x, 0) = f(x)$$
$$u(0, t) = T_1, u(L, t) = T_2,$$

   where T_1 and T_2 are constants. Therefore, if we can solve the heat problem with *homogeneous* Dirichlet conditions, then we know how to solve it when it has *nonhomogeneous*, constant Dirichlet conditions.

8. a) Do the same as in Exercise 7 for the problem with mixed conditions

$$u_t = 3u_{xx}$$
$$u(x, 0) = f(x)$$
$$u(0, t) = 12, \, u_x(5, t) + u(5, t) = -2.$$

b) Do the same for the problem

$$u_t = 4u_{xx}$$
$$u(x, 0) = f(x)$$
$$u_x(0, t) = 5, \, u_x(2, t) = 3.$$

What's the trouble in this case?

9. Suppose that we have a rod immersed in a large container of water which is held at a temperature of $0°$, and suppose that the rod *is not insulated at all*. Then, it turns out, the temperature function satisfies a PDE of the form

$$u_t = \alpha^2 u_{xx} - \beta u.$$

Briefly explain where the term $-\beta u$ comes from.

2.5   Laplace's Equation—The Potential Equation

In Chapter 9, when we generalize the heat and wave equations to higher dimensions, we'll get

two-dimensional heat: $u_t = \alpha^2(u_{xx} + u_{yy})$,

three-dimensional heat: $u_t = \alpha^2(u_{xx} + u_{yy} + u_{zz})$,

two-dimensional wave: $u_{tt} = c^2(u_{xx} + u_{yy})$,

three-dimensional wave: $u_{tt} = c^2(u_{xx} + u_{yy} + u_{zz})$.

It should not be hard to imagine, then, that the expression on the right side of these equations is an important one. Indeed, it may be the most important such expression in all of applied mathematics and mathematical physics.

Definition 2.2 *Given the function $u = u(x, y)$, the function*

$$u_{xx} + u_{yy}$$

*is called the two-dimensional **Laplacian** of u. (Similarly, the one-dimensional Laplacian is u_{xx}, and the three-dimensional Laplacian is $u_{xx} + u_{yy} + u_{zz}$.)*

The *formal* notation for the Laplacian is

$$\nabla^2 u = \nabla \cdot \nabla u = u_{xx} + u_{yy},^*$$

where, from vector analysis, ∇ is the operator

$$\nabla = \hat{i}\frac{\partial}{\partial x} + \hat{j}\frac{\partial}{\partial y},$$

so that the gradient of the scalar field $u(x, y)$ is written

$$\textbf{grad } u = \nabla u = u_x \hat{i} + u_y \hat{j},$$

while the divergence of the vector field $\vec{F} = F_1(x, y)\hat{i} + F_2(x, y)\hat{j}$ is written

$$\text{div } \vec{F} = \nabla \cdot \vec{F} = F_{1x} + F_{2y}.$$

From now on, we may write the heat equation as

$$u_t = \alpha^2 \nabla^2 u$$

and the wave equation as

$$u_{tt} = c^2 \nabla^2 u,$$

where the number of space dimensions should be obvious from the setting.

Now, when Fourier began to study the heat equation, he first looked for *steady state* solutions of the two-dimensional heat equation. These are time-independent solutions of the heat equation and represent, in some sense, the "final state" of the heat problem, after it has "settled down." These solutions will satisfy $u_t = 0$, so they also must satisfy

$$u_{xx} + u_{yy} = \nabla^2 u = 0. \tag{2.17}$$

The same is true, of course, for steady state solutions of the three-dimensional heat equation, as well as the two- and three-dimensional wave equations.

Equation (2.17), the third of the Big Three PDEs, is known as **Laplace's equation**[†] (in two dimensions). Its appearance as a *limiting case* of the heat and wave equations is interesting, but of much greater importance is the role that Laplace's equation plays in *potential theory*.

---

*There is no ambiguity in the notation $\nabla \cdot \nabla = \nabla^2$, since, of the three possibilities $\nabla(\nabla u)$, $\nabla \times (\nabla u)$ and $\nabla \cdot (\nabla u)$, the first makes no sense and the second is identically zero. By the way, the Laplacian shows up often in the *theory of elasticity*. For example, the PDE for a vibrating *plate* is $(\nabla^2)^2 u = \nabla^4 u = -cu_{tt}$, while for a vibrating so-called shallow spherical shell we have $\nabla^6 u - c_1 \nabla^2 u = -c_2 u_{tt}$. One often uses the notation $\Delta = \nabla^2$. Here, the Laplacian Δ sometimes is called the *harmonic operator*, and $\Delta^2 = \nabla^4$ is called the *biharmonic operator*.

[†]Solutions of Laplace's equation are called **harmonic functions**.

Again, from vector analysis, remember that a *potential* for a vector function $\vec{F}$ is a scalar function ϕ for which

$$\mathbf{grad}\ \phi = \nabla\phi = \vec{F}.$$

In physical problems, involving gravitation, electricity and magnetism, fluid flow and the like, it is very convenient if we can find a potential for a force field or velocity field because

(1) It reduces the number of functions that we must deal with.

(2) It makes it *very* easy to calculate line integrals of the vector field.

Suppose, for example, that we'd like to find the electric field $\vec{E} = \vec{E}(x, y, z) = E_1\hat{\imath} + E_2\hat{\jmath} + E_3\hat{k}$ due to a certain distribution of electric charges. To be more specific, suppose that electric charge is distributed along the boundary of a rectangular box, and we'd like to calculate $\vec{E}$ inside the box. We will need two of the four famous *Maxwell's*[‡] *equations* from the theory of electricity and magnetism. (See Appendix D.)

One of Maxwell's equations says that

$$\mathbf{curl}\ \vec{E} = \vec{\nabla} \times \vec{E} = (E_{3y} - E_{2z})\hat{\imath} + (E_{1z} - E_{3x})\hat{\jmath} + (E_{2x} - E_{1y})\hat{k} = \vec{0}.$$

This, in turn, implies that $\vec{E}$ does, indeed, have a potential, $\phi = \phi(x, y, z)$, so

$$-\mathbf{grad}\ \phi = -\nabla\phi = \vec{E}.^{[§]}$$

Then, another of Maxwell's equations says that, in regions where there is no charge, we must have

$$\mathrm{div}\ \vec{E} = \nabla \cdot \vec{E} = 0. \tag{2.18}$$

Hence,

$$-\mathrm{div}\ \vec{E} = \nabla \cdot (\nabla\phi) = \nabla^2\phi = 0,$$

and ϕ satisfies Laplace's equation. We may then solve this equation and determine ϕ from its given values on the boundary. The function ϕ is called the **electric potential**.

Not *every* potential function ϕ satisfies Laplace's equation, but since the potential functions in many important applied problems *do* satisfy it, the Laplace equation often is referred to as the **potential equation**.

We'll concentrate on solving the two-dimensional Laplace equation on a rectangle; in Chapter 9 we will solve it inside a circle.

[‡]James Clerk Maxwell (1831–1879), probably the greatest theoretical physicist of the 19th century, and one of the greatest of all time.

[§]From a theorem in vector analysis: This will follow if the region throughout which $\nabla \times \vec{E} = \vec{0}$ is simply-connected. The minus sign is a convention; see footnote on page 70.

BOUNDARY CONDITIONS FOR LAPLACE'S EQUATION

Since time is not involved, we have only *Laplace boundary-value problems.* As with the heat and wave equations, at this point we are interested in solving Laplace's equation on finite domains and, in particular, on a rectangle (we shall treat the circle in Chapter 9, where we consider polar coordinates). The examples below are typical Laplace boundary-value problems on rectangular domains.

Example 1

$$u_{xx} + u_{yy} = 0, \qquad 0 < x < 3, 0 < y < 2,$$
$$u(x,0) = 0, \qquad 0 < x < 3,$$
$$u_y(x,2) = 0, \qquad 0 < x < 3,$$
$$u_x(0,y) = y^2, \qquad 0 < y < 2,$$
$$u(3,y) = 2, \qquad 0 < y < 2.$$

See Figure 2.12.

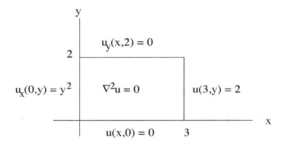

FIGURE 2.12
The Laplace equation boundary-value problem from Example 1.

Example 2

$$u_{xx} + u_{yy} = 0, \qquad 0 < x < 4, 0 < y < 5,$$
$$u_y(x,0) - 2u(x,0) = 2x,$$
$$u(x,5) = 3,$$
$$u(0,y) = y^2,$$
$$u_x(4,y) + 3u(4,y) = 0.$$

In practice, we often encounter Laplace's equation in situations where we have just one type of boundary condition along the entire boundary. These problems are so important that they are just called **the Dirichlet Problem** or **the Neumann Problem** (or, sometimes, the **Interior** or **Exterior**

Dirichlet Problem, etc., depending on whether we are solving the PDE on the interior or the exterior of the boundary).

Finally, going back to the problem of finding the electric potential, if we are doing so throughout a region that *does* contain electric charge, then we must use the more general version of Maxwell's equation (2.18), which is

$$\text{div } \vec{E} = \nabla \cdot \vec{E} = 4\pi\rho,$$

where $\rho = \rho(x, y, z)$ is the density of electric charge in the region. Then, we have

$$-\nabla \cdot \vec{E} = \nabla^2\phi = -4\pi\rho.$$

This is the *nonhomogeneous* Laplace equation. The general equation of this form,

$$\nabla^2 u = f,\P$$

is called **Poisson's**‖ **equation**.

Exercises 2.5

1. Find all $u = u(x)$ which are solutions of Laplace's equation $u_{xx} + u_{yy} = 0$.

2. Find all polynomials of the form $ax^2 + bxy + cy^2 + dx + fy + g$ which satisfy Laplace's equation.

3. Show that the function $u = -\ln\sqrt{x^2 + y^2}$ satisfies Laplace's equation (except at the origin, of course). This function is called the **logarithmic potential**.

4. Show that if u is a solution of Poisson's equation $u_{xx} + u_{yy} = F(x, y)$, then so is $u + c_1 xy + c_2 x + c_3 y + c_4$ for *any* choice of the constants c_1, c_2, c_3 and c_4.

5. In the field of *complex analysis*, we consider *complex-valued functions* of the form $f(x, y) = u(x, y) + iv(x, y)$. In order to decide if f is differentiable, we need to see if u and v satisfy the *Cauchy–Riemann equations* $u_x = v_y$ and $u_y = -v_x$. Show that if u and v satisfy these equations, then u and v both are solutions of Laplace's equation. (Assume that u and v have continuous second derivatives, so that the order of differentiation doesn't matter.)

¶Actually, we usually write $\nabla^2 u = -f$ or, commonly in more advanced texts, $-\Delta u = f$. There are various reasons: Signs of eigenvalues, the fact that an electron is negatively charged, etc.

‖After Siméon-Denis Poisson (1781–1840), known also for his contributions to probability theory.

6. a) Show that the two-dimensional Laplacian is *translation-invariant*, that is, show that if the independent variables undergo a *translation* to the new variables $x' = x + a$, $y' = y + b$, where a and b are constants, then

$$u_{xx} + u_{yy} = u_{x'x'} + u_{y'y'}.$$

   b) Show that the two-dimensional Laplacian is *rotation-invariant*, that is, show that if the independent variables undergo a *rotation* through angle α to the new variables $x' = x \cos \alpha + y \sin \alpha$, $y' = -x \sin \alpha + y \cos \alpha$, where α is a constant, then

$$u_{xx} + u_{yy} = u_{x'x'} + u_{y'y'}.$$

7. Show that Poisson's equation in polar coordinates is

$$u_{rr} + \frac{1}{r} u_r + \frac{1}{r^2} u_{\theta\theta} = F(r, \theta),$$

   where $F(r, \theta) = f(r \cos \theta, r \sin \theta)$.

8. Use the polar form from Exercise 7 to find all θ-independent solutions of Laplace's equation, that is, all solutions $u = u(r)$.

9. Show that, by separating the time variable in the two-dimensional heat and wave equations, that is, by looking for product solutions of the form $u(x, y, t) = T(t)\phi(x, y)$, we are led to the **Helmholtz equation,** $\nabla^2 \phi + \lambda \phi = 0$.

2.6 Using Separation of Variables to Solve the Big Three PDEs

We will solve the Big Three PDEs as we started to solve the heat equation in Sections 1.6 and 1.7, that is, we'll separate the PDE and boundary conditions and then see what happens. Let's look, again, at an example of the heat problem.

Example 1 Try to solve the heat initial-boundary-value problem

$$\left.\begin{array}{l} u_t = u_{xx} \\ u(x, 0) = f(x) \\ u(0, t) = u(\pi, t) = 0 \end{array}\right\} \tag{2.19}$$

(we'll look at specific cases of the function $f(x)$ when the time is right).

Again, we separate the PDE and boundary conditions. We let $u(x,t) = X(x)T(t)$, and the PDE becomes

$$XT' = X''T$$

or

$$\frac{T'}{T} = \frac{X''}{X} = -\lambda, \quad \text{constant}$$

or

$$T' + \lambda T = 0, X'' + \lambda X = 0 \quad \text{(for the same } \lambda\text{)}.$$

The boundary conditions become

$$X(0)T(t) = X(\pi)T(t) = 0,$$

and, as before, since $T(t) \equiv 0$ leads to the zero-solution, we have

$$X(0) = X(\pi) = 0.$$

We first solve the boundary-value problem

$$X'' + \lambda X = 0$$
$$X(0) = X(\pi) = 0.$$

Proceeding as in Example 1 in Section 1.7, we find that the eigenvalues and eigenfunctions are

$$\lambda_n = n^2, \quad X_n(x) = \sin nx, \quad n = 1, 2, 3, \ldots .$$

(You should carry out the calculations yourself.) Therefore, these λ_n are the only values of λ for which the X-boundary-value problem has nontrivial solutions, *so we need only solve the T-equation for these values of λ!* We have

$$T' + \lambda_n T = 0$$

or

$$T' + n^2 T = 0,$$

which has solution

$$T_n(t) = e^{-n^2 t}, \quad n = 1, 2, 3, \ldots .$$

Therefore, the nontrivial product solutions of the original problem, including the boundary conditions (but *not* the initial condition), are

$$u_n(x, t) = e^{-n^2 t} \sin nx, \qquad n = 1, 2, 3, \ldots .$$

Also, since the PDE *and* boundary conditions are homogeneous, any linear combination of these functions is again a solution. So any function of the form

$$u(x, t) = \sum_{n=1}^{\infty} c_n e^{-n^2 t} \sin nx \qquad (2.20)$$

$$= c_1 e^{-t} \sin x + c_2 e^{-4t} \sin 2t + c_3 e^{-9t} \sin 3t + \cdots$$

is a solution (as long as it "converges nicely enough," which turns out to be the case).

Now, at this point we have found only linear combinations of product solutions of the system

$$u_t = u_{xx}$$
$$u(0, t) = u(\pi, t) = 0.$$

However, surprisingly, we'll see in the next chapter that, for just about *any* function $f(x)$, the solution of the original problem (2.19) must be of the form (2.20). More precisely, for "any" function $f(x)$, we'll be able to determine the constant c_n so that

$$u(x, 0) = f(x) = \sum_{n=1}^{\infty} c_n \sin nx$$

is satisfied! That being the case, we will call (2.20) the **general solution** of the heat initial-boundary-value problem (2.19) (realizing, of course, that, since (2.19) is well-posed, there really will be only one solution).

The question we must answer, then, is: Can we find values of c_n, $n = 1, 2, 3, \ldots$, for which

$$f(x) = \sum_{n=1}^{\infty} c_n \sin nx?$$

The answer will have to wait until Chapter 3 for most functions f. However, there are some special cases for which we already can find the values of c_n.

Example 2 Suppose $f(x) = 2 \sin x - 5 \sin 3x$ in the previous example. Then, we'd like to find the values of c_n for which

$$2 \sin x - 5 \sin 3x = \sum_{n=1}^{\infty} c_n \sin nx$$

$$= c_1 \sin x + c_2 \sin 2x + c_3 \sin 3x + \cdots .$$

It should be clear that $c_1 = 2$, $c_2 = 0$, $c_3 = -5$, $c_4 = c_5 = \cdots = 0$ does the job.* Therefore, the unique solution to (2.19), with $f(x) = 2\sin x - 5\sin 3x$, is just (2.20) with these values plugged in, that is,

$$u(x,t) = 2e^{-t}\sin x + 0e^{-4t}\sin 2x - 5e^{-9t}\sin 3x$$
$$+ 0e^{-16t}\sin 4t + 0 + 0 + \cdots$$
$$= 2e^{-t}\sin x - 5e^{-9t}\sin 3x.$$

Example 3 Solve the initial-boundary-value problem

$$u_t = 3u_{xx}$$
$$u(x,0) = 17\sin \pi x$$
$$u(0,t) = u(4,t) = 0.$$

We begin by separating the PDE and the boundary conditions. Letting $u(x,t) = X(x)T(t)$, the PDE becomes

$$XT' = 3X''T \quad \text{or} \quad \frac{X''}{X} = \frac{T'}{3T} = -\lambda$$

so the separated ODEs become

$$X'' + \lambda X = 0, \quad T' + 3\lambda T = 0.$$

Separating the boundary conditions gives us

$$X(0) = X(4) = 0.$$

We now solve the X-boundary-value problem. The result is (see Exercise 1) that the eigenvalues and corresponding eigenfunctions are

$$\lambda_n = \frac{n^2\pi^2}{16}, \quad X_n(x) = \sin \frac{n\pi x}{4}, \quad n = 1,2,3,\ldots .$$

Then, solving the T-equation for each λ_n gives us

$$T_n(t) = e^{-3\lambda_n t} = e^{-\frac{3n^2\pi^2 t}{16}}, \quad n = 1,2,3,\ldots .$$

So the general solution is

$$u(x,t) = \sum_{n=1}^{\infty} c_n e^{-\frac{3n^2\pi^2 t}{16}} \sin \frac{n\pi x}{4} \tag{2.21}$$

---

*Are these the *only* numbers c_n that do the job? The answer is *yes*—we'll say more in Chapter 3.

and, applying the initial condition, we see that we need to find the values of c_n, $n = 1, 2, 3, \ldots$, for which

$$u(x, 0) = 17 \sin \pi x = \sum_{n=1}^{\infty} c_n \sin \frac{n \pi x}{4}.$$

We see that we get the term $\sin \pi x$ on the right side when $n = 4$. Therefore, $c_4 = 17$, $c_1 = c_2 = c_3 = c_5 = c_6 = \cdots = 0$, and our solution is

$$u(x, t) = 17 e^{-3\pi^2 t} \sin \pi x$$

(which, as before, is the general solution (2.21), with the particular values of c_n plugged in).

Exercises 2.6

1. Solve the X-eigenvalue problems in Examples 1 and 3, as was done in Section 1.7.

2. Solve the heat equation initial-boundary-value problem

$$u_t = 2u_{xx}$$
$$u(x, 0) = -\sin 3\pi x + \frac{1}{4} \sin 6\pi x$$
$$u(0, t) = u(1, t) = 0.$$

(Refer to Example 1 in Section 1.7.)

3. Solve the heat equation initial-boundary-value problem

$$u_t = u_{xx}$$
$$u(x, 0) = 3 + \cos 2\pi x$$
$$u_x(0, t) = u_x(3, t) = 0.$$

(Refer to Example 2 in Section 1.7.)

4. Solve the heat equation initial-boundary-value problem

$$u_t = u_{xx}$$
$$u(x, 0) = 7 \cos \frac{5x}{2}$$
$$u_x(0, t) = u(\pi, t) = 0.$$

(Refer to Exercise 3 in Section 1.7.)

5. Find the general solution of the heat equation $u_t = \alpha^2 u_{xx}$, subject to the boundary conditions $u(0, t) = u(L, t) = 0$.

6. Do the same as in Exercise 5, but subject to $u_x(0,t) = u_x(L,t) = 0$.

7. Find the general solution of the wave equation $u_{tt} = u_{xx}$, subject to the boundary conditions $u(0,t) = u(1,t) = 0$. (Refer to Example 1 in Section 1.7.)

8. Find the unique solution of the wave equation problem in Exercise 7, subject to the initial conditions

$$u(x,0) = 2\sin 3\pi x \quad \text{and} \quad u_t(x,0) = 5\sin \pi x.$$

9. Solve the wave equation initial-boundary-value problem

$$u_{tt} = 4u_{xx}$$
$$u(x,0) = 5\sin 2x - 7\sin 4x$$
$$u_t(x,0) = 0$$
$$u(0,t) = u(\pi,t) = 0.$$

10.   a) In Exercise 27 of Section 1.6, you were asked to find all product solutions of Laplace's equation $u_{xx} + u_{yy} = 0$. Now, find which of the product solutions also satisfy the boundary conditions $u(0,y) = u(\pi,y) = 0$, and use these to form the general solution of Laplace's equation subject to these conditions.

   b) Solve the Laplace boundary-value problem

$$u_{xx} + u_{yy} = 0$$
$$u(x,0) = \sin 3x$$
$$u(x,1) = \sin x$$
$$u(0,y) = u(\pi,y) = 0.$$

11. **MATLAB:** Plot the solutions of the heat problems in Exercises 2, 3 and 4. You should plot them in two ways:

   i) As surfaces $u = u(x,t)$ in three-dimensional space

   ii) As functions of x in the x-u plane, for various (fixed) values of t

   In each case, investigate the behavior as $t \to \infty$, that is, the *steady state* behavior.

12. **MATLAB:** Plot the solution of Exercise 9 in the x-u plane, for various values of t.

Prelude to Chapter 3

In this chapter we show that many of the "usual" functions with which we are familiar or which occur in physical problems can be rewritten as—that is, *expanded in*—infinite series of cosines and/or sines. Thus, the Big Three PDEs, as well as other important PDEs, may be solved using separation of variables for a wide variety of initial conditions.

These trigonometric series are called *Fourier series*, as they were investigated by Fourier in conjunction with his application of separation of variables to the heat equation. In fact, trigonometric series of this form had been used by a number of mathematicians throughout the latter half of the 18th century, including Euler, Daniel Bernoulli in his work on the vibrating string, and Alexis-Claude Clairaut (1713–1765) in his study of the motion of the planets.

However, Fourier, in his 1807 paper on heat conduction, made the claim that "any function" can be expanded in a trigonometric series and showed how to do so. This paper was rejected by its reviewing committee—consisting of Laplace, Joseph-Louis Lagrange (1736–1813), Gaspard Monge (1746–1818) and Francois Lacroix (1765–1843), the first three of whom we shall meet again soon. However, they encouraged Fourier to continue to develop these ideas, and he responded with an award winning revision in 1811. Finally, in 1822 Fourier published his masterpiece, *La Théorie Analytique de Chaleur* (The Analytic Theory of Heat).

Fourier and his series were here to stay. However, there still was no rigorous proof of the convergence of these series. More to the point, there still was no precise definition of function, without which rigor is impossible. Much of what was needed was provided by Dirichlet. In fact, mathematicians were realizing that calculus needed a complete overhaul, in order for its arguments to be mathematically justified. This overhaul continued unabated into the 20th century, by which time we had a new theory of integration (initially spurred, by the way, by the integral coefficients in Fourier's series), a precise definition of *real number* and of *limit*, and a "new" calculus based on these notions. This whole process—this "arithmetization of analysis" as it is sometimes called—had its roots in the trigonometric series of Joseph Fourier and his contemporaries.

3

Fourier Series

3.1 Introduction

At the end of the previous chapter, we were given an arbitrary function $f(x)$ defined on $0 \leq x \leq L$, and we wanted to know if there is a linear combination (possibly with infinitely many terms) of the functions $\sin \frac{n\pi x}{L}$ that is identical to $f(x)$ on $0 \leq x \leq L$. That is, can we find coefficients c_n, $n = 1, 2, 3, \ldots$, such that

$$f(x) = \sum_{n=1}^{\infty} c_n \sin \frac{n\pi x}{L} \tag{3.1}$$

for $0 \leq x \leq L$?

Fourier's affirmative answer to this question in 1807, while not quite correct, marks a pivotal moment in the history of mathematics. Indeed, this surprising and profound result not only disturbed Fourier's mathematical contemporaries but, ultimately, rocked the very foundations of mathematical analysis.

What we would *like* to do is to assume that (3.1) is true and actually find the values of the coefficients c_n for which it is, in fact, true. Let's do this and worry about serious mathematical issues afterwards. (We say that we proceed *formally*.) To this end, we will fix positive integer N, multiply both sides of (3.1) by $\sin \frac{N\pi x}{L}$ and then integrate both sides from $x = 0$ to $x = L$, resulting in

$$\int_0^L f(x) \sin \frac{N\pi x}{L} dx = \sum_{n=1}^{\infty} c_n \int_0^L \sin \frac{n\pi x}{L} \sin \frac{N\pi x}{L} dx. \tag{3.2}$$

Then, it is easy enough to calculate the integrals in (3.2). It turns out that

$$\int_0^L \sin \frac{n\pi x}{L} \sin \frac{N\pi x}{L} dx = \begin{cases} 0, & \text{if } n \neq N, \\ \frac{L}{2}, & \text{if } n = N, \end{cases} \tag{3.3}$$

so (3.2) becomes

$$\int_0^L f(x) \sin \frac{N\pi x}{L} dx = c_1 \cdot 0 = c_2 \cdot 0 + \cdots + c_{N-1} \cdot 0 + c_N \cdot \frac{L}{2} + c_{N+1} \cdot 0 + \cdots$$

or

$$c_N = \frac{2}{L} \int_0^L f(x) \sin \frac{N\pi x}{L} dx. \tag{3.4}$$

Therefore, we have determined the values of c_n for which (3.1) is true. Or have we?

Many readers will notice that we *assumed* the truth of (3.1) and then calculated the values that the coefficients must take on. This, of course, says nothing about the truth of (3.1). Further, the series in (3.1) is an *infinite series of functions*. As we know, an infinite series of constants may not even converge; an infinite series of functions may, then, converge for some values of x and diverge for others.

Finally, there is a more subtle problem with the above argument. We actually skipped a step in going from (3.1) to (3.2)—we assumed that we could integrate the right side *term-by-term*, i.e., that

$$\int_0^L \left(\sum_{n=1}^{\infty} c_n \sin \frac{n\pi x}{L} \sin \frac{N\pi x}{L} \right) dx = \sum_{n=1}^{\infty} c_n \int_0^L \sin \frac{n\pi x}{L} \sin \frac{N\pi x}{L} dx.$$

It turns out that, even if an infinite series of functions converges for all appropriate values of x, it *still* may not have this property.

Fortunately, for "most" of the functions that we deal with in calculus and, in particular, for functions which arise in physical problems, we will see that we need not be concerned with these issues.

Now, instead of working directly with (3.1), or with the analogous series involving cosines, we will first consider the following related question:

> Given $f(x)$ on $-L \leq x \leq L$, is it possible to find constants c_n and d_n, $n = 0, 1, 2, \ldots$, so that
>
> $$f(x) = \sum_{n=0}^{\infty} \left(c_n \cos \frac{n\pi x}{L} + d_n \sin \frac{n\pi x}{L} \right)$$
>
> $$= c_0 + \sum_{n=1}^{\infty} \left(c_n \cos \frac{n\pi x}{L} + d_n \sin \frac{n\pi x}{L} \right) \qquad (3.5)$$
>
> on $-L \leq x \leq L$?

(This series will be called the **Fourier series** of $f(x)$ on $-L \leq x \leq L$; similarly, (3.1) will be called the **Fourier sine series** of $f(x)$ on $0 \leq x \leq L$.)

3.2   Properties of Sine and Cosine

We start by looking at the properties of sine and cosine that will be relevant to this discussion. There are three such properties: periodicity, symmetry and orthogonality.

PERIODICITY

Remember that the functions $y = \sin px$ and $y = \cos px$ are **periodic**, as can be seen from their graphs in Figure 3.1. We see there that, for any value of x,

$$\sin px = \sin p \left(x + \frac{2\pi}{p} \right) = \sin p \left(x + \frac{4\pi}{p} \right) = \cdots$$

$$= \sin p \left(x - \frac{2\pi}{p} \right) = \sin p \left(x - \frac{4\pi}{p} \right) = \cdots$$

and, similarly, for $\cos px$.

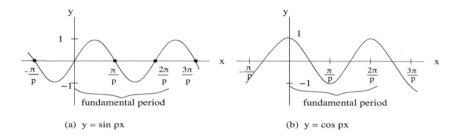

(a) y = sin px   (b) y = cos px

FIGURE 3.1
Graphs of $y = \sin px$ and $y = \cos px$, showing the fundamental period $\frac{2\pi}{p}$.

Definition 3.1 *If there exists a number $T \neq 0$ for which $f(x+T) = f(x)$ for all x in the domain of f, then we say that f is **periodic of period** T. The smallest positive period of f is called its **fundamental period**.*

The fundamental period of $\sin px$, as well as of $\cos px$, is $T = \frac{2\pi}{p}$ (see Exercise 5). Now, since we are interested in linear combinations of functions of this form, let's look at a few examples of such.

Example 1 Is $f(x) = 2 + 3\sin x$ periodic? If so, what are its periods? We know that 2π is the fundamental period of $3\sin x$. Also, the constant function 2 seems to be periodic, with every possible period! (Hence, it has no fundamental period. Why not?) So we suspect that the fundamental period of f is 2π. Let's see:

$$f(x + 2\pi) = 2 + 3\sin(x + 2\pi)$$
$$= 2 + 3\sin x.$$

Therefore, f has period $2k\pi$ for any nonzero integer k. It is easy to show that these are the only periods of f.

Example 2 Do the same for $f(x) = \sin 2x + \cos 4x$. The first function has fundamental period π; the second, $\frac{\pi}{2}$. Therefore, f has fundamental period π (why π and not $\frac{\pi}{2}$?). See Figure 3.2.

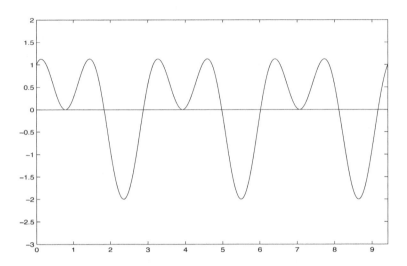

FIGURE 3.2
MATLAB graph of $y = \sin 2x + \cos 4x$, with fundamental period π.

It follows that $\sin\frac{n\pi x}{L}$ and $\cos\frac{n\pi x}{L}$ each has fundamental period $T = \frac{2L}{n}$. Also, since $\sin\frac{\pi x}{L}$ and $\cos\frac{\pi x}{L}$ have fundamental period $2L$, and since any integral multiple of a period also is a period, we find that any *finite* series

$$F_N(x) = c_0 + \sum_{n=1}^{N}\left(c_n\cos\frac{n\pi x}{L} + d_n\sin\frac{n\pi x}{L}\right)$$

has period $2L$. It follows that, if the *infinite* series in (3.4) converges, it also must be periodic of period $2L$. (This is not hard to show and should be "relatively obvious.")

SYMMETRY

We notice that the graph of $y = \cos\frac{n\pi x}{L}$ is symmetric with respect to the y-axis, while that of $y = \sin\frac{n\pi x}{L}$ is symmetric about the origin.

Definition 3.2 $f(x)$ *is* **even** *if its graph is symmetric with respect to the y-axis, that is, if $f(-x) = f(x)$ for all x in the domain of f; $f(x)$ is* **odd** *if its graph is symmetric with respect to the origin, that is, if $f(-x) = -f(x)$ for all x in the domain of f. (See Figure 3.3.)*

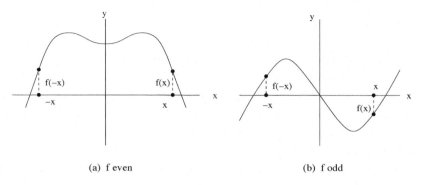

(a) f even   (b) f odd

FIGURE 3.3
Typical even and odd functions.

Example 3 $1, x^2, x^4, \ldots$ are even functions (hence the name "even").

Example 4 $x, x^3, x^5, \ldots$ are odd functions (hence the name "odd").

(Note that the same is true for negative powers, on the restricted domain $x \neq 0$.)

It is easy to show that any linear combination of even/odd functions is even/odd. What about products and quotients? Suppose $f(x)$ and $g(x)$ are even functions, and look at the product $h(x) = f(x)g(x)$. We have

$$h(-x) = f(-x)g(-x)$$
$$= f(x)g(x)$$
$$= h(x) \quad \text{for every } x \text{ in the domain of } h.$$

(Similarly for the quotient of two even functions.) Thus,

$$f(x), g(x) \text{ even} \implies f(x)g(x) \text{ even}.$$

A similar argument (see Exercise 12) gives us

$$f(x), g(x) \text{ odd} \implies f(x)g(x) \text{ even}$$
$$f(x) \text{ even}, g(x) \text{ odd} \implies f(x)g(x) \text{ odd}.$$

Now, back in (3.2), we needed to integrate products of functions. Similarly, when finding the coefficients in (3.5), we will be integrating such products, but on the interval $-L \leq x \leq L$. Even and odd symmetries not only will make our life easier now but, more importantly, will allow us to make an easy jump from Fourier series to Fourier sine and cosine series. So, looking at Figure 3.4, the following simplification should be "obvious":

$$f \text{ even on } -L \leq x \leq L \implies \int_{-L}^{L} f(x)dx = 2\int_{0}^{L} f(x)dx = 2\int_{-L}^{0} f(x)dx$$

$$\tag{3.6}$$

$$f \text{ odd on } -L \leq x \leq L \implies \int_{-L}^{L} f(x)dx = 0.$$

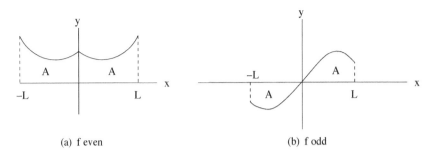

(a) f even   (b) f odd

FIGURE 3.4
Integrals of even and odd functions on $-L \leq x \leq L$.

Let's prove the second statement—the first is left to Exercise 13. Suppose f is integrable and odd on $-L \leq x \leq L$. Then

$$\int_{-L}^{L} f(x)dx = \int_{-L}^{0} f(x)dx + \int_{0}^{L} f(x)dx$$

$$= -\int_{-L}^{0} f(-x)dx + \int_{0}^{L} f(x)dx, \text{ since } f \text{ is odd}$$

$$= \int_{L}^{0} f(u)du + \int_{0}^{L} f(x)dx, \text{ from the substitution } u = -x$$

$$= -\int_{0}^{L} f(u)du + \int_{0}^{L} f(x)dx$$

$$= 0.$$

ORTHOGONALITY

Back in equation (3.3) we mentioned that

$$\int_{0}^{L} \sin\frac{n\pi x}{L} \sin\frac{m\pi x}{L} dx = 0 \quad \text{if } n \neq m.$$

We say that the functions $\sin\frac{n\pi x}{L}$ are orthogonal on $0 \leq x \leq L$. But let us be more precise, while introducing some new terminology.

Definition 3.3 *Given two functions f and g which are integrable on $a \leq x \leq b$, their* **inner product** *on $a \leq x \leq b$ is the real number*

$$\langle f, g \rangle = \int_{a}^{b} f(x)g(x)dx.$$

The inner product is motivated by the fact that Riemann sums of this integral are of the form

$$\Delta x \sum_{i=1}^{n} f(x_i)g(x_i),$$

which has the same form as the dot product of two n-dimensional vectors. Indeed, the inner product is an infinite-dimensional analog of the dot product in $\mathbb{R}^n$. Then, analogous to the concept of perpendicularity, we have the following definition.

Definition 3.4 *If $\langle f, g \rangle = 0$, we say that f and g are* **orthogonal** *on $a \leq x \leq b$. Given a set of functions $\{f_n(x)\}$ on $a \leq x \leq b$, if*

$$\langle f_n, f_m \rangle = 0 \quad \text{whenever} \quad n \neq m,$$

we say that the functions $f_n(x)$ are **pairwise orthogonal** *and that the set $\{f_n(x)\}$ is an* **orthogonal set**.

Now, just as we needed the fact that the functions in (3.1) form an orthogonal set on $0 \leq x \leq L$ (as we saw from (3.3)), so we will need to show that the functions appearing on the right-hand side of (3.5) are pairwise orthogonal on $-L \leq x \leq L$, that is, that the functions

$$1, \cos\frac{\pi x}{L}, \cos\frac{2\pi x}{L}, \ldots, \sin\frac{\pi x}{L}, \sin\frac{2\pi x}{L}, \ldots$$

form an orthogonal set on that interval.

For example, let's look at

$$\left\langle \sin\frac{n\pi x}{L}, \sin\frac{m\pi x}{L} \right\rangle = \int_{-L}^{L} \sin\frac{n\pi x}{L} \sin\frac{m\pi x}{L} \, dx. \tag{3.7}$$

Since $\sin\frac{n\pi x}{L}\sin\frac{m\pi x}{L}$ is even (why?), we can write

$$\left\langle \sin\frac{n\pi x}{L}, \sin\frac{m\pi x}{L} \right\rangle = 2\int_{0}^{L} \sin\frac{n\pi x}{L} \sin\frac{m\pi x}{L} \, dx. \tag{3.8}$$

To deal with these integrals, we will need the trigonometric identities

$$\sin(a + b) = \sin a \cos b + \cos a \sin b, \tag{3.9a}$$
$$\sin(a - b) = \sin a \cos b - \cos a \sin b, \tag{3.9b}$$
$$\cos(a + b) = \cos a \cos b - \sin a \sin b, \tag{3.9c}$$
$$\cos(a - b) = \cos a \cos b + \sin a \sin b. \tag{3.9d}$$

Now, the integrand in (3.8) is of the form $\sin a \sin b$, so we use identities (3.9c) and (3.9d); specifically, subtracting (3.9c) from (3.9d) gives us

$$\sin a \sin b = \frac{1}{2}[\cos(a - b) - \cos(a + b)].$$

Therefore, (3.7) becomes

$$\left\langle \sin\frac{n\pi x}{L}, \sin\frac{m\pi x}{L} \right\rangle = \int_0^L \left[\cos\left(\frac{n\pi x}{L} - \frac{m\pi x}{L}\right) - \cos\left(\frac{n\pi x}{L} + \frac{m\pi x}{L}\right) \right] dx$$

$$= \begin{cases} \int_0^L \left[\cos\frac{(n-m)\pi x}{L} - \cos\frac{(n+m)\pi x}{L} \right] dx, & \text{if } n \neq m, \\ \int_0^L \left[1 - \cos\frac{2n\pi x}{L} \right] dx, & \text{if } n = m \end{cases}$$

$$= \begin{cases} 0, & \text{if } n \neq m \\ L, & \text{if } n = m, \text{ for } n = 1, 2, \dots, m = 1, 2, \dots. \end{cases}$$

(3.10a)

For example, see Figure 3.5 where we look at $\langle \sin 2x, \sin 4x \rangle$ on $-\pi \leq x \leq \pi$.

Similarly (see Exercise 14),

$$\left\langle 1, \cos\frac{n\pi x}{L} \right\rangle = 0, \text{ for } n = 1, 2, \dots \tag{3.10b}$$

$$\left\langle 1, \sin\frac{n\pi x}{L} \right\rangle = 0, \text{ for } n = 1, 2, \dots \tag{3.10c}$$

$$\left\langle \sin\frac{n\pi x}{L}, \cos\frac{m\pi x}{L} \right\rangle = 0, \text{ for } n = 1, 2, \dots; m = 1, 2, \dots \tag{3.10d}$$

$$\left\langle \cos\frac{n\pi x}{L}, \cos\frac{m\pi x}{L} \right\rangle = \begin{cases} 0, & \text{if } n \neq m, \\ L, & \text{if } n = m, \text{ for } n = 1, 2, \dots; m = 1, 2, \dots \end{cases} \tag{3.10e}$$

and, of course,

$$\langle 1, 1 \rangle = 2L. \tag{3.10f}$$

So we see that the functions in (3.5) do, indeed, form an orthogonal set.

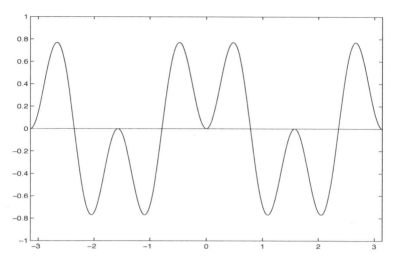

FIGURE 3.5
MATLAB graph showing the orthogonality of the functions $\sin 2x$ and $\sin 4x$ on $-\pi \le x \le \pi$: $\langle \sin 2x, \sin 4x \rangle = \int_{-\pi}^{\pi} \sin 2x \sin 4x \, dx = 0.$

Exercises 3.2

In Exercises 1–4, determine whether or not $f(x)$ is periodic. If so, find its fundamental period.
MATLAB: Afterwards, check your answer by graphing the function.

1. $f(x) = 2 \sin \pi x - 3 \cos \frac{\pi x}{2} + 7$

2. $f(x) = \cos \frac{x}{4} + \sin x$

3. $f(x) = x \sin x$

4. $f(x) = \cos 4x + \cos 3x$

5. a) Use the identities in (3.9) to prove that $\sin px$ and $\cos px$ are periodic of period $T = \frac{2k\pi}{p}$ for any nonzero integer k.

    b) Use the fact that $\sin px$ and $\cos px$ are linearly independent to prove that these functions have no other periods.

6. True or false? Prove, or provide a counterexample.

    a) If $f(x)$ and $g(x)$ are periodic of period T, then so is $c_1 f(x) + c_2 g(x)$ for any choice of constants c_1 and c_2.

    b) If $f(x)$ and $g(x)$ are periodic of period T, then so are $f(x)g(x)$ and $f(x)/g(x)$ (on their domains).

c) If $f(x)$ and $g(x)$ are periodic with fundamental period T, then $f(x)g(x)$ is periodic with fundamental period T.

7. Use mathematical induction to prove that if $f(x)$ has period T, then $f(x)$ also has period kT for any nonzero integer k.

In Exercises 8–11, determine if $f(x)$ is even, odd or neither.

8. $f(x) = 5x^4 - 9x^2 + 6$

9. $f(x) = 2x^5 + 5x^3 - 2$

10. $f(x) = \cos 3 \left(x - \frac{\pi}{4} \right)$

11. $f(x) = x^2 \sin x$

12. Prove:

   a) If $f(x)$ and $g(x)$ are odd, then $f(x)g(x)$ is even.

   b) If $f(x)$ is even and $g(x)$ is odd, then $f(x)g(x)$ is odd.

13. Prove statement (3.6).

14. Fill in the details in the derivation of (3.10a), then derive equations (3.10b) through (3.10f).

15.   a) Show that the functions $f(x) = x^n$ and $g(x) = x^m$, where $n, m \in \mathbb{N}$, are orthogonal on the interval $-L \le x \le L$ if and only if n is even and m is odd (or vice versa, of course!).

   b) Show that the functions $\{1, x, x^2, x^3, \ldots\}$ do not form an orthogonal set on $-1 \le x \le 1$.

16.   a) Show that the functions 1 and x are orthogonal on $-1 \le x \le 1$.

   b) Find a quadratic polynomial which is orthogonal to both 1 and x, on $-1 \le x \le 1$.

   c) Show that the quadratic in part (b) actually is orthogonal to any linear function $f(x) = ax + b$, on $-1 \le x \le 1$.

   d) Find a cubic polynomial which is orthogonal to $1, x$ and the quadratic from part (b), on $-1 \le x \le 1$. (If we continue in this manner, we generate a set of orthogonal polynomials called the **Legendre polynomials**, which we treat in Chapter 7.)

Suppose we are given a (fixed) function $w(x) > 0$ on $a < x < b$. We may generalize the idea of inner product by defining

$$\langle f, g \rangle = \int_a^b f(x)g(x)w(x)dx.$$

This is called the *inner product with respect to the* **weight function** $w(x)$ and, if $\langle f, g \rangle = 0$, we say that f and g are *orthogonal with respect to w* on $a \leq x \leq b$. (If $w(x) = 1$, then we're back to the original definition of orthogonal—in the present context, we will then say that f and g are *simply orthogonal*.)

17. Find the inner product of f and g, with respect to the weight function w, on the given interval.

    a) $f(x) = x, g(x) = x^2, w(x) = \sqrt{x}, \quad 0 \leq x \leq 4$

    b) $f(x) = 1 + x, g(x) = 2 + x, w(x) = x^2, \quad 0 \leq x \leq 1$

18. a) Show that the functions $1, x, 2x^2 - 1$ form an orthogonal set on $-1 \leq x \leq 1$, with respect to the weight function $w(x) = \frac{1}{\sqrt{1-x^2}}$. (Hint: Use a trigonometric substitution.) These are the first three **Chebyshev polynomials of the first kind**.

    b) Show that the functions $1, 1-x, 1-2x+\frac{1}{2}x^2$ form an orthogonal set on $[0, \infty)$, with respect to the weight function $w(x) = e^{-x}$. These are the first three **Laguerre polynomials**.

3.3   The Fourier Series

Now we are ready to look at the question from Section 3.1, namely, given any function $f(x)$ on $-L \leq x \leq L$, can we find coefficients c_n, $n = 0, 1, 2, \ldots$, and d_n, $n = 1, 2, \ldots$, such that

$$f(x) = c_0 + \sum_{n=1}^{\infty} \left(c_n \cos \frac{n\pi x}{L} + d_n \sin \frac{n\pi x}{L} \right) \tag{3.11}$$

on $-L \leq x \leq L$? Fourier asserted that the answer is yes for *any* function $f(x)$ and, thus, that the corresponding series always converges, and he provided formulas for the coefficients. Actually, the coefficients already had been determined much earlier by Euler and Clairaut, with Euler employing the now-standard method that we use here. However, as mentioned, Euler, Clairaut and others were more circumspect with regard to which functions $f(x)$ could be so expanded.

All of this talk ultimately led to a more precise idea of what we mean by a function and to a much closer look at what it means for a function to be integrable. Also, as we shall soon see, Fourier was overly ambitious in his claims, although he was "more-or-less correct" if $f(x)$ is the kind of function that we are familiar with from calculus.

Again, let us proceed formally, assuming that there are constants c_n, d_n for which (3.11) is true, and go about finding these constants.

So, in order to determine a coefficient, we multiply (3.11) by the function associated with that coefficient. Then, we integrate both sides from $x = -L$ to $x = L$, assuming that we may do so term-by-term. For example, to find c_N, for $N \geq 1$, we multiply through by $\cos \frac{N\pi x}{L}$ and integrate, resulting in

$$
\int_{-L}^{L} f(x) \cos \frac{N\pi x}{L} dx = c_0 \int_{-L}^{L} \cos \frac{N\pi x}{L} dx
$$
$$
+ \sum_{n=1}^{\infty} c_n \int_{-L}^{L} \cos \frac{n\pi x}{L} \cos \frac{N\pi x}{L} dx
$$
$$
+ \sum_{n=1}^{\infty} d_n \int_{-L}^{L} \sin \frac{n\pi x}{L} \cos \frac{N\pi x}{L} dx
$$
$$
= c_0 \left\langle 1, \cos \frac{N\pi x}{L} \right\rangle + \sum_{n=1}^{\infty} c_n \left\langle \cos \frac{n\pi x}{L}, \cos \frac{N\pi x}{L} \right\rangle
$$
$$
+ \sum_{n=1}^{\infty} d_n \left\langle \sin \frac{n\pi x}{L}, \cos \frac{N\pi x}{L} \right\rangle
$$
$$
= c_0 \cdot 0 + c_1 \cdot 0 + \cdots + c_{N-1} \cdot 0 + c_n \cdot L + c_{N+1} \cdot 0 +
$$
$$
\cdots + d_1 \cdot 0 + d_2 \cdot 0 + \cdots
$$

and, solving for c_N, we have

$$
c_N = \frac{1}{L} \int_{-L}^{L} f(x) \cos \frac{N\pi x}{L} dx.
$$

Similarly, we can find c_0 and each d_N:

$$
c_0 = \frac{1}{2L} \int_{-L}^{L} f(x) dx,
$$
$$
d_N = \frac{1}{L} \int_{-L}^{L} f(x) \sin \frac{n\pi x}{L} dx.
$$

We summarize these results in the following definition.

Definition 3.5 *Given a function $f(x)$ on $-L \leq x \leq L$, the **Fourier series** of f on $-L \leq x \leq L$ is the series*

$$
F(x) = \frac{a_0}{2} + \sum_{n=1}^{\infty} \left(a_n \cos \frac{n\pi x}{L} + b_n \sin \frac{n\pi x}{L} \right), \qquad (3.12)
$$

where the coefficients are given by

$$a_n = \frac{1}{L} \int_{-L}^{L} f(x) \cos \frac{n\pi x}{L} dx, \qquad n = 0, 1, 2, \ldots$$

$$b_n = \frac{1}{L} \int_{-L}^{L} f(x) \sin \frac{n\pi x}{L} dx, \qquad n = 1, 2, \ldots$$

(assuming these integrals exist).

By definition, **the Fourier series of a function is unique,** and from the above it follows that, if f can be expanded in such a series of sines and cosines, then this series must be its Fourier series.* (Also, see Exercise 14.)

Actually, we shall see later (in Chapter 8) that it is possible to expand f in series involving functions other than sine and cosine. It is because of these **generalized Fourier series** that we often refer to (3.12) as the **trigonometric Fourier series** of f on $-L \leq x \leq L$.

At this point, we have not answered the questions: Does the Fourier series actually converge? If so, does it converge to $f(x)$? In fact, we really should not even write $F(x) = \cdots$ in (3.12) until we know that the series does, indeed, converge. Therefore, we write

$$f(x) \sim \frac{a_0}{2} + \sum_{n=1}^{\infty} \left(a_n \cos \frac{n\pi x}{L} + b_n \sin \frac{n\pi x}{L} \right). \qquad (3.13)$$

Let's compute some. Calculate the Fourier series for the given function on the given interval.

Example 1 $f(x) = x$ on $-\pi \leq x \leq \pi$
Our coefficients are

$$a_0 = \frac{1}{\pi} \int_{-\pi}^{\pi} x \, dx = 0,$$

$$a_n = \frac{1}{\pi} \int_{-\pi}^{\pi} x \cos nx \, dx = 0, \text{ since } x \cos nx \text{ is odd,}$$

$$b_n = \frac{1}{\pi} \int_{-\pi}^{\pi} x \sin nx \, dx$$

$$= \frac{2}{\pi} \int_{0}^{\pi} x \sin nx \, dx, \text{ since } x \sin nx \text{ is even}$$

$$= \frac{2}{\pi} \left[-\frac{1}{n} x \cos nx \Big|_{0}^{\pi} + \frac{1}{n} \int_{0}^{\pi} \cos nx \, dx \right],$$

---

*To be precise, this is true so long as term-by-term integration of the series is justified. This certainly is the case if the series is *finite* (and, more generally, if it *converges uniformly* to f on $-L \leq x \leq L$. See Appendix A.)

where we have integrated by parts

$$= \frac{2}{\pi} \left[-\frac{1}{n} \pi \cos n\pi \right] = -\frac{2}{n}(-1)^n$$

and we have

$$x \sim \sum_{n=1}^{\infty} \left(-\frac{2}{n} \right) (-1)^n \sin nx = 2 \sum_{n=1}^{\infty} \frac{(-1)^{n+1}}{n} \sin nx.$$

Example 2

$$f(x) = \begin{cases} 0, & \text{if } -2 \leq x < 0, \\ 1, & \text{if } 0 \leq x \leq 2 \end{cases}$$

The coefficients are

$$a_0 = \frac{1}{2} \int_{-2}^{2} f(x)dx = \frac{1}{2} \int_{-2}^{0} f(x)dx + \frac{1}{2} \int_{0}^{2} f(x)dx$$

$$= \frac{1}{2} \int_{-2}^{0} 0 \, dx + \frac{1}{2} \int_{0}^{2} 1 \, dx$$

$$= 0 + 1 = 1,$$

$$a_n = \frac{1}{2} \int_{-2}^{2} f(x) \cos \frac{n\pi x}{2} dx$$

$$= \frac{1}{2} \int_{-2}^{0} 0 \cos \frac{n\pi x}{2} dx + \frac{1}{2} \int_{0}^{2} 1 \cos \frac{n\pi x}{2} dx$$

$$= 0 + \frac{1}{n\pi} \sin \frac{n\pi x}{2} \Big|_{0}^{2} = 0,$$

$$b_n = \frac{1}{2} \int_{-2}^{2} f(x) \sin \frac{n\pi x}{2} dx$$

$$= \frac{1}{2} \int_{-2}^{0} 0 \sin \frac{n\pi x}{2} dx + \frac{1}{2} \int_{0}^{2} 1 \sin \frac{n\pi x}{2} dx$$

$$= 0 - \frac{1}{n\pi} \cos \frac{n\pi x}{2} \Big|_{0}^{2} = \frac{1 - (-1)^n}{n\pi}$$

and

$$f(x) \sim \frac{1}{2} + \frac{1}{\pi} \sum_{n=1}^{\infty} \frac{1 - (-1)^n}{n} \sin \frac{n\pi x}{2}$$

$$= \frac{1}{2} + \frac{2}{\pi} \sum_{n=0}^{\infty} \frac{1}{2n+1} \sin \frac{(2n+1)\pi x}{2}.$$

We'll wait until the next section to answer the big questions about convergence of the Fourier series.

Exercises 3.3

In Exercises 1–12, calculate the Fourier series of $f(x)$ on the given interval.

1. $f(x) = \begin{cases} 2, & \text{if } -1 \le x \le 0 \\ 0, & \text{if } 0 < x \le 1 \end{cases}$

2. $f(x) = \begin{cases} -1, & \text{if } -3 \le x < 0 \\ 2, & \text{if } 0 \le x \le 3 \end{cases}$

3. $f(x) = x$, on $-5 \le x \le 5$

4. $f(x) = 2x + 1$, on $-\pi \le x \le \pi$

5. $f(x) = x^2$, on $-1 \le x \le 1$

6. $f(x) = \begin{cases} x^2, & \text{if } -2\pi \le x < 0 \\ -x^2, & \text{if } 0 \le x \le 2\pi \end{cases}$

7. $f(x) = |x|$, on $-1 \le x \le 1$

8. $f(x) = x^2 + x$, on $-\pi \le x \le \pi$

9. $f(x) = \begin{cases} 0, & \text{if } -\pi \le x < \frac{\pi}{2} \\ 1, & \text{if } \frac{\pi}{2} \le x \le \pi \end{cases}$

10. $f(x) = \begin{cases} 0, & \text{if } -\pi \le x < -\frac{\pi}{2} \\ 1, & \text{if } -\frac{\pi}{2} < x < \frac{\pi}{2} \\ 0, & \text{if } \frac{\pi}{2} \le x \le \pi \end{cases}$

11. $f(x) = 2 + 3\sin 2x - 5\cos 4x$, on $-\pi \le x \le \pi$

12. $f(x) = \sin x$, on $-\frac{\pi}{2} \le x \le \frac{\pi}{2}$

13. In Example 1, we saw that, for the odd function $f(x) = x$, the Fourier series contains only sine terms, i.e., we had $a_n = 0$, $n = 0, 1, 2, \ldots$. Such a Fourier series is called a **pure sine series**.

   a) Show that the Fourier series of any function which is odd on $-L \le x \le L$ is, in fact, a pure sine series.

   b) Show that the Fourier series of any function which is even on $-L \le x \le L$ is a **pure cosine series**, i.e., that $b_n = 0$, $n = 1, 2, \ldots$.

   c) Is the function in Example 2 even or odd? Is its Fourier series a pure cosine or pure sine series?

14. a) Explain why the functions $1, \cos\frac{n\pi x}{L}, \sin\frac{n\pi x}{L}$, $n = 1, 2, 3, \ldots$, are linearly independent on $-L \le x \le L$.

b) Show that, if

$$\frac{a_0}{2} + \sum_{n=1}^{\infty} \left(a_n \cos \frac{n\pi x}{L} + b_n \sin \frac{n\pi x}{L} \right)$$

$$= \frac{c_0}{2} + \sum_{n=1}^{\infty} \left(c_n \cos \frac{n\pi x}{L} + d_n \sin \frac{n\pi x}{L} \right),$$

we must have $a_n = c_n$, $n = 0, 1, 2, \ldots$, and $b_n = d_n$, $n = 1, 2, \ldots$.

15. Given $f(x)$ on $-L < x \le L$, let $z = ax$, where a is a constant, and $g(z) = f(z/a)$.

   a) Find a so that $g(z)$ has domain $-\pi \le z \le \pi$.

   b) Using the value for a, compute the Fourier series for g on $-\pi \le z \le \pi$. Then, use this series to recover the Fourier series of f on $-L \le x \le L$.

   Thus, WLOG, we need only know how to compute Fourier series on $-\pi \le x \le \pi$.

16. **Fourier Series on Other Intervals**

   a) Show that the functions $1, \cos \frac{n\pi x}{L}$, $\sin \frac{n\pi x}{L}$, $n = 1, 2, \ldots$, are pairwise orthogonal on any interval of length $2L$.

   b) Show that any piecewise smooth function $f(x)$ on $a \le x \le b$ can be expanded into a series

$$f(x) = \frac{a_0}{2} + \sum_{n=1}^{\infty} \left[a_n \cos \frac{n\pi x}{L} + b_n \sin \frac{n\pi x}{L} \right],$$

   where $L = \frac{b-a}{2}$; in particular, find expressions for the constants a_n, b_n.

   c) Compute this series for $f(x) = x$ on $1 \le x \le 4$.

17. Complex Fourier Series: Using Euler's formula, $e^{i\theta} = \cos\theta + i\sin\theta$, we can rewrite the Fourier series in the form

$$f(x) \sim \sum_{k=-\infty}^{\infty} c_k e^{\frac{ik\pi x}{L}}.$$

Using $L = \pi$,

   a) Find the coefficients c_k in terms of the coefficients a_n and b_n of the original Fourier series (where $k = \pm n$).

It turns out that, when complex functions are involved, we must change the definition of inner product, so that it has certain desirable mathematical properties which are satisfied by the *real* inner product and the vector dot product. Actually, this complex inner product turns out to be

$$\langle f(x), g(x) \rangle = \int_{-L}^{L} \overline{f(x)} g(x) \, dx,$$

where $\overline{f(x)}$ is the *complex conjugate* of $f(x)$. Note that if f and g both are real, then this turns out to be the same as the inner product from Definition 3.3.

b) Show that $\overline{e^{ikx}} = e^{-ikx}$.

c) Assume that $f(x) = \sum_{k=-\infty}^{\infty} c_k e^{ikx}$, on $-\pi \leq x \leq \pi$. Find the coefficients by way of the complex inner product $\langle e^{inx}, f(x) \rangle$ on $-\pi \leq x \leq \pi$.

3.4 The Fourier Series, Continued

Now we are ready to deal with the question of convergence for Fourier series. Specifically, we'll state in this section, and prove in the next section, the famous theorem of Dirichlet[†] from 1829, in which he proved that if $f(x)$ consists of a finite number of smooth arcs, then the Fourier series converges for all x, and it converges to f except, possibly, at a finite number of x-values. But let's be more precise.

Definition 3.6 *Given $f(x)$ and a point $x = x_0$, not necessarily in the domain of f, we define*

$$f(x_0+) = \lim_{x \to x_0^+} f(x)$$

$$f(x_0-) = \lim_{x \to x_0^-} f(x).$$

See Figure 3.6.

[†] Actually, the theorem we state is not exactly Dirichlet's Theorem. For a very nice historical treatment of the latter, see David Bressoud's *A Radical Approach to Real Analysis*.

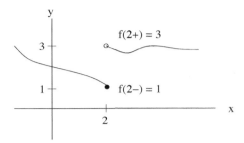

FIGURE 3.6
A typical jump discontinuity.

Definition 3.7 *Given $f(x)$ with domain $a \leq x \leq b$, we say that f is* **piecewise continuous** *on $a \leq x \leq b$ if*

   1) *f has a finite number of discontinuities in $a \leq x \leq b$.*

   2) *At each point of discontinuity x_0, $a < x_0 < b$, $f(x_0+)$ and $f(x_0-)$ both exist (and, therefore, are finite).*

   3) *$f(a+)$ and $f(b-)$ both exist (and are finite).*

(Actually, f need not exist at the points of discontinuity.) If $f'(x)$ also is piecewise continuous on $a \leq x \leq b$, we say that f is **piecewise smooth** *there.*

Example 1 $f(x) = \begin{cases} 1, & \text{if } -2 \leq x \leq 3, \\ 5, & \text{if } 3 < x \leq 6 \end{cases}$

is piecewise smooth on $-2 \leq x \leq 6$.

Example 2 $f(x) = x^2$ is piecewise smooth on any interval $a \leq x \leq b$ (in fact, it is **smooth** on any interval $a \leq x \leq b$).

Example 3

$$f(x) = \begin{cases} x, & \text{if } 1 \leq x \leq 2, \\ \frac{1}{x-2}, & \text{if } 2 < x \leq 3 \end{cases}$$

is *not* piecewise continuous on $1 \leq x \leq 3$, since $f(2+) = \infty$. See Figure 3.7.

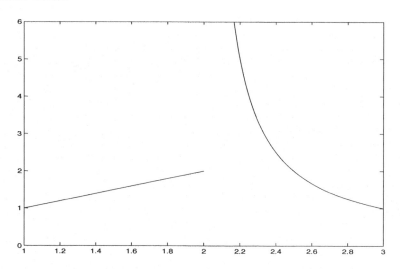

FIGURE 3.7
MATLAB graph of the function from Example 3.

Example 4

$$f(x) = \begin{cases} 1, & \text{if } x = -4, \\ x & \text{if } -4 < x < 1, \\ 2, & \text{if } x = 1, \\ -2, \text{if } 1 < x \leq 4 \end{cases}$$

is piecewise smooth on $-4 \leq x \leq 4$. See Figure 3.8.

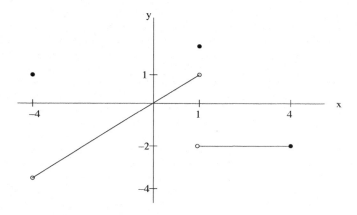

FIGURE 3.8
Graph of the function from Example 4.

Example 5 $f(x) = \sqrt[3]{x}$ is piecewise continuous but *not* piecewise smooth on any interval containing $x = 0$, since $f'(0+) = f'(0-) = \infty$. See Figure 3.9.

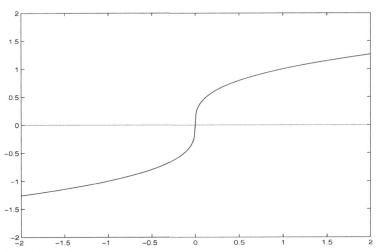

FIGURE 3.9
MATLAB graph of the function $f(x) = \sqrt[3]{x}$.

From calculus, a very important property of piecewise continuous functions is that they are *integrable*. More precisely, if $f(x)$ is piecewise continuous on $a \leq x \leq b$, with discontinuities at $x_1, \ldots, x_{n-1}$ (and, possibly, at $x_0 = a$ or $x_n = b$), then

$$\int_a^b f(x)dx = \sum_{i=1}^{n} \int_{x_{i-1}}^{x_i} f(x)dx.$$

(A "typical" example is shown in Figure 3.10.) Also, many of the theorems on continuous functions have their analogs for piecewise continuous/smooth function, e.g., if f and g are piecewise continuous/smooth, then so is their sum, their product and the like.

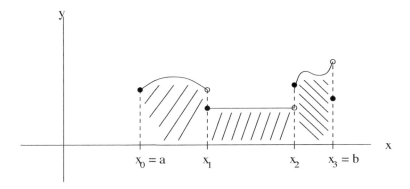

FIGURE 3.10
Evaluating the definite integral of a piecewise continuous function.

Now we are ready to state the big theorem.

Theorem 3.1 *If $f(x)$ is piecewise smooth[‡] on $-L \leq x \leq L$, then its Fourier series, $F(x)$, converges on $-L \leq x \leq L$, and*

   1) $F(x) = f(x)$ for all x in $-L < x < L$ where f is continuous.

   2) $F(x) = \frac{f(x+)+f(x-)}{2}$ for all x in $-L < x < L$ where f is discontinuous.

   3) $F(-L) = F(L) = \frac{f(-L+)+f(L-)}{2}$.

(The proof will be given in Section 3.5.) It also can be shown that the Fourier series converges *uniformly*, and that's all we need to be able to integrate it term-by-term. See Appendix A.

Let's get right to some examples. In each case, we draw the graph of $y = f(x)$ and the graph of its Fourier series, $y = F(x)$.

Example 6 $f(x) = x, -2 \leq x \leq 2$.

See Figure 3.11.

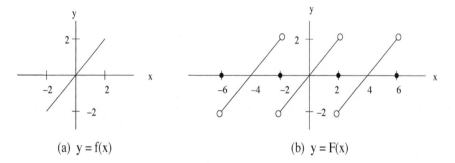

       (a) y = f(x)   (b) y = F(x)

FIGURE 3.11
The function from Example 6 and its Fourier series.

Example 7
$$f(x) = \begin{cases} 2, & \text{if } -2 \leq x < 1, \\ x - 1, & \text{if } 1 \leq x \leq 2. \end{cases}$$

See Figure 3.12.

[‡]It actually turns out that the assumption that f be piecewise smooth can be relaxed, without *too* much difficulty.

We note that the Fourier series is "blind" to the discontinuous points of f, i.e., if we change the value of f at a discontinuous point, it does *not* change the Fourier series. In fact, we may change f at a finite number of points *without* effecting a change in the Fourier series. The reason for the behavior is that the coefficients are determined by *integrals* involving f, and the value of an integral is unaffected by changing the values of the integrand at finitely many points.

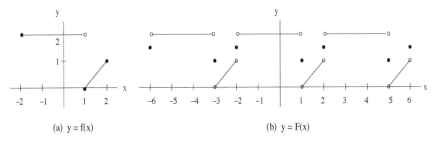

(a) y = f(x)   (b) y = F(x)

FIGURE 3.12
The function from Example 7 and its Fourier series.

Now, when we graphed the Fourier series above, we graphed them on $-L \leq x \leq L$. However, the domain of each term of the Fourier series is all real numbers. Also, we mentioned earlier that **the Fourier series is periodic of period 2L**. Hence the actual graph of the Fourier series is the graph on $-L \leq x \leq L$ extended periodically.

Definition 3.8 *Given a function $f(x)$ defined on $a \leq x \leq b$, the function $g(x)$ defined by*

$$g(x) = f(x), \quad a < x < b, \quad and$$
$$g(x + T) = g(x), \quad for\ all\ x,\ where\ T = b - a,$$

is called the **periodic extension** *of f (of period T). See Figure 3.13. (We're being a bit sloppy here, since we may have $f(a) \neq f(b)$. Again, though, we're not so concerned about the points of discontinuity.)*

So the Fourier series of f actually is the periodic extension of f (of period $2L$, of course), with the possible exception of what happens at the points where f is discontinuous, as well as the endpoints $x = \pm L$.

Finally, we may restate Theorem 3.1 more succinctly. First, if x is a point in $-L < x < L$ where f is continuous, then

$$f(x) = \frac{f(x+) + f(x-)}{2}.$$

Next, letting $f_p(x)$ be the periodic extension of f, we can write

$$f(-L+) = f_p(L+)$$

and

$$f(L-) = f_p(-L-).$$

We then have the following corollary.

Corollary 3.1 *If $f(x)$ is piecewise smooth on $-L \leq x \leq L$, and $f_p(x)$ is the periodic extension of $f(x)$, then the Fourier series $F(x)$ converges for all x and*

$$F(x) = \frac{f_p(x+) + f_p(x-)}{2} \quad \text{for all} \quad x.$$

Exercises 3.4

In Exercises 1–6, determine if $f(x)$ is piecewise continuous and if it is piecewise smooth (assume that $L > 0$). (If f is actually continuous or smooth, say so.)

1. $f(x) = |x|$, on $-L \leq x \leq L$

2. $f(x) = \begin{cases} -x^2, & \text{if } -L \leq x < 0, \\ x^2, & \text{if } 0 \leq x \leq L \end{cases}$

3. $f(x) = x^{2/3}$, on $-L \leq x \leq L$

4. $f(x) = x^{4/3}$, on $-L \leq x \leq L$

5. $f(x) = \begin{cases} \tan x, & \text{if } -\frac{\pi}{2} < x < \frac{\pi}{2}, \\ 0, & \text{if } x = \pm\frac{\pi}{2} \end{cases}$

6. $f(x) = \begin{cases} 2, & \text{if } x = -3, \\ x, & \text{if } -3 < x \leq 1, \\ \frac{1}{x}, & \text{if } 1 < x \leq 3 \end{cases}$

7. $f(0) = 0$ and, for $x \neq 0, -L \leq x \leq L$,

   a) $f(x) = \sin\frac{1}{x}$

   b) $f(x) = x\sin\frac{1}{x}$

   c) $f(x) = x^3\sin\frac{1}{x}$

In Exercises 8–13, sketch three periods of the graph of the Fourier series of $f(x)$.

8. $f(x) = x^2$ on $-2 \leq x \leq 2$

9. $f(x) = x + 1$ on $-2 \leq x \leq 2$

10. $f(x) = \begin{cases} 0, & \text{if } -3 \leq x < 0, \\ 1 - x, & \text{if } 0 \leq x \leq 3 \end{cases}$

11. $f(x) = |3 - x|$ on $-1 \leq x \leq 1$

12. $f(x) = \begin{cases} 2 + x, & \text{if } -3 \leq x < -1, \\ 4, & \text{if } -1 \leq x < 2, \\ 6 - x, & \text{if } 2 \leq x \leq 3 \end{cases}$

13. $f(x) = \begin{cases} x, & \text{if } x = -1, -\frac{4}{5}, -\frac{3}{5}, -\frac{2}{5}, -\frac{1}{5}, 0, \frac{1}{5}, \frac{2}{5}, \frac{3}{5}, \frac{4}{5}, 1, \\ 2, & \text{otherwise} \end{cases}$
    on $-1 \leq x \leq 1$

14. **MATLAB:** Redo Exercises 10 and 12 using MATLAB.

In Exercises 15–17, decide if the statement is true or false. Assume that f is piecewise smooth on $-L \leq x \leq L$.

15. If f is continuous, then so is its Fourier series.

16. If f is discontinuous, then so is its Fourier series.

17. If f is even and continuous, then its periodic extension is identical to its Fourier series.

18. Although each term in the Fourier series is a continuous function for all x, the (infinite!) sum of these terms can be (and "usually" is) a function which is not continuous. This was surprising and unacceptable to mathematicians of the 18th century. Looked at more precisely, we have a sequence of continuous functions

$$F_n(x) = \frac{a_0}{2} + \sum_{n=1}^{N} \left(a_n \cos \frac{n\pi x}{L} + b_n \sin \frac{n\pi x}{L} \right)$$

which converges pointwise to a discontinuous function. Actually, there are examples of this phenomenon which are closer to home. Consider the functions

$$f_n(x) = x^n \quad \text{on} \quad 0 \leq x \leq 1, n = 1, 2, \ldots .$$

What function does this sequence tend to as $n \to \infty$?

19. In the 17th and 18th centuries, starting even before the discovery/invention of calculus, finding the sum of various infinite series was a hot topic. For example, as early as 1736, Euler quite cleverly came up with the sum

$$\sum_{n=1}^{\infty} \frac{1}{n^2} = \frac{\pi^2}{6},$$

a sum that other famous mathematicians had been unable to evaluate, the most notable being the Swiss mathematician James Bernoulli (1654–1705). (By the way, the Bernoulli family is to mathematics as the Bach family is to music. The most well-known of the Bernoullis are James; his brother John who, incidentally, had been Euler's teacher; and John's son, Daniel, the last two of whom we have already met.) Around this same time, Euler also came up with

$$\sum_{n=1}^{\infty} \frac{1}{n^4} = \frac{\pi^4}{90} \quad \text{and} \quad \sum_{n=1}^{\infty} \frac{1}{n^6} = \frac{\pi^6}{945}.$$

a) Many of these infinite series are much easier to deal with using Fourier series. Use the Fourier series for $f(x) = x^2$ on $-1 \le x \le 1$ and Theorem 3.1 to derive Euler's sum

$$\sum_{n=1}^{\infty} \frac{1}{n^2} = 1 + \frac{1}{4} + \frac{1}{9} + \cdots = \frac{\pi^2}{6}.$$

b) Use the same Fourier series to find the sum

$$\sum_{n=1}^{\infty} \frac{(-1)^{n+1}}{n^2} = 1 - \frac{1}{4} + \frac{1}{9} - \frac{1}{16} + \cdots,$$

that is, the alternating version of the sum in (a).

c) Even before he discovered/invented calculus, Gottfried Wilhelm Leibniz (1646–1716) was working on summing infinite series; indeed, this work led him to his development of calculus. Along the way, in 1673, Leibniz used a beautiful geometric argument to conclude that

$$\sum_{k=1}^{\infty} \frac{(-1)^{k+1}}{(2k-1)} = 1 - \frac{1}{3} + \frac{1}{5} - \frac{1}{7} + \cdots = \frac{\pi}{4}.$$

Post-calculus, there are various ways to determine this result. Use the Fourier series for $f(x) = x$ on $-\pi \le x \le \pi$ to prove it.

20. Establish the result

$$\frac{1}{1^2} + \frac{1}{3^2} + \frac{1}{5^2} + \cdots = \frac{\pi^2}{8}$$

two different ways.

a) Using the Fourier series for $f(x) = |x|$ on $-1 \le x \le 1$.

b) Algebraically, using the results from Exercises 19a and 19b.

21. **MATLAB: Gibbs phenomenon.** If you plan on skipping the following section, read Example 1 of that section. Then do the same for the following functions.

a) $f(x) = x, -\pi \le x \le \pi$

b) $f(x) = x^2, -1 \le x \le 1$

c) $f(x) = \begin{cases} 2, & \text{if } -2 \le x < 1, \\ x - 1, & \text{if } 1 \le x \le 2 \end{cases}$

d) $f(x) = \begin{cases} 1 + x, & \text{if } -1 \le x \le 0, \\ 1 - x, & \text{if } 0 < x \le 1 \end{cases}$

e) $f(x) = \begin{cases} -1, & \text{if } -3 \le x < -2, \\ 1, & \text{if } -2 \le x \le 2, \\ -1, & \text{if } 2 < x \le 3 \end{cases}$

3.5   The Fourier Series—Proof of Pointwise Convergence[§]

We now prove the results given in the previous section. We do so for the interval $-\pi \le x \le \pi$, that is, for $L = \pi$, realizing that the proof for arbitrary L proceeds similarly (or realizing that we may transform a problem on $-L \le x \le L$ to a problem on $-\pi \le x \le \pi$). Further, we may assume at the start that $f(x)$ already has been extended to a function on the real line, with period 2π—in essence, then, we are proving Corollary 3.1.

We proceed as follows. First, we list the steps required to prove Corollary 3.1 for those x where f is continuous. Next, we fill in the proof of these steps. Finally, we modify the proof in order to deal with those values of x where f is discontinuous.

PROOF that $f(x) = F(x)$ at all points where f is continuous

For each such x we wish to prove that the nth partial sum

$$s_n(x) = \frac{a_0}{2} + \sum_{k=1}^{n} (a_k \cos kx + b_k \sin kx)$$

converges to $f(x)$, that is, that

$$\lim_{n \to \infty} [s_n(x) - f(x)] = 0$$

for each (fixed) x at which f is continuous. To this end we will

---

[§]This section may be skipped without loss of continuity (no pun intended).

[1] Rewrite $s_n(x)$ as

$$s_n(x) = \frac{1}{2\pi} \int_{-\pi}^{\pi} \left[1 + 2 \sum_{k=1}^{n} \cos k(t-x) \right] f(t)dt.$$

[2] Show that

$$1 + 2 \sum_{k=1}^{n} \cos k(t-x) = \frac{\sin \frac{2n+1}{2}(t-x)}{\sin \frac{t-x}{2}} \quad * \quad (\text{for } -\pi \le t - x \le \pi, t \ne x).$$

[3] Use a substitution to rewrite $s_n(x)$ as

$$s_n(x) = \frac{1}{2\pi} \int_{-\pi}^{\pi} \frac{\sin \frac{2n+1}{2} u}{\sin \frac{u}{2}} f(x+u)du.$$

[4] Show that

$$\frac{1}{2\pi} \int_{-\pi}^{\pi} \frac{\sin \frac{2n+1}{2} u}{\sin \frac{u}{2}} du = 1$$

so that we may write

$$s_n(x) - f(x) = \frac{1}{2\pi} \int_{-\pi}^{\pi} \frac{\sin \frac{2n+1}{2} u}{\sin \frac{u}{2}} f(x+u)du - f(x)$$

$$\cdot \frac{1}{2\pi} \int_{-\pi}^{\pi} \frac{\sin \frac{(2n+1)}{2} u}{\sin \frac{u}{2}} du$$

$$= \frac{1}{2\pi} \int_{-\pi}^{\pi} [f(x+u) - f(x)] \frac{\sin \frac{2n+1}{2} u}{\sin \frac{u}{2}} du.$$

[5] Show that this last integral $\to 0$ as $n \to \infty$.

PROOF of [1]   Remember that the Fourier coefficients are

$$a_0 = \frac{1}{\pi} \int_{-\pi}^{\pi} f(t)dt, \quad a_k = \frac{1}{\pi} \int_{-\pi}^{\pi} f(t) \cos kt \, dt,$$

$$b_k = \frac{1}{\pi} \int_{-\pi}^{\pi} f(t) \sin kt \, dt, \quad k = 1, 2, \ldots, n.$$

---

*This is $2D_n(t-x)$, where $D_n(x)$ is the well-known **Dirichlet kernel**.

Therefore,

$$s_n(x) = \frac{a_0}{2} + \sum_{k=1}^{n}(a_k \cos kx + b_k \sin kx)$$

$$= \frac{1}{2\pi}\int_{-\pi}^{\pi} f(t)dt + \sum_{k=1}^{n}\left[\cos kx \cdot \frac{1}{\pi}\int_{-\pi}^{\pi} f(t)\cos kt\ dt\right.$$

$$\left. + \sin kx \cdot \frac{1}{\pi}\int_{-\pi}^{\pi} f(t)\sin kt\ dt\right]$$

$$= \frac{1}{2\pi}\int_{-\pi}^{\pi}\left[1 + 2\sum_{k=1}^{n}(\cos kx \cos kt + \sin kx \sin kt)\right]f(t)dt$$

$$= \frac{1}{2\pi}\int_{-\pi}^{\pi}\left[1 + 2\sum_{k=1}^{n}\cos k(t - x)\right]f(t)dt$$

where, in the last step, we have used the trigonometric identity

$$\cos(A - B) = \cos A \cos B + \sin A \sin B.$$

∎

PROOF of [2]   We need to show that

$$1 + 2\sum_{k=1}^{n}\cos k\theta = \frac{\sin\frac{2n+1}{2}\theta}{\sin\frac{\theta}{2}}, \quad \text{for} \quad -\pi \le \theta \le \pi, \theta \ne 0.$$

Now, we could use trig identities to do this, but this approach presupposes that we know what the right side should be. Instead, as is the case with so many situations which involve trig identities, life is much easier if we get things in terms of complex exponentials. Remember Euler's formula

$$e^{i\theta} = \cos\theta + i\sin\theta.$$

Replacing θ with $-\theta$ gives us

$$e^{-i\theta} = \cos(-\theta) + i\sin(-\theta) = \cos\theta - i\sin\theta.$$

Finally, adding these two equations, we get

$$\cos\theta = \frac{e^{i\theta} + e^{-i\theta}}{2}.$$

Also, subtracting the second from the first gives us

$$\sin\theta = \frac{e^{i\theta} - e^{-i\theta}}{2i}.$$

Then,

$$1 + 2 \sum_{k=1}^{n} \cos k\theta = 1 + \sum_{k=1}^{n} \left(e^{ik\theta} + e^{-ik\theta} \right)$$

$$= e^{-in\theta} + e^{-i(n-1)\theta} + \ldots + e^{-i\theta} + 1 + e^{i\theta} + \ldots + e^{in\theta}$$

$$= e^{-in\theta} \left(1 + e^{i\theta} + e^{2i\theta} + \ldots + e^{2ni\theta} \right). \tag{3.14}$$

The sum in parentheses is geometric, of the form

$$1 + r + r^2 + \ldots + r^{2n} = \frac{1 - r^{2n+1}}{1 - r}.$$

Therefore, (3.14) becomes

$$= e^{-in\theta} \frac{1 - e^{(2n+1)i\theta}}{1 - e^{i\theta}}$$

$$= \frac{e^{-in\theta} - e^{(n+1)i\theta}}{1 - e^{i\theta}},$$

and, multiplying top and bottom by $e^{\frac{-i\theta}{2}}$, we get

$$= \frac{e^{(n+\frac{1}{2})i\theta} - e^{-(n+\frac{1}{2})i\theta}}{e^{\frac{i\theta}{2}} - e^{\frac{-i\theta}{2}}} = \frac{\sin \frac{2n+1}{2}\theta}{\sin \frac{\theta}{2}}.$$

∎

PROOF of [3]   Letting $u = t - x$, $du = dt$, we have

$$s_n(x) = \frac{1}{2\pi} \int_{-\pi}^{\pi} \frac{\sin \frac{2n+1}{2}(t-x)}{\sin \frac{t-x}{2}} f(t) dt$$

$$= \frac{1}{2\pi} \int_{-\pi-x}^{\pi-x} \frac{\sin \frac{2n+1}{2}u}{\sin u/2} f(x+u) du.$$

But f has period 2π, as does $\frac{\sin \frac{2n+1}{2}u}{\sin u/2}$ (why?), so we may write the above as

$$s_n(x) = \frac{1}{2\pi} \int_{-\pi}^{\pi} \frac{\sin \frac{2n+1}{2}u}{\sin u/2} f(x+u) du.$$

∎

PROOF of [4]

$$\int_{-\pi}^{\pi} \frac{\sin \frac{2n+1}{2} u}{\sin u/2} du = \int_{-\pi}^{\pi} \left[1 + 2 \sum_{k=1}^{n} \cos ku \right] du$$

$$= 2 \int_{0}^{\pi} \left[1 + 2 \sum_{k=1}^{n} \cos ku \right] du$$

and

$$\int_{0}^{\pi} \left[1 + 2 \sum_{k=1}^{n} \cos ku \right] du = \int_{0}^{\pi} du + 2 \sum_{k=1}^{n} \int_{0}^{\pi} \cos ku \, du$$

$$= \pi + 2 \sum_{k=1}^{n} \frac{\sin ku}{k} \Big|_{0}^{\pi}$$

$$= \pi.$$

∎

PROOF of [5]   We must show that

$$\lim_{n \to \infty} \int_{-\pi}^{\pi} [f(x+u) - f(x)] \frac{\sin \frac{2n+1}{2} u}{\sin \frac{u}{2}} du = 0.$$

In order to do this, we'll need the following well known lemma.

Lemma (Riemann–Lebesgue)[†] *If g is piecewise continuous on $[a, b]$, then*

$$\lim_{\lambda \to \infty} \int_{a}^{b} g(x) \sin \lambda x \, dx = 0.$$

Then, we need only show that

$$\frac{f(x+u) - f(x)}{\sin u/2}$$

is piecewise continuous (in u, of course—x is fixed) on $-\pi \le u \le \pi$.

PROOF of Lemma   Since g is piecewise continuous on $[a, b]$, we can write

$$\int_{a}^{b} g(x) \sin \lambda x \, dx = \sum_{i=1}^{n} \int_{x_{i-1}}^{x_i} g_i(x) \sin \lambda x \, dx,$$

[†]This is the Riemann–Lebesgue Lemma for the sine function.

where $g_i(x) = g(x)$ on (x_{i-1}, x_i) and g_i is continuous on $[x_{i-1}, x_i]$, $i = 1, \ldots, n$. It follows that each g_i is *uniformly* continuous on $[x_{i-1}, x_i]$.

We now wish to prove that

$$\lim_{\lambda \to \infty} \left| \int_{x_{i-1}}^{x_i} g_i(x) \sin \lambda x \, dx \right| = 0, \quad i = 1, \ldots, N.$$

First, letting $x = z + \frac{\pi}{\lambda}$ and noting that $\sin \lambda x = -\sin \lambda z$, we can write

$$I = \int_{x_{i-1}}^{x_i} g_i(x) \sin \lambda x \, dx = -\int_{x_{i-1} - \frac{\pi}{\lambda}}^{x_i - \frac{\pi}{\lambda}} g_i\left(z + \frac{\pi}{\lambda}\right) \sin \lambda z \, dz$$

so that

$$2I = \int_{x_{i-1}}^{x_i} g_i(x) \sin \lambda x \, dx - \int_{x_{i-1} - \frac{\pi}{\lambda}}^{x_i - \frac{\pi}{\lambda}} g_i\left(z + \frac{\pi}{\lambda}\right) \sin \lambda z \, dz$$

$$= -\int_{x_{i-1} - \frac{\pi}{\lambda}}^{x_{i-1}} g_i\left(x + \frac{\pi}{\lambda}\right) \sin \lambda x \, dx + \int_{x_i - \frac{\pi}{\lambda}}^{x_i} g_i(x) \sin \lambda x \, dx$$

$$+ \int_{x_{i-1}}^{x_i - \frac{\pi}{\lambda}} \left[g_i(x) - g_i\left(x + \frac{\pi}{\lambda}\right) \right] \sin \lambda x \, dx$$

$$= -I_1 + I_2 + I_3.$$

Now, g_i continuous on $[x_{i-1}, x_i]$ implies that $g_i(x)$ is bounded there. Thus, there exists M such that

$$|g_i(x)| \leq M, \quad x_{i-1} \leq x \leq x_i$$

and

$$\left| g_i\left(x + \frac{\pi}{\lambda}\right) \right| \leq M, \quad x_{i-1} - \frac{\pi}{\lambda} \leq x \leq x_i - \frac{\pi}{\lambda}.$$

Then, $2|I| \leq |I_1| + |I_2| + |I_3|$ and

$$|I_1| \leq \int_{x_{i-1} - \frac{\pi}{\lambda}}^{x_{i-1}} \left| g_i\left(x + \frac{\pi}{\lambda}\right) \right| |\sin \lambda x| \, dx$$

$$\leq \int_{x_{i-1} - \frac{\pi}{\lambda}}^{x_{i-1}} M \, dx \quad \text{(why?)}$$

$$\leq \frac{M\pi}{\lambda} \quad \text{(why?)}.$$

Similarly, $|I_2| \leq \frac{M\pi}{\lambda}$. Thus, $I_1 \to 0$ and $I_2 \to 0$ as $\lambda \to \infty$.

Finally, we must show that, for every $\epsilon > 0$, there is a constant k such that

$$\lambda > k \Rightarrow |I_3| < \epsilon.$$

Now, as g is uniformly continuous on $[x_{i-1}, x_i]$, we know that for all $\epsilon_1 > 0$, there is a $\delta > 0$ such that

$$\frac{\pi}{\lambda} < \delta \Rightarrow \left| g_i(x) - g_i\left(x + \frac{\pi}{\lambda}\right) \right| < \epsilon_1$$

for all $x \in [x_{i-1}, x_i]$. So, given ϵ above, let $\epsilon_1 = \frac{\epsilon}{x_i - x_{i-1}}$ and let $\delta = \delta_1$ be the corresponding value of δ. Then,

$$\lambda > \frac{\pi}{\delta_1} \Rightarrow \frac{\pi}{\lambda} < \delta_1 \Rightarrow \left| g_i(x) - g_i\left(x + \frac{\pi}{\lambda}\right) \right| < \frac{\epsilon}{x_i - x_{i-1}}$$

for all $x \in [x_{i-1}, x_i]$

$$\Rightarrow |I_3| \leq \int_{x_{i-1}}^{x_i - \frac{\pi}{\lambda}} \left| g_i(x) - g_i\left(x + \frac{\pi}{\lambda}\right) \right| dx$$

$$< \int_{x_{i-1}}^{x_i - \frac{\pi}{\lambda}} \frac{\epsilon}{x_i - x_{i-1}} dx < \epsilon.$$

∎

Now, all that's left is to show that $\frac{f(x+u)-f(x)}{\sin u/2}$ is piecewise smooth. Since f is piecewise smooth and the sine is smooth, we need only be concerned when the denominator vanishes, which occurs only at $u = 0$ (since we're restricted to the interval $-\pi \leq u \leq \pi$).

The trick here is to use the following two facts:

1) $f'(x+)$ and $f'(x-)$ exist (since f is piecewise smooth) and
2) $\lim\limits_{x \to 0} \frac{\sin x}{x} = 1$ (using l'Hôpital's rule—but see Exercise 5).

With these in mind, we rewrite

$$g(u) = \frac{f(x+u) - f(x)}{\sin u/2} = \frac{f(x+u) - f(x)}{u} \cdot \frac{u}{\sin \frac{u}{2}}.$$

Then,

$$\lim_{u \to 0^+} g(u) = \lim_{u \to 0^+} \frac{f(x+u) - f(x)}{u} \cdot \lim_{u \to 0^+} \frac{u}{\sin \frac{u}{2}}$$

$$= f'(x+) \cdot 2$$

and

$$\lim_{u \to 0^-} g(u) = f'(x-) \cdot 2,$$

each of which exists. Therefore, g is piecewise continuous, and we are done!

∎

Now, what happens at a point x where f is *not* continuous? Life is a bit more complicated, since at least one of

$$\lim_{u \to 0^+} \frac{f(x+u) - f(x)}{u}, \quad \lim_{u \to 0^-} \frac{f(x+u) - f(x)}{u}$$

will not exist. (In fact, we would like to be able to include points where $f(x)$ doesn't even exist.) However, we need only look at the expression

$$s_n(x) - \frac{1}{2}[f(x+) + f(x-)],$$

which we may rewrite as

$$\frac{1}{2\pi} \int_{-\pi}^{\pi} \frac{\sin \frac{2n+1}{2} u}{\sin \frac{u}{2}} f(x+u) du - \frac{1}{2}[f(x-) + f(x+)]$$

$$= \frac{1}{2\pi} \int_{-\pi}^{0} \frac{\sin \frac{2n+1}{2} u}{\sin \frac{u}{2}} f(x+u) du - \frac{1}{2} f(x-)$$

$$+ \frac{1}{2\pi} \int_{0}^{\pi} \frac{\sin \frac{2n+1}{2} u}{\sin \frac{u}{2}} f(x+u) du - \frac{1}{2} f(x+)$$

$$= \frac{1}{2\pi} \int_{-\pi}^{0} \frac{\sin \frac{2n+1}{2} u}{\sin \frac{u}{2}} [f(x+u) - f(x-)] du$$

$$+ \frac{1}{2\pi} \int_{0}^{\pi} \frac{\sin \frac{2n+1}{2} u}{\sin \frac{u}{2}} [f(x+u) - f(x+)] du,$$

where we have used the fact that

$$\frac{1}{2\pi} \int_{-\pi}^{0} \frac{\sin \frac{2n+1}{2} u}{\sin \frac{u}{2}} du = \frac{1}{2\pi} \int_{0}^{\pi} \frac{\sin \frac{2n+1}{2} u}{\sin \frac{u}{2}} du = \frac{1}{2} \text{ (why is this true?)}.$$

We proceed pretty much as we did above and use the Riemann–Lebesgue Lemma to show that each of the integrals $\to 0$ as $n \to \infty$. We do so here for the second integral (the first is dealt with in the same way).

As above, we write

$$\int_{0}^{\pi} \frac{\sin \frac{2n+1}{2} u}{\sin \frac{u}{2}} [f(x+u) - f(x+)] du = \int_{-\pi}^{\pi} g(u) \sin \frac{2n+1}{2} u \, du,$$

where

$$g(u) = \begin{cases} 0, & \text{if } -\pi \le u \le 0 \\ \dfrac{f(x+u) - f(x+)}{u} \cdot \dfrac{u}{\sin \frac{u}{2}}, & \text{if } 0 < u \le \pi. \end{cases}$$

We need only show that g is piecewise continuous on $-\pi \le u \le \pi$, and, as before, we need only be concerned with what happens as $u \to 0^+$. But

$$\lim_{u \to 0^+} g(u) = \lim_{u \to 0^+} \frac{f(x+u) - f(x+)}{u} \cdot \lim_{u \to 0^+} \frac{u}{\sin \frac{u}{2}} = 2f'(x+),$$

which exists as f is piecewise smooth.[‡] So we are done!

Now that we know *what* the Fourier series of a function converges to, we may ask *how* it converges (that is, does it converge "quickly" or "slowly," and what do the partial sums look like as the value of n increases?). It turns out that the Fourier series of f converges "nicely" to f at those points where f is continuous. However, remembering that each sine and cosine function in a Fourier series is continuous for all values of x, it should come as no surprise that the Fourier series may behave somewhat strangely near those x-values where f has a jump.

Example 1 The Fourier series for

$$f(x) = \begin{cases} 0, \text{ if } -\pi \leq x < 0 \\ 1, \text{ if } 0 \leq x \leq \pi \end{cases}$$

is

$$F(x) = \frac{1}{2} + \frac{2}{\pi} \sum_{n=1}^{\infty} \frac{\sin(2n-1)x}{2n-1}.$$

In Figure 3.13 we have graphed the function

$$s_n(x) = \frac{1}{2} + \frac{2}{\pi} \sum_{k=1}^{n} \frac{\sin(2k-1)x}{2k-1} \quad \text{[§]}$$

for $n = 5, n = 20$ and $n = 200$, respectively. We can see, in each case, that the function $s_n(x)$ exhibits a noticeable *overshoot* just to the right (and left) of $x = 0$ (with similar behavior near $x = \pm\pi$). We also see that, while most of the humps in the graph of $s_n(x)$ tend to flatten out as n increases, this first overshoot, although also shrinking, seems to remain fairly large. So what happens as $n \to \infty$?

[‡]Note that $f'(x+)$ is not the same as the right-hand derivative that you learned in calculus. The latter,

$$\lim_{h \to 0^+} \frac{f(x+h) - f(x)}{h},$$

need not exist even if f is piecewise smooth and, therefore, is less interesting and less useful than the former (which, by the way, is the "old" right-hand derivative of the function whose value at x is $f(x+)$). See Exercises 6–9 for a more in-depth look.

[§]To be more precise, we should have $s_{2n-1}(x) = s_{2n}(x)$, instead of $s_n(x)$.

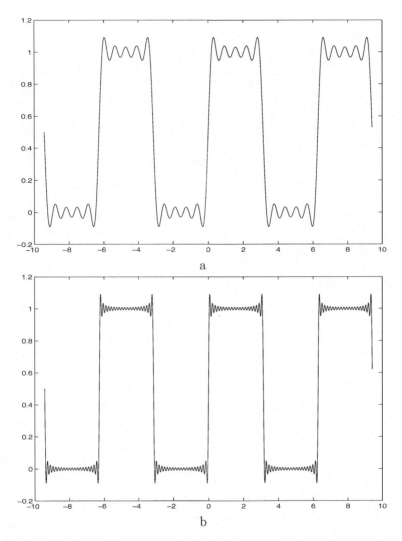

FIGURE 3.13
MATLAB graphs of the truncated Fourier series for the function from Example 1, for (a) $n = 5$, (b) $n = 20$ and (c) $n = 200$.

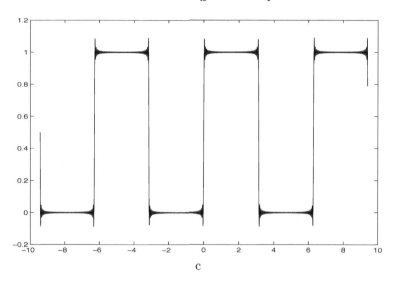

c

FIGURE 3.13 continued

This overshoot is known as the **Gibbs phenomenon**, after Josiah Willard Gibbs[¶] who, in 1899, pointed out this behavior and stated that the overshoot approaches a quantity involving $\int_0^\pi \frac{\sin x}{x} dx$ as $n \to \infty$. We'll examine the Gibbs phenomenon as it applies to this particular example.[‖]

So, let's look at the function

$$g_n(x) = s_n(x) - f(x),$$

find its least positive critical point $x = x_n$, show that $g_n(x_n)$ is a relative maximum and, finally, relate $\lim_{n \to \infty} g_n(x_n)$ to $\int_0^\pi \frac{\sin x}{x} dx$.

First, we restrict ourselves to the interval $0 < x < \pi$, where

$$g(x) = \frac{2}{\pi} \sum_{k=1}^{n} \frac{\sin(2k-1)x}{2k-1} - \frac{1}{2}.$$

Hence, on $0 < x < \pi$,

$$g_n'(x) = \frac{2}{\pi} \sum_{k=1}^{n} \cos(2k-1)x.$$

[¶] Josiah Willard Gibbs (1839–1903) is often considered to be the first American mathematician of note. Gibbs actually treated the case where $f(x) = x$, and he gave no proof of his claims. Later, in 1906, Maxime Bôcher proved Gibbs's statement and showed that this phenomenon occurs at any jump discontinuity. (Actually, a British mathematician named Wilbraham had discovered the phenomenon in 1848! Thus, it is sometimes called the Wilbraham–Gibbs phenomenon.)

[‖] For a more detailed treatment, see, e.g., *Introduction to the Theory of Fourier's Series and Integrals* by H. S. Carslaw.

Next, we may use an argument similar to the proof of [2], above (see Exercise 1), to show that

$$g_n'(x) = \frac{2}{\pi} \frac{\sin 2nx}{\sin x} \quad \text{on} \quad 0 < x < \pi.$$

The critical points are those values of x for which $2nx$ is an integral multiple of π, and the least such value is $x_n = \frac{\pi}{2n}$. Further, g_n attains a relative maximum here (see Exercise 2).

We would like to relate $\lim_{n\to\infty} g_n\left(\frac{\pi}{2n}\right)$ to $\int_0^\pi \frac{\sin x}{x} dx$. The easiest way to see this is to notice that the sum

$$g_n\left(\frac{\pi}{2n}\right) = \frac{2}{\pi} \sum_{k=1}^n \frac{1}{2k-1} \sin \frac{(2k-1)\pi}{2n} - \frac{1}{2}$$

can be made to look like a Riemann sum for the integral $\int_0^\pi \frac{\sin x}{x} dx$. To that end, break the interval $[0, \pi]$ into n equal subdivisions and, on each subinterval, choose x_i = midpoint of that subinterval. We then have $\Delta x = \frac{\pi}{n}$, $x_i = \frac{2i-1}{2}\frac{\pi}{n}$, $i = 1, 2, \ldots, n$ and

$$\int_0^\pi \frac{\sin x}{x} dx = \lim_{n\to\infty} \sum_{i=1}^n \frac{1}{\frac{(2i-1)\pi}{2n}} \cdot \sin \frac{(2i-1)\pi}{2n} \cdot \frac{\pi}{n}$$

$$= \lim_{n\to\infty} 2 \sum_{i=1}^n \frac{1}{2i-1} \sin \frac{(2i-1)\pi}{2n}.$$

It follows that

$$\lim_{n\to\infty} g_n\left(\frac{\pi}{2n}\right) = \lim_{n\to\infty} \frac{2}{\pi} \sum_{k=1}^n \frac{1}{2k-1} \sin \frac{(2k-1)\pi}{2n} - \frac{1}{2}$$

$$= \frac{1}{\pi} \int_0^\pi \frac{\sin x}{x} dx - \frac{1}{2},$$

which turns out to be approximately .09 (see Exercise 4). Therefore, the Gibbs overshoot here is approximately 9% of the jump in the graph of f. More generally, it can be shown that this is *always* the case (for functions on $-\pi \leq x \leq \pi$).

The true mathematical significance of the Gibbs phenomenon is to show that if f has any jump discontinuities, then its Fourier series will not converge *uniformly* to f. We discuss this important type of convergence in Appendix B.

Exercises 3.5

1. **MATLAB:** Do Exercise 22 of the previous section.

2. Proceed as in the proof of [2] to show that

$$\sum_{k=1}^n \cos(2k-1)x = \frac{\sin 2nx}{\sin x} \quad \text{on} \quad 0 < x < \pi.$$

3. Show that the function

$$\sum_{k=1}^{n} \frac{1}{2k-1} \sin \frac{(2k-1)\pi}{2n}$$

has $2n - 1$ critical values in the interval $0 < x < \pi$, that the least of these corresponds to a relative maximum and that each thereafter is, alternatively, a relative min or a relative max.

4. Evaluate $\int_0^\pi \frac{\sin x}{x} \, dx$ by expanding $\sin x$ in its Maclaurin series. Then show that $\frac{1}{\pi} \int_0^\pi \frac{\sin x}{x} \, dx - \frac{1}{2} \approx .09$.

5. Explain why $2 \sum_{k=1}^{n} \frac{1}{2k-1} \sin \frac{(2k-1)\pi}{2n}$ converges *down* to $\int_0^\pi \frac{\sin x}{x} \, dx$, i.e., why the sequence is monotonic decreasing.

6. a) Use L'Hôpital's rule to show that $\lim_{x \to 0} \frac{\sin x}{x} = 1$.

    b) As you did in calculus, show that

$$\frac{d}{dx}(\sin x) = \lim_{h \to 0} \cos x \cdot \frac{\sin h}{h} + \sin x \frac{\cos h - 1}{h}.$$

(Therefore, in order to apply L'Hôpital's rule to $\frac{\sin x}{x}$, you need to compute $\frac{d}{dx}(\sin x)$. However, in order to compute $\frac{d}{dx}(\sin x)$, you already need to know $\lim_{x \to 0} \frac{\sin x}{x}$! Is this circular reasoning?)

In Exercises 7–9 we look more closely at the difference between $f'(x_0+)$ and the right-hand derivative of f at x_0, $f'_R(x_0)$. Remember, the latter is just

$$f'_R(x_0) = \lim_{h \to 0^+} \frac{f(x_0 + h) - f(x_0)}{h},$$

while the former is

$$f'(x_0+) = \lim_{x \to x_0^+} f'(x).$$

(Of course, we can deal with

$$f_L(x_0) = \lim_{h \to 0^-} \frac{f(x_0 + h) - f(x_0)}{h}$$

and $f'(x_0-)$ in a similar manner.)

7. For each function below, compute $f'_R(0), f'_L(0), f'(0+), f'(0-)$ and $f'(0)$, if they exist.

a) $f(x) = x^3 + 5x$

b) $f(x) = \begin{cases} 2x + 3, & \text{if } x < 0 \\ x - 1, & \text{if } x \geq 0 \end{cases}$

c) $f(x) = \begin{cases} x^2, & \text{if } x \leq 0 \\ x^2 + 1, & \text{if } x > 0 \end{cases}$

8. In Exercise 7, we've seen that it's possible that $f'(x_0+)$ exists while $f'_R(x_0)$ doesn't exist. But the opposite can occur, too! Do the same as in Exercise 7 for each function below.

a) $f(x) = \begin{cases} x \sin \frac{1}{x}, & \text{if } x \neq 0 \\ 0, & \text{if } x = 0 \end{cases}$

b) $f(x) = \begin{cases} x^2 \sin \frac{1}{x}, & \text{if } x \neq 0 \\ 0, & \text{if } x = 0 \end{cases}$

c) $f(x) = \begin{cases} x^3 \sin \frac{1}{x}, & \text{if } x \neq 0 \\ 0, & \text{if } x = 0 \end{cases}$

9. Decide if the statement is true or false. In each case, f is piecewise smooth.

a) If $f'(x_0+) = f'(x_0-)$, then $f'(x_0)$ exists and $f'(x_0) = f'(x_0+)$.

b) If $f'_R(x_0) = f'_L(x_0)$, then $f'(x_0)$ exists and $f'(x_0) = f'_R(x_0)$.

c) If $f'_R(x_0) = f'(x_0+)$ and $f'_L(x_0) = f'(x_0-)$, then $f'(x_0)$ exists.

3.6 Fourier Sine and Cosine Series

At last we are in a position to answer the questions posed at the end of the previous chapter.

Given a function $f(x)$ on $0 \leq x \leq L$, is it possible to find constants c_n, $n = 0, 1, 2, \ldots$, and d_n, $n = 0, 1, 2, \ldots$, so that

$$f(x) = \sum_{n=0}^{\infty} c_n \cos \frac{n\pi x}{L} = c_0 + \sum_{n=1}^{\infty} c_n \cos \frac{n\pi x}{L}$$

and

$$f(x) = \sum_{n=0}^{\infty} d_n \sin \frac{n\pi x}{L} = \sum_{n=1}^{\infty} d_n \sin \frac{n\pi x}{L}$$

on $0 \leq x \leq L$?

What we do is this: If f is piecewise smooth, then we extend it to a piecewise smooth function on $-L \leq x \leq L$. Then, we can find the Fourier series of the latter, which will be identical to $f(x)$ on $0 \leq x \leq L$ (except possibly, of course, at finitely many points).

But—how do we extend f? Further, since the Fourier series contains both sines *and* cosines, how will this answer our question, anyway? Our key can be found in Exercise 13 of Section 3.3:

> If $f(x)$ is *even* on $-L \leq x \leq L$, then its Fourier series is a *pure cosine series*.
>
> If $f(x)$ is *odd* on $-L \leq x \leq L$, then its Fourier series is a *pure sine series*.

Obviously, if we want to expand f in a cosine series, we need only extend it to an *even function* on $-L \leq x \leq L$; similarly for a sine series, we extend it to an *odd function* on $-L \leq x \leq L$.

Definition 3.9 *Given $f(x)$ on $0 \leq x \leq L$, the even function*

$$g(x) = \begin{cases} f(x), & \text{if } 0 \leq x \leq L, \\ f(-x), & \text{if } -L \leq x < 0 \end{cases}$$

is called the **even extension of f to $-L \leq x \leq L$.** *The odd function*

$$h(x) = \begin{cases} f(x), & \text{if } 0 \leq x \leq L, \\ -f(-x), & \text{if } -L \leq x < 0 \end{cases}$$

is called the **odd extension of f to $-L \leq x \leq L$.** *(See Figure 3.14.)*

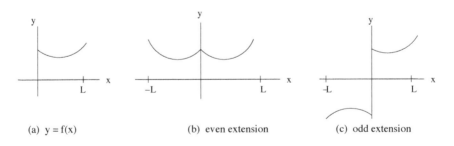

(a) y = f(x)   (b) even extension   (c) odd extension

FIGURE 3.14
Even and odd extensions.

Note that if f is piecewise smooth, then so are g and h. Also, technically, $h(x)$ is not an odd function unless $f(0) = 0$. However, remember that what happens at one point has no effect on the Fourier series of the function.

Let us calculate the Fourier series of g and h, with an eye toward getting everything in terms of f. For g we get the series

$$G(x) = \frac{a_0}{2} + \sum_{n=1}^{\infty} \left(a_n \cos \frac{n\pi x}{L} + b_n \sin \frac{n\pi x}{L} \right),$$

where

$$b_n = 0, \quad n = 1, 2, \ldots,$$

$$\begin{aligned} a_n &= \frac{1}{L} \int_{-L}^{L} g(x) \cos \frac{n\pi x}{L} \, dx \\ &= \frac{2}{L} \int_{0}^{L} g(x) \cos \frac{n\pi x}{L} \, dx, \text{ since the integrand is even} \\ &= \frac{2}{L} \int_{0}^{L} f(x) \cos \frac{n\pi x}{L} \, dx, \text{ since } g(x) = f(x) \text{ on } 0 \le x \le L, n = 0, 1, 2, \ldots. \end{aligned}$$

Furthermore,

$$G(x) = g(x) \text{ }^{*} \quad \text{on} \quad -L \le x \le L$$

and so

$$G(x) = f(x) \quad \text{on} \quad 0 \le x \le L.$$

We can do the same for the function $h(x)$, resulting in the series

$$H(x) = \sum_{n=1}^{\infty} b_n \sin \frac{n\pi x}{L},$$

where

$$b_n = \frac{2}{L} \int_{0}^{L} f(x) \sin \frac{n\pi x}{L} \, dx, \qquad n = 1, 2, 3, \ldots$$

(see Exercise 13).

Definition 3.10 *Given $f(x)$ on $0 \le x \le L$, the series*

$$F_c(x) = \frac{a_0}{2} + \sum_{n=1}^{\infty} a_n \cos \frac{n\pi x}{L}, \text{ where } a_n = \frac{2}{L} \int_{0}^{L} f(x) \cos \frac{n\pi x}{L} \, dx$$

is called the **Fourier cosine series** *of f on $0 \le x \le L$. The series*

$$F_s(x) = \sum_{n=1}^{\infty} b_n \sin \frac{n\pi x}{L}, \text{ where } b_n = \frac{2}{L} \int_{0}^{L} f(x) \sin \frac{n\pi x}{L} \, dx$$

is called the **Fourier sine series** *of f on $0 \le x \le L$.*

---

*Here, and following, we write "=," realizing, of course, that the two functions may differ at finitely many points.

Corollary 3.2 *If f is piecewise smooth on* $0 \leq x \leq L$, *we have*

$$F_c(x) = f(x) \text{ on } 0 \leq x \leq L$$

and

$$F_s(x) = f(x) \text{ on } 0 \leq x \leq L.$$

Of course, we can be precise—as we were in Theorem 3.1—as to the value of $F_c(x)$ and $F_s(x)$ at the discontinuous points of f and the endpoints. Further, since $F_c(x)$ and $F_s(x)$ are also *Fourier* series, as in Section 3.4, each is periodic of period $2L$. See Figure 3.15.

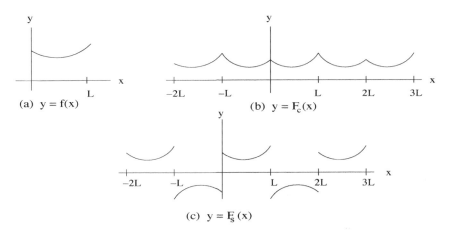

(a) y = f(x)

(b) y = F$_c$(x)

(c) y = F$_s$(x)

FIGURE 3.15
The graphs of the Fourier cosine and sine series (ignoring the points of discontinuity).

Example 1 Calculate the Fourier sine and Fourier cosine series of $f(x) = 3$ on $0 \leq x \leq \pi$.

$$F_s(x) = \sum_{n=1}^{\infty} b_n \sin nx, \text{ where } b_n = \frac{2}{\pi} \int_0^{\pi} 3 \cdot \sin nx \, dx$$

$$= \begin{cases} 0, & \text{if } n \text{ is even} \\ \frac{12}{n\pi}, & \text{if } n \text{ is odd.} \end{cases}$$

So

$$F_s(x) = \frac{12}{\pi} \sin x + 0 + \frac{12}{3\pi} \sin 3x + 0 + \cdots$$

$$= \frac{12}{\pi} \sum_{k=0}^{\infty} \frac{1}{2k+1} \sin(2k+1)x.$$

Also,

$$F_c(x) = \frac{a_0}{2} + \sum_{n=1}^{\infty} a_n \cos nx, \quad \text{where } a_n = \frac{2}{\pi} \int_0^{\pi} 3 \cos nx \, dx.$$

Now, we *could* do the calculation and find that

$$a_0 = 6, \quad a_n = 0 \quad \text{for} \quad n = 1, 2, 3, \ldots \ .$$

However, life is much easier if we notice that $f(x)$ *already is in the same form as* $F_c(x)$ (compare Exercise 9 in Section 3.3) and, in light of Corollary 3.2, we must have

$$F_c(x) = 3 = f(x).$$

Example 2 Calculate the Fourier sine and cosine series of $f(x) = x$ on $0 \le x \le 1$.

$$F_s(x) = \sum_{n=1}^{\infty} b_n \sin n\pi x, \quad \text{where } b_n = 2 \int_0^1 x \sin n\pi x \, dx = \frac{2(-1)^{n+1}}{n\pi}$$

$$F_c(x) = \frac{a_0}{2} + \sum_{n=1}^{\infty} a_n \cos n\pi x, \quad \text{where } a_n = 2 \int_0^1 x \cos n\pi x \, dx,$$

so

$$a_0 = 2 \int_0^1 x \, dx = 1,$$

$$
\begin{aligned}
a_n = 2 \int_0^1 x \cos n\pi x \, dx &= \frac{2}{n^2 \pi^2} [\cos n\pi - 1] \\
&= \frac{2}{n^2 \pi^2} [(-1)^n - 1] \\
&= \begin{cases} 0, & \text{if } n \text{ is even,} \\ -\frac{4}{n^2 \pi^2}, & \text{if } n \text{ is odd,} \end{cases}
\end{aligned}
$$

and

$$
\begin{aligned}
F_c(x) &= \frac{1}{2} - \frac{4}{\pi^2} \cos \pi x - \frac{4}{3^2 \pi^2} \cos 3\pi x - \frac{4}{5^2 \pi^2} \cos 5\pi x - \cdots \\
&= \frac{1}{2} - \frac{4}{\pi^2} \sum_{k=1}^{\infty} \frac{1}{(2k-1)^2} \cos(2k-1)\pi x.
\end{aligned}
$$

Example 3 Draw the graph of $y = f(x)$, along with the graph of its Fourier cosine series, $y = F_c(x)$, and the graph of its Fourier sine series, $y = F_s(x)$, for $f(x) = 1 - x$ on $0 \le x \le 1$. See Figure 3.16.

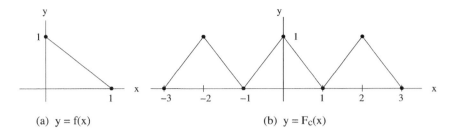

(a) y = f(x)   (b) y = F_c(x)

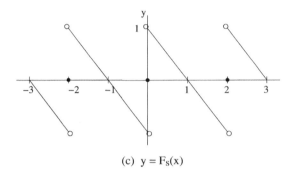

(c) y = F_s(x)

FIGURE 3.16
The graphs for Example 3.

Exercises 3.6

In Exercises 1–3, calculate the Fourier sine and Fourier cosine series of $f(x)$.

1. $f(x) = \begin{cases} 1, & \text{if } 0 \leq x < 2, \\ 0, & \text{if } 2 \leq x \leq 4 \end{cases}$

2. $f(x) = x^2, 0 \leq x \leq \pi$

3. $f(x) = \sin x, 0 \leq x \leq \pi$

In Exercises 4–7, proceed as in Example 3.

4. $f(x) = x^2 + 1$ on $0 \leq x \leq 1$

5. $f(x) = x - 1$ on $0 \leq x \leq 2$

6. $f(x) = \begin{cases} 3, & \text{if } x = 0 \text{ or } x = 1, \\ x + 1, & \text{otherwise, on } 0 \leq x \leq 2 \end{cases}$

7. $f(x) = \begin{cases} 1, & \text{if } 0 \leq x \leq 3, \\ -2, & \text{if } 3 < x \leq 5 \end{cases}$

8. **MATLAB:** Following Example 1, Section 3.5, use MATLAB to graph the truncated Fourier sine and cosine series, for various values of n, for the functions in Exercises 1–7. When is the Gibbs phenomenon exhibited?

In Exercises 9–12, decide if the statement is true or false. Assume that f is piecewise smooth on $0 \le x \le L$.

9. $F_s(kL) = 0$ for every integer k.

10. $F_c(2kL) = f(0)$ for every integer k.

11. $F_c(2kL) = f(0+)$ for every integer k.

12. The constant term $\frac{a_0}{2}$ in $F_c(x)$ actually is the average value of f on $0 \le x \le L$. (Similarly, true or false? The constant term $\frac{a_0}{2}$ in the Fourier series is the average value of piecewise smooth f on $-L \le x \le L$.)

13. Derive the Fourier sine series summation and coefficients, as we did for the Fourier cosine series.

14. a) Show that the functions $1, \cos \frac{n\pi x}{b-a}$, $n = 1, 2, \ldots$, are pairwise orthogonal on the interval $a \le x \le b$. Do the same for the functions $\sin \frac{n\pi x}{b-a}$, $n = 1, 2 \ldots$.

    b) Given piecewise smooth $f(x)$ on $a \le x \le b$, show that it can be expanded in a series of functions $1, \cos \frac{n\pi x}{b-a}$, $n = 1, 2, \ldots$, and also in a series of the functions $\sin \frac{n\pi x}{b-a}$, $n = 1, 2, \ldots$. (See Exercise 18, Section 3.4.)

    c) Compute both series from part b, for the function $f(x) = x$ on $1 \le x \le 4$.

15. Suppose we wanted to do this whole process the other way around, that is, suppose we start with a theorem which tells us that every piecewise smooth function on $0 \le x \le L$ has a Fourier sine and Fourier cosine series, and we want to show that it follows that any piecewise smooth $f(x)$ on $-L \le x \le L$ has a Fourier series. Show that this can be accomplished using the fact that $f(x) = g(x) + h(x)$, where

$$g(x) = \frac{f(x) + f(-x)}{2} \text{ is even}$$

and

$$h(x) = \frac{f(x) - f(-x)}{2} \text{ is odd}.$$

3.7   Completeness

So far, in our discussion of the various types of Fourier series, we've had to qualify all of our convergence statements with "except, possibly, at finitely many points." So, although the Fourier series of a function is unique, many different functions—infinitely many, of course—have the same Fourier series. While not a problem for physical applications, this state of affairs is not at all satisfying from a mathematical standpoint.

In order to remedy the situation, mathematicians introduce a new setting, based on a different type of convergence (which we'll study in Chapter 8, when we look at generalized Fourier series). In this *weaker* setting, two functions which differ from each other at finitely many points are considered to be the "same" function. Thus, we are able to say that the Fourier series of f converges to f, without qualification.

Although it is more appropriate to wait and discuss the idea of completeness in this new setting, we briefly introduce it now, as it is such an important concept. Thus, we will use this weaker definition of "=" throughout this section.

So, as we have seen, given any piecewise smooth function $f(x)$ on $0 \leq x \leq L$, we can find constants $b_1, b_2, b_3, \ldots$, such that

$$f(x) = \sum_{n=1}^{\infty} b_n \sin \frac{n\pi x}{L}.$$

Therefore, the functions $\sin \frac{n\pi x}{L}$, $n = 1, 2, 3, \ldots$, essentially span the space of piecewise smooth functions on $0 \leq x \leq L$. We say, then, that the set

$$\left\{ \sin \frac{n\pi x}{L} \right\}_{n=1}^{\infty} = \left\{ \sin \frac{\pi x}{L}, \sin \frac{2\pi x}{L}, \sin \frac{3\pi x}{L}, \ldots \right\} \tag{3.15}$$

is **complete** in this space of functions. Further, as we have seen, these functions form an orthogonal set. We say, then, that (3.15) forms a **complete orthogonal set** (in the space of piecewise smooth functions on $0 \leq x \leq L$).

Similarly, the functions

$$\left\{ \cos \frac{n\pi x}{L} \right\}_{n=0}^{\infty} = \left\{ 1, \cos \frac{\pi x}{L}, \cos \frac{2\pi x}{L}, \ldots \right\} \tag{3.16}$$

form a complete orthogonal set (in the same space).

Now, let's go back to Chapter 1, where we solved the eigenvalue problem

$$y'' + \lambda y = 0$$
$$y(0) = y(L) = 0$$

and got (3.15) as our set of eigenfunctions. Similarly, (3.16) are the eigenfunctions of the problem

$$y'' + \lambda y = 0$$
$$y'(0) = y'(L) = 0.$$

It is natural to ask if the set of eigenfunctions of such an eigenvalue problem *always* forms a complete orthogonal set in some space of functions on the interval in question. Well, the answer is "sometimes," and we'll give a precise treatment in Chapter 11. However, there are two other eigenvalue problems which need to be addressed now, as they also arise in connection with the heat, wave and Laplace's equations.

In Exercise 15b of Section 1.7, we solved the problem

$$y'' + \lambda y = 0$$
$$y(0) = y'(L) = 0$$

and found that the eigenfunctions are

$$\left\{ \sin \frac{(2n-1)\pi x}{2L} \right\}_{n=1}^{\infty} = \left\{ \sin \frac{\pi x}{2L}, \sin \frac{3\pi x}{2L}, \sin \frac{5\pi x}{2L}, \ldots \right\}. \qquad (3.17)$$

Similarly, the eigenfunctions for

$$y'' + \lambda y = 0$$
$$y'(0) = y(L) = 0$$

and (Exercise 15c, Section 1.7)

$$\left\{ \cos \frac{(2n-1)\pi x}{2L} \right\}_{n=1}^{\infty} = \left\{ \cos \frac{\pi x}{2L}, \cos \frac{3\pi x}{2L}, \cos \frac{5\pi x}{2L}, \ldots \right\}. \qquad (3.18)$$

Does each of (3.17) and (3.18) form a complete orthogonal set in the space of piecewise smooth functions on $0 \le x \le L$? We'll prove that the answer is "yes," in the exercises (although we'll see that this affirmative answer also follows from a general result of Chapter 11—also, orthogonality was already established in Exercise 22 of Section 1.7).

As a result, we'll see:

1) That any piecewise smooth function $f(x)$ on $0 \le x \le L$ can be expanded in series of the form

$$f(x) = \sum_{n=1}^{\infty} c_n \sin \frac{(2n-1)\pi x}{2L}$$

and

$$f(x) = \sum_{n=1}^{\infty} d_n \cos \frac{(2n-1)\pi x}{2L}$$

2) That the coefficients c_n and d_n are given by

$$c_n = \frac{2}{L} \int_0^L f(x) \sin \frac{(2n-1)\pi x}{2L} dx$$

and

$$d_n = \frac{2}{L} \int_0^L f(x) \cos \frac{(2n-1)\pi x}{2L} dx, \qquad n = 1, 2, 3, \dots .$$

Exercises 3.7

1. Prove directly by integration that each of the sets $\left\{ \sin \frac{(2n-1)x}{2} \right\}_{n=1}^{\infty}$ and $\left\{ \cos \frac{(2n-1)x}{2} \right\}_{n=1}^{\infty}$ forms an orthogonal set on $0 \le x \le \pi$.

2. Prove that the set of functions

$$\left\{ \sin \frac{(2n-1)x}{2} \right\}_{n=1}^{\infty} = \left\{ \sin \frac{x}{2}, \sin \frac{3x}{2}, \sin \frac{5x}{2}, \dots \right\}$$

forms a complete set in the space of piecewise smooth functions on $0 \le x \le \pi$, as follows.

We must show that there exist constants $c_1, c_2, c_3, \dots$, such that

$$f(x) = \sum_{n=1}^{\infty} c_n \sin \frac{(2n-1)x}{2} \tag{3.19}$$

$$= c_1 \sin \frac{x}{2} + c_2 \sin \frac{3x}{2} + \cdots \qquad 0 \le x \le \pi.$$

In order to accomplish this,

a) Let $F(x)$ be the even extension of $f(x)$ to $0 \le x \le 2\pi$, that is, extend f so that F is symmetric about the line $x = \pi$.

b) Find the Fourier sine series of F on $0 \le x \le 2\pi$.

c) Show that (3.19) follows; what are the constants, c_n?

3. Prove that the set of functions

$$\left\{ \cos \frac{(2n-1)x}{2} \right\}_{n=1}^{\infty} = \left\{ \cos \frac{x}{2}, \cos \frac{3x}{2}, \cos \frac{5x}{2}, \dots \right\}$$

forms a complete set in the space of piecewise smooth functions on $0 \le x \le \pi$.

Prelude to Chapter 4

Now, with the introduction of Fourier's sine and cosine series, we are able to solve the Big Three PDEs, along with many others, for fairly arbitrary initial and boundary conditions. So, in this chapter, we solve the one-dimensional homogeneous heat equation for a finite rod, the one-dimensional homogeneous wave equation for a finite string and the two-dimensional Laplace's equation on a rectangle, in each case with homogeneous boundary conditions. After that, we consider how to deal with nonhomogeneous boundary conditions. Finally, we treat the case where the PDE itself is nonhomogeneous.

As we have seen, the nonhomogeneous Laplace's equation actually is called Poisson's equation. Laplace had been under the mistaken impression that the gravitational potential of, say, a planet must satisfy Laplace's equation everywhere—in particular, he thought it must be satisfied in the interior of the attracting body. In 1813, Siméon-Denis Poisson (1781–1840) pointed out Laplace's error and showed that the PDE must be nonhomogeneous. Although we solve Poisson's equation in this chapter, we must wait until Chapter 8 in order to treat this gravitational problem, as it occurs most naturally in the setting of spherical coordinates.

And so, without further ado, ...

4

Solving the Big Three PDEs
on Finite Domains

4.1 Solving the Homogeneous Heat Equation for a Finite Rod

Finally we are in a position to solve the PDE problems which were derived in Chapter 2. We start with some examples involving the heat equation, leaving the general case for the exercises.

Example 1 Solve the heat equation initial-boundary-value problem

$$u_t = 3u_{xx},$$
$$u(x, 0) = x(\pi - x),$$
$$u(0, t) = u(\pi, t) = 0.$$

As before, we separate the PDE:

$$u(x, t) = X(x)T(t) \Rightarrow \frac{T'}{3T} = \frac{X''}{X} = -\lambda$$
$$\Rightarrow X'' + \lambda X = 0, \quad T' + 3\lambda T = 0.$$

Next, separate the boundary conditions:

$$u(0, t) = 0 = X(0)T(t) \Rightarrow X(0) = 0,$$
$$u(\pi, t) = 0 = X(\pi)T(t) \Rightarrow X(\pi) = 0.$$

So we have

$$X'' + \lambda X = 0, \quad T' + 3\lambda T = 0,$$
$$X(0) = X(\pi) = 0.$$

Now we solve the X-eigenvalue problem. We get (see Example 3, Section 2.6)

eigenvalues: $\quad \lambda_n = n^2$
eigenfunctions: $X_n(x) = \sin nx, \qquad n = 1, 2, 3, \dots$.

Next, we go back and solve the T-equation for $\lambda = \lambda_n$:

$$T' + 3n^2 T = 0 \Rightarrow T_n(t) = e^{-3n^2 t}.$$

Then, for each eigenvalue λ_n, we form the product solution $X_n T_n$, and use them to form the general solution

$$u(x,t) = \sum_{n=1}^{\infty} c_n e^{-3n^2 t} \sin nx. \tag{4.1}$$

Finally, we determine the coefficients from the initial condition,

$$u(x,0) = x(\pi - x) = \sum_{n=1}^{\infty} c_n \sin nx, \quad \text{on } 0 \leq x \leq \pi. \tag{4.2}$$

Now, we know that the Fourier sine series for $f(x) = x(\pi - x)$ on $0 \leq x \leq \pi$ is

$$F_s(x) = \sum_{n=1}^{\infty} b_n \sin nx, \quad b_n = \frac{2}{\pi} \int_0^{\pi} x(\pi-x) \sin nx \, dx = \begin{cases} 8/\pi n^3, & \text{if } n \text{ is odd,} \\ 0, & \text{if } n \text{ is even.} \end{cases}$$

In other words, we must have

$$c_n = b_n = \begin{cases} 8/\pi n^3, & \text{if } n \text{ is odd,} \\ 0, & \text{if } n \text{ is even,} \end{cases}$$

that is, (4.2) must be the Fourier sine series for $f(x) = x^2$! Our final solution is the general solution (4.1), with these particular values of the c_n:

$$u(x,t) = \frac{8}{\pi} \sum_{k=1}^{\infty} \frac{1}{(2k-1)^3} e^{-3(2k-1)^2 t} \sin(2k-1)x \quad \text{(why?)}.$$

(Of course, we can evaluate the integrals using integration by parts or an integral table.) Figure 4.1 shows the solution for various values of t. For more, see Exercises 4 and 9. Note that we're actually plotting the *truncated* solution (using $\sum_{k=1}^{10}$).

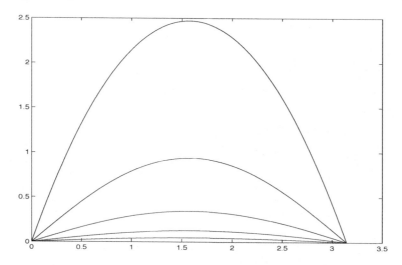

FIGURE 4.1
MATLAB graphs of the solution of Example 1, in the *x-u* plane, for
t = 0, 1, 2, 3 and 4 (from top to bottom). We can see how quickly
the solution approaches the *steady state solution* $u \equiv 0$. (We have
used the truncated solution, with $n = 10$ terms.)

Example 2 Do the same for

$$u_t = u_{xx},$$
$$u(x,0) = x,$$
$$u_x(0,t) = u_x(3,t) = 0.$$

Again, we separate:

$$X'' + \lambda X = 0, \quad T' + \lambda T = 0,$$
$$X'(0) = X'(3) = 0.$$

Solve the X-problem (see Example 2 in Section 1.7):

eigenvalues: $\quad \lambda_0 = 0, \quad \lambda_n = \dfrac{n^2 \pi^2}{9}, \qquad n = 1, 2, 3, \ldots$

eigenfunctions: $X_0(x) = 1, \quad X_n(x) = \cos \dfrac{n\pi x}{3}, \qquad n = 1, 2, 3, \ldots$.

Solve the T-equation for each eigenvalue:

$\lambda_0 = 0: \qquad T' = 0 \Rightarrow T_0(t) = 1$

$\lambda_n = \dfrac{n^2 \pi^2}{9} : T' + \dfrac{n^2 \pi^2}{9} T = 0 \Rightarrow T_n(t) = e^{-\frac{n^2 \pi^2 t}{9}}, \qquad n = 1, 2, 3, \ldots$.

Form the product solutions:

$$u_0(x,t) = X_0(x)T_0(t) = 1$$

$$u_n(x,t) = X_n(x)T_n(t) = e^{-\frac{n^2\pi^2 t}{9}}\cos\frac{n\pi x}{3}, \qquad n = 1,2,3,\ldots .$$

Form the general solution:

$$u(x,t) = c_0 \cdot 1 + c_1 e^{-\frac{\pi^2 t}{9}}\cos\frac{\pi x}{3} + c_2 e^{-\frac{4\pi^2 t}{9}}\cos\frac{2\pi x}{3} + \cdots$$

$$= c_0 + \sum_{n=1}^{\infty} c_n e^{-\frac{n^2\pi^2 t}{9}}\cos\frac{n\pi x}{3}. \tag{4.3}$$

Finally, apply the initial condition:

$$u(x,0) = x = c_0 + \sum_{n=1}^{\infty} c_n \cos\frac{n\pi x}{3},$$

and the right side must be the *Fourier cosine series* of $f(x) = x$ on $0 \leq x \leq 3$. (Alternatively, expand x in its Fourier cosine series on $0 \leq x \leq 3$, then equate corresponding coefficients.) So we have

$$c_0 = \frac{a_0}{2}, \quad c_n = a_n, \qquad n = 1,2,3,\ldots,$$

where

$$a_n = \frac{2}{3}\int_0^3 x\cos\frac{n\pi x}{3}dx, \qquad n = 0,1,2,\ldots .$$

Therefore, our solution is (4.3) with these coefficients plugged in:

$$u(x,t) = \frac{a_0}{2} + \sum_{n=1}^{\infty} a_n e^{-\frac{n^2\pi^2 t}{9}}\cos\frac{n\pi x}{3} = 3 - \frac{12}{\pi^2}\sum_{k=1}^{\infty}\frac{1}{(2k-1)^2}e^{-\frac{(2k-1)^2\pi^2 t}{9}}\cos\frac{n\pi x}{3}.$$

See Figure 4.2 for plots of the solution for various values of t. Here, we use $\sum_{k=1}^{50}$.

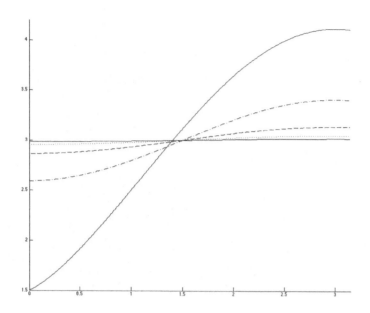

FIGURE 4.2
MATLAB graphs of the truncated solution from Example 2, using
$n = 50$ terms, in the x-u plane for $t = 0$, 1, 2, 3 and 4. Again, we
see its very fast approach to the steady state solution $u \equiv 3$. (Note
the not-so-good Fourier approximation to the initial straight line
$u(x, 0) = x$.)

Example 3 Do the same for

$$u_t = 2u_{xx},$$
$$u(x, 0) = x + 1,$$
$$u(0, t) = u_x(4, t) = 0.$$

First, separate:

$$X'' + \lambda X = 0, \quad T' + 2\lambda T = 0,$$
$$X(0) = X'(4) = 0.$$

Then, find the eigenvalues and eigenfunctions for the X-system (see Exercise
4 in Section 1.7):

$$\text{eigenvalues:} \quad \lambda_n = \frac{(2n - 1)^2 \pi^2}{64}$$
$$\text{eigenfunctions: } X_n(x) = \sin \frac{(2n - 1)\pi x}{8}.$$

Solve the T-equation for each λ_n:

$$T' + \frac{(2n-1)^2\pi^2}{32}T = 0 \Rightarrow T_n(t) = e^{-\frac{(2n-1)^2\pi^2 t}{32}}.$$

Form the product solutions and the general solution:

$$u(x,t) = \sum_{n=1}^{\infty} c_n e^{-\frac{(2n-1)^2\pi^2 t}{32}} \sin\frac{(2n-1)\pi x}{8}.$$

Finally, apply the initial condition:

$$u(x,0) = x + 1 = \sum_{n=1}^{\infty} c_n \sin\frac{(2n-1)\pi x}{8}.$$

From the discussion in Section 3.7—that is, that the functions $\left\{\sin\frac{(2n-1)\pi x}{8}\right\}_{n=1}^{\infty}$ form a *complete orthogonal set* on $0 \le x \le 4$—we are guaranteed that there are coefficients that satisfy this equality (from completeness) and that they are (from orthogonality)

$$c_n = \frac{2}{4}\int_0^4 (x+1)\sin\frac{(2n-1)\pi x}{8}\,dx.$$

How do we deal with more complicated heat problems, for example, those with nonhomogeneous boundary conditions or nonhomogeneous PDEs? We'll take these up in Section 4.4.

Exercises 4.1

1. Solve the heat equation $u_t = 2u_{xx}$ for a rod of length L with both ends held at $0°$, if

   a) $L = \pi, u(x,0) = 20$

   b) $L = 1, u(x,0) = x$ (See Example 2, Section 3.6.)

   c) $L = 2, u(x,0) = \begin{cases} 20, & \text{if } 0 \le x < 1 \\ 0, & \text{if } 1 \le x \le 2 \end{cases}$

2. Solve the heat equation $u_t = 4u_{xx}$ for a rod of length L with both ends insulated, if

   a) $L = \pi, u(x,0) = x^2$ (See Exercise 2, Section 3.6.)

   b) $L = 1, u(x,0) = 10$

   c) $L = 2, u(x,0) = \begin{cases} 10, & \text{if } 0 \le x \le 1 \\ 0, & \text{if } 1 < x \le 2 \end{cases}$

3. Solve the heat equation $u_t = u_{xx}$ for a rod of length π, subject to

    a) $u(0,t) = u_x(\pi,t) = 0, u(x,0) = 100$

    b) $u_x(0,t) = u(\pi,t) = 0, u(x,0) = x$

4. **MATLAB:** Plot the truncated solution for each heat problem, in the x-u plane, for various values of t. What happens as $t \to \infty$?

    a) Exercise 1c

    b) Exercise 2a

    c) Exercise 2c

    d) Exercise 3b

    e) Exercise 3c

5. Solve the general heat equation $u_t = \alpha^2 u_{xx}$ subject to initial condition $u(x,0) = f(x)$ and to the boundary conditions

    a) $u(0,t) = u(L,t) = 0$

    b) $u_x(0,t) = u_x(L,t) = 0$

    c) $u(0,t) = u_x(L,t) = 0$

    d) $u_x(0,t) = u(L,t) = 0$

    e) $u(0,t) = u(L,t) + u_x(L,t) = 0^*$

    f) $u_x(0,t) = u(L,t) + u_x(L,t) = 0^*$

6. In quantum mechanics, if we have a particle of mass m, then its **wave function** $u = u(x,y,z,t)$ satisfies the famous **Schrödinger's equation**

$$-i\hbar u_t = \frac{\hbar^2}{2m}\nabla^2 u - V(x,y,z)u.$$

Here, i is the imaginary number, $\hbar$ is Planck's constant divided by 2π and V is a potential for the force acting on the particle.

For now let's consider the case of a "particle in a box, with zero potential," where the box is long and narrow enough to be considered one-dimensional. In this case, $\nabla^2 u = u_{xx}$, and the PDE becomes

$$u_t = \frac{i\hbar}{2m}u_{xx},$$

which looks suspiciously like the heat equation! If the wave function is zero at both ends of the box, then we have the initial-boundary-value

---

*You should assume that we have completeness and orthogonality here. This will be justified in Chapter 8.

problem

$$u_t = \frac{i\hbar}{2m} u_{xx},$$
$$u(x,0) = f(x),$$
$$u(0,t) = u(\pi,t) = 0,$$

where $f(x)$ is the *initial state* and L is the length of the box. Solve this problem.

7. In Exercise 9 of Section 2.4, we saw that a rod whose sides are *not* insulated satisfies the PDE $u_t = \alpha^2 u_{xx} - \beta u$. Solve the problem

$$u_t = u_{xx} - u,$$
$$u(x,0) = f(x),$$
$$u(0,t) = u(\pi,t) = 0.$$

8. If a pollutant is spilled into a still body of water, it will *diffuse* throughout the water and, thus, its concentration will satisfy the heat/diffusion equation. Suppose, instead, that it is spilled into a moving stream. The pollutant is then carried downstream; this process is called *convection* or *advection*. If there is no diffusion, then its concentration satisfies the *convection* or *advection equation* (discussed in Section 5.1). Finally, if the pollutant undergoes both diffusion and convection—which is what we would expect—its concentration will satisfy the **diffusion-convection equation** $u_t = \alpha^2 u_{xx} - \nu u_x$. Here, ν is the velocity of the stream (ν may depend on x or t).

Solve the diffusion-convection problem

$$u_t = u_{xx} + u_x, \qquad 0 < x < \pi,$$
$$u(x,0) = f(x)$$

subject to the (not very realistic!) boundary conditions

$$u(0,t) = u(\pi,t) = 0.$$

9. Remember that the steady state temperature of a rod is the time-independent function which solves the problem and which represents the temperature distribution of the rod "after a long time."

   a) Find the steady state temperature of the system

$$u_t = u_{xx}, \qquad 0 < x < L,$$
$$u(x,0) = f(x),$$
$$u(0,t) = u(L,t),$$

in two different ways:

(1) Letting $u_t \equiv 0$ and solving the PDE $u_{xx} = 0$

(2) Using the Fourier method of this section, and then allowing $t \to \infty$ (you may assume that you may interchange the sum and the limit, that is, you may assume that

$$\lim_{t \to \infty} \sum_{n=0}^{\infty} g_n(x, t) = \sum_{n=0}^{\infty} \lim_{t \to \infty} g_n(x, t))$$

What role does the initial temperature distribution play in the result? What's happening, physically?

b) Show that the steady state temperature of the system

$$u_t = u_{xx},$$
$$u(x, 0) = f(x),$$
$$u_x(0, t) = u_x(L, t)$$

is just the (constant) *average value* of the initial temperature distribution. Explain what's happening, physically.

c) What happens when the boundary conditions are mixed, that is, when we have $u(0) = u'(L) = 0$ or $u'(0) = u(L) = 0$? Again, what's happening, physically?

10. a) Given the heat problem

$$u_t = \alpha^2 u_{xx},$$
$$u(x, 0) = f(x),$$
$$u(0, t) = u(L, t) = 0,$$

show that, when we change variables to $s = \frac{\pi x}{L}$ and $\tau = \frac{\alpha^2 \pi^2}{L^2} t$, the PDE and boundary conditions become

$$v_\tau = v_{ss},$$
$$v(0, \tau) = v(\pi, \tau) = 0,$$

where v is the new dependent variable, $v(s, \tau) = u\left(\frac{L}{\pi} s, \frac{L^2}{\alpha^2 \pi^2} \tau\right)$. Thus, we need only know how to solve the heat problem on $0 \le x \le \pi$, with $\alpha^2 = 1$. What is the new initial condition?

b) Redo each part of Exercise 1 but, this time, first solve the v-problem in part (a), and then transform back to $u(x, t)$.

4.2   Solving the Homogeneous Wave Equation for a Finite String

The solution of the wave equation is quite similar to that of the heat equation.

Example 1 Solve the wave equation initial-boundary-value problem

$$u_{tt} = 4u_{xx},$$
$$u(x,0) = x(1-x),$$
$$u_t(x,0) = \cos x,$$
$$u(0,t) = u(1,t) = 0.$$

We begin by separating the PDE:

$$u(x,t) = X(x)T(t) \Rightarrow XT'' = 4X''T$$
$$\Rightarrow \frac{T''}{4T} = \frac{X''}{X} = -\lambda, \text{ constant}$$
$$\Rightarrow X'' + \lambda X = 0, \quad T'' + 4\lambda T = 0.$$

Then, separating the boundary conditions gives us

$$X(0) = X(1) = 0.$$

Next, solve the X-boundary-value problem (see Example 1, Section 1.7):

eigenvalues: $\lambda_n = n^2\pi^2$
eigenfunctions: $X_n = \sin n\pi x, \qquad n = 1,2,3,\dots$.

Now we solve the T-equation for $\lambda = \lambda_n$, and it is here that we find that the wave equation's solution differs from that of the heat equation:

$$T'' + 4\lambda_n T = 0 \Rightarrow T'' + 4n^2\pi^2 T = 0$$
$$\Rightarrow T(t) = c_1 \cos 2n\pi t + c_2 \sin 2n\pi t.$$

Since we must do this for each positive integer n, we write

$$T_n(t) = c_n \cos 2n\pi t + d_n \sin 2n\pi t, \qquad n = 1,2,3,\dots .$$

Our product solutions, then, are

$$u_n(x,t) = \sin n\pi x (c_n \cos 2n\pi t + d_n \sin 2n\pi t), \qquad n = 1,2,3,\dots,$$

so the general solution is the linear combination

$$u(x,t) = \sum_{n=1}^{\infty} \sin n\pi x (c_n \cos 2n\pi t + d_n \sin 2n\pi t). \tag{4.4}$$

Finally, we apply *both* initial conditions. First, we need to calculate u_t:

$$u_t(x,t) = \sum_{n=1}^{\infty} \sin n\pi x(-2n\pi c_n \sin 2n\pi t + 2n\pi d_n \cos 2n\pi t).$$

Then, we have

$$u(x,0) = x(1-x) = \sum_{n=1}^{\infty} c_n \sin n\pi x, \quad \text{on } 0 \le x \le 1 \qquad (4.5)$$

and

$$u_t(x,0) = \cos x = \sum_{n=1}^{\infty} 2n\pi d_n \sin n\pi x, \quad \text{on } 0 \le x \le 1. \qquad (4.6)$$

So the right-hand side of (4.5) must be the Fourier sine series for the function $x(1-x)$ on $0 \le x \le 1$ and similarly for (4.6) and the function $\cos x$. Or, if you prefer, expand $x(1-x)$ and $\cos x$ into their Fourier sine series on $0 \le x \le 1$, and (4.5) and (4.6) become, respectively,

$$\sum_{n=1}^{\infty} b_n \sin n\pi x = \sum_{n=1}^{\infty} c_n \sin n\pi x,$$

where

$$b_n = \frac{2}{1} \int_0^1 x(1-x) \sin n\pi x \; dx, \qquad n = 1,2,3,\ldots,$$

and

$$\sum_{n=1}^{\infty} b_n \sin n\pi x = \sum_{n=1}^{\infty} 2n\pi d_n \sin n\pi x,$$

where

$$b_n = \frac{2}{1} \int_0^1 \cos x \sin n\pi x \; dx, \qquad n = 1,2,3,\ldots .$$

Therefore,

$$c_n = 2 \int_0^1 x(1-x) \sin n\pi x \; dx, \qquad n = 1,2,3,\ldots,$$

$$2n\pi \, d_n = 2 \int_0^1 \cos x \sin n\pi x \; dx, \qquad n = 1,2,3,\ldots,$$

and our solution is just the general solution (4.4) with these values for the coefficients, that is,

$$u(x,t) = \sum_{n=1}^{\infty} \sin n\pi x(c_n \cos 2n\pi t + d_n \sin 2n\pi t),$$

$$c_n = 2 \int_0^1 x(1-x) \sin n\pi x \; dx, \qquad n = 1,2,3,\ldots,$$

$$d_n = \frac{1}{n\pi} \int_0^1 \cos x \sin n\pi x \; dx, \qquad n = 1,2,3,\ldots .$$

Example 2 Solve the system. Describe the string's motion.

$$u_{tt} = u_{xx},$$
$$u(x,0) = 0,$$
$$u_t(x,0) = 1,$$
$$u_x(0,t) = u_x(\pi,t) = 0.$$

Physically, it looks like there will be nothing to cause the string to vibrate, and it should just continue moving vertically at the initial velocity, without changing shape (not very realistic!). Let's make sure that the Fourier method actually gives us this solution.

So, first, separate:

$$X'' + \lambda X = 0, \quad T'' + \lambda T = 0,$$
$$X'(0) = X'(\pi) = 0.$$

Solve the X-problem:

eigenvalues: $\lambda_0 = 0, \quad \lambda_n = n^2, \quad n = 1,2,3,\ldots,$
eigenfunctions: $X_0(x) = 1, \quad X_n(x) = \cos nx, \quad n = 1,2,3,\ldots$.

Solve the T-equation for $\lambda = \lambda_n$:

$$\lambda_0 = 0: \quad T'' = 0 \Rightarrow T_0(t) = c_0 + d_0 t$$
$$\lambda_n = n^2: T'' + n^2 T = 0 \Rightarrow T_n(t) = c_n \cos nt + d_n \sin nt, \quad n = 1,2,3,\ldots$$.

Form the product solutions:

$$u_0(x,t) = X_0(x)T_0(t) = c_0 + d_0 t$$
$$u_n(x,t) = X_n(x)T_n(t)$$
$$= \cos nx (c_n \cos nt + d_n \sin nt), \quad n = 1,2,3,\ldots$$.

Then the general solution is

$$u(x,t) = c_0 + d_0 t + \sum_{n=1}^{\infty} \cos nx (c_n \cos nt + d_n \sin nt). \tag{4.7}$$

Now, the initial shape gives us

$$u(x,0) = 0 = c_0 + \sum_{n=1}^{\infty} c_n \cos nx,$$

which means that we have

$$c_n = 0, \quad n = 0,1,2,\ldots$$.

The initial velocity is

$$u_t(x,0) = 1 = d_0 + \sum_{n=1}^{\infty} n d_n \cos nx,$$

and, remembering that the Fourier cosine series for 1 on $0 \leq x \leq \pi$ is

$$1 = \frac{a_0}{2} + \sum_{n=1}^{\infty} a_n \cos nx,$$

where

$$a_n = \frac{2}{\pi} \int_0^{\pi} 1 \cdot \cos nx \; dx, \qquad n = 0, 1, 2, \ldots,$$

we have

$$d_0 = \frac{a_0}{2} = \frac{1}{\pi} \int_0^{\pi} 1 \; dx = 1$$

and

$$d_n = \frac{1}{n} a_n = \frac{2}{n\pi} \int_0^{\pi} \cos nx \; dx = 0, \qquad n = 1, 2, 3, \ldots .$$

Therefore, our solution is the general solution (4.7) with the above values of the coefficients plugged in:

$$u(x,t) = t,$$

so the string does *not* vibrate but, instead, retains its initial shape and continues to move upward at a velocity of 1.

Of course, we can't look at steady state solutions of the wave equation, but we *can* get some very important physical (and mathematical) information if we use our Fourier series solution to *decompose* u into its various *vibration modes*.

So, setting $g(x) \equiv 0$ for the sake of convenience (and it won't make a difference as far as what we'd like to show), we look at the solution

$$u(x,t) = \sum_{n=1}^{\infty} a_n \sin \frac{n\pi x}{L} \cos \frac{n\pi ct}{L}$$

of the string problem

$$u_{tt} = c^2 u_{xx}, \qquad 0 < x < L, t > 0,$$
$$u(x,0) = f(x),$$
$$u_t(x,0) = 0,$$
$$u(0,t) = u(L,t) = 0.$$

The individual product solution

$$u_n(x,t) = \sin\frac{n\pi x}{L}\cos\frac{n\pi ct}{L}, \qquad n = 1, 2, 3, \ldots,$$

of the PDE and boundary conditions, is called the nth **normal mode of vibration** for the problem. What do these modes look like?

1. The functions
$$u_n(x,0) = X_n(x) = \sin\frac{n\pi x}{L}$$

   are, of course, standard sine waves. The first few can be seen in Figure 4.3. The points where each curve intersects the x-axis (including the endpoints, for the string which is nailed down) are called its **nodes**. More on these below.

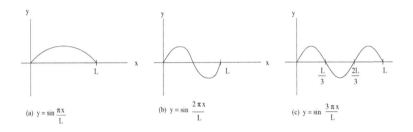

(a) y = sin $\frac{\pi x}{L}$ (b) y = sin $\frac{2\pi x}{L}$ (c) y = sin $\frac{3\pi x}{L}$

FIGURE 4.3
First three vibration modes for the vibrating string.

2. The function
$$T_n(t) = \cos\frac{n\pi ct}{L}$$

   is what tells us how the nth mode *vibrates*. Figure 4.4 gives snapshots of the second mode each at various time t. Note that the nodes remain fixed!

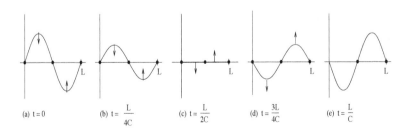

(a) t = 0   (b) t = $\frac{L}{4C}$ (c) t = $\frac{L}{2C}$ (d) t = $\frac{3L}{4C}$ (e) t = $\frac{L}{C}$

FIGURE 4.4
Vibration of the second mode. The three *nodes* remain fixed.

The time that it takes for the nth mode to go through one cycle of vibration—the vibration **period** of the mode—is just the (least positive) period of the cosine function. Thus, it's the value t_n for which $\frac{n\pi c t_n}{L} = 2\pi$, or

$$\text{period of } n^{\text{th}} \text{ mode: } t_n = \frac{2L}{nc}.$$

Then, the mode's vibration **frequency**, or number of vibrations per unit time, is just

$$\text{frequency of } n^{\text{th}} \text{ mode: } \nu_n = \frac{1}{t_n} = \frac{nc}{2L}.$$

We see that each frequency is an integral multiple of the **fundamental frequency**

$$\nu_1 = \frac{c}{2L},$$

the frequency of the first or **fundamental mode**. The complete set of frequencies is called the **frequency spectrum** of the string, and it is a *discrete*, as opposed to *continuous*, spectrum (the separation between consecutive frequencies being the fundamental frequency—we say they are "spaced according to n").

Thus, regardless of the initial displacement and velocity of the string, its motion can be decomposed into these modes and frequencies, and, since each frequency is an integral multiple of the fundamental frequency, the string actually vibrates at this frequency (why?). So, what tells us the difference between two different vibrations?

3. The coefficients a_n, of course! The number $|a_n|$ gives us the **amplitude** of the corresponding mode and, since the Fourier series solution converges, we must have $a_n \to 0$ as $n \to \infty$. Thus, the lowest modes contribute most to the motion of the string. (One may also show that the total energy of the vibrating string is equal to the sum of the energies of the individual modes and that the latter $\to 0$ as $a_n \to 0$, so as $n \to \infty$.)

A nice real-world illustration of the situation can be seen (and heard!) in the case of a violin string. Every note played by the violinist corresponds to a specific vibration frequency of the string, and the higher the frequency, the higher the note. For example, "concert A" (the A above "middle C") corresponds to a frequency of 440 hertz. So, suppose she plays this A—what do we actually hear? We hear the A, of course! The string is vibrating at 440 hertz. However, it's virtually impossible for anyone to cause *any* string to vibrate in a single mode. Thus, we should be able to decompose the motion into many—infinitely many—frequencies and modes. It turns out that 440 hertz is the fundamental frequency (which makes sense, since the A turns out to be the *loudest* note we hear) and that, in principle, the string also sounds the notes corresponding to all integral multiples of this frequency. In practice, we actually *do* hear the first few of these higher notes, with the various contributions dying out and becoming inaudible as the frequencies increase (again, the lower modes contribute more than the higher modes).

It is the *relative* contributions of the notes of various frequencies that give the violin string its unique sound. (To be more specific, striking the string differently will lead to different relative contributions, and we can hear these differences. More generally, though, the possible relative contributions at various frequencies for a violin string are different from those of other instruments. If we play the same A on a piano and a violin, and if we arrange things so that someone with a trained ear hears them only *after* they've been struck, then that person still can tell the difference between the two.)

In our example, we call A the **fundamental** and the higher tones the **overtones** of the particular string. As it turns out, the first few overtones turn out to be

2nd: A (an octave above the original, 880 hertz)
3rd: E (above the second A, 1320 hertz)*
4th: A (two octaves above the original, 1760 hertz)
5th: D (above this last A, 2200 hertz).*

These notes are harmonically *consonant* with the original—when sounded together, they are pleasing to the ear. Hence, the fundamental and overtones have come to be known as the **harmonics** of the particular string (although, if we go high enough, we start to encounter overtones which form a *dissonance* with the original A; however, these generally are inaudible).

Exercises 4.2

1. Solve the wave equation $u_{tt} = 5u_{xx}$ for a string of length L with both ends nailed down along the x-axis, if

    a) $L = \pi, u(x,0) = 3\sin 2x, u_t(x,0) = \sin x - 7\sin 4x$

    b) **(the plucked string)** $L = 4, u(x,0) = \begin{cases} x, & \text{if } 0 \le x \le 2, \\ 4 - x, & \text{if } 2 \le x \le 4, \end{cases}$

    $u_t(x,0) = 0$

    c) $L = 2, u(x,0) = 0, u_t(x,0) = 3$

2. Solve the wave equation $u_{tt} = 4u_{xx}$ for a string of length π, subject to the boundary conditions $u_x(0,t) = u_x(\pi,t) = 0$ and to the initial conditions

---

*For the equal-tempered scale, these values are approximately 1318.51 and 2349.32, respectively.

  a) $u(x,0) = 4\cos 3x \quad u_t(x,0) = 6\cos 2x - \cos 5x$

  b) $u(x,0) = \sin x \quad u_t(x,0) = 0$ (See Exercise 3, Section 3.6.)

  c) $u(x,0) = 1 \quad u_t(x,0) = x$

3. Solve the wave equation $u_{tt} = u_{xx}$ subject to

  a) $u(0,t) = u_x(1,t) = 0, \quad u(x,0) = 0, \quad u_t(x,0) = 1$

  b) $u_x(0,t) = u(\pi,t) = 0, \quad u(x,0) = \pi^2 - x^2, \quad u_t(x,0) = 0$

4. Solve the general wave equation $u_{tt} = c^2 u_{xx}$ subject to the initial conditions $u(x,0) = f(x)$ and $u_t(x,0) = g(x)$ and to the boundary conditions

  a) $u(0,t) = u(L,t) = 0$

  b) $u_x(0,t) = u_x(L,t) = 0$

  c) $u(0,t) = u_x(L,t) = 0$

  d) $u_x(0,t) = (L,t) = 0.$

5. In Exercise 4 in Section 2.3, we derived the damped wave equation $u_{tt} = c^2 u_{xx} - \beta u_t$.

  a) Solve the initial-boundary-value problem

$$u_{tt} = u_{xx} - 4u_t,$$
$$u(x,0) = 1,$$
$$u_t(x,0) = 0,$$
$$u(0,t) = u(\pi,t) = 0.$$

  What happens as $t \to \infty$?

  b) Solve the initial-boundary-value problem

$$u_{tt} = u_{xx} - 2u_t,$$
$$u(x,0) = 5\sin x - 3\sin 4x,$$
$$u_t(x,0) = 0,$$
$$u(0,t) = u(\pi,t) = 0.$$

  Again, what happens as $t \to \infty$?

  c) Solve the same problem as in part (b), but with initial conditions $u(x,0) = 0$, $u_t(x,0) = \sin x + \sin 2x$. Once more, what happens as $t \to \infty$?

  d) Remember that $u_{tt} = c^2 u_{xx} - \beta u_t - \gamma u$ is the **telegraph equation**. Solve the initial-boundary-value problem

$$u_{tt} = u_{xx} - 2u_t - 3u,$$
$$u(x,0) = x,$$
$$u_t(x,0) = 0,$$
$$u(0,t) = u(\pi,t) = 0.$$

6. **MATLAB:** Plot the truncated (unless it's not necessary to truncate) solution for each problem, in the x-u plane, for various values of t.

   a) Exercise 1a
   b) Exercise 1b
   c) Exercise 1c
   d) Exercise 3a
   e) Exercise 5b
   f) Exercise 5c

7. **MATLAB:** Plot the first four modes of the vibrating string of length 1 with boundary conditions $u(0,t) = u_x(1,t) = 0$.

8. Using trigonometric identities (as in Section 3.2), show that the solution of the wave initial-boundary problem

$$u_{tt} = c^2 u_{xx},$$
$$u(x,0) = f(x),$$
$$u_t(x,0) = g(x),$$
$$u(0,t) = u(L,t) = 0$$

   can be written as

$$u(x,t) = \frac{1}{2}[F(x+ct) + F(x-ct)] + \frac{1}{2c}\int_{x-ct}^{x+ct} G(s)ds,$$

   where $F(x)$ and $G(x)$ are the odd periodic extensions of $f(x)$ and $g(x)$, respectively.

9. **Vibrating Euler–Bernoulli beam:** From Appendix D, we know that a simply-supported vibrating E–B beam, given an initial displacement with zero initial velocity, is

$$u_{tt} + a^4 u_{xxxx} = 0, \qquad 0 < x < \pi, t > 0,$$
$$u(x,0) = f(x),$$
$$u_t(x,0) = 0,$$
$$u(0,t) = u_{xx}(0,t) = u(\pi,t) = u_{xx}(\pi,t) = 0.$$

   a) Solve this problem. (Hint: Let the separation constant be $\lambda = -k^4, 0, k^4$, for $k > 0$. For each eigenvalue λ, the corresponding k is called the *wave number*.)

   b) What is the vibration spectrum? Are the frequencies of the overtones integral multiples of the fundamental frequency? Conversely, is every integral multiple of the fundamental also the frequency of an overtone?

   c) Which musical instrument is, essentially, a series of such beams?

4.3 Solving the Homogeneous Laplace's Equation on a Rectangular Domain

Let's start by looking at a particular case of the Dirichlet problem on a rectangle:

$$u_{xx} + u_{yy} = 0, \qquad 0 < x < a, 0 < y < b$$
$$u(x,0) = f(x),$$
$$u(x,b) = g(x),$$
$$u(0,y) = u(a,y) = 0.$$

(See Figure 4.5.) First, separate the PDE:

$$u(x,y) = X(x)Y(y) \Rightarrow \frac{X''}{X} = -\frac{Y''}{Y} = -\lambda$$
$$\Rightarrow X'' + \lambda X = 0, \quad Y'' - \lambda Y = 0.$$

Next, separate the left and right side boundary conditions, as we've been doing. We now have

$$X'' + \lambda X = 0, \quad Y'' - \lambda Y = 0,$$
$$X(0) = X(a) = 0.$$

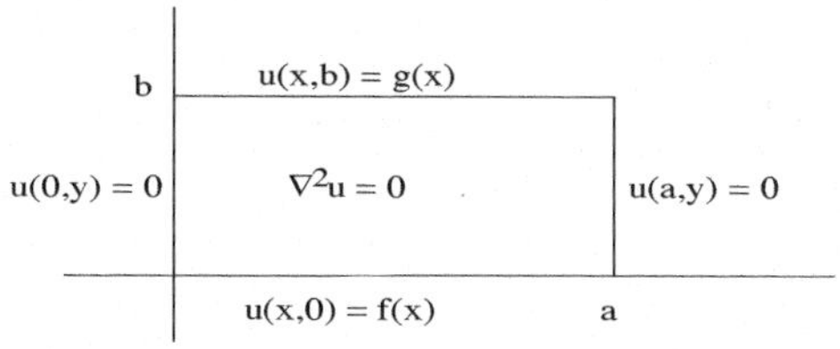

FIGURE 4.5
The Dirichlet problem on a rectangle (with homogeneous BCs along left and right edges).

Solving the X-eigenvalue problem, we get

eigenvalues: $\lambda_n = \frac{n^2\pi^2}{a^2}$,
eigenfunctions: $X_n(x) = \sin\frac{n\pi x}{a}$, $\qquad n = 1, 2, 3, \ldots$.

Again, we now must solve the Y-equation for $\lambda = \lambda_n$:

$$Y'' - \frac{n^2\pi^2}{a^2}Y = 0 \Rightarrow Y_n(y) = c_n \cosh\frac{n\pi y}{a} + d_n \sinh\frac{n\pi y}{a}.$$

The penultimate step, as usual, is to form the product solutions and use them to form the general solution:

$$u(x,y) = \sum_{n=1}^{\infty} \sin \frac{n\pi x}{a} \left(c_n \cosh \frac{n\pi y}{a} + d_n \sinh \frac{n\pi y}{a} \right).$$

Finally, we use the remaining two boundary conditions to determine the coefficients c_n, d_n, $n = 1, 2, 3, \ldots$:

$$u(x,0) = f(x) = \sum_{n=1}^{\infty} c_n \sin \frac{n\pi x}{a} \Rightarrow c_n = \frac{2}{a} \int_0^a f(x) \sin \frac{n\pi x}{a} dx, \quad n = 1, 2, 3, \ldots,$$

$$u(x,b) = g(x) = \sum_{n=1}^{\infty} \sin \frac{n\pi x}{a} \left(c_n \cosh \frac{n\pi b}{a} + d_n \sinh \frac{n\pi b}{a} \right)$$

$$\Rightarrow d_n \sinh \frac{n\pi b}{a} = \frac{2}{a} \int_0^a g(x) \sin \frac{n\pi x}{a} dx - c_n \cosh \frac{n\pi b}{a}$$

or

$$d_n = \frac{2}{a \sinh \frac{n\pi b}{a}} \int_0^a g(x) \sin \frac{n\pi x}{a} dx - c_n \coth \frac{n\pi b}{a}, \quad n = 1, 2, 3, \ldots .^{\dagger}$$

Of course, this example is by no means the most general Dirichlet problem. Specifically, the conditions involving $u(0,y)$ and $u(a,y)$ need not be homogeneous. We'll look more closely in the exercises.

Suppose, instead, we have the Neumann problem

$$u_{xx} + u_{yy} = 0,$$
$$u_y(x,0) = u_y(x,b) = 0,$$
$$u_x(0,y) = f(y),$$
$$u_x(a,y) = g(y).$$

Again, we separate, but we do so with an eye toward the fact that the homogeneous boundary conditions are $u_y(x,0) = u_y(x,b) = 0$:

$$u(x,y) = X(x)Y(y) \Rightarrow \frac{Y''}{Y} = -\frac{X''}{X} = -\lambda$$
$$\Rightarrow Y'' + \lambda Y = 0, \quad X'' - \lambda X = 0.$$

Now, separating these two boundary conditions, we see that we must solve

$$Y'' + \lambda Y = 0, \quad X'' - \lambda X = 0,$$
$$Y'(0) = Y'(b) = 0.$$

---

†Remember, the hyperbolic cotangent is $\coth x = \frac{\cosh x}{\sinh x}$.

As before, we see that the eigenvalues and eigenfunctions of the Y-problem are:

$$\text{eigenvalues:} \quad \lambda_0 = 0, \quad \lambda_n = \frac{n^2\pi^2}{b^2}, \qquad n = 1, 2, 3, \ldots$$
$$\text{eigenfunctions:} \; Y_0(y) = 1, \quad Y_n(y) = \cos\frac{n\pi y}{b}, \qquad n = 1, 2, 3, \ldots.$$

Solving the X-ODE for each of these eigenvalues, we have

$$\lambda_0 = 0: \; X_0(x) = c_0 + d_0 x$$

$$\lambda_n = \frac{n^2\pi^2}{b^2}: \; X_n(x) = c_n \cosh\frac{n\pi x}{b} + d_n \sinh\frac{n\pi x}{b}, \qquad n = 1, 2, 3, \ldots,$$

and the general solution becomes

$$u(x, y) = c_0 + d_0 x + \sum_{n=1}^{\infty} \cos\frac{n\pi y}{b} \left(c_n \cosh\frac{n\pi x}{b} + d_n \sinh\frac{n\pi x}{b} \right).$$

Finally, the other boundary conditions determine the constants:

$$u_x(0, y) = f(y) = d_0 + \sum_{n=1}^{\infty} \frac{n\pi}{b} d_n \cos\frac{n\pi y}{b}, \qquad 0 \le y \le b$$

$$\Rightarrow d_0 = \frac{1}{b} \int_0^b f(y)\,dy,$$

$$d_n = \frac{2}{n\pi} \int_0^b f(y) \cos\frac{n\pi y}{b}\,dy, \qquad n = 1, 2, 3, \ldots,$$

$$u_x(a, y) = g(y) = d_0 + \sum_{n=1}^{\infty} \frac{n\pi}{b} \cos\frac{n\pi y}{b} \left(c_n \sinh\frac{n\pi a}{b} + d_n \cosh\frac{n\pi a}{b} \right)$$

$$\Rightarrow d_0 = \frac{1}{b} \int_0^b g(y)\,dy,$$

$$\frac{n\pi}{b} \left(c_n \sinh\frac{n\pi a}{b} + d_n \cosh\frac{n\pi a}{b} \right) = \frac{2}{b} \int_0^b g(y) \cos\frac{n\pi y}{b}\,dy$$

or

$$c_n = \frac{2}{n\pi \sinh\frac{n\pi a}{b}} \int_0^b g(y) \cos\frac{n\pi y}{b}\,dy - \frac{n\pi d_n}{b} \cosh\frac{n\pi a}{b}, \qquad n = 1, 2, 3, \ldots.$$

Now, we notice two important differences between this example and the previous one:

1) c_0 is arbitrary! Therefore, there are infinitely many solutions, so this Neumann problem is not well-posed. In fact, this is a property of all Neumann problems.

2) There seem to be *two* expressions for d_0. Indeed, f and g must satisfy the **compatibility condition**

$$\int_0^b f(y)dy = \int_0^b g(y)dy$$

or else the problem has *no* solution. See Exercises 12–14.

Again, we'll look at problems with more than two nonhomogeneous boundary conditions in the exercises.

Exercises 4.3

In Exercises 1–10, solve Laplace's equation subject to the given boundary conditions. Work each problem out completely, rather than referring to the solutions in this section.

1. $u(x,0) = 0$, $\quad u(x,2) = 10$, $\quad u(0,y) = u(1,y) = 0$

2. $u(x,0) = 3\sin \pi x$, $\quad u(x,3) = 0$, $\quad u(0,y) = u(4,y) = 0$

3. $u(x,0) = u(x,\pi) = 0$, $\quad u(0,y) = y$, $\quad u(1,y) = 0$

4. $u(x,0) = \cos x$, $\quad u(x,1) = \sin \frac{x}{2} - \sin 2x$, $\quad u(0,y) = u(\pi,y) = 0$

5. $u(x,0) = u(x,2) = 0$, $\quad u(0,y) = y$, $\quad u(1,y) = 2y$

6. $u(x,0) = u(x,\pi) = 0$, $\quad u(0,y) = 0$, $\quad u_x(5,y) = 3\sin y - 5\sin 4y$

7. $u_y(x,0) = 0$, $\quad u(x,1) = x$, $\quad u(0,y) = u(1,y) = 0$

8. $u_y(x,0) = u_y(x,\pi) = 0$, $\quad u_x(0,y) = 0$, $\quad u_x(2\pi,y) = \cos 5y$

9. $u_y(x,0) = 2$, $\quad u_y(x,\pi) = 3$, $\quad u_x(0,y) = u_x(2\pi,y) = 0$

10. **MATLAB:** Plot the (truncated) solution of the given problem. Where do the maximum and minimum values of u seem to occur?

    a) Exercise 3

    b) Exercise 6

11.   a) Use the solution to Exercise 6 to solve

$$u_{xx} + u_{yy} = 0,$$
$$u(x,0) = 0,$$
$$u_y(x,5) = 3\sin x - 5\sin 4x,$$
$$u(0,y) = u(\pi,y) = 0.$$

b) Use the solution to the first example in this section to solve

$$u_{xx} + u_{yy} = 0,$$
$$u(x,0) = u(x,b) = 0,$$
$$u(0,y) = f(y),$$
$$u(a,y) = g(y).$$

Hint: Don't work too hard—look at part (a)!

c) More generally, solve

$$u_{xx} + u_{yy} = 0,$$
$$u(x,0) = u(x,b) = 0,$$
$$u(0,y) = f(y),$$
$$u(a,y) = g(y).$$

12. a) Use the solutions to Exercises 1 and 5 of the Dirichlet problems

$$u_{xx} + u_{yy} = 0, \qquad\qquad u_{xx} + u_{yy} = 0,$$
$$u(x,0) = u(x,2) = 0, \qquad u(x,0) = 0,$$
$$u(0,y) = y, \qquad\qquad u(x,2) = 10,$$
$$u(1,y) = 2y, \qquad\qquad u(0,y) = u(1,y) = 0,$$

to solve

$$u_{xx} + u_{yy} = 0,$$
$$u(x,0) = 0,$$
$$u(x,2) = 10,$$
$$u(0,y) = y,$$
$$u(1,y) = 2y.$$

b) More generally, what is the solution of the Dirichlet problem

$$u_{xx} + u_{yy} = 0,$$
$$u(x,0) = f_1(x),$$
$$u(x,b) = f_2(x),$$
$$u(0,y) = g_1(y),$$
$$u(a,y) = g_2(y)?$$

c) What is the solution of the Neumann problem

$$u_{xx} + u_{yy} = 0,$$
$$u_y(x,0) = f_1(x),$$
$$u_y(x,b) = f_2(x),$$
$$u_x(0,y) = g_1(y),$$
$$u_x(a,y) = g_2(y)?$$

Note: There may be compatibility conditions. See the following two exercises.

13. Solve the problem

$$u_{xx} + u_{yy} = 0,$$
$$u(0, y) = u(a, y) = 0,$$

subject to

a) $u_x(x, 0) = f(x), u_x(x, b) = g(x)$. What restrictions, if any, are there on the functions f and g?

b) $u(x, 0) = f(x), u_x(x, b) = g(x)$. Again, what restrictions, if any, are there on f and g?

14. a) Show that the Neumann problem

$$u_{xx} + u_{yy} = 0, \qquad 0 < x < a, 0 < y < b,$$
$$u_y(x, 0) = f_1(x),$$
$$u_y(x, b) = f_2(x),$$
$$u_x(0, y) = g_1(y),$$
$$u_x(a, y) = g_2(y),$$

must satisfy the compatibility condition

$$\int_0^a [f_2(x) - f_1(x)]dx + \int_0^b [g_2(y) - g_1(y)]dy = 0,$$

as follows: write

$$0 = u_{xx} + u_{yy} \Rightarrow 0 = \int_0^b \int_0^a (u_{xx} + u_{yy})dxdy$$
$$= \int_0^b \int_0^a u_{xx}\,dxdy + \int_0^a \int_0^b u_{yy}\,dydx$$

and then integrate.

b) Instead, use Green's Theorem on

$$0 = \int_a^b \int_0^a (u_{xx} + u_{yy})dxdy$$

to arrive at the same result.

c) More generally, show that, if D and C are well-enough behaved, Green's Theorem says that

$$\iint_D \nabla^2 u\, dA = \oint_C \frac{\partial u}{\partial n}ds,$$

where D is a two-dimensional region, C is its boundary curve and $\frac{\partial u}{\partial n}$ is the outward normal derivative of u along C. Therefore, the above result is just a special case of the more general compatibility condition

$$\oint_C \frac{\partial u}{\partial n} ds = 0.$$

This says, of course, that the total flux across C is zero, which makes sense as solutions of Laplace's equation can be looked at as steady state temperatures.

4.4   Nonhomogeneous Problems

The heat and wave equation problems can be nonhomogeneous in one of two ways: the PDE itself may be nonhomogeneous or the boundary conditions may be nonhomogeneous. Let's first consider the *simplest* nonhomogeneous boundary conditions, leaving the more general case for the exercises.

NONHOMOGENEOUS BOUNDARY CONDITIONS

We start with the heat equation, with each end held at a constant temperature (see Exercise 7 in Section 2.4):

$$u_t = \alpha^2 u_{xx},$$
$$u(x,0) = f(x),$$
$$u(0,t) = T_1, \quad u(L,t) = T_2,$$

where T_1 and T_2 are not necessarily zero. We plan to transform the problem into a *homogeneous* problem in a new unknown $w(x,t)$. In order to accomplish this, we wish to find the "simplest" function $v(x,t)$ so that $w = u - v$ satisfies

$$w_t = \alpha^2 w_{xx}$$

and

$$w(0,t) = w(L,t) = 0.$$

The first condition certainly will be satisfied if $v_t = v_{xx} = 0$, and this will require v to be a linear function in x,

$$v = v(x) = c_1 x + c_2.$$

Then, if $w(x,t) = u(x,t) - (c_1 x + c_2)$, we have

$$w(0,t) = u(0,t) - c_2$$
$$= T_1 \quad - c_2$$

and

$$w(L,t) = u(L,t) - (c_1 L + c_2)$$
$$= T_2 - c_1 L - c_2.$$

Solving the two equations

$$T_1 - c_2 = 0,$$
$$T_2 - c_1 L - c_2 = 0$$

for c_1 and c_2 gives us

$$c_1 = \frac{T_2 - T_1}{L},$$
$$c_2 = T_1,$$

so our new unknown w is given by

$$w(x,t) = u(x,t) + \frac{T_1 - T_2}{L} x - T_1.$$

See Figure 4.6.

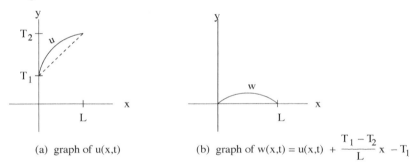

(a) graph of u(x,t)   (b) graph of $w(x,t) = u(x,t) + \dfrac{T_1 - T_2}{L} x - T_1$

FIGURE 4.6
Transforming nonhomogeneous boundary conditions into homogeneous BCs.

Finally, the initial condition in w becomes

$$w(x,0) = u(x,0) + \frac{T_1 - T_2}{L} x - T_1$$
$$= f(x) + \frac{T_1 - T_2}{L} x - T_1.$$

Therefore, we need only solve the problem

$$w_t = \alpha^2 w_{xx},$$
$$w(x,0) = f(x) + \frac{T_1 - T_2}{L} x - T_1,$$
$$w(0,t) = w(L,t) = 0,$$

and the solution to the original problem will be

$$u(x,t) = w(x,t) + v(x),$$
$$= w(x,t) + \frac{T_2 - T_1}{L}x + T_1.$$

Of course, the wave equation with nonhomogeneous boundary conditions can be treated in the same way. Also, it is not difficult to extend this method to more complicated boundary conditions.

Example 1 Transform the wave equation initial-boundary-value problem

$$u_{tt} = u_{xx},$$
$$u(x,0) = f(x),$$
$$u_t(x,0) = g(x),$$
$$u_x(0,t) = 10, \quad u_x(2,t) - u(2,t) = 3$$

into one with homogeneous boundary conditions. Again, we try $w(x,t) = u(x,t) - c_1 x - c_2$. The boundary conditions become

$$w_x(0,t) = u_x(0,t) - c_1 = 10 - c_1,$$
$$w_x(2,t) - w(2,t) = u_x(2,t) - u(2,t) - c_1 + 2c_1 + c_2 = 3 + c_1 + c_2,$$

which become homogeneous in w if and only if $c_1 = 10$ and $c_2 = -13$. So our transformation is

$$w(x,t) = u(x,t) - 10x + 13.$$

Finally, the new initial conditions are

$$w(x,0) = u(x,0) - 10x + 3 = f(x) - 10x + 13,$$
$$w_t(x,0) = u_t(x,0) = g(x).$$

In the exercises we treat more general boundary conditions, including those which involve functions of t.

NONHOMOGENEOUS PDEs

A more difficult question is how to solve an initial-boundary-value problem when the PDE is nonhomogeneous. Again, let's begin by considering the heat problem

$$u_t = u_{xx} + F(x,t), \tag{4.8}$$
$$u(x,0) = f(x), \tag{4.9}$$
$$u(0,t) = u(L,t) = 0. \tag{4.10}$$

Now, if the problem were, instead, homogeneous, it would have a solution of the form

$$u(x,t) = \sum_{n=1}^{\infty} T_n(t) \sin \frac{n\pi x}{L}, \tag{4.11}$$

where we have $\sin \frac{n\pi x}{L}$ because of the boundary conditions (4.10) (and, of course, in the homogeneous case, $T_n(t) = b_n e^{-\frac{n^2\pi^2 t}{L^2}}$, where the constants b_n are the Fourier sine coefficients of $f(x)$ on $0 \le x \le L$). These boundary conditions, then, suggest that the solution of the *nonhomogeneous* problem also will have the form (4.11) (for different functions $T_n(t)$, of course). Our goal is to determine the functions $T_n(t)$ so that (4.11) is the solution of the problem (4.8)–(4.10).

To this end, for each value of t we expand $F(x,t)$ into its Fourier sine series on $0 \le x \le L$. Thus, we have

$$F(x,t) = \sum_{n=1}^{\infty} F_n(t) \sin \frac{n\pi x}{L}$$

where

$$F_n(t) = \frac{2}{L} \int_0^L F(x,t) \sin \frac{n\pi x}{L} dx. \tag{4.12}$$

Next, we substitute (4.11) into the PDE (4.8) and, assuming we may differentiate term-by-term, we get

$$\sum_{n=1}^{\infty} T_n'(t) \sin \frac{n\pi x}{L} = -\sum_{n=1}^{\infty} \frac{n^2\pi^2}{L^2} T_n(t) \sin \frac{n\pi x}{L} + \sum_{n=1}^{\infty} F_n(t) \sin \frac{n\pi x}{L},$$

$$\sum_{n=1}^{\infty} T_n'(t) \sin \frac{n\pi x}{L} = \sum_{n=1}^{\infty} \left[-\frac{n^2\pi^2}{L^2} T_n(t) + F_n(t) \right] \sin \frac{n\pi x}{L}.$$

Equating coefficients of $\sin \frac{n\pi x}{L}$ (which, by the way, is equivalent to forming the inner product of each side with $\sin \frac{n\pi x}{L}$, as these sine functions are orthogonal on $0 \le x \le L$) we have

$$T_n'(t) + \frac{n^2\pi^2}{L^2} T_n(t) = F_n(t), \qquad n = 1, 2, \ldots .$$

Again, we have reduced a PDE to a problem involving ODEs. We may solve each ODE by multiplying through by the integrating factor $e^{\frac{n^2\pi^2 t}{L^2}}$, resulting in

$$\frac{d}{dt} \left[e^{\frac{n^2\pi^2 t}{L^2}} T_n(t) \right] = e^{\frac{n^2\pi^2 t}{L^2}} F_n(t)$$

or

$$e^{\frac{n^2\pi^2 t}{L^2}} T_n(t) = \int_0^t e^{\frac{n^2\pi^2 \tau}{L^2}} F_n(\tau) d\tau + k_n$$

or

$$T_n(t) = e^{-\frac{n^2\pi^2 t}{L^2}} \int_0^t e^{\frac{n^2\pi^2 \tau}{L^2}} F_n(\tau)d\tau + k_n e^{-\frac{n^2\pi^2 t}{L^2}}, \quad n = 1, 2, 3, \dots . \quad (4.13)$$

Note that we could have chosen *any* constant as the value of the lower limit of integration. We choose it to be zero for the sake of convenience, because the initial condition is given at time $t = 0$.

Now, what's the value of each k_n? The initial condition (4.9) implies that

$$u(x, 0) = f(x) = \sum_{n=1}^{\infty} T_n(0) \sin \frac{n\pi x}{L}$$

$$= \sum_{n=1}^{\infty} k_n \sin \frac{n\pi x}{L}$$

and, since this must hold for all x in $0 \le x \le L$, we must have

$$k_n = \frac{2}{L} \int_0^L f(x) \sin \frac{n\pi x}{L} dx, \quad (4.14)$$

that is, the constants k_n must be the Fourier sine coefficients of $f(x)$ on $0 \le x \le L$.

So, our solution is (4.11), with each $T_n(t)$ given by (4.13), each k_n by (4.14) and each function $F_n(t)$ by (4.12).

Again, we may treat the nonhomogeneous wave equation, as well as the nonhomogeneous Laplace's equation (i.e., Poisson's equation), in a similar manner (and we do so in the exercises).

Exercises 4.4

1. Suppose we are given the wave equation initial-boundary-value problem $u_{tt} = u_{xx}$, $u(x, 0) = f(x)$, $u_t(x, 0) = g(x)$, subject to the given nonhomogeneous boundary conditions. Transform each problem to one with *homogeneous* boundary conditions.

   a) $u(0, t) = u(L, t) = T$
   b) $u(0, t) = T, u_x(L, t) = a$
   c) $u_x(0, t) = a, u(L, t) = T$
   d) $u_x(0, t) = u_x(L, t) = a$

   (In each case, T and a are constants.)

2. Do the same as in Exercise 1 for the boundary conditions $u_x(0, t) = a \ne u_x(L, t) = b$. (Note: The PDE in the transformed system will be *non*homogeneous.)

3. Consider the general heat problem

$$u_t = u_{xx} + F(x,t),$$
$$u(x,0) = f(x),$$
$$u(0,t) = g(t),$$
$$u(L,t) = h(t).$$

Determine functions $A(t)$ and $B(t)$ so that the transformation

$$w(x,t) = u(x,t) - A(t) - xB(t)$$

leads to a problem with homogeneous boundary conditions. Write down the initial-boundary-value problem for w.

In Exercises 4–10, solve the nonhomogeneous heat initial-boundary-value problem. In each case, rather than following the example in the text, start from scratch by assuming a solution of the correct form, as in (4.11). Also, calculate the steady state temperature, $\lim_{t\to\infty} u(x,t)$; if the limit doesn't exist, describe the system's behavior as $t \to \infty$. (Hint: Use l'Hôpital's rule and the fundamental theorem of calculus.)

4. $u_t = u_{xx} + x,$
   $u(x,0) = \sin 2x,$
   $u(0,t) = u(\pi,t) = 0.$

5. $u_t = u_{xx} + 10,$
   $u(x,0) = 3\sin x - 4\sin 2x + 5\sin 3x,$
   $u(0,t) = u(\pi,t) = 0.$

6. $u_t = u_{xx} + 10,$
   $u(x,0) = 50,$
   $u(0,t) = u(\pi,t) = 0.$

7. $u_t = u_{xx} + x + t,$
   $u(x,0) = x,$
   $u(0,t) = u(\pi,t) = 0.$

8. $u_t = u_{xx} + \cos 2x - \cos 5x,$
   $u(x,0) = \cos 3x,$
   $u_x(0,t) = u_x(\pi,t) = 0.$

9. $u_t = u_{xx} + 10,$
   $u(x,0) = -15,$
   $u_x(0,t) = u_x(\pi,t) = 0.$

10. $u_t = u_{xx},$
    $u(x,0) = \frac{10x^2}{2\pi},$

$u_x(0, t) = 0,$

$u_x(\pi, t) = 10.$

(Hint: See Exercise 2.)

In Exercises 11–15, solve the nonhomogeneous wave initial-boundary-value problem. In each case, start by letting $u(x, t) = \sum_{n=1}^{\infty} T_n(t) \sin nx$ and proceed from there.

11. $u_{tt} = u_{xx} + \sin x,$

$u(x, 0) = \sin 3x,$

$u_t(x, 0) = \sin 5x,$

$u(0, t) = u(\pi, t) = 0.$

12. $u_{tt} = u_{xx} + x,$

$u(x, 0) = 1,$

$u_t(x, 0) = 0,$

$u(0, t) = u(\pi, t) = 0.$

13. $u_{tt} = u_{xx} + \sin x \sin t,$

$u(x, 0) = 0,$

$u_t(x, 0) = \sin 3x,$

$u(0, t) = u(\pi, t) = 0.$

(Explain why this system is said to exhibit *resonance*.)

14. **MATLAB:** Plot snapshots of the (truncated) solution in the *x*-*u* plane, for various values of t, for the following problems. What happens as $t \to \infty$?

   a) Exercise 5

   b) Exercise 9

   c) Exercise 11

   d) Exercise 15

15. Show that the solution of the general nonhomogeneous wave initial-boundary-value problem

$$u_{tt} = u_{xx} + F(x, t),$$
$$u(x, 0) = f(x),$$
$$u_t(x, 0) = g(x),$$
$$u(0, t) = u(\pi, t) = 0$$

is

$$u(x, t) = \sum_{n=1}^{\infty} \sin nx [c_n \cos nt + d_n \sin nt + p_n(t)],$$

where, if $F(x,t) = \sum_{n=1}^{\infty} F_n(t) \sin nx$, then $p_n(t)$ is any particular solution of the ODE

$$T_n'' + n^2 T_n = F_n,$$

and where

$$c_n = \frac{2}{\pi} \int_0^{\pi} f(x) \sin nx \ dx - p_n(0)$$

and

$$d_n = \frac{2}{n\pi} \int_n^{\pi} g(x) \sin nx \ dx - \frac{1}{n} p_n'(0), \quad n = 1, 2, 3, \dots .$$

In Exercises 16 and 17, solve the Poisson's equation boundary-value problem. You will want to let $u(x,y) = \sum_{n=1}^{\infty} X_n(x) \sin ny$ or $u(x,y) = \sum_{n=1}^{\infty} Y_n(y) \sin nx$, depending on the boundary conditions.

16. $u_{xx} + u_{yy} = \sin 2x,$
    $u(x,0) = \sin 3x,$
    $u(x,\pi) = 0,$
    $u(0,y) = u(\pi,y) = 0.$

17. $u_{xx} + u_{yy} = x^2 y^2,$
    $u(x,0) = u(x,\pi) = 0,$
    $u(0,y) = u(\pi,y) = 0.$

Prelude to Chapter 5

This chapter is devoted to the idea of *characteristics of a PDE*. For our purposes, these characteristics are curves in the domain of the PDE—here, the x-y plane—along which solutions of the PDE have certain "nice" properties. For example, for some equations, solutions are constant along these characteristics.

We begin with first-order linear equations, which were studied by many notable mathematicians in the late 18th century, particularly Lagrange, who, as in all of his work, used an analytical approach, and Monge, whose geometrical ideas were quite fruitful.

The *method of characteristics* can be generalized to equations of higher order. We do so only for second-order linear equations with constant coefficients, starting with Jean Le Rond d'Alembert's (1717–1783) famous solution for the infinite vibrating string, which he actually derived around 1746, *before* any of the important work on characteristics. Although he did not call them such, it is implicit in d'Alembert's work that he understood the behavior of the characteristics of the wave equation. Daniel Bernoulli, Euler and Lagrange also made important contributions in this area.

Finally, we classify second-order linear PDEs according to how many families of characteristics they have. It is here that we see that the wave, heat and Laplace equations have a mathematical importance far beyond their connection with specific physical problems.

5

Characteristics

5.1 First-Order PDEs with Constant Coefficients

In Chapter 1, we saw that there are many first-order PDEs which we can already solve. In particular, the linear equation

$$a(x, y)u_x + b(x, y)u = f(x, y) \tag{5.1}$$

can be treated as an ODE. What about more general linear, first-order PDEs? And why would we be interested in these equations?

The PDE $au_x + bu_y = 0$ often is called the **convection equation** (as in "convey") or **advection equation** (where *advection* is a synonym of the noun *transport*). Imagine a very narrow stream flowing at constant velocity v. Suppose there is a chemical that has polluted the stream and that this chemical is carried downstream without diffusing at all. Let

$$u(x, t) = \text{concentration of chemical per unit length, at point } x$$
$$\text{along the stream, at time } t$$

(we are assuming that the stream is narrow enough so that the concentration is the same along any line across its width, so we may treat the stream as one-dimensional). See Figure 5.1. What can we say about the function u? Consider a short length of the stream, from x to $x + \Delta x$, and calculate the change in the amount of chemical present from time t to time $t + \Delta t$. First, the amount present at time t is approximately

$$u(x, t)\Delta x,$$

so the change is approximately

$$[u(x, t + \Delta t) - u(x, t)]\Delta x.$$

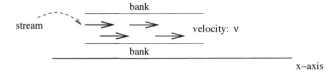

FIGURE 5.1
Narrow stream flowing at constant velocity ν.

Now, during the time interval Δt, a particle in the stream travels a distance $v\Delta t$. Therefore, the amount of material passing a point x during this time interval is approximately

$$u(x,t)v\Delta t,$$

so the net change in the amount of chemical in our length Δx, during the interval Δt, is approximately

$$- \text{(Amount leaving right end)}$$
$$+ \text{(Amount entering left end)}$$
$$= [-u(x+\Delta x, t) + u(x,t)]v\Delta t.$$

Divide through by $\Delta x \Delta t$ and take limits as Δt and Δx go to 0, and we get

$$u_t + vu_x = 0$$

(see Figure 5.2).

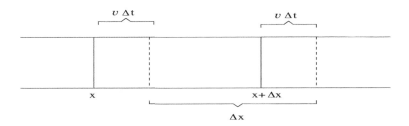

FIGURE 5.2
Differential element of stream, "before" and "after."

So how do we solve the convection equation? Let's look at some examples.

Example 1 Find all solutions of the PDE

$$2u_x + 3u_y = 0. \tag{5.2}$$

We'll try to reduce this equation, via a transformation of the independent variables, to one which we know how to solve, namely, to (5.1). Let

$$\xi = x$$
$$\eta = Ax + By$$

and let's choose the constants A and B so that the transformed PDE has no u_η term. We have

$$u_x = u_\xi + Au_\eta$$
$$u_y = Bu_\eta$$

and the PDE becomes

$$2u_\xi + (2A + 3B)u_\eta = 0.$$

We may choose, for example, $A = 3$ and $B = -2$, and the transformation

$$\begin{aligned}\xi &= x \\ \eta &= 3x - 2y\end{aligned} \quad \text{or} \quad \begin{aligned}x &= \xi \\ y &= \tfrac{3}{2}\xi - \tfrac{1}{2}\eta\end{aligned}$$

reduces our PDE to

$$2u_\xi = 0.$$

Integrating by ξ (and treating η as a constant), we have our solution

$$\begin{aligned}u &= g(\eta) \\ &= g(3x - 2y),\end{aligned} \tag{5.3}$$

where g is any arbitrary function (with certain restrictions). We can always check our solution:

$$u_x = 3g'(3x - 2y), \quad u_y = -2g'(3x - 2y)$$

and

$$2u_x + 3u_y = 0.$$

Note that the above implies that g must be differentiable. (We may refer to (5.3) as the **general solution** of the PDE.)

First, you should convince yourself that we could have interchanged the roles of x and y and used a transformation

$$\begin{aligned}\xi &= Cx + Dy \\ \eta &= y\end{aligned}$$

to eliminate the u_ξ term and arrive at the same solution. Now, let's look more carefully at the solution. Notice that, for all points on the line $3x - 2y = c$, where c is a given constant, we have

$$\begin{aligned}u(x, y) &= g(3x - 2y) \\ &= g(c),\end{aligned}$$

that is, u is constant along each of the lines $3x - 2y = c$. These important lines are called the **characteristics** or **characteristic curves** of the PDE, while ξ

and η are called **characteristic coordinates**. (We will give the "official" definition of characteristics in Section 5.2. See Definition 5.1.) Essentially, what has happened is that the PDE (5.2) has been turned into an ODE "along" these characteristics (an ODE which, in this particular case, has solution $u = constant$ along them).

Example 1 (cont.) Continuing with our example, suppose that the PDE is to be solved subject to the additional condition

$$u(x,0) = \sin x.$$

Then,

$$u(x,0) = g(3x) = \sin x$$

and, letting $z = 3x, x = \frac{1}{3}z$, we have

$$g(z) = \sin\left(\frac{1}{3}z\right).$$

So the *unique* solution to the system

$$2u_x + 3u_y = 0,$$
$$u(x,0) = \sin x \tag{5.4}$$

is

$$u(x,y) = \sin\frac{1}{3}(3x - 2y)$$
$$= \sin\left(x - \frac{2}{3}y\right).$$

The condition $u(x,0) = \sin x$ is called a **side condition** or, in the case where y is replaced by the time variable, t, an **initial condition**. We shall use the latter, even when the variable does not necessarily represent time. Then, the curve along which the condition is given will be called the **initial curve**, and the system (5.4) is called an **initial-value problem**.*

Looking at Figure 5.3, we see the relationship between the solution in the transformed coordinates and the actual solution. In the ξ-η plane, u is constant along the characteristic $\eta = constant$. To get the graph of the actual solution, we "tilt and stretch" the latter until they coincide with the lines $3x - 2y = constant$.

---

*Alternatively, the system often is called a **Cauchy problem**, after Augustin-Louis Cauchy (1789–1857), and the initial condition is called the **Cauchy data**.

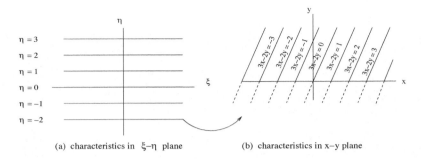

(a) characteristics in ξ–η plane (b) characteristics in x–y plane

FIGURE 5.3
Characteristics for Example 1 in (a) *characteristic coordinates* and (b) Cartesian coordinates.

Also, we have a nice physical interpretation of the solution if we replace y by the time variable, t. Then, if we take "snapshots" of the solution $u(x,t) = \sin\left(x - \frac{2}{3}t\right)$ at various times t_0, we see that we can think of our solution as an initial curve $u = \sin x$ which moves to the right at constant velocity. See Exercise 11. Here, we can see the phenomenon in three dimensions in Figure 5.4, which graphs the solution in x-t-u space.

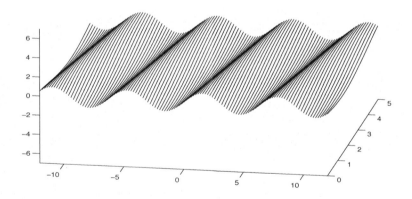

FIGURE 5.4
MATLAB graph of the function $u(x,t) = \sin\left(x - \frac{2}{3}t\right)$, for $t \geq 0$. We see that the cross sections $t = constant$ are copies of the initial shape, traveling to the right.

More generally, given the first-order, linear PDE

$$au_x + bu_y + cu = f(x,y), \tag{5.5}$$

where a, b and c are constant, we may always proceed as above. We find that the transformation

$$\begin{aligned}\xi &= x \\ \eta &= bx - ay\end{aligned} \quad \text{or} \quad \begin{aligned}x &= \xi \\ y &= \tfrac{b}{a}\xi - \tfrac{1}{a}\eta\end{aligned}$$

(for example) reduces the PDE (5.5) to

$$au_\xi + cu = F(\xi, \eta),$$

where $F(\xi, \eta)$ is just the function $f(x, y)$ with x and y replaced by $x = \xi$ and $y = \tfrac{b}{a}\xi - \tfrac{1}{a}\eta$. Rather than memorizing, let's apply the basic principle to some examples.

Example 2 Solve $u_x - 4u_y + u = 0$

$$u(0, y) = \cos 3y.$$

Again, we let

$$\begin{aligned}\xi &= x \\ \eta &= Ax + By\end{aligned}$$

and the transformed PDE becomes

$$u_\xi + (A - 4B)u_\eta + u = 0.$$

We may choose $A = 4, B = 1$, so that

$$\begin{aligned}\xi &= x \\ \eta &= 4x + y\end{aligned}$$

and the PDE becomes

$$u_\xi + u = 0.$$

Its solution is

$$u = g(\eta)e^{-\xi} \quad \text{or} \quad u = g(4x + y)e^{-x},$$

where g is any function. Then, applying the initial condition, we have

$$u(0, y) = g(y) = \cos 3y,$$

so our final, unique solution is

$$u(x, y) = e^{-x}\cos 3(4x + y).$$

Figure 5.5a shows the graph of $u = e^{-\xi}\cos 3\eta$. Along each characteristic $\eta = constant$, we have the exponential solution $u = e^{-\xi} \cdot constant$. Then, tilting these lines clockwise, and stretching them until they coincide with the lines $4x+y = constant$, gives us the graph of our actual solution in Figure 5.5b.

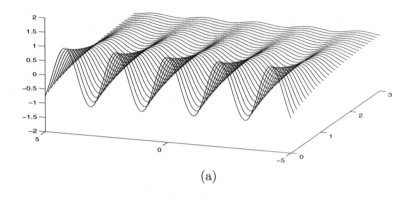

(a)

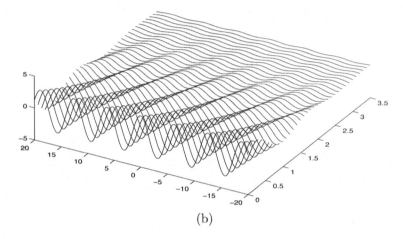

(b)

FIGURE 5.5
MATLAB graphs of (a) $u = e^{-\xi}\cos 3\eta$ and (b) $u = e^{-x}\cos 3(4x + y)$.

The initial condition need not be given along one of the coordinate axes. In fact, as we shall see shortly, it can be given along almost any curve. For example, the values of our solution u could be given along the curve $y = x^3$, in which case the side condition would be of the form

$$u(x, x^3) = f_1(x)$$

or

$$u(\sqrt[3]{y}, y) = f_2(y),$$

where f_1 and f_2 are given functions.

Example 3 Solve

$$u_x - u_y - 2y = 0,$$
$$u(x, 2x + 1) = e^x.$$

We see that the transformation

$$\begin{array}{cc} \xi = x & x = \xi \\ \eta = x + y & \text{or} \quad y = \eta - \xi \end{array}$$

reduces the PDE to

$$u_\xi = 2\eta - 2\xi.$$

Integrating both sides with respect to ξ, we have

$$u = 2\xi\eta - \xi^2 + g(\eta),$$

or

$$u(x, y) = 2x(x + y) - x^2 + g(x + y)$$
$$= x^2 + 2xy + g(x + y).$$

Then,

$$u(x, 2x + 1) = e^x = x^2 + 2x(2x + 1) + g(x + 2x + 1),$$

so

$$g(3x + 1) = e^x - 6x^2 - 2x.$$

Letting $z = 3x + 1, x = \frac{z-1}{3}$, we have

$$g(z) = e^{\frac{z-1}{3}} - 6\left(\frac{z-1}{3}\right)^2 - 2\left(\frac{z-1}{3}\right),$$

and our solution becomes

$$u(x, y) = x^2 + 2xy + e^{\frac{x+y-1}{3}} - 6\left(\frac{x+y-1}{3}\right)^2 - 2\left(\frac{x+y-1}{3}\right).$$

Example 4 Solve

$$u_x - u_y - u = 0,$$
$$u(x, -x) = \sin x.$$

As in Example 3, we use the transformation

$$\begin{array}{cc} \xi = x, & x = \xi, \\ \eta = x + y, & \text{or} \quad y = \eta - \xi \end{array}$$

to reduce the PDE to

$$u_\xi - u = 0.$$

The solution is

$$u = g(\eta)e^{\xi}$$
$$= g(x+y)e^{x}.$$

Then,

$$u(x, -x) = \sin x = g(0)e^{x}.$$

But $g(0)$ is a constant. Rather than determining g, the initial condition seems to require that $\sin x$ be a multiple of e^x for all x, which is impossible (why?). Therefore, it seems that this problem has no solution. Why did this happen?

Notice that the characteristic curves are the lines $x + y = constant$. Also, notice that the initial condition is given along the line $y = -x$, which happens to be a characteristic of the PDE.

Basically, what's going on is this. As mentioned before, the PDE becomes a first-order ODE along each of the characteristic curves. In order to have a unique solution, the initial condition must specify the value of u at exactly one point on each of these characteristics. Therefore, it seems that the curve along which the side condition u is given must intersect each characteristic at exactly one point. This, indeed, is the case. We state the following theorem without proof.

Theorem 5.1 *Given the initial-value problem*

$$au_x + bu_y + cu = f(x, y)$$
$$u(x, f_1(x)) = f_2(x),$$

where a, b and c are constant, suppose that

1) f_x, f_y, f_1' and f_2' are continuous.

2) Each characteristic of the PDE intersects the initial curve $y = f_1(x)$ exactly once.

3) No characteristic is tangent to the initial curve.

Then, the initial-value problem has a unique solution u, with the property that u_x and u_y are continuous.

Note that, by interchanging the roles of x and y, the theorem is also true for initial conditions of the form $u(f_1(y), y) = f_2(y)$. Further, it is easy to generalize the theorem to include parametric initial curves which are continuously differentiable.

We generalize the convection equation in Exercise 18.

Exercises 5.1

In Exercises 1–10, first find all possible solutions of the PDE, and check your answer. Then, solve the initial-value problem, and check your answer, or say why the initial-value problem cannot be solved.

1. $5u_x - 7u_y = 0, u(x, 0) = 4\sin 5x$

2. $3u_x + u_y = 0, u(4, y) = e^{-y^2}$

3. $u_x - 2u_y = 0, u(x, -2x + 4) = x^2 + 3x - 1$

4. $-2u_x + 6u_y = 0, u(4y, y) = 2y + 1$

5. $u_x - u_y + 2u = 0, u(x, 3) = x^2$

6. $2u_x + u_y - 5u = 0, u(y^2, y) = 3(-y^2 + 2y)e^{\frac{5y^2}{2}}$

7. $2u_x + u_y = 12x(1 + y), u(x, -1) = 4x$

8. $u_x + 4u_y - 2u = e^{x+y}, u(x, 0) = \cos x$

9. $u_x + u_y + u = y, u(0, y) = \sin y$

10. $4u_x - 3u_y + 5u = 0, u\left(-\frac{4}{3}y, y\right) = \frac{1}{1+y^2}$

11. In the remarks following Example 1, it is mentioned that the solution $u(x, t) = \sin\left(x - \frac{2}{3}t\right)$ can be viewed as a sine wave moving to the right at constant speed. Why is this so? What is this speed?

12. **MATLAB:** Plot the solution, in x-y-u space, of the given problem.

    a) Exercise 2

    b) Exercise 5

    c) Exercise 9

13. **MATLAB:** Plot snapshots of the solution in the x-u plane, for various versions of t, for the given problem. In each case, what is the velocity of the moving wave?

    a) Exercise 1 (first, change x to t and y to x)

    b) Exercise 2 (change x to t and y to x)

    c) Exercise 5 (change y to t)

    d) Exercise 9 (change x to t and y to x)

14. If we solve the PDE $u_x - u_y + u = 0$, we get $u = g(x + y)e^{-x}$, for arbitrary g. However, if, instead, we let $\xi = Ax + By$ and $\eta = y$ and solve, we get $u = g(x + y)e^y$. What's going on?

In Exercises 15 and 16, solve each initial-value problem two ways. First, solve each as we've been doing; then, transform both the PDE *and* the initial condition in terms of ξ and η; solve that problem; and substitute back. (Make sure you get the same answer!) Hint: When using the second method, it may help to rename u, i.e., write $u(x,y) = v(\xi, \eta)$ and work with v.

15. $u_x - u_y = 0, u(x, 3x) = x^2$

16. $u_x + u_y = u, u(x, x^3) = 5(x - x^3)e^x$

17. Solve the initial-value problem $u_x - u_y = 0, u(x, x^3 - x) = x^2$. Where is the solution not differentiable? Why does this example not contradict Theorem 5.1?

18. a) In the derivation of the convection equation, suppose that the velocity of the stream depends on position, that is, that $v = v(x)$. Show that the PDE becomes

$$u_t + v(x)u_x = 0.$$

   b) If, in addition, there is a *source* of pollutant of the form

$$f(x,t) = \text{amount added at point } x \text{ at time } t,$$
$$\text{per unit length per unit time,}$$

   show that we get the *non*homogeneous equation

$$u_t + v(x)u_x = f(x,t).$$

19. In Example 1, we solved $2u_x + 3u_y = 0$ using the transformation

$$\xi = x \quad \eta = 3x - 2y.$$

Solve this same equation using various other transformations

$$\xi = Ax + By \quad \eta = 3x - 2y.$$

Which transformations "don't work," that is, for which values of A and B are we unable to solve the problem?

20. a) Solve the problem

$$\begin{aligned} au_x + bu_y &= 0 & x > 0, y > 0, \\ u(x,0) &= f(x), & x \geq 0, \\ u(0,y) &= g(y), & y \geq 0, \end{aligned}$$

   where $a > 0$ and $b > 0$ are constants, and where $f(0) = g(0)$.

   Hint: Follow the characteristics.

b) Solve the problem

$$u_x - u_y = u, \qquad x > 0, y > 0,$$
$$u(x,0) = \sin 2x, \quad x \geq 0,$$
$$u(0,y) = \sin 3y, \quad y \geq 0.$$

5.2   First-Order PDEs with Variable Coefficients

In Exercise 18 of the previous section, we derived the more general version of the convection equation

$$u_t + v(x)u_x = f(x,t).$$

Can we extend the method of Section 5.1 to deal with equations where the coefficients are not all constant? Let's see.

Example 1 Try to solve $u_x + yu_y = 0$.
   We let

$$\xi = x$$
$$\eta = Ax + By$$

and the u_η coefficient becomes $A + By$. However, it is impossible to choose constants A and B which will make $A + By = 0$ for all y (why?). We meet with a similar fate if we try

$$\xi = Ax + By$$
$$\eta = y.$$

What can we do?
   In the previous section, our transformation was chosen so that the characteristics were the curves $\eta = constant$. This suggests that we try the same thing for equations with variable coefficients. The obvious question, then, is, "What are the characteristics?"
   Remember that the characteristics were curves along which the PDE could be treated as an ODE. Let's go back to Example 1 of Section 5.1 and see if we can look at that problem in a different way.

Example 2 $2u_x + 3u_y = 0$.
   Notice that we can write this equation as

$$(2\hat{\imath} + 3\hat{\jmath}) \cdot (u_x\hat{\imath} + u_y\hat{\jmath}) = 0$$

or

$$(2\hat{\imath} + 3\hat{\jmath}) \cdot \mathbf{grad}\ u = 0.$$

The left side of the PDE is just a constant multiple of the directional derivative, $\frac{du}{ds}$, of u, in the direction $2\hat{\imath} + 3\hat{\jmath}$, so the PDE says that

$$\frac{du}{ds} = 0$$

in this direction. More precisely, u is constant along curves which have tangent vector $2\hat{\imath} + 3\hat{\jmath}$ at each point, i.e., along curves with slope

$$\frac{dy}{dx} = \frac{3}{2}.$$

These curves are, of course, the characteristics

$$y = \frac{3}{2}x + c.$$

Therefore, the value of u at any point (x, y) depends only on the characteristic on which (x, y) lies; in other words, u depends only on the value of $\frac{3}{2}x - y$, i.e.,

$$u = g\left(\frac{3}{2}x - y\right)$$

for any arbitrary function g. Notice that this says the same thing as does equation (5.3).

This approach is a nice geometric way of looking at the problem. However, it may seem somewhat ad hoc, and it may not be clear how it generalizes to equations with more terms. For example, what, exactly, does an equation like

$$\frac{du}{ds} + u = 0$$

mean, since we expect PDEs to have at least two independent variables?

We will recast this geometric method slightly, so that it looks like what we have been doing in Section 5.1. To that end, remember, again, that we chose the transformation so that the characteristics were given by $\eta = c$ in the transformed coordinates. We shall make the same choice for η in equations with variable coefficients, and we shall see that ξ plays a role similar to that of the arc length variable s in the notation for the directional derivative of u.

Example 3 Find all solutions of $u_x + 4xu_y - u = 0$. Again, the term $u_x + 4xu_y$ is the directional derivative of u in the direction of the vector $\hat{\imath} + 4x\hat{\jmath}$. We expect curves with this vector as tangent vectors to be the characteristics. They satisfy

$$\frac{dy}{dx} = \frac{4x}{1},$$

so the characteristics are the curves

$$y = 2x^2 + c \quad \text{or} \quad 2x^2 - y = constant.$$

Now, let

$$\begin{aligned} \xi &= x \\ \eta &= 2x^2 - y \end{aligned} \quad \text{or} \quad \begin{aligned} x &= \xi \\ y &= 2\xi^2 - \eta. \end{aligned}$$

Then

$$u_x = u_\xi + 4x u_\eta$$
$$u_y = -u_\eta$$

and the transformed PDE is

$$u_\xi - u = 0$$

with solution

$$\begin{aligned} u &= g(\eta)e^\xi \\ &= g(2x^2 - y)e^x \end{aligned}$$

for arbitrary function g. So the curves $2x^2 - y = c$ are, indeed, characteristic. More precisely, given the first-order linear PDE

$$a(x, y)u_x + b(x, y)u_y + c(x, y)u = f(x, y),$$

we *define* the characteristic curves to be those curves satisfying

$$\frac{dy}{dx} = \frac{b(x, y)}{a(x, y)}$$

or, as is traditionally written,

$$\frac{dx}{a(x, y)} = \frac{dy}{b(x, y)}. \tag{5.6}$$

Definition 5.1 *The **characteristics** or **characteristic curves** of the first-order linear PDE*

$$a(x, y)u_x + b(x, y)u_y + c(x, y)u = f(x, y)$$

are those curves satisfying the ODE

$$\frac{dx}{a(x, y)} = \frac{dy}{b(x, y)}.$$

Then, supposing that the ODE (5.6) has general solution $h(x,y) = c$, we make the transformation

$$\xi = x$$
$$\eta = h(x,y).$$

In this case, we have

$$u_x = u_\xi + u_\eta h_x$$
$$u_y = u_\eta h_y$$

and

$$au_x + bu_y = au_\xi + (ah_x + bh_y)u_\eta.$$

But

$$\frac{dx}{a} = \frac{dy}{b} \Rightarrow h(x,y) = c$$
$$\Rightarrow dh = 0 = h_x \, dx + h_y \, dy$$
$$= dx\left(h_x + h_y\frac{dy}{dx}\right)$$
$$= dx\left(h_x + h_y\frac{b}{a}\right)$$
$$= \frac{dx}{a}(ah_x + bh_y)$$

and we have

$$au_x + bu_y = au_\xi,$$

so the PDE has been reduced to an ODE.

We haven't been very rigorous here. For example, what happens at points where $a = 0$ or $b = 0$? Theorem 5.1 can be extended somewhat to the case of variable coefficients.[†] It turns out that a necessary condition for existence and uniqueness of a solution throughout a neighborhood of a point is that $a \neq 0$ or $b \neq 0$ at that point.

Example 4 Solve

$$u_x + yu_y = x,$$
$$u(1, y) = \cos y.$$

The characteristics are given by

$$\frac{dx}{1} = \frac{dy}{y}$$

[†]See, e.g., *Introduction to Partial Differential Equations with Applications* by E. C. Zachmanoglou and Dale W. Thoe.

with solution

$$y = ce^x \quad \text{or} \quad ye^{-x} = c.$$

Then, our transformation is

$$\xi = x, \qquad \qquad x = \xi,$$
$$\eta = ye^{-x}, \quad \text{or} \quad y = \eta e^{\xi}$$

and our PDE becomes

$$u_{\xi} = \xi$$

with solution

$$u = \frac{\xi^2}{2} + g(\eta),$$

$$= \frac{x^2}{2} + g(ye^{-x}).$$

Finally,

$$u(1,y) = \cos y = \frac{1}{2} + g(ye^{-1}),$$

and, letting $z = \frac{y}{e}, y = ez$, we get

$$g(z) = \cos ez - \frac{1}{2}.$$

Therefore, our unique solution is

$$u = \frac{x^2}{2} + \cos(ye^{1-x}) - \frac{1}{2}.$$

Exercises 5.2

In Exercises 1–6, first find all possible solutions of the PDE, and check your answer. Then, solve the initial-value problem, and check your answer. Also, sketch the initial curve and some of the characteristics.

1. $yu_x - xu_y = 0, u(x, 2x) = x^4$

2. $(1 + x^2)u_x + u_y = 0, u(1, y) = \cos y$

3. $u_x + 3x^2 u_y = 0, u(0, y) = \sin 3y$

4. $(1 + y^2)u_x + u_y = 0, u(x, 0) = e^{-x^2}$

5. $u_x + 3x^2 u_y - u = 0, u(2, y) = 3y + 1$

6. $u_x + xu_y = x^2 y, u\left(x, 2x + \frac{x^2}{2}\right) = 5x$

7. **MATLAB:** Plot the solution of the given problem in x-y-u space. On a separate graph, plot some of the characteristic curves.

a) Exercise 1

b) Exercise 3

8. Solve the problem $u_t + xu_x = 0$, $u(x,0) = \sin x$; draw the characteristics in the x-t plane (the first quadrant will suffice); and compare the solution with the solution of $u_t + vu_x = 0$, $u(x,0) = \sin x$, where v is a constant. Specifically, describe what happens to the initial sine wave in both cases.

We may extend the methods of Sections 5.1 and 5.2 to equations in higher dimensions. For example, for the PDE

$$au_x + bu_y + cu_z + du = f,$$

where a, b, c, d and f are functions of x, y and z, it can be shown that the characteristics can be determined by the three ODEs (one of which is redundant)

$$\frac{dx}{a} = \frac{dy}{b} = \frac{dz}{c}.$$

More precisely, solving any two of the ODEs gives us two families of surfaces, the intersections of which give the characteristic curves. This suggests solving, say, $\frac{dx}{a} = \frac{dy}{b}$, resulting in $h_1(x,y) = c$, and $\frac{dx}{a} = \frac{dz}{c}$, resulting in $h_2(x,z) = c_2$, and transforming via the **characteristic coordinates**

$$\xi = x$$
$$\eta = h_1(x,y)$$
$$\zeta = h_2(x,z).$$

Use this method to solve the following PDEs, making sure to check your answers, then solve each initial value problem, again checking your answers.

9. $u_x + 2u_y + 3u_z = 0, u(x,y,0) = xy$

10. $5u_x - 3u_y + u_z = 0, u(1,y,z) = y\cos z$

11. $u_x + 3x^2 u_y - u_z = 0, u(x,0,z) = x^4 + x^3 z$

12. $2u_x + u_y - 4u_z = 0, u(x,y,0) = y$

An important type of first-order PDE in physical applications is the equation

$$a(x,y,u)u_x + b(x,y,u)u_y = c(x,y,u).$$

This equation is, of course, nonlinear, but it *is* linear in the derivatives u_x and u_y. It is called a **quasi-linear** equation. A class of quasi-linear equations which describe certain **conservation laws** consists of PDEs of the form

$$u_t + \frac{\partial}{\partial x}[F(u)] = u_t + f(u)u_x = 0,$$

where F is a given function with $F' = f$. Lagrange, in the late 1700s, devised a method of transforming the general quasi-linear equation into a *linear* PDE involving three independent variables. Instead, we will solve the simpler conservation equation as a special case. In fact, it can be shown that the characteristic equations are given by

$$\frac{dt}{1} = \frac{dx}{f(u)},$$

where u is treated formally as a constant.

13. a) Given Burger's equation from the study of gas dynamics,

$$u_t + uu_x = 0,$$
$$u(x,0) = g(x),$$

show that the characteristics actually must be the *straight lines*

$$x - tg(x_0) = x_0,$$

that is, show that the solution is constant along these lines.

b) Using part (a), and given the Burger's problem

$$u_t + uu_x = 0,$$
$$u(x,0) = 2x,$$

(a rather unrealistic initial condition!), compute $u(6,8)$, $u(0,5)$.

c) More generally, what is $u(x,t)$ for any point (x,t)?

Note that the characteristics do *not* intersect each other in this particular example. They *will* for "most" functions $g(x)$, in which case life becomes more complicated (specifically, the solution develops what we call *shock waves*, or just *shocks*).

5.3   The Infinite String

We have solved first-order equations by considering curves along which the PDE and its solutions have certain nice properties. The obvious question now is whether we can extend this idea of characteristics to higher order equations.

To this end, let's look at how d'Alembert cleverly solved the wave equation for an infinitely long string. The well-posed system for the infinite string is

$$u_{tt} = c^2 u_{xx}, \qquad -\infty < x < \infty, t > 0, \qquad (5.7)$$
$$u(x,0) = f(x), \qquad (5.8)$$
$$u_t(x,0) = g(x), \qquad (5.9)$$

where each term has the same meaning as in Chapter 3. There are no boundary conditions, as we are supposing the string to have infinite length. (However, it turns out that we *do* need certain "conditions at infinity." We'll say more, later.)

D'Alembert suggested the following change in coordinates:

$$\xi = x + ct$$
$$\eta = x - ct,$$

in which case we have (see Exercise 1)

$$u_{tt} = c^2(u_{\xi\xi} - 2u_{\xi\eta} + u_{\eta\eta})$$
$$u_{xx} = u_{\xi\xi} + 2u_{\xi\eta} + u_{\eta\eta}.$$

The PDE in the transformed coordinates becomes

$$u_{\xi\eta} = 0, \tag{5.10}$$

which can be solved by integrating twice! Doing so leads to

$$u = \phi(\xi) + \psi(\eta) \tag{5.11}$$
$$= \phi(x + ct) + \psi(x - ct),$$

where ϕ and ψ are arbitrary functions (again, with certain restrictions).

Now, if we choose $\psi \equiv 0$, we have that

$$u = \phi(x + ct)$$

is a solution for any ϕ. These solutions, of course, will be constant along the lines

$$x + ct = constant,$$

so that these lines behave very much like the characteristics we have already met. Similarly, if $\phi \equiv 0$, our solutions are constant along the lines

$$x - ct = constant.$$

We call these lines the **characteristics** for the wave equation. Of course, if neither ϕ nor ψ is the zero-function, we do not expect the solution to be constant along the characteristics.

Before we finish solving the initial-value problem and looking more closely at the characteristics, let's look at the physical interpretation of our solution.

Suppose, first, that $\phi \equiv 0$. Then, again, we have the solution

$$u = \psi(x - ct).$$

Further, suppose the initial shape $u(x, 0) = \psi(x)$ has the graph given in Figure 5.6a, and let $x = x_0$ be the point where the "hump" is, initially. What will the shape be at any time $t > 0$? For starters, consider what happens at time $t = 1$, when the shape is $u(x,1) = \psi(x - c)$. As can be seen in Figure 5.6b, this is just the initial curve shifted c units to the right. More generally, $u(x,t_0) = \psi(x - ct_0)$ is the initial curve moved ct_0 units to the right. Therefore, our solution represents a wave whose shape is unchanged and which moves to the right with velocity c. (Compare with Exercise 11 in Section 5.1.)

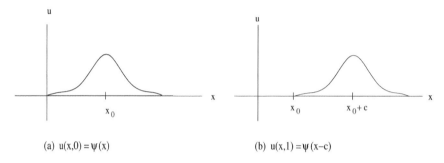

(a) $u(x,0) = \psi(x)$ (b) $u(x,1) = \psi(x-c)$

FIGURE 5.6
Hump moves to the right at velocity c.

Similarly, $u = \phi(x + ct)$ is a wave moving to the left with velocity c. It is this physical interpretation that gives the wave equation its name.

Finally, let's apply the initial conditions. Equation (5.8) becomes

$$u(x,0) = f(x) = \phi(x) + \psi(x). \tag{5.12}$$

As for (5.9), we need to calculate u_t first. Using the chain rule, we have

$$u_t = c\phi'(x + ct) - c\psi'(x - ct),$$

and (5.9) becomes

$$u_t(x,0) = g(x) = c\phi'(x) - c\psi'(x).$$

We must solve for ϕ and ψ in terms of f and g. Therefore, we first differentiate (5.12) with respect to x, and we solve

$$f'(x) = \phi'(x) + \psi'(x)$$
$$g(x) = c\phi'(x) - c\psi'(x).$$

With a little algebra, we get

$$\phi'(x) = \frac{1}{2}f'(x) + \frac{1}{2c}g(x)$$
$$\psi'(x) = \frac{1}{2}f'(x) - \frac{1}{2c}g(x),$$

and, upon integrating, we have

$$\phi(x) = \frac{1}{2}f(x) + \frac{1}{2c}\int_0^x g(z)dz + c_1 \tag{5.13}$$

$$\psi(x) = \frac{1}{2}f(x) - \frac{1}{2c}\int_0^x g(z)dz + c_2.$$

(Remember that the Fundamental Theorem of Calculus says that

$$\frac{d}{dx}\int_a^x g(z)dz = g(x)$$

for any constant a. We choose $a = 0$ for the sake of convenience.) Are c_1 and c_2 arbitrary constants? Well, (5.12) and (5.13) say that

$$f(x) = \phi(x) + \psi(x)$$
$$= \frac{1}{2}f(x) + \frac{1}{2c}\int_0^x g(z)dz + c_1 + \frac{1}{2}f(x) - \frac{1}{2c}\int_0^x g(z)dz + c_2$$
$$= f(x) + c_1 + c_2,$$

so we must have

$$c_1 + c_2 = 0.$$

Our solution, then, is

$$u = \phi(x + ct) + \psi(x - ct)$$
$$= \frac{1}{2}f(x + ct) + \frac{1}{2c}\int_0^{x+ct} g(z)dz + \frac{1}{2}f(x - ct) - \frac{1}{2c}\int_0^{x-ct} g(z)dz$$

or

$$u = \frac{1}{2}[f(x + ct) + f(x - ct)] + \frac{1}{2c}\int_{x-ct}^{x+ct} g(z)dz. \tag{5.14}$$

We can check this function to show that it does, indeed, satisfy the original system (see Exercise 2).

Example 1 Solve

$$u_{tt} = 5u_{xx},$$
$$u(x,0) = \sin x,$$
$$u_t(x,0) = \sin 3x.$$

We have

$$u = \frac{1}{2}[\sin(x + \sqrt{5}t) + \sin(x - \sqrt{5}t)] + \frac{1}{10}\int_{x-\sqrt{5}t}^{x+\sqrt{5}t} \sin 3z \; dz$$

$$= \frac{1}{2}[\sin(x + \sqrt{5}t) + \sin(x - \sqrt{5}t)] - \frac{1}{30}\cos 3z \Big|_{z=x-\sqrt{5}t}^{z=x+\sqrt{5}t}$$

$$= \frac{1}{2}[\sin(x + \sqrt{5}t) - \sin(x - \sqrt{5}t)] - \frac{1}{30}[\cos 3(x + \sqrt{5}t) - \cos 3(x - \sqrt{5}t)].$$

Example 2 Let's look graphically at the solution of

$$u_{tt} = u_{xx},$$
$$u(x,0) = f(x),$$
$$u_t(x,0) = 0,$$

where

$$f(x) = \begin{cases} 2, & \text{if } -3 \le x \le 3, \\ 0, & \text{if } |x| > 3, \end{cases}$$

for various times t. Note that the function $f(x)$ is known as a **square wave** (although maybe it should be called a rectangular wave!). Of course, a real string can only approximate such a shape; however, square waves are very important in many wave phenomena, in fields like electronics and acoustics. Also, this system technically does not have a solution, since f' (and, therefore, f'') does not exist at $x = \pm 3$. However, d'Alembert's solution still "works," and the square wave makes it easy to see what's going on in these problems.

Now, our solution is

$$u = \frac{1}{2}[f(x+t) + f(x-t)].$$

Therefore, at any time t, the graph of u is the superposition of two identical waves, one shifted t units to the right, the other shifted t units to the left, and each a copy of the initial shape but half its height. So, for example, we have (see Figure 5.7)

$$t = 0: \quad u(x,0) = f(x),$$
$$t = 1: \quad u(x,1) = \frac{1}{2}[f(x+1) + f(x-1)],$$
$$t = 2: \quad u(x,2) = \frac{1}{2}[f(x+1) + f(x-2)],$$
$$t = 3: \quad u(x,3) = \frac{1}{2}[f(x+3) + f(x-3)],$$
$$t = 4: \quad u(x,4) = \frac{1}{2}[f(x+4) + f(x-4)]$$

and, by this point in time, we clearly see that the initial shape has broken into two half-waves, traveling in opposite directions. We continue this example below.

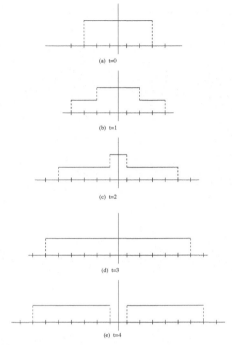

(a) t=0

(b) t=1

(c) t=2

(d) t=3

(e) t=4

FIGURE 5.7
Breaking up of square wave into two half-waves, traveling in oppo-site directions.

Now let's look more carefully at the characteristic curves. We ask the following question: How do the initial conditions affect the solution at points (x_0, t_0), where $t_0 > 0$? We will treat $u(x, 0)$ and $u_t(x, 0)$ separately (more precisely, we know that we may deal with the initial conditions

$$u(x, 0) = f(x),$$
$$u_t(x, 0) = 0$$

and

$$u(x, 0) = 0,$$
$$u_t(x, 0) = g(x)$$

separately because of the principle of superposition).

The contribution of f to our solution u gives us

$$u(x_0, t_0) = \frac{1}{2}[f(x_0 + ct_0) + f(x_0 - ct_0)]$$
$$= \frac{1}{2}[u(x_0 + ct_0, 0) + u(x_0 - ct_0, 0)].$$

In other words, the value of $u(x_0, t_0)$ depends only on the values of f at $x = x_0 - ct_0$ and $x = x_0 + ct_0$. We have a nice geometrical interpretation of this fact in Figure 5.8a. There are two characteristics which pass through the point (x_0, t_0), and these intersect the initial line $t = 0$ at the above-mentioned x-values. We can think of the **initial disturbance** of the string at these points as **propagating "along" the characteristics** to the point (x_0, t_0), with velocity c.

From $g(x)$, we have

$$u(x_0, t_0) = \frac{1}{2c} \int_{x_0 - ct_0}^{x_0 + ct_0} g(z)\,dz,$$

which tells us that the solution $u(x_0, t_0)$ depends on the values of g along the interval $x_0 - ct_0 \le x \le x_0 + ct_0$. This fact is illustrated in Figure 5.8b.

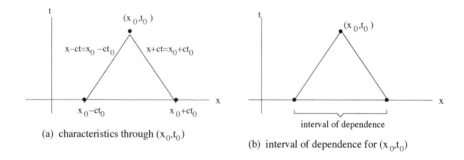

(a) characteristics through (x_0, t_0)

interval of dependence

(b) interval of dependence for (x_0, t_0)

FIGURE 5.8
The *characteristics* and *interval of dependence* of the point (x_0, t_0).

This interval often is called the **interval of dependence** or **domain of dependence** of the point (x_0, t_0). More generally, though, we may think of all points in the triangle in Figure 5.9 as having an effect on the solution at (x_0, t_0). That is why this whole triangle, and not just its base, is sometimes referred to as the *domain of dependence*.

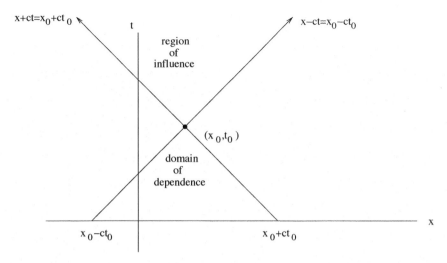

FIGURE 5.9
Domain of dependence and *region of influence* of the point (x_0, t_0).

With the latter idea in mind, we may turn the question around and ask, instead, at which points (x, t) is the solution affected by what has "happened" at the point (x_0, t_0)? Of course, (x, t) will be such a point if and only if (x_0, t_0) is in the domain of dependence of (x, t). Therefore, this **region of influence** of the point (x_0, t_0) will consist of the (infinite) region in Figure 5.9.

Example 2 (cont.) Let's go back to Example 2, and use these ideas to get a different look at our solution. Indeed, we see that we may represent our solution in the x-t plane as given below. In Figure 5.10, we take three representative points and use the characteristics and initial data to show that

$$u(x_0, t_0) = \frac{1}{2}[f(x_0 + ct_0) + f(x_0 - ct_0)] = 2,$$
$$u(x_1, t_1) = 1,$$

and

$$u(x_2, t_2) = 0.$$

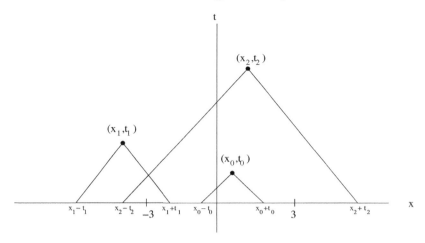

FIGURE 5.10
Graphical solution of Example 2.

Then, using these ideas, we see in Figure 5.11 a graphical solution of the problem on its domain. We leave the case where the initial *velocity* is a square wave to Exercise 13.

So we have shown that a solution of the wave equation can be thought of as a superposition of two waves (5.11) traveling in opposite directions, each with velocity c. That they do not interfere with each other is implicit in (5.11)—we say that the waves do not **scatter** each other. This is a consequence of the fact that the equation is linear.

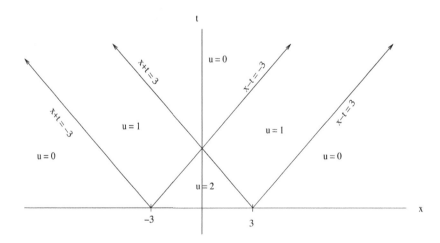

FIGURE 5.11
Complete graphical solution of Example 2.

Also, these waves—and all disturbances—travel at the same speed. This need *not* be the case for linear PDEs with wave solutions. When wave solutions move at different speeds, the PDE is said to exhibit the property of **dispersion**. (See Exercise 14.)

Exercises 5.3

1. Carry through the computations that transform PDE (5.7) into (5.10).

2. Show that the function u given by (5.14) does, indeed, satisfy equations (5.7), (5.8) and (5.9).
   In Exercises 3 and 4, find the solution of

$$u_{tt} = c^2 u_{xx}, \qquad -\infty < x < -\infty,$$
$$u(x,0) = f(x),$$
$$u_t(x,0) = g(x),$$

where c, f and g are given by

3. $c = 1$, $f(x) = 3e^{-x^2}$, $g(x) = 0$

4. $c = 3$, $f(x) = 0$, $g(x) = xe^{-x^2}$
   In Exercises 5–7,

   a) Graph, by hand, snapshots of the solution in the x-u plane, for various times t, as in Example 2.

   b) **MATLAB:** Plot the solution in x-t-u space—make sure you can see the "characteristic directions."

5. Use $f(x) = 3e^{-x^2}$.

6. Use
$$f(x) = \begin{cases} |4 - x|, & \text{if } -4 \le x \le 4, \\ 0, & \text{if } |x| > 4. \end{cases}$$

7. Use
$$f(x) = \begin{cases} 2, & \text{if } -4 \le x \le 0, \\ 3, & \text{if } 4 \le x \le 8, \\ 0, & \text{otherwise.} \end{cases}$$

8. a) Show that if $f(x)$ and $g(x)$ are both even, then the solution (5.14) also is even.

   b) Show that if $f(x)$ and $g(x)$ are both odd, then the solution (5.14) also is odd.

In Exercises 9 and 10, find and sketch the (triangular) domain of dependence and the region of influence of the given point (x_0, t_0) for the given PDE.

9. $u_{tt} = 4u_{xx}, (x_0, t_0) = (5, 4)$

10. $u_{tt} = 3u_{xx}, (x_0, t_0) = (0, 5)$

11. For the problem in Example 2, use only characteristics to find the following: $u(0, 2)$, $u(0, 4)$, $u(5, 5)$, $u(10, 6)$, $u(-5, 3)$. See Figure 5.10. By the way, again, notice that, in those cases where $u_t(x, 0) = 0$, the value of $u(x_0, t_0)$ depends only on the values of u along the two characteristics containing (x_0, t_0). This means that a disturbance at some point on the string arrives at, and leaves from, the point $x = x_0$ instantaneously. In this particular example, a square wave hits the point $x = x_0$, moves through it and leaves as sharply as it arrived—we say that it has sharp leading and trailing edges. When *all* disturbances propagate in this manner, the system is said to satisfy **Huygens's‡ Principle**.

12. We wish to show that Huygens's Principle is *not* satisfied when $u_t(x, 0)$ is not the zero-function. Specifically, consider the problem in Example 2, but with $f(x)$ and $g(x)$ interchanged. Show that the solution u is as given in Figure 5.12, and then describe what the graph of u looks like for various times t.

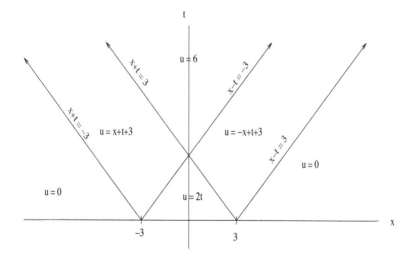

FIGURE 5.12
Graphical solution of the problem in Exercise 12.

---

‡Named after Christiaan Huygens (1629–1695), famous Dutch scientist and contemporary of Newton.

13. **Green's Theorem and the nonhomogeneous wave equation:** There's another, elegant method for deriving d'Alembert's solution for the infinite string, a method which can be extended to solve the nonhomogeneous equation, as well. Suppose that we're given the problem

$$u_{tt} = c^2 u_{xx}, \quad -\infty < x < \infty, t > 0,$$
$$u(x,0) = f(x),$$
$$u_t(x,0) = g(x).$$

a) Integrate both sides of the PDE over any simple closed region D in the upper half of the x-t plane, then use Green's Theorem to conclude that we must have

$$\oint_{\partial D} (c^2 u_x \, dt + u_x \, dx),$$

where ∂D is the boundary of D with positive (counterclockwise) orientation.

Now, which region D will do something for us? Given a point (x_0, t_0), it seems the most logical choice would be the *domain of dependence* (or *characteristic triangle* or *past history*) of (x_0, t_0), given in Figure 5.9.

b) Show that the line integral on the bottom line segment gives us

$$\int_{c_1} (c^2 u_x \, dt + u_t \, dx) = \int_{x_0 - ct_0}^{x_0 + ct_0} g(z)dz.$$

c) Parametrize c_2 as

$$t = t,$$
$$x = x_0 + ct_0 - ct, \quad 0 \le t \le t_0,$$

to show that

$$\int_{c_2} (c^2 u_x \, dt + u_t \, dx) = cu(x_0, t_0) - cf(x_0 + ct_0).$$

d) Similarly, show that

$$\int_{c_3} (c^2 u_x \, dt + u_t \, dx) = cu(x_0, t_0) - cf(x_0 - ct_0),$$

and conclude that we, indeed, end up with d'Alembert's formula for the solution $u(x_0, t_0)$.

e) Proceed similarly to show that the solution of the nonhomogeneous problem

$$u_{tt} = c^2 u_{xx} + F(x,t), \qquad -\infty < x < \infty, t > 0,$$
$$u(x,0) = f(x),$$
$$u_t(x,0) = g(x)$$

is

$$u(x,t) = \frac{1}{2}[f(x+ct) + f(x-ct)] + \frac{1}{2c}\int_{x-ct}^{x+ct} g(z)dz$$
$$+ \frac{1}{2c}\iint_D F(x,t)dxdt,$$

where, again, D is the *domain of dependence* of the point (x,t).

14. a) Find all solutions of the wave equation of the form $\sin(\kappa x - \omega t)$ and $\cos(\kappa x - \omega t)$, then use this information to find all *complex* solutions of the form $e^{i(\kappa x - \omega t)}$, where κ and ω are constants (remember **Euler's formula:** $e^{i\theta} = \cos\theta + i\sin\theta$). The number κ is called the **wave number** of such a solution, while ω is called its **angular frequency**. Show that, regardless of the wave number κ, these solutions all travel at velocity c.

  b) Remember the **Euler–Bernoulli beam equation**

$$u_{tt} + u_{xxxx} = 0.$$

Find all solutions of this equation of the form $e^{i(\kappa x - \omega t)}$. What is the velocity of a solution with given wave number κ? (Solutions of this PDE exhibit **dispersion**, and we say that the Euler–Bernoulli beam equation is a **dispersive equation**.)

5.4 Characteristics for Semi-Infinite and Finite String Problems

Now let's see how characteristics relate to the wave equation on more restricted intervals. We begin with the interval $x \geq 0$.

SEMI-INFINITE STRING

We would like to solve

$$u_{tt} = c^2 u_{xx}, \qquad x > 0, t > 0,$$
$$u(x,0) = f(x),$$
$$u_t(x,0) = g(x),$$
$$u(0,t) = 0$$

and, proceeding as in the previous section, we arrive at the same solution *so long as* $x - ct \geq 0$. So if $x_0 \geq ct_0$, we have

$$u(x_0,t_0) = \frac{1}{2}[f(x_0 + ct_0) + f(x_0 - ct_0)]$$
$$+ \frac{1}{2c} \int_{x_0-ct_0}^{x_0+ct_0} g(z)dz,$$

and the domain of dependence is the same as earlier. The problem arises, of course, if $x_0 < ct_0$, as $f(x)$ and $g(x)$ are not defined for $x < 0$. In that case, we need to do something with the term $\psi(x - ct)$ in (5.11). Now, note that we still have the boundary condition at our disposal. So, using (5.11), we have

$$u(0,t) = \phi(ct) + \psi(-ct) = 0 \quad \text{for all} \quad t \geq 0,$$

on

$$\psi(z) = -\phi(-z) \quad \text{for all} \quad z < 0.$$

Then, from (5.13),

$$\psi(x_0 - ct_0) = -\phi(ct_0 - x_0)$$
$$= -\frac{1}{2}f(ct_0 - x_0) - \frac{1}{2c} \int_0^{ct_0-x_0} g(z)dz$$

and, combining, we have

$$u(x_0,t_0) = \frac{1}{2}[f(x_0 + ct_0) - f(ct_0 - x_0)]$$
$$+ \frac{1}{2c} \int_{ct_0-x_0}^{x_0+ct_0} g(z)dz, \qquad x_0 < ct_0.$$

We see that the solution depends on the initial interval $ct_0 - x_0 \leq x \leq x_0 + ct_0$. In the x-t plane, we see that the characteristic

$$x - ct = x_0 - ct_0$$

comes about via a *reflection of the characteristic*

$$x + ct = ct_0 - x_0$$

against the t-axis. The two-dimensional domain of dependence no longer is a triangle, as can be seen in Figure 5.13. Further, the region of influence of *any* point (x_0,t_0) is no longer a triangle (see Figure 5.14).

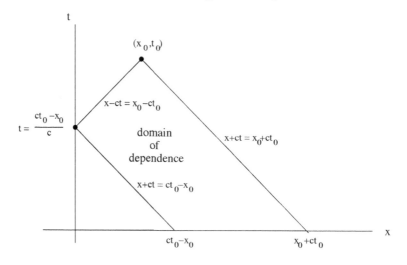

FIGURE 5.13
Domain of dependence for the semi-infinite string, for a point (x_0, t_0)
"above" the line $x = ct$.

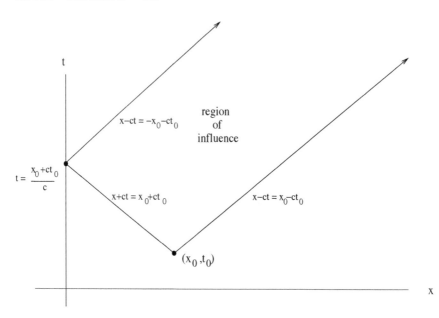

FIGURE 5.14
Region of influence for the semi-infinite string, for any point (x_0, t_0).

It's interesting to see that the word *reflection* is no misnomer, as the initial
disturbance actually does reflect at the boundary. We look closely at this
phenomenon in Exercises 4 and 5.

Example 1 Let's look at an analog of the problem in Example 2 of the previous section,

$$u_{tt} = u_{xx}, \qquad x > 0, t > 0,$$

$$u(x, 0) = \begin{cases} 2, & \text{if } 2 \le x \le 3, \\ 0, & \text{otherwise,} \end{cases}$$

$$u_t(x, 0) = 0,$$

$$u(0, t) = 0.$$

We follow the characteristics and their reflections and arrive at the x-t representation of the solution given in Figure 5.15. To be sure, look at the shaded region. Here, we have $x < t$ and

$$2 < x + t < 3,$$
$$2 < t - x < 3,$$

so the solution at any point here is

$$u(x, t) = \frac{1}{2}[f(x + t) - f(t - x)]$$
$$= \frac{1}{2}(2 - 2).$$

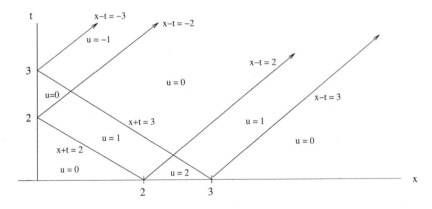

FIGURE 5.15
Graphical solution of the problem in Example 1.

Before moving on, we mention that there's a more elegant approach to problems on semi-infinite intervals—the so-called **method of images**, which we deal with in Exercises 10 and 11.

FINITE STRING

We can extend this principle of reflection to a *finite* string easily enough. Consider the problem

$$u_{tt} = c^2 u_{xx}, \qquad 0 < x < L,$$
$$u(x,0) = f(x),$$
$$u_t(x,0) = g(x),$$
$$u(0,t) = u(L,t) = 0.$$

Here, now, the original d'Alembert's solution is only good for those points (x_0, t_0) for which

$$0 \le x_0 + ct_0 \le L$$

and

$$0 \le x_0 - ct_0 \le L,$$

and we must reflect the characteristics at *both* boundaries in order to find the solution elsewhere. Let's begin by looking at a point (x_0, t_0) situated as in Figure 5.16. We see that

$$0 \le x_0 - ct_0 \le L,$$

but

$$L \le x_0 + ct_0 \le 2L.$$

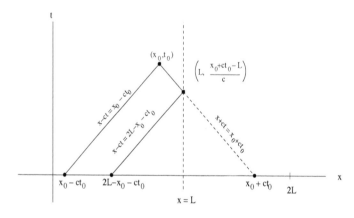

FIGURE 5.16
Reflection of characteristic for the finite string.

Therefore, we use $u(L,t) = 0$ to give us

$$\phi(L + ct) + \psi(L - ct) = 0$$

or

$$\phi(z) = -\psi(2L - z) \quad \text{for} \quad L \le z \le 2L \quad \text{(why?)}.$$

So we have

$$u(x_0, t_0) = \phi(x_0 + ct_0) + \psi(x_0 - ct_0)$$
$$= -\psi(2L - x_0 - ct_0) + \psi(x_0 - ct_0)$$
$$= \frac{1}{2}[f(x_0 - ct_0) - f(2L - x_0 - ct_0)]$$
$$+ \frac{1}{2c}\int_{x_0-ct_0}^{2L-x_0-ct_0} g(z)dz.$$

Now, let's take a point that requires *two* reflections. Here, we must bounce off both ends; since $x_0 - ct_0 < 0$, while $x_0 + ct_0 > L$, we use

$$\psi(z) = -\phi(-z), \qquad -L \leq z \leq 0$$

and

$$\phi(z) = \psi(2L - z), \qquad L \leq z \leq 2L$$

in order to write

$$u(x_0, t_0) = \phi(x_0 + ct_0) + \psi(x_0 - ct_0)$$
$$= -\psi(2L - x_0 - ct_0) - \phi(ct_0 - x_0)$$
$$= -\frac{1}{2}[f(ct_0 - x_0) + f(2L - x_0 - ct_0)]$$
$$+ \frac{1}{2c}\int_{ct_0-x_0}^{2L-x_0-ct_0} g(z)dz.$$

Figure 5.17 shows the domain of dependence of the point (x_0, t_0) (although we need not always have $ct_0 - x_0 < 2L - x_0 - ct_0$, as we do in this particular case).

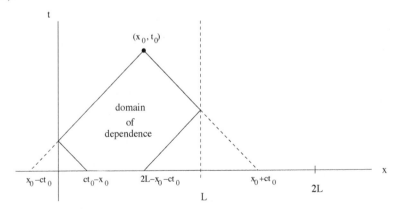

FIGURE 5.17
Reflection of two characteristics for the finite string.

We may, of course, continue the process indefinitely. In order to get a simple picture of what's going on, let's look at the following example.

Example 2 Suppose that we have the problem

$$u_{tt} = u_{xx}, \qquad 0 < x < 3, t > 0,$$
$$u_t(x, 0) = 0,$$
$$u(0, t) = u(3, t) = 0,$$

where the initial shape $u(x, 0)$ is given by a "point-impulse" of "magnitude 2" at $x = 1$. (These ideas will be made more precise in Section 6.5, where we introduce the *Dirac delta function*, $\delta(x)$. Then, this initial shape would be written as $u(x, 0) = 2\delta(x - 1)$.) We should then have two pulses, each of magnitude 1, traveling in opposite directions from the point $x = 1$. In the x-t plane, each disturbance travels along a characteristic and, as earlier, each undergoes a change of sign as it reflects off either boundary. Thus, the solution will be $u = 0$ everywhere except along the characteristics $x \pm t =$ constant, with values as given in Figure 5.18.

We investigate more complicated problems in the exercises.

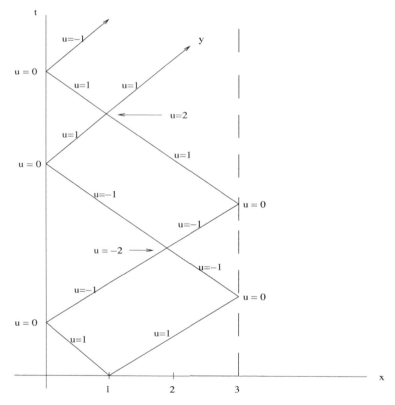

FIGURE 5.18
Graphical solution of the problem in Example 2.

Exercises 5.4

In Exercises 1–3, find the solution of

$$u_{tt} = c^2 u_{xx}, \qquad x > 0, t > 0,$$
$$u(x, 0) = f(x),$$
$$u_t(x, 0) = g(x),$$
$$u(0, t) = 0,$$

where c, f and g are given by

1. $c = 2$, $f(x) = 3e^{-x}$, $g(x) = 0$

2. $c = 1$, $f(x) = \begin{cases} 2, & \text{if } 0 \le x \le 1, \\ 0, & \text{otherwise,} \end{cases}$ $g(x) = 0$.

3. $c = 4$, $f(x) = 0$, $g(x) = e^{-x}$

In Exercises 4 and 5, proceed as in Exercises 5–7 of the previous section for the problem

$$u_{tt} = u_{xx}, \qquad x > 0, t > 0,$$
$$u(x, 0) = f(x),$$
$$u_t(x, 0) = 0,$$
$$u(0, t) = 0.$$

That is,

a) Draw snapshots of the solution for various time t.

b) Graph the solution in x-t-u space—make sure you can see the "characteristic directions," both before and after reflection.

4. For $f(x) = \begin{cases} 2, & \text{if } 2 \le x \le 5, \\ 0, & \text{otherwise.} \end{cases}$

5. For $f(x) = \begin{cases} |12 - x|, & \text{if } 4 \le x \le 8, \\ 0, & \text{otherwise.} \end{cases}$

In Exercises 6 and 7, proceed as in Exercises 4 and 5 for the finite string problem on $0 \le x \le 8$.

6. For $f(x) = \begin{cases} 2, & \text{if } 2 \le x \le 4, \\ 0, & \text{otherwise.} \end{cases}$

7. For $f(x) = \begin{cases} 2, & \text{if } 0 \le x \le 4, \\ 0, & \text{otherwise.} \end{cases}$

8. a) Proceed as in Example 1 to find a graphical representation of the solution of that same problem, but with the initial displacement changed to

$$u(x,0) = \begin{cases} (2-x)(3-x), & \text{if } 2 \le x \le 3, \\ 0, & \text{otherwise.} \end{cases}$$

b) Do the same, but with initial conditions

$$u(x,0) = 0,$$

$$u_t(x,0) = \begin{cases} 2, & \text{if } 2 \le x \le 3, \\ 0, & \text{otherwise.} \end{cases}$$

9. Do the same as in Exercise 8, but for the finite string problems in Exercises 6 and 7. You need only go "up" to two reflections.

10. **Method of images:** Given the semi-infinite string problem

$$u_{tt} = c^2 u_{xx}, \qquad x > 0, t > 0,$$
$$u(x,0) = f(x),$$
$$u_t(x,0) = g(x),$$
$$u(0,t) = 0,$$

solve, instead, the *infinite* string problem

$$u_{tt} = c^2 u_{xx}, \qquad -\infty < x < \infty, t > 0,$$
$$u(x,0) = F(x),$$
$$u_t(x,0) = G(x),$$

where F and G are the odd extensions of f and g, respectively. Then, show that your solution, when restricted to $x \ge 0$, is the solution of the original problem on $x \ge 0$.

11. **Semi-infinite string with Neumann condition**

a) Solve the problem with *free end*

$$u_{tt} = c^2 u_{xx}, \qquad x > 0, t > 0,$$
$$u(x,0) = f(x),$$
$$u_t(x,0) = g(x),$$
$$u_x(0,t) = 0$$

as in the text.

b) Solve the same problem, but use the *method of images*.

12. Use Green's Theorem, as in Exercise 13 of the previous section, to solve the nonhomogeneous semi-infinite string problem

$$u_{tt} = c^2 u_{xx} + F(x,t), \qquad x > 0, t > 0,$$
$$u(x,0) = f(x),$$
$$u_t(x,0) = g(x),$$
$$u(0,t) = 0.^{\S}$$

5.5 General Second-Order Linear PDEs and Characteristics

The general second-order linear PDE in x and y is

$$a u_{xx} + b u_{xy} + c u_{yy} + d u_x + f u_y + g u + h(x,y) = 0, \qquad (5.15)$$

where a, b, c, d, f, g and h are functions of x and y. For now, let's assume that they are constant. The question we would like to answer is: Is there a transformation similar to that in Section 5.3 which will reduce PDE (5.15) to some simple form? Let's try. First, if any two of a, b and c are zero, then the PDE already is in the simplest form. So we assume this is not the case, and we set

$$\xi = Ax + By, \quad \eta = Cx + Dy,$$

where A, B, C and D are constants to be determined. Proceeding as in Exercise 1 from Section 5.3, we have

$$u_x = u_\xi \xi_x + u_\eta \eta_x$$
$$= A u_\xi + C u_\eta$$

and, similarly,

$$u_y = B u_\xi + D u_\eta.$$

Then,

$$u_{xx} = (u_x)_x = (A u_\xi + C u_\eta)_x$$
$$= A(A u_\xi + C u_\eta)_\xi + C(A u_\xi + C u_\eta)_\eta$$
$$= A^2 u_{\xi\xi} + 2AC u_{\xi\eta} + C^2 u_{\eta\eta}.$$

Similarly,

$$u_{xy} = AB u_{\xi\xi} + (BC + AD) u_{\xi\eta} + CD u_{\eta\eta} \qquad (5.16)$$
$$u_{yy} = B^2 u_{\xi\xi} + 2BD u_{\xi\eta} + D^2 u_{\eta\eta}.$$

---

§It turns out that we need to assume that $f(0) = g(0) = 0$.

The second-derivative part of PDE (5.15) then transforms to

$$(A^2a + ABb + B^2c)u_{\xi\xi} + [2ACa + (BC + AD)b + 2BDc]u_{\xi\eta}$$
$$+ (C^2a + CDb + D^2c)u_{\eta\eta}. \tag{5.17}$$

Now, notice that the $u_{\xi\xi}$ and $u_{\eta\eta}$ coefficients are such that if we can eliminate one, then we can eliminate both. For what values of a, b and c will we be able to do this? We are considering only cases where $a \neq 0$ or $c \neq 0$; suppose the former (the latter case will proceed similarly). Then, $a \neq 0$ implies that $B \neq 0$ (because, if $B = 0$, we would need $A = 0$ as well, to eliminate $u_{\xi\xi}$) and $D \neq 0$. Dividing the $u_{\xi\xi}$- and $u_{\eta\eta}$-coefficients by B^2 and D^2, respectively, we see that we need to solve the equations

$$a\left(\frac{A}{B}\right)^2 + b\left(\frac{A}{B}\right) + c = 0$$

$$a\left(\frac{C}{D}\right)^2 + b\left(\frac{C}{D}\right) + c = 0,$$

that is, we need to look at the quadratic equation

$$az^2 + bz + c = 0,$$

which has roots

$$z = \frac{-b \pm \sqrt{b^2 - 4ac}}{2a}.$$

Case 1: $b^2 - 4ac > 0$

In this case, the quadratic has two distinct real roots, so we may take

$$\frac{A}{B} = \frac{-b + \sqrt{b^2 - 4ac}}{2a} \quad \text{and}$$
$$\frac{C}{D} = \frac{-b - \sqrt{b^2 - 4ac}}{2a}. \tag{5.18}$$

So, for example, we may take

$$\xi = (-b + \sqrt{b^2 - 4ac})x + 2ay \tag{5.19}$$
$$\eta = (-b - \sqrt{b^2 - 4ac})x + 2ay,$$

in which case the PDE (5.15) reduces to its **standard** or **canonical form**

$$u_{\xi\eta} + lower\ order\ terms = 0.$$

In particular, the PDE

$$au_{xx} + bu_{xy} + cu_{yy} = 0$$

reduces to

$$u_{\xi\eta} = 0$$

and, thus, has solution

$$u = \phi(\xi) + \psi(\eta) \tag{5.20}$$
$$= \phi[(-b + \sqrt{b^2 - 4ac})x + 2ay] + \psi[(-b - \sqrt{b^2 - 4ac})x + 2ay]$$

for arbitrary ϕ and ψ. Actually, we will show that, in this case, equation (5.15) reduces to the alternate canonical form

$$u_{\xi\xi} = \kappa^2 u_{\eta\eta} + lower\ order\ terms = 0$$

(see Exercise 11).

Definition 5.2 *If $b^2 - 4ac > 0$, we say that PDE (5.15) is* **hyperbolic.**

Examples

1. The wave equation $u_{tt} - c^2 u_{xx} = 0$ is hyperbolic

2. The **telegraph equation** or **dissipative wave equation** $u_{tt} - c^2 u_{xx} + \gamma u_t = 0$, where $\gamma > 0$ is a constant, is hyperbolic.

3. Consider the equation $2u_{xx} - 5u_{xy} + 2u_{yy} = 0$. We have $b^2 - 4ac = 9$, so the equation is hyperbolic. In order to reduce it to standard form, we solve the quadratic

$$2z^2 - 5z + 2 = 0$$

which gives us

$$z = \frac{5 \pm \sqrt{9}}{2} = \frac{5}{2} \pm \frac{3}{2}.$$

Then, we determine the coefficients of our transformation by

$$\frac{A}{B} = \frac{5}{2} + \frac{3}{2} = 4, \quad \frac{C}{D} = \frac{5}{2} - \frac{3}{2} = 1,$$

so we may choose $A = 4, B = C = D = 1$, arriving at

$$\xi = 4x + y$$
$$\eta = x + y.$$

Our solution, then, is

$$u = \phi(\xi) + \psi(\eta)$$
$$= \phi(4x + y) + \psi(x + y).$$

Often, PDEs which model vibrations and wave phenomena are hyperbolic, and *the wave equation is a "standard example" of a hyperbolic equation.*

Notice that, in Example 3, the lines

$$4x + y = constant, \quad x + y = constant$$

play the same role as did the lines $x \pm ct = constant$ for the wave equation. In general, this will be true of the lines

$$\xi = constant, \quad \psi = constant$$

for hyperbolic equations and, thus, *these lines are called the* **characteristics** *of the PDE.* In general:

Hyperbolic equations have two families of characteristics.

Case 2: $b^2 - 4ac = 0$

In this case, the quadratic has one repeated root, $z = -\frac{b}{2a}$. The problem here is that, although we may choose

$$\frac{A}{B} = -\frac{b}{2a}$$

and eliminate the $u_{\xi\xi}$ term, we may not choose

$$\frac{C}{D} = -\frac{b}{2a}$$

since this choice will make η a multiple of ξ and, thus, the new variables will not be independent (we say that the transformation to ξ, η is **singular** because it cannot be inverted to give us x and y in terms of ξ and η).

However, by choosing only the former, for example,

$$A = b, \quad B = -2a \quad \text{and} \quad \xi = bx - 2ay, \tag{5.21}$$

we get lucky—the coefficient of $u_{\xi\eta}$ becomes

$$2ACa + (BC + AD)b + 2BDC$$
$$= 2Cab + (-2aC + bD)b - 4Dac$$
$$= (b^2 - 4ac)D = 0$$

and, by letting $\eta = x$ (or anything else that is not a constant multiple of ξ), the PDE (5.15) reduces to the **standard form**

$$u_{\eta\eta} + lower\ order\ terms = 0.$$

In this case, the PDE

$$au_{xx} + bu_{xy} + cu_{yy} = 0$$

transforms into

$$u_{\eta\eta} = 0$$

with solution

$$u = \eta\phi(\xi) + \psi(\xi) \tag{5.22}$$
$$= x\phi(bx - 2ay) + \psi(bx - 2ay)$$

for any functions ϕ and ψ.

Definition 5.3 *If $b^2 - 4ac = 0$, we say that PDE (5.15) is* **parabolic**.

Examples

4. The heat equation $u_t - a^2 u_{xx} = 0$ is parabolic.

5. Consider the equation $9u_{xx} + 12u_{xy} + 4u_{yy} = 0$. We have $b^2 - 4ac = 0$, so it is parabolic. Let's reduce it to standard form:

$$9z^2 + 12z + 4 = 0$$

has the repeated root $z = -\frac{2}{3}$. We choose A and B to satisfy

$$\frac{A}{B} = -\frac{2}{3}$$

so the transformation

$$\xi = 2x - 3y, \quad \eta = x$$

reduces the PDE to canonical form

$$u_{\eta\eta} = 0$$

with solution

$$u = \eta\phi(\xi) + \psi(\xi)$$
$$= x\phi(2x - 3y) + \psi(2x - 3y).$$

In general, PDEs which model heat flow and other dissipative phenomena are parabolic. *The heat equation is a "standard example" of a parabolic equation.*

In Example 5, the lines $2x - 3y = constant$ play the role of characteristics of the PDE and, in general, the lines

$$\xi = constant$$

are the characteristics of a parabolic PDE. The lines $\eta = constant$ are *not* characteristic, for various reasons, one of which is the fact that our choice of η is arbitrary (with minimal restriction). In fact, in many parabolic problems, the line $\eta = 0$ will correspond to the initial line $t = 0$. Thus,

Parabolic equations have one family of characteristics.

Case 3: $b^2 - 4ac < 0$

In this case the quadratic has no real roots and there is no transformation

$$\xi = Ax + By$$
$$\eta = Cx + Dy$$

(with A, B, C and D real) that will eliminate the $u_{\xi\xi}$ and $u_{\eta\eta}$ terms. Can we simplify the PDE, nonetheless? Let's try

$$\eta = x,$$

i.e.,

$$C = 1 \quad \text{and} \quad D = 0,$$

and see what that does to (5.17). We get

$$(A^2 a + ABb + B^2 c)u_{\xi\xi} + (2Aa + Bb)u_{\xi\eta} + au_{\eta\eta}.$$

Next, choosing $Bb = -2Aa$ eliminates the middle term, resulting in

$$\left(-A^2 a + \frac{4A^2 a^2 c}{b^2}\right)u_{\xi\xi} + au_{\eta\eta}$$

or

$$\frac{A^2 a}{b^2}(4ac - b^2)u_{\xi\xi} + au_{\eta\eta}.$$

(Here we have assumed that $b \neq 0$. If $b = 0$, things are much easier.) Then, choosing

$$A^2 = \frac{b^2}{4ac - b^2}$$

or

$$A = \frac{b}{\sqrt{4ac - b^2}}$$

reduces (5.17) to

$$a(u_{\xi\xi} + u_{\eta\eta}),$$

the Laplace operator! Therefore, the transformation

$$\xi = \frac{b}{\sqrt{4ac - b^2}}x - \frac{2a}{\sqrt{4ac - b^2}}y$$
$$\eta = x$$

reduces the PDE (5.15) to its canonical form

$$u_{\xi\xi} + u_{\eta\eta} + \textit{lower order terms} = 0.$$

Definition 5.4 *If $b^2 - 4ac < 0$, we say that PDE (5.15) is* **elliptic**.

Examples

6. The Laplace equation $u_{xx} + u_{yy} = 0$ is elliptic.

7. The equation $4u_{xx} - 5u_{xy} + 3u_{yy} = 0$ is elliptic.

8. The **Helmholtz equation** or **reduced wave equation** $u_{xx} + u_{yy} + \lambda u = 0$ is very important in the study of wave motion and shows up in the fields of elasticity theory, electricity and magnetism, acoustics and quantum mechanics. It is elliptic.

Elliptic PDEs arise in steady state problems and in other problems involving two (or more) space dimensions in which the time behavior is neglected or has been separated from the original PDE. *The Laplace equation is a "standard example"* of an elliptic equation.

Unfortunately, there are no characteristic coordinates, like (5.19) and (5.21), for elliptic equations, so elliptic equations cannot be integrated as we did to get (5.20) and (5.22). (More precisely, the transformation

$$\frac{A}{B} = \frac{-b + \sqrt{b^2 - 4ac}}{2a}, \quad \frac{C}{D} = \frac{-b - \sqrt{b^2 - 4ac}}{2a}$$

leads to a *complex* transformation and, hence, to *complex* characteristics—see Exercise 13.)

Elliptic equations have no (real) characteristics.

Now, what if the coefficients in (5.13) are not constant? Well, pretty much the same as what we've been doing, except that, since the quantity $b^2 - 4ac$ is now a function of x and y, it may have different signs in different parts of the x-y plane.

Example 9 What can we say about the PDE $u_{xx} + x u_{xy} + y^2 u_{yy} = 0$? We have

$$b^2 - 4ac = x^2 - 4y^2 = (x - 2y)(x + 2y),$$

so, looking at Figure 5.19, we see that the equation is hyperbolic in the region marked H, elliptic in the region marked E and parabolic along the boundary of these regions.

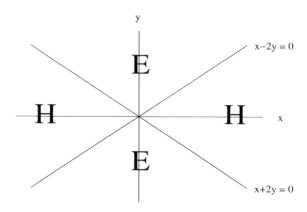

FIGURE 5.19
Regions where the PDE of Example 9 is hyperbolic (H) and elliptic (E).

Example 10 An important equation from hydrodynamics is the **Tricomi equation,** $yu_{xx} + u_{yy} = 0$. We have

$$b^2 - 4ac = -4y,$$

so this equation is hyperbolic in the lower half-plane elliptic in the upper half-plane, and parabolic along the x-axis.

Although we mentioned that each of these last two equations is parabolic along certain curves, in practice we are only interested in two-dimensional regions throughout which an equation is hyperbolic, parabolic or elliptic.

Exercises 5.5

In Exercises 1–6, determine if the PDE is hyperbolic, parabolic or elliptic. Then, transform the equation to canonical form. If hyperbolic or parabolic, find its solution.

1. $u_{xx} + 8u_{xy} + 16u_{yy} = 0$

2. $21u_{xx} - 10u_{xy} + u_{yy} = 0$

3. $6u_{xx} - u_{xy} - u_{yy} = 0$

4. $25u_{xx} - 30u_{xy} + 9u_{yy} = 0$

5. $u_{xx} + 4u_{xy} + 5u_{yy} = 0$

6. $2u_{xx} + 5u_{xy} + u_{yy} = 0$

7. Solve $u_{xx} - u_{xy} - 6u_{yy} = 0$, $u(0, y) = \sin 3y$, $u_x(0, y) = \sin 2y$.

8. Solve $u_{xx} - 4u_{xy} + 4u_{yy} = 0$, $u(0, y) = y^2$, $u_x(0, y) = 1 - 3y$.

9. Derive formulas (5.16).

10. Prove that the classification of the PDE

$$au_{xx} + bu_{xy} + cu_{yy} + lower\ order\ terms = 0,$$

where a, b and c are constant, is invariant with respect to coordinate transformations

$$\xi = Ax + By, \quad \eta = Cx + Dy,$$

where A, B, C and D are constant. In other words, prove that the transformed equation

$$a_1 u_{\xi\xi} + b_1 u_{\xi\eta} + c_1 u_{\eta\eta} + \cdots = 0$$

has the same classification as the original PDE.

11. We showed that the hyperbolic PDE

$$au_{xx} + bu_{xy} + cu_{yy} = 0, a, b, c\ \text{constant},$$

can always be transformed to the PDE

$$u_{\xi\eta} = 0.$$

Show that the latter equation can always be transformed into the equation

$$u_{tt} - \kappa^2 u_{ss} = 0, \kappa\ \text{constant}$$

and, therefore, that the original also can be put in this form. In other words, the original PDE is just the wave equation in disguise and *the equation $u_{\xi\xi} - \kappa^2 u_{\eta\eta} +$ lower order terms $= 0$ is an alternate canonical form for any hyperbolic equation.*

12. What is the graph, in the x-y plane, of the equation

$$ax^2 + bxy + cy^2 = 0$$

when $b^2 - 4ac > 0$?
when $b^2 - 4ac = 0$?
when $b^2 - 4ac < 0$?

13. Show that, by way of the transformation

$$\xi = x + iy, \quad \eta = x - iy,$$

the Laplace equation has solution

$$u = \phi(x + iy) + \psi(x - iy).$$

We say that the "lines" $x + iy = constant$ and $x - iy = constant$ are the *complex* characteristics of the Laplace equation. More generally, for any elliptic equation

$$au_{xx} + bu_{xy} + cu_{yy} = 0,$$

show that we may use a complex version of equations (5.18) which reduces the PDE to

$$u_{\xi\eta} = 0.$$

(Hint: If $x < 0$, then $\sqrt{x} = \sqrt{(-1)(-x)} = i\sqrt{-x}$, where i, as always, has the property that $i^2 = -1$.)

Find and sketch the regions in the x-y plane where the PDE is hyperbolic, parabolic or elliptic.

14. $xu_{xx} - 4u_{xy} + 2yu_{yy} = 0$

15. $u_{xx} + y^2 u_{yy} + u = 0$

16. $xu_{xx} + 2u_{xy} + xu_{yy} - xu_x + x^2 yu_y = \sin y$

Prelude to Chapter 6

Although the previous chapter was somewhat of a detour from our Fourier path, we did meet PDEs which were to be solved on unbounded intervals. There, we had no problem solving the infinite and semi-infinite string problems, but what happens with *parabolic* or *elliptic* equations on such domains?

Specifically, we ask whether we can extend the idea of the Fourier series to functions on unbounded domains, and we see that an affirmative answer was given by Fourier, Cauchy and Poisson in the second decade of the 19th century. We now call this representation the *Fourier integral*, although all three seemed to have discovered it more or less independently. (Of course, they were all in Paris at the time and knew each other—the idea was *in the air*.)

The Fourier integral immediately gives us the *Fourier transform* which, like the *Laplace transform* (introduced by Laplace in 1782), turns ODEs into algebraic equations (and, as we'll see, PDEs into ODEs). In turn, we use the Fourier transform to solve the infinite and semi-infinite heat equation. Further, it gives us the perfect opportunity to introduce the theory of *generalized functions* or *distributions*, which began its life as the *operational calculus* of the British scientist Oliver Heaviside (1850–1925) and, ultimately, was given a firm mathematical footing around 1950, in the setting of Fourier transforms.

6

Integral Transforms

6.1   The Laplace Transform for PDEs

In Chapter 4, we used separation of variables and Fourier series to solve PDEs on *finite* intervals. Then, in Sections 5.3 and 5.4, we solved the wave equation on *unbounded* intervals using the method of characteristics. An obvious question is, how do we solve more general PDEs on unbounded intervals?

We assume that the reader has seen the Laplace transform in connection with the solution of ODEs (specifically, with initial-value problems). It turns out that these and other *integral transforms* play a crucial role in the study of PDEs, on both bounded and unbounded domains.

Since we mean to concentrate on the *Fourier* transform, the natural extension of the Fourier series to functions on unbounded domains, we use the Laplace transform only as a brief introduction to the application of integral transforms to PDEs.

Actually, to get an idea of how transforms work, one need only look at the logarithm function. Back in the old days, before computers and calculators were available, mathematicians made extensive *log tables*. Then, supposing one of these mathematicians wanted to do a quick calculation of the product of two numbers, a and b, she would start by writing

$$P = ab.$$

Then, she would take the log of both sides:

$$\log_{10} P = \log_{10} ab,$$

the point being to use the property of logs:

$$\log_{10} ab = \log_{10} a + \log_{10} b.$$

Thus, *transforming* both sides of the equation would result in a multiplication problem being *transformed* into an (easier) addition problem. Next, she would look up the values of $\log_{10} a$ and $\log_{10} b$ and add them, resulting in

$$\log_{10} P = C.$$

The last step is to *transform back*, that is, to find in the table the number P whose log is C.

Now, let's define the Laplace transform and give some of its properties, for functions $u(x, t)$. We write

$$\mathcal{L}[u(x,t)] = \int_0^\infty e^{-st}u(x,t)dt = U(x,s), \quad s > 0.$$

Sufficient conditions for $U(x, s)$ to exist are that, for each x,

1) $u(x, t)$ is piecewise continuous on any interval $0 \le t \le T$.

2) $u(x, t)$ is of exponential order as $t \to \infty$, i.e., there exist constants α and M such that $u(x, t)$ does not grow more rapidly than $Me^{\alpha t}$ as $t \to \infty$. (We write $u = O(e^{\alpha t})$, and we say that u is "big-oh" of $e^{\alpha t}$.)

Now, when we applied the Laplace transform to ODEs, in order to *transform back* to the solution we needed to have at our disposal a table of *inverse transforms*. (Actually, there is an integral formula that gives the inverse transform for "any" function $F(s)$. However, we need to know some complex analysis in order to evaluate this integral.*) We provide a short table of these inverse transforms before the exercises (Table 6.1).

Let's look at two examples involving the temperature distribution in a *semi-infinite rod*. In the process, we'll introduce two important functions often arising in problems in applied mathematics, and then, in the second example, we'll use a special case of what is known as *Duhamel's Principle*.

Example 1

$$u_t = u_{xx}, \quad x > 0, t > 0,$$
$$u(x,0) = 0, \quad x > 0,$$
$$u(0,t) = 1, \quad t > 0,$$
$$\lim_{x \to \infty} u(x,t) = 0, \quad t > 0.$$

Here, u represents the temperature of a very long rod which, initially, is at a temperature of $0°$ and for which, at time $t = 0$, the one end that is "near" us is raised to, and held thereafter at, $1°$. Note that we require $u \to 0$ as $x \to \infty$.

We transform the PDE and, using the initial condition and Property 1 in Table 6.1, we get

$$sU(x, s) = U_{xx}(x, s).^\dagger$$

---

*Due to Poisson. See, e.g., Ruel V. Churchill's excellent text, *Operational Mathematics*, for a detailed treatment of the Laplace transform.

$\dagger$So long as $\frac{\partial^2}{\partial x^2} \int_0^\infty e^{-st}u(x,t)dt = \int_0^\infty e^{-st}u_{xx}(x,t)dt.$

This equation can be treated as an ODE and has general solution

$$U(x,s) = C_1(s)e^{x\sqrt{s}} + C_2(s)e^{-x\sqrt{s}}.$$

Next, let's apply the "boundary condition at $x = \infty$." We know, via hindsight, that if we take the inverse transform of the first term, we will get a function that grows without bound as $t \to \infty$. Therefore, we must have $C_1(s) = 0$. What we have done, essentially, is *formally* to transform the "boundary condition at $x = \infty$" to $\lim_{x\to\infty} U(x,s) = 0$. Then, applying the other boundary condition, we have

$$U(0,s) = \mathcal{L}[u(0,t)] = \mathcal{L}[1] = \frac{1}{s} = C_2(s),$$

i.e.,

$$U(x,s) = \frac{1}{s}e^{-x\sqrt{s}}.$$

Finally, transforming back (using Property 13 in Table 6.1), our solution is

$$u(x,t) = \frac{2}{\sqrt{\pi}} \int_{\frac{x}{2\sqrt{t}}}^{\infty} e^{-z^2}\, dz.$$

Actually, this is a well-known function. First, let us look at a very important function, the so-called **error function** or **probability integral**

$$\mathrm{erf}(x) = \frac{2}{\sqrt{\pi}} \int_0^x e^{-z^2}\, dz.$$

The latter name is based on the fact that the function

$$f(x) = \frac{1}{\sqrt{\pi}}e^{-x^2}$$

is related to the *normal density function*, or *Gaussian*, from probability (the famous *bell-shaped curve*). As such, we must have $\mathrm{erf}(\infty) = \frac{2}{\sqrt{\pi}}\int_0^\infty e^{-z^2}dz = \frac{1}{\sqrt{\pi}}\int_{-\infty}^\infty e^{-z^2}dz = 1$ (see Exercise 6).
Then, the function

$$\mathrm{erfc}(x) = 1 - \mathrm{erf}(x) = \frac{2}{\sqrt{\pi}} \int_x^\infty e^{-z^2}\, dz$$

is called the **complementary error function**. So the solution to our problem is

$$u(x,t) = \mathrm{erfc}\left(\frac{x}{2\sqrt{t}}\right).$$

We graph both functions in Figure 6.1.

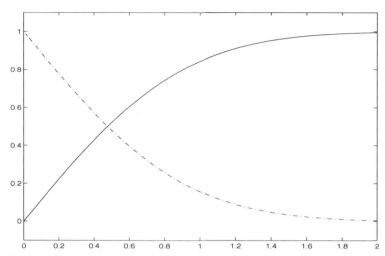

FIGURE 6.1
MATLAB plots of the error function (solid) and the complementary error function (dashed).

Now let's change the previous example slightly by considering, instead, a left end condition which varies with time.

Example 2

$$w_t = w_{xx}$$
$$w(x,0) = 0$$
$$w(0,t) = f(t)$$
$$\lim_{x \to \infty} w(x,t) = 0.$$

The solution is identical to that in Example 1, until we get to the left end boundary condition:

$$W(0,s) = \mathcal{L}[w(0,t)] = \mathcal{L}[f(t)] = F(s) = C_2(s).$$

In this case, we have

$$W(x,s) = \mathcal{L}[w(x,t)] = F(s)e^{-x\sqrt{s}}.$$

Rather than performing the inverse transform directly, we, instead, try to use the solution $U(x,s)$ from Example 1:

$$W(x,s) = F(s)e^{-x\sqrt{s}} = sF(s)\frac{1}{s}e^{-x\sqrt{s}}$$
$$= F(s) \cdot sU(x,s).$$

Now, from Property 1 in Table 6.1, $\mathcal{L}[u_t(x,t)] = sU(x,s) - u(x,0) = sU(x,s)$. Then, from Property 3, our solution is the (**finite**) **convolution** of $f(t)$ and $u_t(x,t)$, that is,

$$w(x,t) = u_t(x,t) * f(t) = \int_0^t f(t-\tau)u_t(x,\tau)d\tau.$$

This method of solution is a special case of what is known as **Duhamel's[‡] Principle**, which, in some cases, involves relating the solution of problems with variable boundary or initial conditions to those with constant boundary or initial conditions. (We'll meet a more important version of this principle in Section 10.5.)

Before getting to the exercises, one may ask why we chose to transform the t, and not the x, in these problems—since both t and x have domain $[0,\infty)$. It's a good exercise to try to do these problems by transforming x, instead. You'll see why we generally choose to transform t when using the Laplace transform.

Finally, most of the exercises will be solved *formally*. We won't be able to check our answers until we've covered Section 6.5

Suppose $\mathcal{L}[g(t)] = G(s)$ and $\mathcal{L}[h(t)] = H(s)$. We have the Laplace transform formulas in Table 6.1.

Exercises 6.1

1. Use Laplace transforms to solve the (unrealistic!) convection problem

   a) $u_t + 2u_x = 0$, $x > 0, t > 0$,
      $u(x,0) = 3$,
      $u(0,t) = 5$.

   b) $u_t + (1 + x^2)u_x = 0$, $x > 0, t > 0$,
      $u(x,0) = 0$,
      $u(0,t) = 1$.

2. Use Laplace transforms to solve the following heat problem:

   a) $u_t = u_{xx}$, $x > 0, t > 0$,
      $u(x,0) = 10e^{-x}$,
      $u(0,t) = 0$,
      $\lim_{x \to \infty} u(x,t) = 0$.

   b) $u_t = u_{xx}$,
      $u(x,0) = 10e^{-x}$,
      $u(0,t) = 10$,
      $\lim_{x \to \infty} u(x,t) = 0$.

[‡] Jean Marie Constant Duhamel (1797–1872).

c) $u_t = u_{xx}$,
   $u(x,0) = 10e^{-x}$,
   $u(0,t) = f(t)$,
   $\lim_{x\to\infty} u(x,t) = 0$.

$f(t) = \mathcal{L}^{-1}[F(s)]$		$F(s) = \mathcal{L}[f(t)]$
1.	$g^{(n)}(t)$	1.	$s^n G(s) - s^{n-1}g(0) - \cdots$ $-sg^{(n-2)}(0) - g^{(n-1)}(0)$
2.	$\int_0^t g(\tau)d\tau$	2.	$G(s)/s$
3.	$g(t)*h(t) = \int_0^t g(\tau)h(t-\tau)d\tau$	3.	$G(s)H(s)$
4.	t^n	4.	$n!/s^{n+1}$
5.	e^{at}	5.	$\frac{1}{s-a}$
6.	$\sin at$	6.	$\frac{a}{s^2+a^2}$
7.	$\cos at$	7.	$\frac{s}{s^2+a^2}$
8.	$e^{at}g(t)$	8.	$G(s-a)$
9.	$H(t-a) = \begin{cases} 1, & \text{if } t \geq a^* \\ 0, & \text{if } t < a \end{cases}$	9.	e^{-as}/s
10.	$f(t) = \begin{cases} g(t-a), & \text{if } t \geq a^* \\ 0, & \text{if } t < a \end{cases}$ $= H(t-a)g(t-a)$	10.	$e^{-as}G(s)$
11.	$\frac{1}{\sqrt{\pi t}}e^{-a^2/4t}$	11.	$e^{-a\sqrt{s}}$
12.	$\mathrm{erf}(t/2a)^{**}$	12.	$e^{a^2 s^2}\mathrm{erfc}(as)/s$
13.	$\mathrm{erfc}(a/2\sqrt{t})^{**}$	13.	$e^{-a\sqrt{s}}/s$

*H is the well-known Heaviside function, named after Oliver Heaviside.
**Again, $\mathrm{erf}(x) = \frac{2}{\sqrt{\pi}}\int_0^x e^{-z^2}\,dz$ and $\mathrm{erfc}(x) = 1 - \mathrm{erf}(x)$.

TABLE 6.1
Laplace transforms.

3. Use Laplace transforms to solve the semi-infinite wave problem

a) $u_{tt} = c^2 u_{xx}$, $x > 0, t > 0$,
   $u(x,0) = u_t(x,0) = 0$,
   $u(0,t) = f(t)$,
   $\lim_{x\to\infty} u(x,t) = 0$.

b) $u_{tt} = c^2 u_{xx} - g$,
   $u(x,0) = u_t(x,0) = 0$,
   $u(0,t) = 0$,
   $\lim_{x\to\infty} u_x(x,t) = 0$.

This, of course, is the equation for a string which is tied down at one end and which falls under its own weight (see equation (2.8), Section 2.3). The constant g is the gravitational constant. (Note the difference between the limit condition here and the one above.)

c) In part (b), replace the constant gravitational force with a force $F(t)$ (units of force per unit mass of string), so that the PDE becomes

$$u_{tt} = c^2 u_{xx} - F(t).$$

Show that the solution is $u(x,t) = H\left(t - \frac{x}{c}\right) G\left(t - \frac{x}{c}\right) - G(t)$, where

$$G(t) = \int_0^t \int_0^\tau F(z)dz \, d\tau.$$

4. Show formally that the solutions in Examples 1 and 2 actually do satisfy the PDE and side conditions.

5. Prove that the finite convolution is commutative, that is, that $f * g = g * f$. (Formula 3 in Table 6.1.)

6. If you haven't done so in multivariable calculus, show that $\text{erf}(0) = 0$ and $\text{erf}(\infty) = 1$. (Hint: If $I = \int_{-\infty}^\infty e^{-x^2} dx$, then $I^2 = \left(\int_{-\infty}^\infty e^{-x^2} dx\right)\left(\int_{-\infty}^\infty e^{-y^2} dy\right)$. Rewrite I^2 as a double integral, and change to polar coordinates.)

7. a) Prove Formula 2 in Table 6.1: $\mathcal{L}\left[\int_0^t f(\tau)d\tau\right] = \frac{1}{s}\mathcal{L}[f(t)]$.

   b) Prove Formula 9 in Table 6.1: $\mathcal{L}[H(t-a)] = e^{-as}/s$.

   c) Prove Formula 10 in Table 6.1.

8. We also may use Laplace transforms to solve problems on finite x-domains, as we did earlier using Fourier series. Use the Laplace transform to solve

   a) $u_t = 3u_{xx}$,
   $u(x,0) = 17\sin\pi x$,
   $u(0,t) = u(4,t) = 0$. (This is Example 3, Section 2.6.)

   b) $u_{tt} = 4u_{xx}$,
   $u(x,0) = 5\sin 2x - 7\sin 4x$,
   $u_t(x,0) = 0$,
   $u(0,t) = u(\pi,t) = 0$. (This is Exercise 9, Section 2.6.)

9. a) Use Laplace transforms to solve
   $u_t = u_{xx}$, $\quad x > 0, t > 0$,
   $u(x,0) = 0$,
   $u_x(0,t) = 1$,
   $\lim_{x\to\infty} u(x,t) = 0$.

b) Use part (a) and Duhamel's Principle to solve

$$w_t = w_{xx}, \quad x > 0, t > 0,$$
$$w(x,0) = 0,$$
$$w_x(0,t) = g(t),$$
$$\lim_{x \to \infty} w(x,t) = 0.$$

(Note: You need not use Duhamel—try to solve it directly, too, and make sure you get the same answer.)

10. **MATLAB:** For each problem,

    i) Plot snapshots of the solution in the *x-u* plane for various time *t*.

    ii) Plot the solution in *x-t-u* space.

    a) Exercise 1a

    b) Exercise 2a

    c) Exercise 2b

    d) Exercise 9a

6.2 Fourier Sine and Cosine Transforms

What has all this to do with Fourier series? Well, Laplace transforming an ODE or PDE can be looked at as multiplying both sides of the equation by e^{-st} and then integrating. Similarly, solving a PDE by Fourier series *can* be made to look like multiplying both sides of the equation by the appropriate trigonometric function and integrating. In fact, the various Fourier coefficients are referred to as *finite Fourier transforms* (while the inverse transform would, in a sense, be the corresponding Fourier series).

As the Laplace transform involves functions on $[0, \infty)$, it is not unnatural to ask if we can extend Fourier's idea to functions on unbounded domains. The answer, of course, is yes (or we wouldn't have asked the question!). So, remember that any piecewise smooth function on a finite domain, or any piecewise smooth periodic function, can be expanded in a Fourier series. Now, what if f is neither? For definiteness, suppose $f(x)$ has domain $[0, \infty)$ and is not periodic. For any $L > 0$, we have

$$f(x) = F_s(x) = \sum_{n=1}^{\infty} b_n \sin \frac{n\pi x}{L} \quad \text{on} \quad 0 \le x \le L, \tag{6.1}$$

where

$$b_n = \frac{2}{L} \int_0^L f(x) \sin \frac{n\pi x}{L} \, dx$$

and, of course, $F_s(x)$ is the Fourier sine series for f on $[0, L]$. Since this statement is *not* true on (L, ∞), let's see what happens as $L \to \infty$. Rewriting (6.1), we have

$$f(x) = \sum_{n=1}^{\infty} \frac{2}{L} \int_0^L f(z) \sin \frac{n\pi z}{L} \, dz \sin \frac{n\pi x}{L}.$$

Since $\frac{2}{L} \to 0$ as $L \to \infty$, this looks a little like a Riemann sum. If we let $\Delta\alpha = \frac{\pi}{L}$, we then have

$$f(x) = \frac{2}{\pi} \sum_{n=1}^{\infty} \int_0^{\pi/\Delta\alpha} f(z) \sin(nz\Delta\alpha) \, dz \sin(nx\Delta\alpha)\Delta\alpha, \qquad (6.2)$$

which looks a lot more like a Riemann sum—specifically, we've broken up the nonnegative α-axis into pieces of length $\Delta\alpha$, and the function inside is evaluated at each grid point $n\Delta\alpha$. Letting $\Delta\alpha \to 0$, it should not be *too* surprising (although we must be very careful with that $\pi/\Delta\alpha$ term) that (6.2) becomes

$$f(x) = \frac{2}{\pi} \int_0^{\infty} \int_0^{\infty} f(z) \sin \alpha z \, dz \sin \alpha x \, d\alpha \qquad (6.3)$$

or

$$f(x) = \int_0^{\infty} F_s(\alpha) \sin \alpha x \, d\alpha \qquad (6.4)$$

with

$$F_s(\alpha) = \frac{2}{\pi} \int_0^{\infty} f(x) \sin \alpha x \, dx. \qquad (6.5)$$

Note the beautiful symmetry! Of course, what we have done *is not a proof*.

Equation (6.3) is called the **Fourier sine integral formula** for $f(x)$; (6.5) is the **Fourier sine transform** of f, while (6.4) is the **inverse transform** or the **Fourier sine inversion formula**. We often will write

$$F_s(\alpha) = \mathcal{F}_s[f(x)], \quad f(x) = \mathcal{F}_s^{-1}[F_s(\alpha)],$$

similar to the notation for the Laplace transform.

One also can show that

$$f(x) = \int_0^{\infty} F_c(\alpha) \cos \alpha x \, d\alpha = \mathcal{F}_c^{-1}[F_c(\alpha)], \qquad (6.6)$$

where

$$F_c(\alpha) = \frac{2}{\pi} \int_0^{\infty} f(x) \cos \alpha x \, dx = \mathcal{F}_c[f(x)]. \qquad (6.7)$$

These, of course, are the **Fourier cosine transform** and **inverse (Fourier cosine) transform**. Note that the placement of the $\frac{2}{\pi}$ is arbitrary; for example, we could, instead, have chosen to write

$$f(x) = \frac{2}{\pi} \int_0^{\infty} F_s(\alpha) \sin \alpha x \, d\alpha, \quad F_s(\alpha) = \int_0^{\infty} f(x) \sin \alpha x \, d\alpha$$

or, to make the symmetry perfect,

$$f(x) = \sqrt{\frac{2}{\pi}} \int_0^\infty F_s(\alpha) \sin \alpha x \, d\alpha, \quad F_s(\alpha) = \sqrt{\frac{2}{\pi}} \int_0^\infty f(x) \sin \alpha x \, d\alpha.$$

It should not be surprising that, analogous to the *Fourier* series of a function on $-L \leq x \leq L$, we have the *Fourier transform* of a function on $-\infty < x < \infty$. We derive the trigonometric form of this Fourier transform in Exercise 10. However, it is *much* more convenient to deal with the *complex* form of the transform (analogous to the complex form of the Fourier series) for which we wait until the next section.

Now, what properties must f possess in order that its Fourier sine and cosine transforms exist and, in each case, for which values of x will we get (6.3) and (6.6)? We have the following theorem, whose proof will be given in Section 6.6.

Theorem 6.1 *Suppose $f(x)$ is piecewise smooth on every interval $[0, L]$, and suppose that $\int_0^\infty |f(x)|dx$ is finite (we say that f is **absolutely integrable**). Then,*

$$\int_0^\infty F_s(\alpha) \sin \alpha x \, d\alpha = \frac{1}{2}[f(x+) + f(x-)]$$

and

$$\int_0^\infty F_c(\alpha) \cos \alpha x \, d\alpha = \frac{1}{2}[f(x+) + f(x-)]$$

for each $x > 0$. (Also, when $x=0$, the first integral is 0, while the second is $f(0+)$.)

At this point, we must mention that, in some cases, these integrals may not converge. We will give a precise treatment in the final section of this chapter, where we prove these *Fourier inversion theorems*. For now, we'll continue to use the notation $\int_0^\infty$ (and, later, $\int_{-\infty}^\infty$), realizing that it may not, technically, be correct.

Also, in practice, it turns out that the condition that f be absolutely integrable is too strict—there are many situations where this is not the case, but where it is useful to be able to talk about f's Fourier transform, anyway.

Therefore, until we get to Section 6.6, we'll take a purely *formal* approach and not concern ourselves with problems of convergence.

Before deciding for which functions f these statements are valid, let's compute some Fourier sine and cosine transforms.

Example 1 Compute the Fourier sine and cosine transforms of $f(x) = e^{-cx}$.

First, $\int_0^\infty e^{-cx}\,dx = \frac{1}{c}$ (why?), so f *is* absolutely integrable. We compute the transforms directly (although there's a simple and more elegant way to do the problem—see Exercise 1):

$$F_s(\alpha) = \frac{2}{\pi}\int_0^\infty e^{-cx}\sin\alpha x\,dx.$$

Integrating by parts twice gives us

$$\frac{2}{\pi}I = \frac{2}{\pi}\left\{ -\frac{1}{\alpha}e^{-cx}\cos\alpha x\Big|_0^\infty - \frac{c}{\alpha^2}e^{-cx}\sin\alpha x\Big|_0^\infty - \frac{c^2}{\alpha^2}I \right\}$$

so

$$F_s(\alpha) = \frac{2}{\pi}\frac{\alpha}{\alpha^2+c^2}.$$

Similarly,

$$F_c(\alpha) = \frac{2}{\pi}\int_0^\infty e^{-cx}\cos\alpha x\,dx = \frac{2}{\pi}\frac{c}{\alpha^2+c^2}.$$

Note that Theorem 6.1 then tells us that

$$e^{-cx} = \frac{2}{\pi}\int_0^\infty \frac{\alpha}{\alpha^2+c^2}\sin\alpha x\,d\alpha = \frac{2}{\pi}\int_0^\infty \frac{c}{\alpha^2+c^2}\cos\alpha x\,d\alpha.$$

Example 2 Compute $\mathcal{F}_s\left(\frac{x}{x^2+c^2}\right)$ and $\mathcal{F}_c\left(\frac{1}{x^2+c^2}\right)$.

It appears that we must compute the difficult integrals

$$\mathcal{F}_c\left(\frac{x}{x^2+c^2}\right) = \frac{2}{\pi}\int_0^\infty \frac{x\sin\alpha x}{x^2+c^2}\,dx$$

and

$$\mathcal{F}_c\left(\frac{1}{x^2+c^2}\right) = \frac{2}{\pi}\int_0^\infty \frac{\cos\alpha x}{x^2+c^2}\,dx.$$

However, notice that we've already done them in the previous example! Therefore,

$$\mathcal{F}_s\left(\frac{x}{x^2+c^2}\right) = e^{-c\alpha}$$

$$\mathcal{F}_c\left(\frac{1}{x^2+c^2}\right) = \frac{1}{c}e^{-c\alpha}.$$

(By the way, note that $f(x) = \frac{x}{x^2+c^2}$ is *not* absolutely integrable.)

What about transforms of derivatives? Suppose that $\mathcal{F}_s[f(x)] = F_s(\alpha)$ and $\mathcal{F}_c[f(x)] = F_c(\alpha)$. Then it is easy to show (see Exercise 3) that

$$\mathcal{F}_s[f'(x)] = -\alpha F_c(\alpha)$$

and

$$\mathcal{F}_c[f'(x)] = \alpha F_s(\alpha) - \frac{2}{\pi} f(0),^{\S}$$

from which it follows that

$$\mathcal{F}_s[f''(x)] = -\alpha^2 F_s(\alpha) + \frac{2}{\pi} f(0)\alpha$$

and

$$\mathcal{F}_c[f''(x)] = -\alpha^2 F_c(\alpha) - \frac{2}{\pi} f'(0)\alpha.$$

As with the Laplace transform, these transforms essentially turn differentiation into multiplication, thereby allowing us to turn PDEs into ODEs, and ODEs into algebraic equations. Let's look at an example of the latter.

Example 3 Solve the boundary-value problem

$$y'' - y = e^{-2x}, \quad 0 \le x < \infty,$$
$$y(0) = 1,$$
$$\lim_{x \to \infty} y(x) = 0.$$

The presence of $y(0)$ suggests we use the sine transform. Letting $F_s(\alpha) = \mathcal{F}_s[y(x)]$ and transforming the ODE, we have

$$-\alpha^2 F_s(\alpha) + \frac{2}{\pi}\alpha - F_s(\alpha) = \frac{2}{\pi} \frac{\alpha}{\alpha^2 + 4}$$

or, after using partial fractions,

$$F_s(\alpha) = \frac{4}{3\pi} \frac{\alpha}{\alpha^2 + 1} + \frac{2}{3\pi} \frac{\alpha}{\alpha^2 + 4}.$$

The inverse transform, then, is

$$y = \frac{2}{3} e^{-x} + \frac{1}{3} e^{-2x}.$$

It is interesting to note that there are infinitely many solutions to $y'' - y = e^{-2x}$, $y(0) = 1$ (the reader should find some), but the method we used seemed to find the only *bounded* solution (note the limit "boundary condition").

---

§ Since f (or, below, f') may not be continuous at $x = 0$, we really should have $f(0+)$ and $f'(0+)$, respectively.

Of course, we would like to solve not ODEs, but PDEs. It turns out that it is easiest to use the Fourier transform, rather than the sine and cosine transforms, both for problems on $-\infty < x < \infty$ *and* for many on $0 \le x < \infty$. We'll look at the semi-infinite heat equation below, then we'll solve it again in Section 6.4 using the Fourier transform.

Here, as with the Laplace transform, we run into the question of which variable to transform. It turns out that, due to the nature of the Fourier transforms and the problems we solve with them, it is the *space* variable(s) that is transformed, instead of time.

Okay, let's solve the heat equation for a semi-infinite rod. Specifically, let's start with the problem in Example 4.

Example 4

$$u_t = u_{xx}, \quad x > 0, t > 0,$$
$$u(x,0) = f(x), x > 0,$$
$$u(0,t) = 0, t > 0,$$
$$\lim_{x \to \infty} u(x,t) = 0.$$

We choose to transform x and, since we're given $u(0,t)$ (as opposed to $u_x(0,t)$), we use the sine transform. Specifically, we define

$$U(\alpha,t) = \mathcal{F}_s[u(x,t)]$$
$$= \frac{2}{\pi} \int_0^\infty u(x,t) \sin \alpha x \, dx, \text{ for each } t.$$

As with Laplace transforms, we will need the property

$$\frac{\partial}{\partial t} U(\alpha,t) = \frac{2}{\pi} \int_0^\infty u_t(x,t) \sin \alpha x \, dx.$$

Then, the transformed PDE is

$$U_t = -\alpha^2 U + \frac{2}{\pi} u(0,t)\alpha$$

or

$$U_t + \alpha^2 U = 0,$$

with solution
$$U(\alpha,t) = e^{-\alpha^2 t} G(\alpha), G \text{ arbitrary.}$$

Then, the initial condition gives

$$U(\alpha,0) = F_s(\alpha)$$

which implies that

$$U(\alpha,t) = e^{-\alpha^2 t} F_s(\alpha).$$

Therefore, we may transform back:

$$u(x,t) = \int_0^\infty e^{-\alpha^2 t} F_s(\alpha) \sin \alpha x \, d\alpha$$

$$= \int_0^\infty e^{-\alpha^2 t} \frac{2}{\pi} \int_0^\infty f(z) \sin \alpha z \, dz \sin \alpha x \, d\alpha.$$

This is our not-so-satisfying solution. However, let's switch the order of integration (if it's actually allowed! More, later.):

$$u(x,t) = \frac{2}{\pi} \int_0^\infty f(z) \int_0^\infty e^{-\alpha^2 t} \sin \alpha z \sin \alpha x \, d\alpha dz$$

$$= \frac{1}{\pi} \int_0^\infty f(z) \int_0^\infty e^{-\alpha^2 t} [\cos \alpha(z-x) - \cos \alpha(z+x)] d\alpha dz.$$

Now, we'll show in the next section that

$$\int_0^\infty e^{-cx^2} \cos cx \, dx = \frac{\sqrt{\pi}}{2\sqrt{k}} e^{-\frac{c^2}{4k}},$$

so we can rewrite u as

$$u(x,t) = \frac{1}{2\sqrt{\pi t}} \int_0^\infty f(z) \left[e^{-\frac{(z-x)^2}{4t}} - e^{-\frac{(z+x)^2}{4t}} \right] dz.$$

So it seems that the sine transform is the transform of choice for the problem with Dirichlet boundary condition. We hope it's relatively obvious that the cosine transform is appropriate for the Neumann condition (see Exercise 5). (However, we also may solve these problems using the full Fourier transform and the *method of images*, which we do in Section 6.4.)

Exercises 6.2

1. Derive the Fourier sine and cosine transforms of $f(x) = e^{-cx}$ by writing $e^{i\alpha x} = \cos \alpha x + i \sin \alpha x$ and computing the integral $\int_0^\infty e^{-cx} e^{i\alpha x} dx$.

2. Find the Fourier sine and cosine transforms of $f(x) = xe^{-x}$.

3. a) Derive the formulas

$$\mathcal{F}_s[f'(x)] = -\alpha F_c(\alpha)$$

and

$$\mathcal{F}_c[f'(x)] = \alpha F_s(\alpha) - \frac{2}{\pi} f(0+).$$

b) Verify the formulas for $\mathcal{F}_s[f''(x)]$ and $\mathcal{F}_c[f''(x)]$, as well.

c) Show that

$$\mathcal{F}_s[f^{(4)}(x)] = \alpha^4 F_s(\alpha) - \frac{2}{\pi}\alpha^3 f(0) + \frac{2}{\pi}\alpha f''(0)$$

and

$$\mathcal{F}_c[f^{(4)}(x)] = \alpha^4 F_c(\alpha) + \frac{2}{\pi}\alpha^2 f'(0) - \frac{2}{\pi}f'''(0).$$

4. Solve the boundary-value problem.

a) $y'' - y = 3e^{-4x}$,
   $y'(0) = 0$,
   $\lim\limits_{x\to\infty} y(x) = 0$.

b) $y'' - 3y = e^{-x}$,
   $y(0) = 4$,
   $\lim\limits_{x\to\infty} y(x) = 0$.

c) $y^{(4)} - 5y'' + 4y = 3e^{-3x}$,
   $y(0) = 0$,
   $y''(0) = 1$,
   $\lim\limits_{x\to\infty} y(x) = \lim\limits_{x\to\infty} y'(x) = 0$.

5. Solve the problem with Neumann BC,

$$u_t = u_{xx}, \quad x > 0, t > 0,$$
$$u(x,0) = f(x),$$
$$u_x(0,t) = 0,$$
$$\lim_{x\to\infty} u(x,t) = 0.$$

6. a) Use Fourier sine or cosine transforms to show that

$$u(x,t) = \frac{2}{\pi}\int_0^\infty \frac{1 - e^{-\alpha^2 t}}{\alpha}\sin\alpha x \, d\alpha$$

is the solution of Example 1 in Section 6.1, i.e., of

$$u_t = u_{xx},$$
$$u(x,0) = 0,$$
$$u_t(x,0) = 1,$$
$$\lim_{x\to\infty} u(x,t) = 0.$$

b) Conclude from part (a) that if we, indeed, have found *the* solution, then

$$\mathcal{F}_s\left[\operatorname{erfc}\frac{x}{2\sqrt{c}}\right] = \frac{2}{\pi}\frac{1 - e^{-c\alpha^2}}{\alpha}.$$

7. Derive the following properties of Fourier sine and cosine transforms.

   a) $\mathcal{F}_s[xf(x)] = -F_c'(\alpha)$

   b) $\mathcal{F}_c[xf(x)] = F_s'(\alpha)$

8. Do the same for the following properties:

   a) $\mathcal{F}_s[f(x)\cos kx] = \frac{1}{2}[F_s(\alpha + k) + F_s(\alpha - k)]$

   b) $\mathcal{F}_c[f(x)\sin kx] = \frac{1}{2}[F_c(\alpha + k) - F_c(\alpha - k)]$

   c) What would the corresponding formula be for

   $$\mathcal{F}_s[f(x)\sin kx] \quad \text{and} \quad \mathcal{F}_c[f(x)\cos kx]?$$

9. **Convolution and Fourier sine and cosine transforms:** We wish to derive the inverse Fourier transforms of a product, that is, if $\mathcal{F}_s[f(x)] = F_s(\alpha)$ and $\mathcal{F}_c[g(x)] = G_s(\alpha)$, we wish to find out how to inverse transform $F_s(\alpha)G_s(\alpha)$. To this end,

   a) Suppose that we're given $f(x)$ on $x \geq 0$, and let $f_1(x)$ be the odd extension of f to the interval $-\infty < x < \infty$. Show that

   $$\mathcal{F}_c\left\{\frac{1}{2}[f_1(x + k) - f_1(x - k)]\right\} = F_s(\alpha)\sin k\alpha$$

   for any constant k.

   b) Show that

   $$2F_s(\alpha)G_s(\alpha) = \int_0^\infty g(y)\int_0^\infty [f_1(x + y) - f_1(x - y)]\cos \alpha x \, dx dy$$

   and, therefore, that

   $$\mathcal{F}_c^{-1}[F_s(\alpha)G_s(\alpha)] = \int_0^\infty g(y)[f_1(x + y) - f_1(x - y)]dy.$$

   Thus, we have found out how to find the inverse *cosine* transform in this case. Similarly, one can show that

   $$\mathcal{F}_s^{-1}[F_s(\alpha)G_c(\alpha)] = \int_0^\infty g(y)[f_1(x + y) + f_1(x - y)]dy$$

   $$= \int_0^\infty f(y)[g(|x - y|) - g(x + y)]dy$$

   and

   $$\mathcal{F}_c^{-1}[F_c(\alpha)G_c(\alpha)] = \int_0^\infty g(y)[f(|x - y|) + f(x + y)]dy.$$

10. Given $f(x)$ piecewise continuous on any interval, and $\int_{-\infty}^{\infty} |f(x)| dx < \infty$, show that we can write

$$f(x) = \int_{-\infty}^{\infty} [A(\alpha) \cos \alpha x + B(\alpha) \sin \alpha x] d\alpha,$$

where

$$A(\alpha) = \frac{1}{\pi} \int_{-\infty}^{\infty} f(x) \cos \alpha x \ dx$$

and

$$B(\alpha) = \frac{1}{\pi} \int_{-\infty}^{\infty} f(x) \sin \alpha x \ dx.$$

(Hint: We can write $f(x) = \frac{f(x)+f(-x)}{2} + \frac{f(x)-f(-x)}{2}$. Then, one of those functions is even and the other is odd. For $x \geq 0$, use the Fourier cosine integral for the even one, and the Fourier sine integral for the odd one; then extend everything to the interval $-\infty < x < \infty$.)

11. **Parseval's equality for the Fourier transform**:

   a) Show formally that if the Fourier sine and cosine transforms of f exist, then

   $$\frac{2}{\pi} \int_0^{\infty} f^2(x) dx = \int_0^{\infty} F_s^2(\alpha) d\alpha = \int_0^{\infty} F_c^2(\alpha) d\alpha.$$

   (Hint: $\int_0^{\infty} f^2(x) dx = \int_0^{\infty} f(x) \int_0^{\infty} F_s(\alpha) \sin \alpha x \ d\alpha dx$; now switch the order of integration.)

   b) Similarly, show formally that if f is as in Exercise 10, then

   $$\frac{1}{\pi} \int_{-\infty}^{\infty} f^2(x) dx = \int_{-\infty}^{\infty} [A^2(\alpha) + B^2(\alpha)] d\alpha.$$

   Each of these is called **Parseval's equality** for the corresponding transforms; compare it to the version of Parseval's equality for Fourier series in Section 8.5 and, particularly, in Example 4 and Exercise 6 of that section.

12. **Sampling:** Given a function $f(x)$ on $x \geq 0$, we say that f is **band-limited** if its Fourier sine transform is 0 except on a finite interval. (Actually, this definition *usually* refers to functions on $-\infty < x < \infty$ and their exponential Fourier transform—see Exercise 16 in the following section.) Specifically, there exists a positive constant L such that

$F_s(\alpha) = 0$ outside of $0 \le \alpha \le L$. The least such $L^\P$ is called the **cut-off frequency** for f.

a) For convenience, suppose that the cutoff frequency for f is $L = \pi$. Then, we may expand $F_s(\alpha)$ in a Fourier sine series on $0 \le \alpha \le \pi$. Do this and conclude that

$$f(x) = \int_0^\pi \sin \alpha x \left[\sum_{n=1}^\infty b_n \sin n\alpha\right] d\alpha,$$

where

$$b_n = \frac{2}{\pi} \int_0^\pi F_s(\alpha) \sin n\alpha \, d\alpha, \qquad n = 1, 2, 3, \ldots .$$

b) Show that we actually have

$$b_n = \frac{2}{\pi} f(n), \qquad n = 1, 2, 3, \ldots !$$

c) Conclude that

$$f(x) = \frac{2}{\pi} \sum_{n=1}^\infty f(n) \int_0^\pi \sin \alpha x \sin n\alpha \, d\alpha$$

and, thus, that if we know beforehand that f is band-limited, then we can construct f by knowing *only* its values on the positive integers! This result is known as the **Sampling Theorem**.

6.3   The Fourier Transform

One either can define the Fourier sine and cosine transforms and then derive from them the Fourier transform for functions on $-\infty < x < \infty$ (as we do here), or go the other way around (as we do in the exercises, and as we did with Fourier series). Our procedure is quite similar, not surprisingly, to what we did in Exercise 14 of Section 3.6.

So, suppose we're given a piecewise smooth function $f(x)$ with domain $-\infty < x < \infty$ (as earlier, *officially* we also would need $\int_{-\infty}^\infty |f(x)| dx$ to be finite). We write f as the sum of an even function and an odd function:

$$f(x) = g(x) + h(x),$$

¶And there always is a least such L—from analysis, it is the *least upper bound* of the set of all values of α for which $F(\alpha) \ne 0$.

where

$$g(x) = \frac{f(x) + f(-x)}{2}, \quad h(x) = \frac{f(x) - f(-x)}{2}.$$

Now, for g we have

$$g(x) = \int_0^\infty G_c(\alpha) \cos \alpha x \, d\alpha, \quad G_c(\alpha) = \frac{2}{\pi} \int_0^\infty g(x) \cos \alpha x \, dx \qquad (6.8)$$

on $0 \le x < \infty$. But the fact that $g(x)$ and $\cos \alpha x$ are even functions (of x) allow us to rewrite the second half of (6.8) as

$$
\begin{aligned}
G_c(\alpha) &= \frac{1}{\pi} \int_{-\infty}^\infty g(x) \cos \alpha x \, dx \\
&= \frac{1}{\pi} \int_{-\infty}^\infty \frac{f(x) + f(-x)}{2} \cos \alpha x \, dx \\
&= \frac{1}{\pi} \int_{-\infty}^\infty f(x) \cos \alpha x \, dx \quad \text{(why?)}.
\end{aligned}
$$

Similarly, for h we have

$$h(x) = \int_0^\infty H_s(\alpha) \sin \alpha x,$$

$$H_s(\alpha) = \frac{2}{\pi} \int_0^\infty h(x) \sin \alpha x \, dx \qquad (6.9)$$

and, as above, we may write

$$
\begin{aligned}
H_s(\alpha) &= \frac{1}{\pi} \int_{-\infty}^\infty h(x) \sin \alpha x \, dx \\
&= \frac{1}{\pi} \int_{-\infty}^\infty \frac{f(x) - f(-x)}{2} \sin \alpha x \, dx \\
&= \frac{1}{\pi} \int_{-\infty}^\infty f(x) \sin \alpha x \, dx \quad \text{(again, why?)}.
\end{aligned}
$$

Combining, we arrive at

$$f(x) = \int_0^\infty [A(\alpha) \cos \alpha x + B(\alpha) \sin \alpha x] d\alpha \qquad (6.10)$$

where

$$A(\alpha) = G_c(\alpha), \quad B(\alpha) = H_s(\alpha).$$

Thus we can write

$$f(x) = \frac{1}{\pi} \int_0^\infty \int_{-\infty}^\infty f(z)[\cos \alpha z \cos \alpha x + \sin \alpha z \sin \alpha x] dz \, d\alpha.$$

This *Fourier integral* representation of f is, of course, the analog of the Fourier series. But we may go still further. In Exercise 17 in Section 3.3, we derived the *complex form* of the Fourier series which, ultimately, is more convenient (and more elegant) than the *real* Fourier series. With regard to Fourier integrals, this is decidedly the case, as well.

Now we note that $A(\alpha)$ and $B(\alpha)$ actually are defined for $-\infty < \alpha < \infty$, and that $A(-\alpha) = A(\alpha)$ and $B(-\alpha) = -B(\alpha)$ for all α. Thus, we may rewrite (6.10) as

$$f(x) = \frac{1}{2} \int_{-\infty}^{\infty} [A(\alpha)\cos\alpha x + B(\alpha)\sin\alpha x]d\alpha. \quad \text{(why?)}$$

Then, remembering that

$$\cos\theta = \frac{e^{i\theta} + e^{-i\theta}}{2}, \quad \sin\theta = \frac{e^{i\theta} - e^{-i\theta}}{2i},$$

we have

$$
\begin{aligned}
f(x) &= \frac{1}{2} \int_{-\infty}^{\infty} \left[A(\alpha)\frac{e^{i\alpha x} + e^{-i\alpha x}}{2} + B(\alpha)\frac{e^{i\alpha x} - e^{-i\alpha x}}{2i} \right] d\alpha \\
&= \frac{1}{4} \int_{-\infty}^{\infty} \{[A(\alpha) - iB(\alpha)]e^{i\alpha x} + [A(\alpha) + iB(\alpha)]e^{-i\alpha x}\}d\alpha \\
&= \frac{1}{4} \int_{-\infty}^{\infty} \{A(\alpha) + A(-\alpha) - i[B(\alpha) - B(-\alpha)]\}e^{i\alpha x}d\alpha \\
&= \frac{1}{2} \int_{-\infty}^{\infty} [A(\alpha) - iB(\alpha)]e^{i\alpha x}d\alpha.
\end{aligned}
$$

We then have

$$
\begin{aligned}
A(\alpha) - iB(\alpha) &= \frac{1}{\pi} \int_{-\infty}^{\infty} f(z)[\cos\alpha z - i\sin\alpha z]dz \\
&= \frac{1}{\pi} \int_{-\infty}^{\infty} f(z)e^{-i\alpha z}dz
\end{aligned}
$$

and it follows that

$$f(x) = \frac{1}{2\pi} \int_{-\infty}^{\infty} \left[\int_{-\infty}^{\infty} f(z)e^{-i\alpha z}dz \right] e^{i\alpha x} \, d\alpha. \quad (6.11)$$

This is the *complex* **Fourier integral** representation of f, and we generally write

$$f(x) = \frac{1}{\sqrt{2\pi}} \int_{-\infty}^{\infty} F(\alpha)e^{i\alpha x} \, d\alpha = \mathcal{F}^{-1}[F(\alpha)],$$

where

$$F(\alpha) = \frac{1}{\sqrt{2\pi}} \int_{-\infty}^{\infty} f(x)e^{-i\alpha x}\ dx = \mathcal{F}[f(x)]$$

is the **Fourier transform** of f. Again, it really doesn't matter how we divvy up the $\frac{1}{2\pi}$, as long as we're consistent. We say that the functions $f(x)$ and $F(\alpha)$ form a **Fourier transform pair**. It can be shown that if $f(x)$ and $F(\alpha)$ form such a pair, then so do $F(x)$ and $f(-\alpha)$ (see Exercise 2).

As before, sufficient conditions for the validity of (6.11) are given by the following theorem.

Theorem 6.2 *Suppose that f is piecewise smooth on every finite interval and $\int_{-\infty}^{\infty} |f(x)|dx$ is finite. Then, for each x,*

$$\frac{1}{2\pi} \int_{-\infty}^{\infty} \int_{-\infty}^{\infty} f(z)e^{-i\alpha z}\ dz\ e^{i\alpha x}\ d\alpha = \frac{1}{2}[f(x+) + f(x-)].$$

This is the theorem that we'll prove in Section 6.6. Again, we must tread lightly when using "$\int_{-\infty}^{\infty}$." Also, as with Theorem 6.1, this theorem actually is too restrictive, so we won't take it *too* seriously.

First, let's compute some transforms and inverse transforms.

Example 1 Find the Fourier transform of the square wave

$$f(x) = \begin{cases} 1, & \text{if } |x| \leq 1, \\ 0, & \text{if } |x| > 1. \end{cases}$$

We have

$$F(\alpha) = \frac{1}{\sqrt{2\pi}} \int_{-\infty}^{\infty} f(x)e^{-i\alpha x}dx$$

$$= \frac{1}{\sqrt{2\pi}} \int_{-1}^{1} e^{-i\alpha x}dx$$

$$= \sqrt{\frac{2}{\pi}}\frac{\sin\alpha}{\alpha}.$$

Of course, this means that

$$\mathcal{F}^{-1}\left[\frac{\sin\alpha}{\alpha}\right] = \sqrt{\frac{\pi}{2}}f(x) \quad \text{(why?)}.$$

Example 2 Find the Fourier transform of $f(x) = e^{-c|x|}$, $c > 0$.

We have

$$F(\alpha) = \frac{1}{\sqrt{2\pi}} \int_{-\infty}^{\infty} e^{-c|x|} e^{-i\alpha x} \, dx$$

$$= \frac{1}{\sqrt{2\pi}} \left[\int_{-\infty}^{0} e^{(c-i\alpha)x} dx + \int_{0}^{\infty} e^{-(c+i\alpha)x} dx \right]$$

$$= \frac{1}{\sqrt{2\pi}} \left[\frac{1}{c - i\alpha} e^{(c-i\alpha)x} \Big|_{x=-\infty}^{0} - \frac{1}{c + i\alpha} e^{-(c+i\alpha)x} \Big|_{x=0}^{\infty} \right]$$

$$= \sqrt{\frac{2}{\pi}} \frac{c}{c^2 + \alpha^2}.$$

Again, we have

$$\mathcal{F}^{-1} \left[\frac{1}{c^2 + \alpha^2} \right] = \sqrt{\frac{\pi}{2}} \frac{e^{-c|x|}}{c}.$$

Example 3 As the normal density function is such an important function in many applications, let's find the Fourier transform of $f(x) = e^{-x^2}$.

We have

$$F(\alpha) = \frac{1}{\sqrt{2\pi}} \int_{-\infty}^{\infty} e^{-x^2} e^{-i\alpha x} \, dx$$

$$= \frac{1}{\sqrt{2\pi}} \int_{-\infty}^{\infty} e^{-(x^2 + i\alpha x)} \, dx.$$

Now, we know that $\int_{-\infty}^{\infty} e^{-x^2} \, dx = \sqrt{\pi}$, so it looks like we need to complete the square in the exponent:

$$x^2 + i\alpha x = \left(x + \frac{i\alpha}{2} \right)^2 + \frac{\alpha^2}{4}.$$

So

$$F(\alpha) = \frac{1}{\sqrt{2\pi}} e^{-\frac{\alpha^2}{4}} \int_{-\infty}^{\infty} e^{-(x + \frac{i\alpha}{2})^2} \, dx$$

$$= \frac{1}{\sqrt{2\pi}} e^{-\frac{\alpha^2}{4}} \int_{-\infty}^{\infty} e^{-u^2} \, du$$

$$= \frac{1}{\sqrt{2}} e^{-\frac{\alpha^2}{4}}.$$

So the Fourier transform of the Gaussian e^{-x^2} is another Gaussian—in fact, we'll see that this is *always* the case (see Exercises 3 and 4). More specifically, we'll see that a sharply peaked Gaussian has a transform which is "lower" and more "spread out," and vice versa. See Figures 6.2 and 6.3.

Each of the Fourier transforms above turned out to be *real*. This is not always the case, as we will see below and in the exercises.

Let's look now at the important properties of the Fourier transforms.

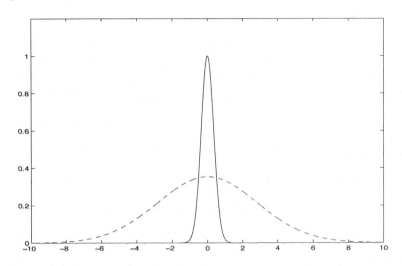

FIGURE 6.2
MATLAB graphs of $f(x) = e^{-4x^2}$ (solid curve) and its Fourier transform $F(\alpha) = \frac{1}{2\sqrt{2}} e^{-\alpha^2/16}$ (dashed curve).

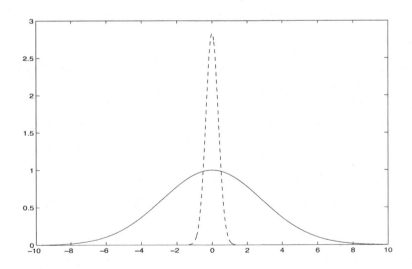

FIGURE 6.3
MATLAB graph of $f(x) = e^{-x^2/16}$ (solid curve) and its Fourier transform $F(\alpha) = 2\sqrt{2} e^{-4\alpha^2}$ (dashed curve).

TRANSFORMS AND DERIVATIVES

If $\mathcal{F}[f(x)] = F(\alpha)$, let's find $\mathcal{F}[f'(x)]$:

$$\mathcal{F}[f'(x)] = \frac{1}{\sqrt{2\pi}} \int_{-\infty}^{\infty} f'(x)e^{-i\alpha x} \, dx$$

$$= \frac{1}{\sqrt{2\pi}} \left[e^{-i\alpha x} f(x) \Big|_{-\infty}^{\infty} + i\alpha \int_{-\infty}^{\infty} f(x)e^{-i\alpha x} \, dx \right] \quad \text{(why?)}$$

$$= i\alpha F(\alpha).$$

So we have

$$\mathcal{F}[f'(x)] = i\alpha \mathcal{F}[f(x)].$$

It follows that

$$\mathcal{F}[f^{(n)}(x)] = (i\alpha)^n \mathcal{F}(f(x))$$

(see Exercise 10a). Similarly, we can show that

$$\mathcal{F}[-ixf(x)] = F'(\alpha)$$

and, more generally, that

$$\mathcal{F}[(-ix)^n f(x)] = F^{(n)}(\alpha).$$

CONVOLUTION

As with Laplace transforms, we often need to find inverse transforms of products. Specifically, if

$$\mathcal{F}[f(x)] = F(\alpha) \quad \text{and} \quad \mathcal{F}[g(x)] = G(\alpha),$$

what is

$$\mathcal{F}^{-1}[F(\alpha)G(\alpha)]?$$

Well, we know that

$$F(\alpha) = \frac{1}{\sqrt{2\pi}} \int_{-\infty}^{\infty} f(x)e^{-i\alpha x} dx$$

and

$$G(\alpha) = \frac{1}{\sqrt{2\pi}} \int_{-\infty}^{\infty} g(y)e^{-i\alpha y} \, dy,$$

so

$$F(\alpha)G(\alpha) = \frac{1}{2\pi} \left[\int_{-\infty}^{\infty} f(x)e^{-i\alpha x} \, dx \right] \left[\int_{-\infty}^{\infty} g(y)e^{-i\alpha y} \, dy \right]$$

$$= \frac{1}{2\pi} \int_{-\infty}^{\infty} g(y) \left[\int_{-\infty}^{\infty} f(x)e^{-i\alpha(x+y)} \, dx \right] dy \quad \text{(why?)}.$$

Now, we substitute $z = x + y$ in the inner integral in order to make it look like a Fourier transform (so $dz = dx$), and we get

$$F(\alpha)G(\alpha) = \frac{1}{2\pi} \int_{-\infty}^{\infty} g(y) \left[\int_{-\infty}^{\infty} f(z-y)e^{-i\alpha z} \, dz \right] dy$$

$$= \frac{1}{\sqrt{2\pi}} \int_{-\infty}^{\infty} \left[\frac{1}{\sqrt{2\pi}} \int_{-\infty}^{\infty} f(x-y)g(y)dy \right] e^{-i\alpha x} \, dx,$$

where we have reversed the order of integration,[∥] and replaced z by (the new variable) x. But this is just the Fourier transform of the function of x in the brackets! So,

$$F(\alpha)G(\alpha) = \mathcal{F}\left[\frac{1}{\sqrt{2\pi}} \int_{-\infty}^{\infty} f(x-y)g(y)dy \right]$$

and, thus,

$$\mathcal{F}^{-1}[F(\alpha)G(\alpha)] = \frac{1}{\sqrt{2\pi}} \int_{-\infty}^{\infty} f(x-y)g(y)dy$$

$$= \frac{1}{\sqrt{2\pi}} f * g,$$

where $f * g$ is called the (infinite) **convolution** of f with g (as long as the integral exists).

Example 4 Find $\mathcal{F}^{-1}\left[\frac{e^{-\frac{\alpha^2}{4}}}{1+\alpha^2} \right]$.

From Examples 2 and 3, $\mathcal{F}[e^{-|x|}] = \sqrt{\frac{2}{\pi}} \frac{1}{1+\alpha^2}$ and $\mathcal{F}[e^{-x^2}] = \frac{1}{\sqrt{2}} e^{-\alpha^2/4}$. Therefore,

$$\mathcal{F}^{-1}\left[\frac{e^{-\alpha^2/4}}{1+\alpha^2} \right] = \sqrt{\pi} \, \mathcal{F}^{-1}\left[\frac{1}{\sqrt{2}} e^{-\alpha^2/4} \cdot \sqrt{\frac{2}{\pi}} \frac{1}{1+\alpha^2} \right]$$

$$= \sqrt{\pi} \, \frac{1}{\sqrt{2\pi}} e^{-x^2} * e^{-|x|}$$

$$= \frac{1}{\sqrt{2}} \int_{-\infty}^{\infty} e^{-(x-y)^2} e^{-|y|} \, dy.$$

[∥]It turns out that, in order to switch the order of integration, we need the inner integral to be a *uniformly continuous* function of y.

TRANSLATION

If $\mathcal{F}[f(x)] = F(\alpha)$, what is $\mathcal{F}[f(x - c)]$, where c is a constant?

$$\mathcal{F}[f(x - c)] = \frac{1}{\sqrt{2\pi}} \int_{-\infty}^{\infty} f(x - c)e^{-i\alpha x} \, dx$$

$$= \frac{1}{\sqrt{2\pi}} \int_{-\infty}^{\infty} f(z)e^{-i\alpha(z+c)} \, dz$$

$$= e^{-ic\alpha} F(\alpha).$$

Example 5 If we would like to find the Fourier transform of

$$g(x) = \begin{cases} 1, \text{ if } -3 \le x \le -1, \\ 0, \text{ otherwise}, \end{cases}$$

we may notice that $g(x) = f(x + 2)$, where f is the function from Example 1. Then,

$$G(\alpha) = e^{2i\alpha} F(\alpha) = \sqrt{\frac{2}{\pi}} e^{2i\alpha} \frac{\sin \alpha}{\alpha}.$$

Let's list these properties again. If $\mathcal{F}[f(x)] = F(\alpha)$ and $\mathcal{F}[g(x)] = G(\alpha)$, then

Derivative property: $\mathcal{F}[f^{(n)}(x)] = (i\alpha)^n F(\alpha)$
Convolution: $\mathcal{F}^{-1}[F(\alpha)G(\alpha)] = \frac{1}{\sqrt{2\pi}} f(x) * g(x)$
Translation: $\mathcal{F}[f(x - c)] = e^{-ic\alpha} F(\alpha)$

(as long as the integrals involved do not give us any problems). Finally, given the symmetry of the various Fourier transform pairs, it should be no surprise that these properties can be "turned around" (as we mentioned when deriving the derivative properties). See Exercises 2 and 9.

Of course, as with the sine and cosine transforms, when using Fourier transforms to solve partial differential equations, the functions no longer are functions of a *single* variable. We treat these problems in the following section.

We include a table of Fourier transforms and properties at the end of this chapter (see Table 6.3).

Exercises 6.3

1. Calculate the Fourier transform of f, then use Theorem 6.2 to describe the function

$$g(x) = \frac{1}{2\pi} \int_{-\infty}^{\infty} e^{i\alpha x} \int_{-\infty}^{\infty} f(z)e^{-i\alpha z} \, dz d\alpha.$$

a) $f(x) = \begin{cases} 1, \text{ if } a \le x \le b \\ 0, \text{ otherwise} \end{cases}$

b) $f(x) = \begin{cases} e^{2x}, & \text{if } x \leq 0 \\ e^{-x}, & \text{if } x > 0 \end{cases}$

c) $f(x) = xe^{-c|x|}, c > 0$

d) $f(x) = e^{-|x|}\cos x$ and $g(x) = e^{-|x|}\sin x$

e) $f(x) = \dfrac{\sin x}{x}$ (see Example 1)

2. Suppose that $f(x)$ and $F(\alpha)$ form a Fourier transform pair. Show that the following also form Fourier transform pairs:

$$f(-x) \quad \text{and} \quad F(-\alpha),$$
$$F(x) \quad \text{and} \quad f(-\alpha),$$
$$F(-x) \quad \text{and} \quad f(\alpha).$$

3. The most general form of the *Gaussian* or *normal density function* is

$$f(x) = \frac{1}{\sqrt{2\pi\sigma^2}} e^{-\frac{(x-m)^2}{2\sigma^2}},$$

where m is the mean and σ is the standard deviation.

a) Compute its Fourier transform, $F(\alpha)$.

b) **MATLAB:** Investigate graphically the behavior of f versus F as σ varies; in particular, justify the statement made in the text, that sharply peaked Gaussian's have flat, spread-out Fourier transforms, and vice versa.

4. Find a function which is its own Fourier transform, i.e., for which $\mathcal{F}[f(x)] = f(\alpha)$.

5. Prove that the (infinite) convolution satisfies the following properties:

a) Commutativity: $f * g = g * f$.

b) Associativity: $f * (g * h) = (f * g) * h$.

c) Distributive law: $f * (c_1 g + c_2 h) = c_1 f * g + c_2 f * h$, for all constants c_1 and c_2.

6. We've been treating all of these transforms as though they're linear without even justifying it! Prove that all of the integral transforms in this chapter are linear, that is, prove that each transform T satisfies

$$T(c_1 f_1 + c_2 f_2) = c_1 T(f_1) + c_2 T(f_2)$$

for all constants c_1 and c_2 and for all functions f_1 and f_2 for which the transform is defined.

7. Use the result of Example 1 to show that $\int_{-\infty}^{\infty} \frac{\sin \alpha}{\alpha} d\alpha = \pi$. Then, find $\mathcal{F}_c \left[\frac{\sin cx}{x} \right]$. (Actually, we'll need the first statement when we prove the Fourier inversion theorem, so in Section 6.5 we must prove $\int_{-\infty}^{\infty} \frac{\sin \alpha}{\alpha} d\alpha = \pi$ *without* using transforms.)

8. Show that the Fourier integral representation also can be written

$$\mathcal{F}[f(x)] = \frac{1}{\pi} \int_0^\infty \int_{-\infty}^\infty f(z) \cos \alpha (x - z) dz d\alpha.$$

9. Prove the following properties of Fourier transforms. In each case, $F(\alpha) = \mathcal{F}[f(x)]$ and $G(\alpha) = \mathcal{F}[g(x)]$.

 a) $\mathcal{F}[f(cx)] = \frac{1}{c} F \left(\frac{\alpha}{c} \right)$

 b) $\mathcal{F}[xf(x)] = iF'(\alpha)$

 c) $\mathcal{F}[f(x)g(x)] = \frac{1}{\sqrt{2\pi}} F(\alpha) * G(\alpha)$

 d) $\mathcal{F}[e^{icx} f(x)] = F(\alpha - c)$

10. a) Use the property $\mathcal{F}[f'(x)] = i\alpha F(\alpha)$ to prove that $\mathcal{F}[f^{(n)}(x)] = (i\alpha)^n F(\alpha)$, $n = 1, 2, 3, \ldots$.

 b) How would you find $\mathcal{F}^{-1} \left[\frac{F(\alpha)}{\alpha} \right]$ and, more generally, $\mathcal{F}^{-1} \left[\frac{F(\alpha)}{\alpha^n} \right]$? (We can't really answer these questions definitively until we get to Section 6.5.)

11. Calculate the Fourier transform of f.

 a) $f(x) = \dfrac{x}{(1 + x^2)^2}$

 b) $f(x) = \displaystyle\int_{-\infty}^\infty \frac{\sin y}{y} e^{-|x-y|} \, dy$

 c) $f(x) = \dfrac{2}{3 + (x - 1)^2}$

 d) $f(x) = e^{-x^2 - |x|}$

 e) $f(x) = x^2 e^{-x^2}$

12. a) Show that if f is even, then $\mathcal{F}[f]$ is even, and that if f is odd, then $\mathcal{F}[f]$ is odd.

 b) Suppose that $f(x)$ and $F(\alpha)$ form a Fourier transform pair. Show that if f is even, then $F(x)$ and $f(\alpha)$ form a Fourier transform pair, while if f is odd, then $F(x)$ and $-f(\alpha)$ form a Fourier transform pair.

13. **The Laplace transform and the Fourier transform:** Of course, the Laplace transform is defined for functions with domain $[0, \infty)$. However, suppose that

$$f(t) = \begin{cases} 0, & \text{if } t < 0, \\ g(t), & \text{if } t \geq 0. \end{cases}$$

Show formally that

$$\mathcal{L}[g(t)] = \sqrt{2\pi} \, F(-i\alpha),$$

where $F(\alpha) = \mathcal{F}[f(t)]$.

14. Suppose that we start with the complex Fourier transform and derive the sine and cosine integral formulas for functions on $x \geq 0$. That is, suppose that

$$f(x) = \frac{1}{2\pi} \int_{-\infty}^{\infty} e^{i\alpha x} \int_{-\infty}^{\infty} f(z) e^{-i\alpha z} \, dz \, d\alpha$$

for any well-behaved function f on $-\infty < x < \infty$, and show that, for any well-behaved function g on $x \geq 0$, we have

$$g(x) = \frac{2}{\pi} \int_0^{\infty} \sin \alpha x \int_0^{\infty} g(z) \sin \alpha z \, dz \, d\alpha$$

$$= \frac{2}{\pi} \int_0^{\infty} \cos \alpha x \int_0^{\infty} g(z) \cos \alpha z \, dz \, d\alpha.$$

15. **Parseval's equality for the complex Fourier transform:** If we allow $f(x)$ and $F(\alpha)$ to be complex-valued functions, then we must alter slightly the form of Parseval's equality. Here it becomes

$$\int_{-\infty}^{\infty} |f(x)|^2 dx = \int_{-\infty}^{\infty} |F(\alpha)|^2 d\alpha,$$

where $|z| = |a + bi| = \sqrt{a^2 + b^2}$ is the modulus of the complex number z. Noticing that $|z|^2 = a^2 + b^2 = (a + bi)(a - bi)$, we see that the above can be rewritten as

$$\int_{-\infty}^{\infty} f(x)\overline{f(x)} \, dx = \int_{-\infty}^{\infty} F(\alpha)\overline{F(\alpha)} \, d\alpha.$$

We prove this as follows:

a) Use Euler's formula to show that $\overline{e^{i\theta}} = e^{-i\theta}$, where θ is real.

b) Show that $\bar{z}_1 \bar{z}_2 = \overline{(z_1 z_2)}$ for complex numbers z_1 and z_2.

c) Follow what you did in Exercise 11 of the previous section and, along the way, use the fact that

$$\int_{-\infty}^{\infty} \overline{f(x)e^{-i\alpha x}} \, dx = \overline{\int_{-\infty}^{\infty} f(x)e^{-i\alpha x} \, dx},$$

to arrive at the above result.

16. **Sampling, revisited:** Exercise 12 of Section 6.2 dealt with the *sampling theorem* for the Fourier sine series. Here, we do the same for the complex Fourier transform and series. So, a function f on $-\infty < x < \infty$ will be **band-limited** if its Fourier transform $F(\alpha) = 0$ outside some interval $-L \leq \alpha \leq L$. The least such (positive) L is f's **cutoff frequency.**

a) Supposing that the cutoff frequency for f is $L = \pi$, expand $F(\alpha)$ in a *complex* Fourier series, as in Exercise 13, Section 3.3, and conclude that

$$f(x) = \frac{1}{\sqrt{2\pi}} \int_{-\infty}^{\infty} e^{i\alpha x} \left[\sum_{n=-\infty}^{\infty} c_n e^{-in\alpha} \right] d\alpha,$$

where

$$c_n = \frac{1}{2\pi} \int_{-\pi}^{\pi} F(\alpha) e^{in\alpha}\, d\alpha, \qquad n = \ldots, -1, 0, 1, \ldots .$$

b) Show that

$$c_n = \frac{1}{\sqrt{2\pi}} f(n), \qquad n = \ldots, -1, 0, 1, \ldots$$

and, thus, that

$$f(x) = \frac{1}{2\pi} \sum_{n=-\infty}^{\infty} f(n) \int_{-\pi}^{\pi} e^{i\alpha(x-n)}\, d\alpha.$$

Again, f is recovered by *sampling* it on the integers!

6.4   The Infinite and Semi-Infinite Heat Equations

We are now in a position to use Fourier transforms to solve the heat equation for an *infinite* rod. So suppose we have the problem

$$u_t = k^2 u_{xx}, \quad -\infty < x < \infty, t > 0,^*$$

$$u(x, 0) = f(x), \quad -\infty < x < \infty,$$

$$\lim_{|x| \to \infty} u(x, t) = 0.$$

---

*We now use k^2, instead of α^2, for the thermal diffusivity, in order to avoid confusion.

Since we have $-\infty < x < \infty$, we'll apply the Fourier transform to the x-variable. The transformed PDE is

$$U_t(\alpha, t) = -\alpha^2 k^2 U(\alpha, t),$$

where $U(\alpha, t) = \mathcal{F}[u(x,t)]$. This is essentially an ODE, with general solution

$$U(\alpha, t) = e^{-\alpha^2 k^2 t} G(\alpha),$$

where G is an arbitrary function of α and may be determined by transforming the initial condition. We arrive at

$$\begin{aligned} U(\alpha, 0) = F(\alpha) &= \mathcal{F}[f(x)] \\ &= e^0 G(\alpha), \end{aligned}$$

so we now have

$$U(\alpha, t) = e^{-\alpha^2 k^2 t} F(\alpha).$$

Therefore, our solution is a convolution. First, then, we need to compute

$$\mathcal{F}^{-1}[e^{-\alpha^2 k^2 t}] = \frac{1}{\sqrt{2\pi}} \int_{-\infty}^{\infty} e^{-\alpha^2 k^2 t} e^{i\alpha x} \, d\alpha. \tag{6.12}$$

We can evaluate this directly from the result of Exercise 4 in the previous section. However, let's compute it here. We know from Example 3 in Section 6.3 that

$$\mathcal{F}^{-1}[e^{-\frac{\alpha^2}{4}}] = \sqrt{2} \, e^{-x^2} = \frac{1}{\sqrt{2\pi}} \int_{-\infty}^{\infty} e^{-\alpha^2/4} e^{i\alpha x} \, d\alpha,$$

so we need only put (6.12) in this form. Remembering that α is the integration variable and, therefore, treating x and t as constants, we let

$$\frac{u}{2} = \alpha k \sqrt{t}, \quad \frac{1}{2} du = k\sqrt{t} \, d\alpha.$$

Then, (6.12) becomes

$$\begin{aligned} \frac{1}{\sqrt{2\pi}} \int_{-\infty}^{\infty} e^{-\frac{u^2}{4}} e^{iu\left(\frac{x}{2k\sqrt{t}}\right)} \frac{du}{2k\sqrt{t}} &= \frac{1}{k\sqrt{2t}} e^{-\left(\frac{x}{2k\sqrt{t}}\right)^2} \\ &= \frac{1}{k\sqrt{2t}} e^{-\frac{x^2}{4k^2 t}}. \end{aligned}$$

So our solution is the convolution (in x, of course)

$$u(x,t) = \frac{1}{k\sqrt{2\pi}} \frac{1}{\sqrt{2t}} e^{-x^2/4k^2 t} * f(x) = \frac{1}{k\sqrt{2\pi}} \frac{1}{\sqrt{2t}} f(x) * e^{-x^2/4k^2 t},$$

which can be written

$$u(x,t) = \frac{1}{2k\sqrt{\pi t}} \int_{-\infty}^{\infty} e^{-\frac{(x-\xi)^2}{4k^2 t}} f(\xi) \, d\xi \tag{6.13}$$

or

$$u(x,t) = \frac{1}{2k\sqrt{\pi t}} \int_{-\infty}^{\infty} f(x-\xi) e^{-\frac{\xi^2}{4k^2 t}} d\xi.$$

For various reasons, the solution usually is written as the former. The function $G(x,t) = \frac{1}{2k\sqrt{\pi t}} e^{-\frac{x^2}{4k^2 t}}$ is called the **heat kernel**, and it's also called the **fundamental solution** for this problem (see Section 10.5). Of course, for fixed t, G is just the normal density function of Exercise 3 in the previous section, with variance $\sigma^2 = 2k^2 t$ (and variance is the square of the standard deviation).

Sadly, (6.12) is not the only solution to this problem. It turns out, however, that it *is* the only continuous, bounded solution (so long as f is bounded).

There is a very nice physical interpretation of this solution, which we'll mention in Chapter 10. Let's look at an example.

Example 1 Solve the heat problem

$$u_t = u_{xx}$$

$$u(x,0) = f(x) = \begin{cases} 1/2a, & \text{if } |x| \le a, \\ 0, & \text{if } |x| > a, a > 0, \end{cases}$$

$$\lim_{|x| \to \infty} u(x,t) = 0.$$

Note that the initial heat content of the bar is the same for any choice of a. Using (6.12), our solution is

$$u(x,t) = \frac{1}{4a\sqrt{\pi t}} \int_{-a}^{a} e^{-\frac{(x-\xi)^2}{4t}} d\xi.$$

It's interesting to analyze what happens at $t \to \infty$, or as a varies. In either case, let's first calculate the *heat content* of the bar (assuming constant cross section, as always); it is proportional to

$$\int_{-\infty}^{\infty} u(x,t) dx = \frac{1}{4a\sqrt{\pi t}} \int_{-\infty}^{\infty} \int_{-a}^{a} e^{-\frac{(x-\xi)^2}{4t}} d\xi dx$$

$$= \frac{1}{4a\sqrt{\pi t}} \int_{-a}^{a} \int_{-\infty}^{\infty} e^{-\frac{(x-\xi)^2}{4t}} dx d\xi.^{\dagger}$$

Now, in order to evaluate the x-integral, we—as usual!—substitute so that we can use $\int_{-\infty}^{\infty} e^{-x^2} dx = \sqrt{\pi}$. Then, everything simplifies to

$$\int_{-\infty}^{\infty} u(x,t) dx = 1,$$

---

†Again, so long as we may reverse the order of integration.

for *any* choice of t! Actually, this is not as surprising as it seems; it's just a statement of the conservation of (heat) energy.

Of course, the last equation says that the heat content does not depend on a, as well. This is because we rigged the problem so that the *initial* heat content is independent of the choice of a.

Figure 6.4 illustrates $u(x, t)$ for $a = 1$ and for various values of t. Figure 6.5, on the other hand, shows $u(x, t)$ for the specific time $t = 1$, but for different values of a.

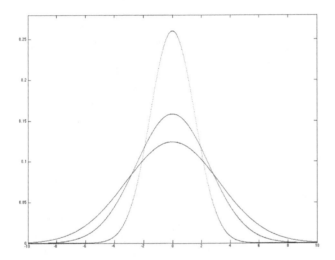

FIGURE 6.4
MATLAB graph of the solution of the problem in Example 1, with
$a = 1$, for $t = 1, 3$ and 5 (from highest peaked to lowest peaked).

We look more closely at the solution $u(x, t)$, and the conservation of energy, in the exercises.

Before leaving this example, we should ask what happens as $a \to 0$. Specifically, let's rewrite some of the functions $f(x)$ by letting $a = \frac{1}{n}$, where n is a natural number:

$$f_n(x) = \begin{cases} n/2, & \text{if } |x| \le \dfrac{1}{n}, \\ 0, & \text{if } |x| > \dfrac{1}{n}. \end{cases}$$

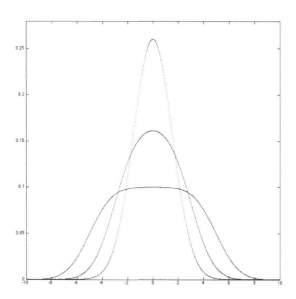

FIGURE 6.5
**MATLAB graph of the solution of the problem in Example 1, at
time $t = 1$, for $a = 1, 3$ and 5 (from highest peaked to lowest peaked).**

As $n \to \infty$, f *seems* to be approaching a "function" which is infinite at the
origin and 0 everywhere else—certainly *not* a function, as far as we're concerned. (See Figure 6.6.) However, in the more general setting of *distributions*
or *generalized functions*, it turns out that it *does* make sense and, in fact, that
it turns out to be one of the most important ideas in mathematics! We write

$$\lim_{n \to \infty} f_n(x) = \delta(x), \ddagger$$

where $\delta(x)$ is called the **Dirac**§ **delta function**, and the functions $f_n(x)$ are
said to form a *delta sequence*.

Also, as $\int_{-\infty}^{\infty} f_n(x)dx = 1$ for each n, it *looks* as though $\delta(x)$ has the
property

$$\int_{-\infty}^{\infty} \delta(x)dx = 1.$$

We'll look more closely at these ideas in the next section.

---

‡Realizing that this is not a true "=."
§After the great quantum physicist P.A.M. Dirac (1902–1984).

Now that we've solved the heat equation for an infinite bar, we turn to the case of the semi-infinite bar. Specifically, we'd like to solve the problem

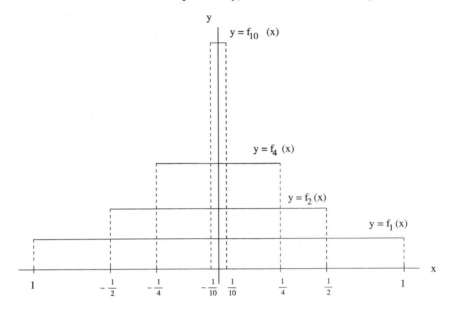

FIGURE 6.6 Graphs of $y = f_n(x)$ for $n = 1, 2, 4$ and 10.

$$u_t = u_{xx}, \quad 0 < x < \infty, t > 0,$$
$$u(x, 0) = f(x), \quad 0 < x < \infty,$$
$$u(0, t) = 0,$$

from Section 6.2. We'll solve it here using the **method of images**. (You may recall that we used the method of images to solve the semi-infinite wave equation in Section 5.4.) The idea is to solve, instead, the problem

$$u_t = u_{xx}, \quad -\infty < x < \infty, t > 0,$$
$$u(x, 0) = g(x), \quad -\infty < x < \infty,$$

where $g(x)$ is either the even or the odd extension of f to the interval $-\infty < x < \infty$. Which one? Since we would like the solution to satisfy $u(0, t) = 0$, we try the *odd* extension

$$g(x) = \begin{cases} f(x), & \text{if } x > 0, \\ -f(-x), & \text{if } x < 0. \end{cases}$$

We've already solved the new problem; from (6.12) we have

$$u(x, t) = \frac{1}{2\sqrt{\pi t}} \int_{-\infty}^{\infty} e^{-\frac{(x-\xi)^2}{4t}} g(\xi) d\xi.$$

Then, since g is odd, we also have

$$u(0, t) = 0 \quad \text{(why?)}.$$

Thus, we have our solution to the original problem. However, we really should be able to write it in terms of the initial function f. So, we write

$$u(x, t) = \frac{1}{2\sqrt{\pi t}} \left[\int_{-\infty}^{0} e^{-\frac{(x-\xi)^2}{4t}} g(\xi) d\xi + \int_{0}^{\infty} e^{-\frac{(x-\xi)^2}{4t}} g(\xi) d\xi \right].$$

Substituting $z = -\xi$ in the first integral turns it into

$$\int_{0}^{\infty} e^{-\frac{(x+z)^2}{4t}} g(-z) dz = -\int_{0}^{\infty} e^{-\frac{(x+z)^2}{4t}} f(z) dz$$

and our solution becomes

$$u(x, t) = \frac{1}{2\sqrt{\pi t}} \int_{0}^{\infty} \left[e^{-\frac{(x-\xi)^2}{4t}} - e^{-\frac{(x+\xi)^2}{4t}} \right] f(\xi) d\xi$$

which, of course, is exactly what we got in Section 6.2 using the sine transform.

One final note: It turns out that the heat/diffusion problem on $-\infty < x < \infty$ is not well-posed! See Exercise 15.

Exercises 6.4

1. Two identical very long rods are at different temperatures, T_1 and T_2. At time $t = 0$ they are attached; then, the system will satisfy the problem

$$u_t = k^2 u_{xx}, \qquad -\infty < x < \infty, t > 0,$$

$$u(x, 0) = \begin{cases} T_1, & \text{if } x < 0 \\ T_2, & \text{if } x > 0. \end{cases}$$

   a) Solve the problem.

   b) **MATLAB:** Letting $k^2 = 1, T_1 = 10°$ and $T_2 = 30°$, graph the solution in the x-u plane, for various values of t. (You'll need to replace $\pm\infty$ by $\pm M$ and to choose M judiciously.)

2. Show formally that (6.13) does, in fact, satisfy the PDE and initial condition. Find another solution to this problem. (Hint: It need not be continuous at $t = 0$.)

3. Use the method of images to solve

$$u_t = u_{xx}, \qquad x > 0, t > 0,$$
$$u(x, 0) = f(x),$$
$$u_x(0, t) = 0,$$
$$\lim_{x \to \infty} u(x, t) = 0,$$

and make sure you get the same answer as before (Exercise 5, Section 6.2).

4. **Uninsulated rod** (In each case, verify formally that your solution satisfies the problem.)

   a) Solve the system

   $$u_t = u_{xx} - u, \quad -\infty < x < \infty, t > 0,$$
   $$u(x, 0) = f(x),$$
   $$\lim_{|x| \to \infty} u(x, t) = 0,$$

   which represents the temperature of an infinite rod whose sides are *not* insulated (as we derived in Exercise 9, Section 2.4).

   b) Now solve the same problem for the *semi*-infinite rod, with Dirichlet condition $u(0, t) = 0$.

   c) Do the same, but now for the semi-infinite rod subject to the Neumann condition $u_x(0, t) = 0$.

   d) Go back and solve parts (a)–(c) via the substitution $u(x, t) = e^{-t}v(x, t)$.

5. **Diffusion-Convection:** Repeat Exercise 3, but for the diffusion-convection equation $u_t = u_{xx} - u_x$ (see Exercise 7, Section 4.1). Again, verify formally that your solutions satisfy the system.

6. **Infinite and semi-infinite Euler–Bernoulli beam:** (Solve each problem, in terms of Fourier integrals.)

   a) The infinite Euler–Bernoulli beam problem

   $$u_{tt} = u_{xxxx}, \quad -\infty < x < \infty, t > 0,$$
   $$u(x, 0) = f(x),$$
   $$u_t(x, 0) = g(x).$$

   b) The semi-infinite E–B beam problem

   $$u_{tt} = u_{xxxx}, \quad x > 0, t > 0$$
   $$u(x, 0) = f(x),$$
   $$u_t(x, 0) = g(x),$$
   $$u_x(0, t) = 0.$$

7. In Section 5.3 we derived d'Alembert's formula

$$u(x,t) = \frac{1}{2}[f(x+ct) + f(x-ct)] + \frac{1}{2c}\int_{x-ct}^{x+ct} g(z)dz$$

for the solution of the infinite wave problem

$$u_{tt} = c^2 u_{xx},$$
$$u(x,0) = f(x),$$
$$u_t(x,0) = g(x).$$

We also may solve this problem using Fourier transforms.

a) Use Fourier transforms to solve

$$u_{tt} = c^2 u_{xx}, \quad -\infty < x < \infty, t > 0,$$
$$u(x,0) = f(x),$$
$$u_t(x,0) = 0,$$

$$\lim_{|x|\to\infty} u(x,t) = 0. \text{ (Hint: Use } \cos\theta = \frac{e^{i\theta} + e^{-i\theta}}{2}.)$$

b) Do the same for

$$u_{tt} = c^2 u_{xx}, \quad -\infty < x < \infty, t > 0,$$
$$u(x,0) = 0,$$
$$u_t(x,0) = g(x),$$
$$\lim_{|x|\to\infty} u(x,t) = 0. \text{ (Hint: Do } not \text{ use the above hint!)}$$

8. Use Fourier transforms to solve Laplace's equation on the half-plane:

$$u_{xx} + u_{yy} = 0, \quad -\infty < x < \infty, y \geq 0,$$
$$u(x,0) = f(x),$$
$$\lim_{|x|,y\to\infty} |u(x,y)| = 0.$$

Then show that you may write this solution explicitly as

$$u(x,y) = \frac{y}{\pi}\int_{-\infty}^{\infty} \frac{f(z)}{y^2 + (x-z)^2}dz.$$

(When you solve the ODE after transforming, take care to solve it for $\alpha \geq 0$ and for $\alpha < 0$ separately.) This is one form of the **Schwarz integral formula**, or **Poisson's integral formula for the upper half-plane**, and the function $\frac{y}{\pi(y^2+x^2)}$ is the **Poisson kernel**.

9. Solve the semi-infinite heat problem

$$u_t = u_{xx}, \quad x > 0, t > 0,$$
$$u(x, 0) = f(x),$$
$$u(0, t) = g(t),$$
$$\lim_{x \to \infty} u(x, t) = 0.$$

(See Example 4 in Section 6.2.)

10. Solve the Laplace equation problem on the first quadrant:

$$u_{xx} + u_{yy} = 0, \quad x > 0, y > 0,$$
$$u(x, 0) = f(x),$$
$$u(0, y) = g(y),$$
$$\lim_{x, y \to \infty} u(x, y) = 0.$$

11. Consider the PDE

$$u_{tt} + 2au_t + bu = c^2 u_{xx}, \quad -\infty < x < \infty, t > 0.$$

a) What kind of equation is this (hyperbolic, parabolic, elliptic)?

b) Show that, if $a^2 - b < 0$, the Fourier transform of the solution is

$$U(\alpha, t) = c_1(\alpha)e^{-at} \cos \sqrt{b + c^2\alpha^2 - a^2}\, t \ + \ c_2(\alpha)e^{-at} \sin \sqrt{b + c^2\alpha^2 - a^2}\, t$$

for arbitrary functions c_1 and c_2.

c) Show that, if $a^2 - b = 0$, we have

$$U(\alpha, t) = c_1(\alpha)e^{-at} \cos c\alpha t + c_2(\alpha)e^{-at} \sin c\alpha t$$

(unless $\alpha = 0$).

d) Show that, if $a^2 - b > 0$, we have

$$U(\alpha, t) = \begin{cases} c_1(\alpha)e^{-at} \cos \sqrt{b + c^2\alpha^2 - a^2}\, t \\ \quad + c_2(\alpha)e^{-at} \sin \sqrt{b + c^2\alpha^2 - a^2}\, t, & \text{if } c^2\alpha^2 \geq a^2 - b, \\ c_3(\alpha)e^{-at} \cosh \sqrt{-b - c^2\alpha^2 + a^2}\, t \\ \quad + c_4(\alpha)e^{-at} \sinh \sqrt{-b - c^2\alpha^2 + a^2}\, t, & \text{if } c^2\alpha^2 < a^2 - b. \end{cases}$$

12. a) Use the results of the preceding exercise to find the solution, in Fourier integral form, of the one-dimensional *telegraph equation*

$$CLu_{tt} + (RC + GL)u_t + RGu = u_{xx}$$

on $-\infty < x < \infty, t > 0$, subject to the conditions

$$u(x,0) = f(x),$$
$$u_t(x,0) = g(x),$$
$$\lim_{|x| \to \infty} u(x,t) = 0.$$

b) A very important PDE from *particle physics* is the (one-dimensional) linearized **Klein–Gordon equation**

$$u_{tt} + m^2 u = c^2 u_{xx},$$

where m is the mass of the given elementary particle and c is the speed of light. Do the same as in part (a) for this equation.

c) Use the fact that

$$\mathcal{F}[J_0(\sqrt{1-x^2})H(1-x^2)] = \sqrt{\frac{2}{\pi}} \frac{\sin\sqrt{\alpha^2+1}}{\alpha^2+1}$$

to show that the solution of part (b), with $f(x) = 0$, is

$$u(x,t) = \frac{1}{2c} \int_{-ct}^{ct} g(x-\xi) J_0(m\sqrt{c^2t^2-\xi^2})d\xi.$$

13. **Fourier transform in higher dimensions; the two-dimensional heat equation on the plane:** As we show in Section 9.6, we may talk about the Fourier transform of a function of several variables. For instance, for $f = f(x,y)$, we have the Fourier transform

$$F(\alpha, \beta) = \frac{1}{2\pi} \int_{-\infty}^{\infty} \int_{-\infty}^{\infty} f(x,y)e^{-i(\alpha x + \beta y)}\,dx\,dy$$
$$= \mathcal{F}[f(x,y)],$$

with the inverse transform of F given by

$$f(x,y) = \frac{1}{2\pi} \int_{-\infty}^{\infty} \int_{-\infty}^{\infty} F(\alpha, \beta)e^{i(\alpha x + \beta y)}\,d\alpha\,d\beta.$$

a) Show that
$$\mathcal{F}[f_x] = i\alpha F \quad \text{and} \quad \mathcal{F}[f_y] = i\beta F.$$

b) Show that the formal solution to the two-dimensional heat equation

$$u_t = u_{xx} + u_{yy}, \qquad -\infty < x < \infty, -\infty < y < \infty,$$
$$u(x,y,0) = f(x,y),$$
$$\lim_{x^2+y^2 \to \infty} u(x,y,t) = 0,$$

is

$$u(x, y, t) = \frac{1}{2\pi} \int_{-\infty}^{\infty} \int_{-\infty}^{\infty} F(\alpha, \beta) e^{-(\alpha^2 + \beta^2)t + i(\alpha x + \beta y)} d\alpha d\beta$$

$$= \mathcal{F}^{-1}[F(\alpha, \beta) e^{-(\alpha^2 + \beta^2)t}].$$

c) **Convolution:** Show that

$$\mathcal{F}^{-1}[F(\alpha, \beta) G(\alpha, \beta)] = \frac{1}{2\pi} \int_{-\infty}^{\infty} \int_{-\infty}^{\infty} f(x - u, y - v) g(u, v) du dv,$$

where $F = \mathcal{F}[f]$ and $G = \mathcal{F}[g]$. Not surprisingly, we call this integral $f * g$, the *convolution* of f with g.

d) Show that the solution to part (b) can be written

$$u(x, y, t) = \frac{1}{4\pi t} \int_{-\infty}^{\infty} \int_{-\infty}^{\infty} e^{-\frac{[(x-u)^2 + (y-v)^2]}{4t}} f(u, v) du dv.$$

14. Use the result of the previous exercise to solve the two-dimensional heat problem on the *half*-plane

$$u_t = u_{xx} + u_{yy}, \qquad -\infty < x < \infty, 0 < y < \infty,$$
$$u(x, y, 0) = f(x, y),$$

subject to the boundary condition

a) $u(x, 0, t) = 0$

b) $u_y(x, 0, t) = 0$

15. Here we consider the well-known function

$$f(x) = \begin{cases} e^{-1/x^2}, & \text{if } x \neq 0, \\ 0, & \text{if } x = 0. \end{cases}$$

a) What is $\lim_{x \to 0} f(x)$?

b) Show that

$$\lim_{x \to 0} \frac{f(x)}{x} = \lim_{x \to 0} \frac{f(x)}{x^2} = 0,$$

then use mathematical induction to show that

$$\lim_{x \to 0} \frac{f(x)}{x^n} = 0, \qquad n = 3, 4, 5, \dots .$$

c) Show that f is infinitely differentiable and that $f^{(n)}(0) = 0$ for $n = 1, 2, 3, \dots$ (Hint: Look at

$$\lim_{h \to 0} \frac{f^{(n-1)}(h) - f^{(n-1)}(0)}{h} = f^{(n)}(0).)$$

d) Show that f is not analytic at $x = 0$. Remember that f is analytic at x_0 if the Taylor series

$$\sum_{n=0}^{\infty} \frac{f^{(n)}(x_0)}{n!}(x - x_0)^n$$

converges to f on some interval $x_0 - L < x < x_0 + L$ $(L > 0)$.

e) Now show, by differentiating term-by-term (which, as it turns out, is legal here), that the function

$$u_0(x, t) = \sum_{n=0}^{\infty} \frac{f^{(n)}(t)}{(2n)!} x^{2n}$$

is a solution of the heat/diffusion problem

$$u_t = u_{xx}, \qquad -\infty < x < \infty, t > 0,$$
$$u(x, 0) = 0.$$

Why does it follow that the problem does not possess a unique solution?

f) Explain why uniqueness must fail for the general problem

$$u_t = u_{xx}, \qquad -\infty < x < \infty, t > 0,$$
$$u(x, 0) = f(x).$$

This famous example is due to A.N. Tychonov (1935).

6.5   Distributions, the Dirac Delta Function and Generalized Fourier Transforms

We have been lax in our treatment of Fourier transforms. Specifically, we have paid almost no attention to the "absolutely integrable" requirement. And now, in the previous section, we introduce the function $\delta(x)$ which, for all intents and purposes, looks like

$$\delta(x) = \begin{cases} 0, & \text{if } x \neq 0, \\ \infty, & \text{if } x = 0! \end{cases}$$

Why? Well, as in most situations of this type, it's *because it works (so very well)*. Remember that Newton, Leibniz, Euler, et al. used the calculus with astounding results without even knowing why it worked. (Newton went so far as to recast all of the proofs in *Principia Mathematica* in terms of the well-established Euclidean geometry.)

Similarly, during the late 19th and early 20th centuries, mathematicians and scientists used the old *classical* analysis on problems involving these new "functions," with resounding success.[*]

Finally, in the early 1950s, the mathematician Laurent Schwartz[†] generalized the idea of function, putting everything into a setting—the theory of **generalized functions** or **distributions**—which made all of the ideas mathematically legitimate.

Schwarz defined these generalized functions and these operations in terms of integrals and, specifically, in terms of inner products. In this setting, our old classical functions still behave as always.

We begin by defining a **test function** to be any function $\phi \colon \mathbb{R} \to \mathbb{R}$ which

1) Has derivatives of all orders.

2) Is zero outside some finite internal (we say that ϕ has **finite support**).

Then, given any function $f \colon \mathbb{R} \to \mathbb{R}$ which is locally integrable on $(-\infty, \infty)$,[‡] the inner product

$$\langle f, \phi \rangle = \int_{-\infty}^{\infty} f(x)\phi(x)dx$$

converges for any test function ϕ. Further, it is *linear* in the sense that

$$\langle f, c_1\phi_1 + c_2\phi_2 \rangle = c_1\langle f, \phi_1 \rangle + c_2\langle f, \phi_2 \rangle$$

for any constants c_1, c_2 and any test functions ϕ_1, ϕ_2. So, if we think of f as fixed, what we actually have is a function of *functions*, which maps each test function ϕ to the real number $\langle f, \phi \rangle$. (Such a function, with domain equal to a set of functions and range consisting of a set of real or complex numbers, is called a **functional**.) Although the notation is not the usual, we can write

$$f[\phi] = \langle f, \phi \rangle,$$

and the linearity property becomes

$$f[c_1\phi_1 + c_2\phi_2] = c_1 f[\phi_1] + c_2 f[\phi_2].$$

---

[*]Heaviside was the "leader of the pack." (As Heaviside said, "Should I refuse a good dinner simply because I do not understand the process of digestion?") The downside of all this is that they could only show *formally* that they had the right answer, and that only by "plugging it back in."

[†]His famous article "Théorie des distributions" appeared in two parts, in 1950 and 1951.

[‡]That is, integrable on any bounded subset of the number line.

Now, to generalize, we look again at the *delta sequence*

$$f_n(x) = \begin{cases} \dfrac{n}{2}, & \text{if } |x| \le \dfrac{1}{n}, \\[2ex] 0, & \text{if } |x| > \dfrac{1}{n}. \end{cases}$$

For any test function ϕ,

$$\langle f_n, \phi \rangle = \frac{n}{2} \int_{-\frac{1}{n}}^{\frac{1}{n}} \phi(x)dx$$

and, *classically*, we may write

$$\lim_{n \to \infty} \langle f_n, \phi \rangle = \phi(0)$$

(see Exercise 4). Then, *formally*,

$$\phi(0) = \lim_{n \to \infty} \langle f_n, \phi \rangle = \lim_{n \to \infty} \int_{-\infty}^{\infty} f_n(x)\phi(x)dx$$

$$= \int_{-\infty}^{\infty} [\lim_{n \to \infty} f_n(x)]\phi(x)dx$$

$$= \int_{-\infty}^{\infty} \delta(x)\phi(x)dx,$$

suggesting that we *define* the **generalized function** $\delta(x)$, the **Dirac delta function**, by

$$\delta[\phi] = \langle \delta(x), \phi(x) \rangle = \phi(0).$$

Certainly, this is defined for any test function ϕ. Further,

$$\delta[c_1\phi_1 + c_2\phi_2] = [c_1\phi_1(x) + c_2\phi_2(x)]|_{x=0}$$
$$= c_1\delta[\phi_1] + c_2\delta[\phi_2],$$

so δ is a linear functional on the space of test functions.

Generally, we define a **generalized function** or **distribution** to be any linear functional on the space of test functions.[§] We use the notation

$$f[\phi] = \langle f, \phi \rangle = \int_{-\infty}^{\infty} f(x)\phi(x)dx,$$

realizing that the integral may not make sense, classically. *It is more than mere notation, though, in the sense that this generalized integral still behaves like a classical integral (so we may substitute, integrate by parts, etc.).*

[§]Actually, we need more—we define what we mean by a *continuous* functional and then define a distribution to be any *continuous*, linear functional on the space of test functions. For a very nice brief treatment, see the excellent book *Mathematical Methods in Physics and Engineering* by John W. Dettman.

MULTIPLICATION OF A DISTRIBUTION BY A FUNCTION

If f and g are integrable, then we have

$$\langle fg, \phi \rangle = \int_{-\infty}^{\infty} f(x)g(x)\phi(x)dx = \langle g, f\phi \rangle.$$

Therefore, we *define* the multiplication of the *distribution* g by the integrable function f by

$$\langle fg, \phi \rangle = \langle g, f\phi \rangle.^{\P}$$

In particular, it's interesting to note that the distribution $x^r \delta(x)$, for any $r > 0$, is given by

$$\langle x^r \delta(x), \phi(x) \rangle = \langle \delta(x), x^r \phi(x) \rangle = 0 \quad \text{(why?)}.$$

Therefore, there are infinitely many distributions which behave as the "zero distribution." In fact, it follows that

$$\langle f, \phi \rangle = \langle f + cx^r \delta(x), \phi \rangle$$

for any constants c and r, with $r > 0$.

TRANSLATION OF A DISTRIBUTION

If $f(x)$ is integrable, then so is $f(x - x_0)$ for any constant x_0. Then

$$\langle f(x - x_0), \phi(x) \rangle = \int_{-\infty}^{\infty} f(x)\phi(x + x_0)dx = \langle f(x), \phi(x + x_0) \rangle.$$

So we *define* the translation of the *distribution* f via

$$\langle f(x - x_0), \phi(x) \rangle = \langle f(x), \phi(x + x_0) \rangle$$

for all test functions ϕ. In particular, we have

$$\langle \delta(x - x_0), \phi(x) \rangle = \langle \delta(x), \phi(x + x_0) \rangle = \phi(x_0),$$

and we write

$$\int_{-\infty}^{\infty} \delta(x - x_0)\phi(x)dx = \phi(x_0).$$

This is called the **sifting property** of δ.$^{\parallel}$

¶Of course, we must show that this is a distribution, too.

$^{\parallel}$We generalize this statement below, to functions ϕ which are not test functions.

DERIVATIVE OF A DISTRIBUTION

Now, given an integrable and differentiable function $f(x)$, what distribution, in terms of f, is defined by f'? Well,

$$\langle f', \phi \rangle = \int_{-\infty}^{\infty} f'(x)\phi(x)dx,$$

and, integrating by parts, we have

$$= -\int_{-\infty}^{\infty} f(x)\phi'(x)dx \quad \text{(why?)}$$

$$= -\langle f, \phi' \rangle.$$

Therefore, we define the derivative of any distribution f to be the distribution

$$f'[\phi] = -\langle f, \phi' \rangle.$$

For the delta function, we have

$$\langle \delta', \phi \rangle = -\langle \delta, \phi' \rangle = -\phi'(0)$$

and, more generally,

$$\langle \delta^{(n)}, \phi \rangle = (-1)^n \phi^{(n)}(0), \qquad n = 1, 2, 3, \ldots$$

(see Exercise 6).

Another important function—and distribution—is the **Heaviside** [**] function

$$H(x) = \begin{cases} 1, & \text{if } x > 0, \\ 0, & \text{if } x < 0. \end{cases}$$

Classically, of course, $H'(x) = 0$ except at $x = 0$, where it doesn't exist. What is H' as a distribution? We have

$$\langle H', \phi \rangle = -\langle H, \phi' \rangle = -\int_0^{\infty} \phi'(x)dx \quad \text{(why?)}$$

$$= \phi(0) \quad \text{(again, why?)},$$

that is, for any test function ϕ,

$$\langle H', \phi \rangle = \langle \delta, \phi \rangle.$$

We say that $H'(x) = \delta(x)$, and, more generally, we can write

$$\langle H^{(n)}(x - x_0), \phi(x) \rangle = \langle (-1)^{n-1} \delta^{(n-1)}(x - x_0), \phi(x) \rangle$$

$$= (-1)^{n-1} \phi^{(n-1)}(x_0)$$

for any test function ϕ.

[**] Again, Oliver.

GENERALIZED FOURIER TRANSFORMS
AND DISTRIBUTIONS

The definition of the Fourier transform is generalized to this setting, as well, so that we may talk about the Fourier transform of any distribution. So, for example, we will throw around integrals like

$$\mathcal{F}[H(x)] = \int_0^\infty e^{-i\alpha x}\, dx$$

which clearly make no sense classically.

Let's begin by looking at the Fourier transform of the function in the delta sequence $f_n(x)$ from earlier. We have

$$\mathcal{F}[f_n(x)] = \frac{1}{\sqrt{2\pi}} \frac{n}{2} \int_{-1/n}^{1/n} e^{-i\alpha x}\, dx$$

$$= \frac{1}{\sqrt{2\pi}} \frac{\sin\alpha/n}{\alpha/n} \qquad \text{(why?)}$$

and, since

$$\lim_{n\to\infty} \frac{\sin\alpha/n}{\alpha/n} = 1,$$

we have no choice but to define

$$\mathcal{F}[\delta(x)] = \frac{1}{\sqrt{2\pi}}.^{\dagger\dagger}$$

This suggests that we would like to be able to say that

$$\int_{-\infty}^\infty \delta(x)f(x)dx = f(0)$$

even when f is *not* a test function. This is accomplished by looking at what is called the *principal value* of the given integral—an idea discussed classically in the following section. In particular, we'll have

$$\int_{-\infty}^\infty \delta(x)dx = 1.$$

More generally, we have (formally)

$$\mathcal{F}[\delta(x - x_0)] = \frac{1}{\sqrt{2\pi}} \int_{-\infty}^\infty \delta(x - x_0)e^{-i\alpha x}\, dx$$

$$= \frac{1}{\sqrt{2\pi}} \int_{-\infty}^\infty \delta(x)e^{-i\alpha(x+x_0)}\, dx$$

$$= \frac{e^{-i\alpha x_0}}{\sqrt{2\pi}},$$

---

††Many texts choose to write $\mathcal{F}[f(x)] = \int_{-\infty}^\infty f(x)e^{-i\alpha x}\, dx$, $\mathcal{F}^{-1}[F(\alpha)] = \frac{1}{2\pi} \int_{-\infty}^\infty F(\alpha)e^{i\alpha x}dx$, so that $\mathcal{F}[\delta(x)] = 1$.

and we have the transform pairs

$$\delta(x - x_0) \longleftrightarrow \frac{e^{-i\alpha x_0}}{\sqrt{2\pi}}$$

$$e^{icx} \longleftrightarrow \sqrt{2\pi}\, \delta(\alpha - c)$$

(see Exercise 7b).

Now, what about the transform of the Heaviside function? Here it turns out that we must proceed with care. We require that Fourier transforms of distributions must satisfy the old identity

$$\mathcal{F}[f'(x)] = i\alpha \mathcal{F}[f(x)],$$

so we must have

$$i\alpha \mathcal{F}[H(x)] = \mathcal{F}[\delta(x)] = \frac{1}{\sqrt{2\pi}}.$$

So it *looks* like we would get

$$\mathcal{F}[H(x)] = \frac{1}{i\alpha\sqrt{2\pi}}.$$

However, consider the function $f(x) = H(x) + H(-x) = 1$. We have, from Exercise 2 of Section 6.3,

$$\mathcal{F}[H(-x)] = -\frac{1}{i\alpha\sqrt{2\pi}},$$

from which it follows that

$$\mathcal{F}[1] = \sqrt{2\pi}\, \delta(\alpha) = \mathcal{F}[H(x)] + \mathcal{F}[H(-x)] = 0,$$

a contradiction! What happened? It turns out that the presence of $\frac{1}{\alpha}$ is what causes the trouble, due to its severe discontinuity at $\alpha = 0$. For this reason, in the distributional setting, $\frac{1}{\alpha}$ is referred to not as a function, but as a *pseudo function*.[‡‡] For our purposes, it suffices to remember that $f(x) + cx\delta(x)$ generates the same distribution as $f(x)$, for any constant c. So we see if we can find c so that

$$\mathcal{F}[H(x)] = \frac{1}{i\alpha\sqrt{2\pi}} + c\delta(\alpha) = \widehat{H}(\alpha).$$

Back to the equation above, we have

$$\sqrt{2\pi}\, \delta(\alpha) = \mathcal{F}[H(x)] + \mathcal{F}[H(-x)]$$
$$= \widehat{H}(\alpha) + \widehat{H}(-\alpha)$$
$$= 2c\delta(\alpha),$$

[‡‡]See, e.g., *Green's Functions and Boundary Value Problems*, 2nd ed., by Ivar Stakgold.

or

$$\mathcal{F}[H(x)] = \frac{1}{i\alpha\sqrt{2\pi}} + \sqrt{\frac{\pi}{2}}\delta(\alpha).$$

Finally, the **sign** or **signum function**

$$\text{sgn}(x) = H(x) - H(-x)$$

turns out to be very important, as we'll see presently. Therefore, let's look at its transform:

$$\mathcal{F}[\text{sgn } x] = \mathcal{F}[H(x)] - \mathcal{F}[H(-x)]$$

$$= \frac{1}{i\alpha}\sqrt{\frac{2}{\pi}}.$$

Now we're finally in a position to deal with $\mathcal{F}^{-1}\left[\frac{F(\alpha)}{i\alpha}\right]$ and the like (which we *tried* to do in Exercise 9b of Section 6.3). To repeat, though, we have the transform pairs

$$H(x) \longleftrightarrow \frac{1}{i\alpha\sqrt{2\pi}} + \sqrt{\frac{\pi}{2}}\,\delta(x)$$

$$\sqrt{\frac{\pi}{2}}\text{sgn } x \longleftrightarrow \frac{1}{i\alpha}.$$

Then, using the convolution formula, we have

$$\mathcal{F}^{-1}\left[\frac{F(\alpha)}{i\alpha}\right] = \frac{1}{\sqrt{2\pi}}\left(\sqrt{\frac{\pi}{2}}\text{ sgn } x\right) * f(x)$$

$$= \frac{1}{2}\int_{-\infty}^{\infty} \text{sgn}(x-y)f(y)dy$$

$$= \frac{1}{2}\left[\int_{-\infty}^{x} f(y)dy - \int_{x}^{\infty} f(y)dy\right] \quad \text{(why?)}.$$

Exercises 6.5

1. Show formally, using integrals, that we have no choice but to treat $\delta(x)$ as an even function.

2. Suppose that $\phi(x)$ and $\psi(x)$ are test functions. Decide if each statement is true or false, and justify your answer.

   a) $\phi(x + x_0)$ is a test function.

   b) $\phi'(x)$ is a test function.

   c) $\int_{-\infty}^{x} \phi(\xi)d\xi$ is a test function.

   d) $\phi(x)\psi(x)$ is a test function.

e) $\phi(x)/\psi(x)$ is a test function.

f) $f(x)\phi(x)$ is a test function for *any* infinitely differentiable function f.

3. a) Show that the function

$$\phi(x) = \begin{cases} e^{-1/x}e^{1/x-1}, & \text{if } 0 < x < 1, \\ 0, & \text{otherwise,} \end{cases}$$

is a test function.

b) **MATLAB:** Graph $\phi(x)$.

c) Using part (a) as a hint, construct a test function on the more general interval $a < x < b$ (i.e., such that $\phi(x) = 0$ outside of $a < x < b$).

4. Given the delta sequence

$$f_n(x) = \begin{cases} \dfrac{n}{2}, & \text{if } -1/n < x < 1/n, \\ 0, & \text{otherwise,} \end{cases}$$

use the *mean value theorem for integrals* to show that

$$\lim_{n\to\infty} \frac{n}{2} \int_{-1/n}^{1/n} f_n(x)\phi(x)dx = \phi(0)$$

for any continuous function ϕ.

5. a) Show that, for any $x \neq 0$,

$$\lim_{n\to\infty} \sqrt{\frac{n}{\pi}}e^{-nx^2} = 0.$$

b) What happens for $x = 0$?

c) Show that, for any $n > 0$,

$$\sqrt{\frac{n}{\pi}} \int_{-\infty}^{\infty} e^{-nx^2} dx = 1.$$

d) **MATLAB:** Plot the graph of $y = \sqrt{\frac{\pi}{n}} e^{-nx^2}$ for various values of n.

e) Describe how you would show that

$$\lim_{n\to\infty} \sqrt{\frac{n}{\pi}} \int_{-\infty}^{\infty} e^{-nx^2}\phi(x)dx = \phi(0)$$

for any continuous, bounded function ϕ. (Therefore, the sequence of continuous function $\sqrt{\frac{n}{\pi}}e^{-nx^2}$ is a delta sequence.)

f) More generally, suppose that $\int_{-\infty}^{\infty} f(x)dx = 1$. Define

$$f_n(x) = \frac{1}{n}f\left(\frac{x}{n}\right), \qquad n = 1, 2, 3, \ldots$$

and show that

i) $\lim_{n\to\infty} f_n(x) = 0$ for any fixed $x \neq 0$.

ii) $\lim_{n\to\infty} \int_{-\infty}^{\infty} f_n(x)dx = 1$.

6. Use mathematical induction to prove that

$$\langle \delta^{(n)}(x - x_0), \phi(x)\rangle = (-1)^n \phi^{(n)}(x_0), \qquad n = 1, 2, 3, \ldots .$$

7. a) Show that $\mathcal{F}[\delta(x + x_0) \pm \delta(x - x_0)] = \begin{cases} 2\cos\alpha x_0 \\ 2i\sin\alpha x_0. \end{cases}$

b) Show that, if

$$\mathcal{F}[\delta(x - x_0)] = \frac{e^{-i\alpha x_0}}{\sqrt{2\pi}},$$

then

$$\mathcal{F}[e^{icx}] = \sqrt{2\pi}\,\delta(\alpha - c).$$

8. a) Although we have the classical formula $\mathcal{F}[f(x-x_0)] = e^{-ix_0\alpha}\mathcal{F}[f(x)]$, compute $\mathcal{F}[H(x - x_0)]$, instead, the same way that we computed $\mathcal{F}[H(x)]$.

b) In Example 1 of Section 6.3, we found the Fourier transform of the square wave

$$f(x) = \begin{cases} 1, & \text{if } 1 < x < 1, \\ 0, & \text{otherwise.} \end{cases}$$

Here, instead, rewrite f as a sum of various Heaviside functions, then take the Fourier transform (and show that we get the same answer!).

c) If $f(x)$ is a function and $g(x)$ is a distribution, we define the product fg by $\langle fg, \phi\rangle = \langle g, f\phi\rangle$ (so long as f is sufficiently well behaved). Show that the old product rule still holds, that is, that $(fg)' = fg' + gf'$, in the distributional sense.

d) **Distributional derivatives of general piecewise smooth functions:** Suppose $f(x)$ is the piecewise smooth function

$$f(x) = \begin{cases} g(x), & \text{if } x < x_0, \\ h(x), & \text{if } x > x_0. \end{cases}$$

Rewrite f using the Heaviside function, as in part (b), then compute its distributional derivative. (Hopefully, it will be identical to the old $f'(x)$ for every $x \neq x_0$.)

e) Generalize part (d) to any piecewise smooth function

$$f(x) = \begin{cases} f_1(x), \text{if } x < x_1, \\ f_2(x), \text{if } x_1 < x < x_2, \\ \quad \vdots \\ f_n(x), \text{ if } x > x_{n-1}. \end{cases}$$

9. Use Exercise 9d of Section 6.3 to find

a) $\mathcal{F}^{-1}\left[\frac{1}{\alpha - c}\right]$

b) $\mathcal{F}^{-1}\left[\frac{1}{\alpha^2 + \alpha - 2}\right]$

10. Consider an electric network consisting of a resistor, an inductor and a capacitor in series. Suppose that the network has a voltage source, so that the voltage at the terminals of the network is given by $E(t)$. See Figure 6.7. Then it turns out that the current $I(t)$ must satisfy the differential equation

$$L\ddot{I} + R\dot{I} + \frac{1}{C}I = \dot{E}(t), \qquad -\infty < t < \infty.$$

Here, L, R and C are the inductance, resistance and capacitance.

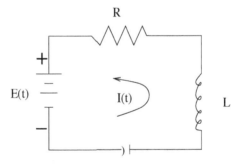

FIGURE 6.7
The electrical network in Exercise 10.

a) Suppose, further, that there is a switch and that the voltage is a constant E_0 when the switch is closed. Conclude that, if the switch is closed at time $t = t_0$, the equation becomes

$$L\ddot{I} + R\dot{I} + \frac{1}{C}I = E_0\delta(t - t_0), \qquad -\infty < t < \infty.$$

b) Use Fourier transforms, as well as Exercise 9 of Section 6.3, to solve this problem from the case $L = 1$, $R = 0$, $C = 1/4$, $E_0 = 1$ and $t_0 = 3$.

11. a) Show that the heat kernel

$$u(x,t) = \frac{1}{2k\sqrt{\pi t}} e^{-\frac{x^2}{4k^2 t}}$$

is the solution of the heat/diffusion problem

$$u_t = k^2 u_{xx}, \qquad -\infty < x < \infty, t > 0,$$
$$u(x,0) = \delta(x).$$

b) Show formally that the *non*homogeneous heat problem

$$w_t = k^2 w_{xx} + \delta(x), \qquad -\infty < x < \infty, t > 0,$$
$$w(x,0) = 0,$$

has solution $w(x,t) = \int_0^t u(x,t)$, where u is the solution from part (a).

12. Suppose we *start out* by defining the Fourier transform/inverse transform formulas for the delta function. That is, formally, we say that

$$\mathcal{F}[\delta(x - x_0)] = \frac{1}{\sqrt{2\pi}} \int_{-\infty}^{\infty} \delta(x - x_0) e^{-i\alpha x}\, dx = \frac{e^{-i\alpha x_0}}{\sqrt{2\pi}}$$

and

$$\delta(x - x_0) = \mathcal{F}^{-1}\left[\frac{e^{-i\alpha x_0}}{\sqrt{2\pi}}\right] = \frac{1}{2\pi} \int_{-\infty}^{\infty} e^{-i\alpha(x - x_0)}\, d\alpha.$$

Then it turns out that we can recover (at least formally) the Fourier transform formulas for arbitrary functions. For example,

a) Given

$$\delta(x - y) = \frac{1}{2\pi} \int_{-\infty}^{\infty} e^{i\alpha(x - y)}\, d\alpha,$$

multiply both sides by $f(y)$, then integrate, to arrive at the Fourier integral representation of f (6.11).

b) Similarly, starting with

$$\delta(\alpha - \beta) = \frac{1}{2\pi} \int_{-\infty}^{\infty} e^{-iy(\alpha - \beta)}\, dy,$$

multiply both sides by $F(\alpha)G(\beta)e^{i\beta x}$, integrate both sides with respect to β and α and then reverse the order of integration on the right side to arrive at the convolution formula

$$\int_{-\infty}^{\infty} F(\alpha)G(\alpha)e^{i\alpha x}\, d\alpha = f(x) * g(x).$$

Here, of course, F and G are the transform of f and g, respectively.

6.6   Proof of the Fourier Integral Formula *

Now we are ready to prove Theorem 6.2 (whence Theorem 6.1 follows). First, let's introduce the idea of the Cauchy principal value.

Definition 6.1 *We define*

$$\lim_{M \to \infty} \int_{-M}^{M} f(x)dx$$

to be the **Cauchy principal value** *or* **Cauchy principal part** *of the (possibly divergent—else, why bother?) integral*

$$\int_{-\infty}^{\infty} f(x)dx.$$

The point is that the integral may diverge, while the Cauchy principal value converges, as we see in the following example. (Also, see Exercise 2.)

Example 1 As we saw in Section 6.2, $\int_{-\infty}^{\infty} \frac{x}{x^2+c^2}dx$ diverges. However, $\lim_{R \to \infty} \int_{-M}^{M} \frac{x}{x^2+c^2}dx = 0$ (why?).

As a result, we needed to be very careful when we talked about

$$\int_{0}^{\infty} \frac{\alpha \sin \alpha x}{\alpha^2 + c^2} d\alpha$$

and the like. In fact, Theorem 6.2 should *really* read as the following.

Theorem 6.2′ *Suppose that f is piecewise smooth on every finite interval, and suppose that $\int_{-\infty}^{\infty} |f(x)|dx < \infty$. Then, for each x,*

$$\frac{1}{2\pi} \lim_{M \to \infty} \int_{-M}^{M} F(\alpha)e^{i\alpha x} \, d\alpha$$

$$= \frac{1}{2\pi} \lim_{M \to \infty} \int_{-M}^{M} \int_{-\infty}^{\infty} f(z)e^{-i\alpha z} \, dz \, e^{i\alpha x} \, d\alpha = \frac{1}{2}[f(x+) + f(x-)].$$

Not surprisingly, there is much similarity between the proof of Theorem 6.2′ and the convergence proof for Fourier *series*, in Section 3.5. Here, we will need first to prove a lemma which will allow us to use the same trick we used in step [4] of that earlier proof. And, in proving this lemma, we'll rely on the proof of step [5], as well.

---

*This section may be skipped without loss of continuity.

Lemma 6.1 *For any real constant $M > 0$,*

$$\int_0^\infty \frac{\sin x}{x}\,dx = \int_0^\infty \frac{\sin Mx}{x}\,dx = \int_{-\infty}^0 \frac{\sin Mx}{x}\,dx = \frac{\pi}{2}.$$

PROOF   The proof entails

[1] Showing that $\int_0^\infty \frac{\sin x}{x}\,dx$ converges;

[2] Rewriting the improper integral as

$$\int_0^\pi \frac{\sin x}{x}\,dx = \lim_{n\to\infty}\left\{\int_0^\pi \left[\frac{\sin u/2}{u/2} - 1\right]\frac{\sin \frac{2n+1}{2}u}{2\sin u/2}\,du \right.$$
$$\left. + \int_0^\pi \frac{\sin \frac{2n+1}{2}u}{2\sin u/\alpha}\,du\right\}$$

(remember that $D_n(u) = \sin \frac{2n+1}{2}u/2\sin \frac{u}{2}$ is the Dirichlet kernel);

[3] Using properties of the Dirichlet kernel to show that the first integral $\to 0$ and the second $\to \frac{\pi}{2}$.

PROOF of [1]   Write

$$\int_0^\infty \frac{\sin x}{x}\,dx = \int_0^1 \frac{\sin x}{x}\,dx + \int_1^\infty \frac{\sin x}{x}\,dx.$$

Since $\lim_{x\to\infty}\frac{\sin x}{x} = 1$, the first integral is finite. For the second, we integrate by parts twice and get

$$\int_1^\infty \frac{\sin x}{x}\,dx = \cos 1 - \lim_{x\to\infty}\frac{\cos x}{x} - \int_1^\infty \frac{\cos x}{x^2}\,dx.$$

The limit on the right-hand side is 0 while, for the last integral, we have

$$\int_1^\infty \left|\frac{\cos x}{x^2}\right|\,dx \le \int_1^\infty \frac{1}{x^2}\,dx = 1.$$

Therefore, the integral is absolutely convergent (which implies it is convergent).   ∎

PROOF of [2] and [3]

$$\int_0^\infty \frac{\sin x}{x}\,dx = \lim_{R\to\infty}\int_0^R \frac{\sin x}{x}\,dx$$
$$= \lim_{n\to\infty}\int_0^{\frac{2n+1}{2}\pi} \frac{\sin x}{x}\,dx, \quad n \in \mathbb{N}$$

(since it converges, we can have the upper limit approach ∞ any way we'd like).

Now, we substitute

$$x = \frac{2n+1}{2}u, \quad dx = \frac{2n+1}{2}du,$$

and we get

$$\int_0^\infty \frac{\sin x}{x}dx = \lim_{n\to\infty}\int_0^\pi \frac{\sin\frac{2n+1}{2}u}{u}du$$

$$= \lim_{n\to\infty}\int_0^\pi \frac{\sin u/2}{u/2}\frac{\sin\frac{2n+1}{2}u}{2\sin u/2}du$$

$$= \lim_{n\to\infty}\int_0^\pi \frac{\sin u/2}{u/2}D_n(u)du,$$

where $D_n(u)$ is, again, the Dirichlet kernel,

$$= \lim_{n\to\infty}\left[\int_0^\pi \left(\frac{\sin u/2}{u/2}-1\right)D_n(u)du + \int_0^\pi D_n(u)du\right].$$

To arrive at the last expression, we just added and subtracted 1 inside the integral. But why? Well, first off, we already computed the second integral in Section 3.5. As for the other, if we consider the continuous function

$$f(x) = \begin{cases} \dfrac{\sin x/2}{x/2}, & \text{if } x \neq 0, \\ 1, & \text{if } x = 0, \end{cases}$$

then

$$\frac{\sin u/2}{u/2} - 1 = f(u) - f(0)$$

and the first integral is almost identical to the one from step [5] in Section 3.5. There, we used the *Riemann–Lebesgue Lemma* to prove that the first integral $\to 0$, and, in step [4], we proved that the second $\to \frac{\pi}{2}$.

Finally, it is straightforward to show that

$$\int_0^\infty \frac{\sin Mx}{x}dx = \int_{-\infty}^0 \frac{\sin Mx}{x}dx = \int_0^\infty \frac{\sin x}{x}dx$$

for any constant M (see Exercise 1).   ∎

Now we prove Theorem 6.2′.

PROOF of Theorem 6.2′   We must prove that

$$\lim_{M\to\infty}\frac{1}{2\pi}\int_{-M}^M\int_{-\infty}^\infty f(z)e^{-i\alpha z}dz\, e^{i\alpha x}d\alpha - \frac{1}{2}[f(x+)+f(x-)] = 0. \quad (6.14)$$

To this end we shall

[1] Justify reversing the order of integration, so that we may rewrite the integral as

$$\int_{-M}^{M} \int_{-\infty}^{\infty} f(z)e^{-i\alpha z} dz \; e^{i\alpha x} dx = \int_{-\infty}^{\infty} \left[\int_{-M}^{M} e^{i\alpha(x-z)} d\alpha \right] f(x) dz$$

$$= \int_{-\infty}^{\infty} \frac{e^{i\alpha(x-z)}}{i(x-z)} \Big|_{\alpha=-M}^{\alpha=M} f(z) dz \; ^\dagger$$

$$= 2 \int_{-\infty}^{\infty} \frac{\sin M(x-z)}{x-z} f(z) dz,$$

(and, letting $u = z - x$)

$$= 2 \int_{-\infty}^{\infty} \frac{\sin Mu}{u} f(x+u) du.$$

[2] Use the lemma to write

$$f(x\pm) = \frac{2}{\pi} \int_{0}^{\infty} \frac{\sin Mu}{u} f(x\pm) du,$$

in which case (6.14) becomes

$$\lim_{M\to\infty} \frac{1}{2\pi} \cdot 2 \int_{-\infty}^{\infty} \frac{\sin Mu}{u} f(x+u) du - \frac{1}{2} \left[\frac{2}{\pi} \int_{0}^{\infty} \frac{\sin Mu}{u} f(x+) du \right.$$

$$\left. + \frac{2}{\pi} \int_{-\infty}^{0} \frac{\sin Mu}{u} f(x-) du \right]$$

$$= \lim_{M\to\infty} \left[\frac{1}{\pi} \int_{0}^{\infty} \frac{f(x+u) - f(x+)}{u} \sin Mu \; du \right.$$

$$\left. + \frac{1}{\pi} \int_{-\infty}^{0} \frac{f(x+u) - f(x-)}{u} \sin Mu \; du \right].$$

[3] Show that each integral $\to 0$.

PROOF of [1] In order to reverse the order of integration, we must be able to show that

$$\int_{-\infty}^{\infty} f(z)e^{i\alpha(x-z)} dz$$

converges *uniformly* as a function of α, on $-R \le \alpha \le R$. While beyond the scope of this book (although we briefly discuss uniform convergence in Appendix B), it turns out that this follows from the fact that

$$\int_{-\infty}^{\infty} |f(z)e^{i\alpha(x-z)}| dz = \int_{-\infty}^{\infty} |f(z)| dz < \infty.$$

---

†Note that, when $x = z$, the inside integral $= 2M$, although this is of no concern.

(Note that the absolute value sign here represents *not* absolute value, but the *modulus* of a complex number. It turns out that $|z_1 z_2| = |z_1||z_2|$ and that $|e^{i\theta}| = 1$ for any real number θ.)   ∎

PROOF of [2]   Easy (why?).   ∎

PROOF of [3]   We prove here that

$$\lim_{M\to\infty} \int_0^\infty \frac{f(x+u) - f(x+)}{u} \sin Mu \, du = 0;$$

the proof for the second integral is almost identical. First, we write

$$\int_0^c \frac{f(x+u) - f(x+)}{u} \sin Mu \, du + \int_c^\infty \frac{f(x+u) - f(x+)}{u} \sin Mu \, du$$

for constant $c > 0$. The only "problem" is what happens near $u = 0$; however, as we've done before, we notice that

$$\lim_{u\to 0} \frac{f(x+u) - f(x+)}{u} = f'(x+)$$

which is finite since f is piecewise smooth. This means that we can apply the Riemann–Lebesgue Lemma to the first integral (again, see Section 3.5), with the result being that, regardless of the choice of c, the first integral $\to 0$ as $M \to \infty$. As for the second integral, we can make it as small as we'd like just by taking c large enough (why?). To be more precise, we may choose c and $M = M_1$ so that the second integral $< \frac{\epsilon}{2}$, and choose $M = M_2$ which makes the first integral $< \frac{\epsilon}{2}$; then, for $M = \max(M_1, M_2)$, the sum is $< \epsilon$.   ∎

And we are finished!

THE FOURIER TRANSFORM, INTUITIVELY

In closing, let's take an intuitive look at the Fourier transform. As the Fourier series resolved a function into its **discrete spectrum**, as well as the contribution *at* each frequency, so the Fourier transform may be looked at as resolving a function into its **continuous spectrum**, with $F(\alpha)$ being a measure of the contribution at each frequency α. Then, we see two phenomena involving the Fourier transform which are analogous to those which we found for the Fourier series:

1) **Functions with sharp spikes require a greater contribution from higher frequencies than do functions without them.** So, we saw that "thin, sharp" Gaussians had transforms which were "wide and flat," and vice versa, culminating in the extreme case involving the delta function. Indeed, $\delta(x)$ requires the *same* contribution from *each* frequency

(since $\mathcal{F}[\delta(x)] = $ constant), while a constant function requires an *infinite* contribution from a *single* frequency.

2) **The smoother a function, the less of a contribution it requires from higher frequencies, and vice versa.** This phenomenon can be seen, for example, in the relationship between the square wave and its Fourier transform, $\frac{\sin \alpha}{\alpha}$, and is obviously related to the *Gibbs phenomenon* for Fourier series. Of course, the delta function represents the extreme case of this phenomenon, as well.

Exercises 6.6

1. Given that $\int_0^\infty \frac{\sin x}{x}dx = \frac{\pi}{2}$, show that $\int_0^\infty \frac{\sin Mx}{x}dx = \int_{-\infty}^0 \frac{\sin Mx}{x}dx = \frac{\pi}{2}$, for any real constant $M > 0$. What can you say about $\int_0^\infty \frac{\cos x}{x}dx$?

2. Give three examples of functions f with the property that $\int_{-\infty}^\infty f(x)dx$ diverges but $\lim_{M\to\infty} \int_{-M}^M f(x)dx$ converges.

3. One also may talk about the principal value of an integral for which the integrand has a singularity. For example, we may have that $\int_{-a}^a f(x)dx$ diverges, but

$$\lim_{\epsilon \to 0+} \left[\int_{-a}^{-\epsilon} f(x)dx + \int_{\epsilon}^a f(x)dx \right]$$

converges, in which case this limit is the principal value of the divergent integral. Give an example of an integral $\int_{-b}^a f(x)dx$ which diverges, but for which the principal value converges.

OTHER INTEGRAL TRANSFORMS

There are many integral transforms, each determined by (a) the function that is multiplied by f (this function is called the **kernel** of the transform) and (b) the interval over which we integrate the product. In general, given $f(x)$ on $a < x < b$, an integral transform of f will be a function

$$F(\alpha) = \int_a^b K(x, \alpha)f(x)dx.$$

Here, K is the kernel. For example, for the Fourier transform, $a = -\infty$, $b = \infty$ and $K(x, \alpha) = \frac{1}{\sqrt{2\pi}}e^{i\alpha x}$. Other transforms are listed in Table 6.2.

		Kernel $K(x,\alpha)$	Interval
1.	Laplace	$e^{-\alpha x}$	$[0,\infty)$
2.	Fourier	$e^{-i\alpha x}$	$(-\infty,\infty)$
3.	Fourier sine	$\sin\alpha x$	$[0,\infty)$
4.	Fourier cosine	$\cos\alpha x$	$[0,\infty)$
5.	Finite Fourier	$K\left(x,\dfrac{n\pi}{L}\right)=e^{\frac{in\pi x}{L}}$	$[-L,L]$
6.	Finite Fourier sine	$K\left(x,\dfrac{n\pi}{L}\right)=\sin\dfrac{n\pi x}{L}$	$[0,L]$
7.	Finite Fourier cosine	$K\left(x,\dfrac{n\pi}{L}\right)=\cos\dfrac{n\pi x}{L}$	$[0,L]$
8.	Mellin	$x^{\alpha-1}$	$[0,\infty)$
9.	Hankel*	$xJ_m(\alpha x)$	$[0,\infty)$
10.	Hilbert	$\dfrac{1}{\pi}\dfrac{1}{x+\alpha}$	$(-\infty,\infty)$
11.	Weierstrass	$\dfrac{1}{2\sqrt{\pi\alpha}}e^{-x^2/4\alpha}$	$(-\infty,\infty)$

* The function J_m is the mth-order Bessel function of the first kind; the Hankel transform sometimes is called the Fourier–Bessel transform.

TABLE 6.2
Some integral transforms.

It turns out that the Mellin transform is very similar to the Fourier transform, as we see in the following exercise.

4. Given $f(x)$ on $[0,\infty)$, we define the *Mellin transform* of f to be

$$\mathcal{M}[f(x)]=F_{\mathcal{M}}(\alpha)=\int_0^\infty f(x)x^{\alpha-1}dx,$$

where α is purely imaginary, i.e., $\alpha=i\beta$, where β is real.

a) Show via the substitution $x=e^{-z}$ that

$$F_{\mathcal{M}}(\alpha)=H(\beta)=\int_{-\infty}^\infty f(e^{-z})e^{-i\beta z}dz.$$

In other words,

$$\mathcal{M}[f(x)]=\sqrt{2\pi}\mathcal{F}[h(x)],\text{ where }h(x)=f(e^{-x}).$$

Therefore, the Mellin transform should have properties similar to those of the Fourier transform.

b) Show that it follows from part (a) that the inversion formula for the Mellin transform is

$$f(x) = \frac{1}{2\pi} \int_{-\infty}^{\infty} x^{-i\beta} H(\beta) d\beta.$$

c) Show formally that

$$\mathcal{M}^{-1}[F_{\mathcal{M}}(\alpha) G_{\mathcal{M}}(\alpha)] = (f \times g)(x),$$

where the *multiplicative convolution* $f \times g$ is defined by

$$(f \times x)(x) = \int_0^{\infty} f(z) g\left(\frac{x}{z}\right) \frac{dz}{z}.$$

5. The Weierstrass transform should look familiar. If we let $\mathcal{W}[f(x)]$ be the Weierstrass transform of f, then describe the pertinence of the function

$$w(y, \alpha) = \mathcal{W}[f(y - x)].$$

(By the way, the Weierstrass kernel sometimes is called the Gauss–Weierstrass kernel.)

INTEGRAL EQUATIONS

Frequently, functions which describe certain physical situations are solutions of what are called **integral equations**; often, a differential equation problem will be recast in the form of an integral equation.

The general linear integral equation in the unknown function $f(x)$ can be written

$$\lambda \int_a^b K(x, y) f(y) dy + g(x) = h(x) f(x),$$

where g, h and K are known functions (K is called the *kernel* of the integral equation), and λ is a parameter (often playing the role of an eigenvalue, as it turns out). If $g(x) \equiv 0$, the equation is *homogeneous*. Often the function K is of the form $K = K(x - y)$.

Solve the following integral equations.

6. $\int_{-\infty}^{\infty} \frac{f(y)}{(x-y)^2 + a^2} dy - \frac{1}{x^2 + b^2} = 0, \ 0 < a < b.$

   This is an example of a so-called **Wiener–Hopf equation**.

7. $\int_{-\infty}^{\infty} e^{-|x-y|} f(y) dy + e^{-x^2/2} + f(x) = 0.$

$f(x) = \frac{1}{\sqrt{2\pi}} \int_{-\infty}^{\infty} F(\alpha)e^{i\alpha x}\, d\alpha$		$F(\alpha) = \frac{1}{\sqrt{2\pi}} \int_{-\infty}^{\infty} f(x)e^{-i\alpha x}\, dx$	
1.	$f'(x)$	1.	$i\alpha F(\alpha)$
2.	$xf(x)$	2.	$iF'(\alpha)$
3.	$f(x-c)$	3.	$e^{-ic\alpha}F(\alpha)$
4.	$e^{icx}f(x)$	4.	$F(\alpha-c)$
5.	$f(cx)$	5.	$\frac{1}{c}F\left(\frac{\alpha}{c}\right)$
6.	$\frac{1}{c}f\left(\frac{x}{c}\right)$	6.	$F(c\alpha)$
7.	$f(x)*g(x)$	7.	$\sqrt{2\pi}\,F(\alpha)G(\alpha)$
8.	$f(x)g(x)$	8.	$\frac{1}{\sqrt{2\pi}}\,F(-\alpha)*G(-\alpha)$
9.	$e^{-c|x|}$	9.	$\sqrt{\frac{2}{\pi}}\,\frac{c}{\alpha^2+c^2}$
10.	$\frac{1}{x^2+c^2}$	10.	$\sqrt{\frac{\pi}{2}}\,e^{-c|\alpha|}/c$
11.	$\exp[-(x-m)^2/2\sigma^2]/\sigma\sqrt{2\pi}$	11.	$\exp\left[-\frac{\sigma^2\alpha^2}{2}-im\alpha\right]\big/\sqrt{2\pi}$
12.	$\begin{cases} 1, & \text{if } |x| \le L, \\ 0, & \text{if } |x| > L. \end{cases}$	12.	$\sqrt{\frac{2}{\pi}}\,\frac{\sin L\alpha}{\alpha}$
13.	$\frac{\sin Lx}{x}$	13.	$\begin{cases} \sqrt{\frac{\pi}{2}}, & \text{if } |\alpha| \le L, \\ 0, & \text{if } |\alpha| > L. \end{cases}$
14.	$\delta(x)$	14.	$\frac{1}{\sqrt{2\pi}}$
15.	1	15.	$\sqrt{2\pi}\,\delta(\alpha)$
16.	$H(x)$	16.	$\frac{1}{i\alpha\sqrt{2\pi}}+\sqrt{\frac{\pi}{2}}\,\delta(\alpha)$
17.	$\operatorname{sgn} x$	17.	$\sqrt{\frac{2}{\pi}}\,\frac{1}{i\alpha}$
18.	$\frac{1}{x}$	18.	$\sqrt{\frac{\pi}{2}}\,\frac{\operatorname{sgn}\alpha}{i}$
19.	$\operatorname{sgn} x * f(x)$	19.	$\frac{2F(\alpha)}{i\alpha}$
20.	$\frac{f(x)}{x}$	20.	$-\frac{i}{2}\operatorname{sgn}\alpha * f(-\alpha)$

TABLE 6.3
Table of Fourier transforms.

Prelude to Chapter 7

Eventually, we'd like to solve PDEs in higher dimensions. As we show in the following chapter, applying separation of variables to many of the important higher-dimensional equations leads to certain special ODEs, the solutions of which are called *the special functions*. (Okay, we never said that mathematicians have much originality when it comes to naming things. To be fair, they're often called *the special functions of mathematical physics*.)

The earliest studied of the special functions are probably the *Bessel functions*. Daniel Bernoulli ran into the first few of these when he solved the *hanging chain problem* in 1733, and Euler encountered the modified Bessel functions around the same time. Then, in 1759, Euler solved the problem of the vibration of a circular drumhead (as we do in Chapter 9), giving us the standard solution in terms of the Bessel functions of the first kind.

However, the Bessel functions are named after the well-known Prussian mathematician and astronomer Friedrich Wilhelm Bessel (1784–1824), who performed the first systematic study of solutions of *Bessel's ODE* during the decade 1815–1825. (Although he made numerous contributions to mathematics and the like, during his lifetime Bessel was best known by virtue of his being the first person to measure the parallax of a star.)

The other special functions we'll look at are the so-called (we'll see why in Chapter 8) *orthogonal polynomials*—the *Legendre, Chebyshev, Hermite and Laguerre polynomials*. The Legendre polynomials seem first to have arisen in the work of Adrien-Marie Legendre (1752–1833) and also of Laplace, in their study of gravitational potentials in the 1780s. (In fact, early on they often were referred to as the *Laplace coefficients*.)

The Russian mathematician Pafnuti Chebyshev (or Tchebycheff or Chebichev!) (1821–1894) derived both sets of polynomials which bear his name, while studying the approximation of functions by these polynomials, in the 1850s. Chebyshev also seems to have been the first to consider orthogonal polynomials as such, in a general setting.

Similarly, the Hermite polynomials were studied by Charles Hermite (1822–1901) in the 1860s, in *his* work on polynomial approximation of functions, although Laplace seems to have encountered them in the early 1800s, when working on probability theory. Incidentally, known for many things, Hermite's most famous contribution is his proof, in 1873, that e is a transcendental number. Finally, the Laguerre polynomials are named after the geometer Edmond Laguerre (1834–1886). The Hermite and Laguerre polynomials are involved in the solution of certain versions of Schrödinger's equation and, thus, play a key role in the study of quantum mechanics.

7

Special Functions and Orthogonal Polynomials

7.1 The Special Functions and Their Differential Equations

The PDEs describing many important physical problems lead, when separated, to ODEs with solutions which are called the **special functions (of mathematical physics)**. These include the Bessel functions and the sets of **orthogonal polynomials**: the Chebyshev, Hermite, Laguerre and Legendre polynomials. (In Chapter 8, we explain why they are called *orthogonal*.)

Here we derive these ODEs in the following exercises, in most cases via separation of variables.

Exercises 7.1

1. **Bessel's ODE:** In Chapter 9, we'll see that the PDE for the motion of a vibrating membrane is

$$u_{tt} = c^2(u_{xx} + u_{yy}).$$

   a) Use polar coordinates to change this into the equation

$$\frac{1}{c^2}u_{tt} = u_{rr} + \frac{1}{r}u_r + \frac{1}{r^2}u_{\theta\theta}$$

   (see Exercise 11, Section 1.6).

   b) Separate variables and show that the separated equations can be written as

$$T'' + c^2\lambda T = 0, \quad \Theta'' + \gamma\Theta = 0, \quad R'' + \frac{1}{r}R' + \left(\lambda - \frac{\gamma}{r^2}\right)R = 0,$$

   where $T = T(t)$, $\Theta = \Theta(\theta)$ and $R = R(r)$, and λ and γ are separation constants. The R-ODE is the eigenvalue version of **Bessel's equation**.

   c) Use Exercise 18 in Section 1.7 to conclude that $\gamma = m^2$, where $m = 0, 1, 2, \ldots$.

d) If $\lambda > 0$, use the change of variable $x = \sqrt{\lambda}\, r$ to rewrite the R-equation in the standard form of **Bessel's equation of order** α,

$$x^2 y'' + xy' + (x^2 - \alpha^2)y = 0$$

(where, for the vibrating membrane, of course, $\alpha = m = 0, 1, 2, \ldots$).

e) If, instead, $\lambda < 0$, use the change of variable $x = \sqrt{-\lambda}\, r$ to rewrite the R-equation as

$$x^2 y'' + xy' - (x^2 + \alpha^2)y = 0.$$

This is the **modified Bessel's equation**, and its solutions are **modified Bessel functions, of order** α.

Note: The only reason to change to polars is if we are looking at a *circular* membrane/drumhead. Now, from the theory of ODEs, the R-equation has a *singular point* (see the following section) at $r = 0$, which means that there may be solutions which "blow up" there. As these solutions are unrealistic, the physical nature of the problem suggests that we must stipulate the additional "boundary condition" that R be bounded as $r \to 0^+$. Similarly, we must have y bounded as $x \to 0^+$ for the equations in parts (d) and (e).

2. **Legendre's ODE:** Laplace's equation in three dimensions is the PDE

$$\nabla^2 u = u_{xx} + u_{yy} + u_{zz} = 0.$$

a) Transform this equation, using spherical coordinates, to

$$(\rho^2 u_\rho)_\rho + \frac{1}{\sin\phi}(u_\phi \sin\phi)_\phi + \frac{1}{\sin^2\phi}u_{\theta\theta} = 0.$$

b) Suppose we are looking for solutions which are θ-independent and ϕ-independent. Show that they are of the form $u = u(\rho)$, where u satisfies the Cauchy–Euler equation

$$\rho^2 u_{\rho\rho} + 2\rho u_\rho = 0.$$

c) If, instead, we want solutions which are only θ-independent, show that they are those functions $u = u(\rho, \phi)$ which satisfy

$$(\rho^2 u_\rho)_\rho + \frac{1}{\sin\phi}(u_\phi \sin\phi)_\phi = 0.$$

d) Why are there no ϕ-independent solutions?

e) Use separation of variables, $u(\rho, \phi) = R(\rho)\Phi(\phi)$, to show that the equation in part (c) can be separated into the ODEs

$$(\rho^2 R')' - \lambda R = 0, \quad (\Phi' \sin\phi)' + \lambda\Phi \sin\phi = 0.$$

Here, λ is the separation constant.

    f) Using the change of variable $x = \cos\phi$, $0 \leq \phi \leq \pi$, show that the ϕ-equation becomes

$$[(1 - x^2)y']' + \lambda y = 0.$$

Here, $y(x) = y(\cos\phi) = \Phi(\phi)$. This ODE is the eigenvalue version of **Legendre's equation**. Note that our change to spherical coordinates has added the artificial-looking requirement that solutions be finite at $\phi = n\pi$, $n = 0, \pm 1, \ldots$, i.e., at $x = \pm 1$. These points correspond, of course, to the north and south poles.

Note: For reasons similar to those above, we also stipulate that y be bounded as $x \to 1^-$ and $x \to -1^+$. It turns out that the only such solutions are polynomials!

3. **Hermite's ODE:** The one-dimensional Schrödinger's equation for a harmonic oscillator is

$$-iu_t = u_{xx} - x^2 u,$$

where $u = u(x,t)$ and $i^2 = -1$. If we solve this equation in $-\infty < x < \infty$, quantum mechanics tells us that the solutions must $\to 0$ as $x \to \pm\infty$.

    a) Show that this equation separates into the ODEs

$$T' + i\lambda T = 0, \quad X'' + (\lambda - x^2)X = 0,$$

    where $u(x,t) = X(x)T(t)$ and λ is the separation constant.

    b) Show that, if $\lambda = 1$, $X = e^{-x^2/2}$ is a solution. What is

$$\lim_{x \to \pm\infty} e^{-x^2/2}?$$

    c) Show that, for arbitrary λ, the change of dependent variable

$$y(x) = e^{\frac{x^2}{2}} X(x)$$

transforms the X-equation into

$$y'' - 2xy' + (\lambda - 1)y = y'' - 2xy' + \lambda_1 y = 0.$$

This is the eigenvalue version of **Hermite's equation**.

Note: Here, from the physics of the problem, it turns out that we need our solutions to have the property

$$\lim_{x \to \pm\infty} e^{-x^2/2} y(x) = 0.^*$$

Again, it turns out that the only viable solutions are polynomials.

---

*Actually, the business of assigning boundary conditions at singular points, especially at $\pm\infty$, is not trivial. For a detailed treatment, see either *Green's Functions and Boundary Value Problems* by Ivar Stakgold or *Ordinary Differential Equations* by E.L. Ince.

4. **Laguerre's ODE: Quantum Mechanics and The Hydrogen Atom:** Quantum mechanics has shown us that we cannot predict where a particle (an electron, say) will be at a given moment. Rather, the best that we can do is to talk about the *probability* of its being at any particular location at any given time. Thus, we look at the so-called **wave function** of the particle,

$\psi(x, y, z, t)$ = probability that the particle is at point (x, y, z) at time t.

In the 1920s, Erwin Schrödinger derived (actually, cobbled together!) the partial differential equation that ψ must satisfy. Thus, we have Schrödinger's equation

$$i\hbar\psi_t = -\frac{\hbar^2}{2m}\nabla^2\psi + V\psi,$$

where

m = mass of particle,

$\hbar(\text{we say } h\text{-}bar) = \dfrac{h}{2\pi}$, where h = Planck's constant,

and

$V = V(x, y, z)$ = potential energy of the force field at point (x, y, z).

We may always choose units so that $m = \hbar = 1$ in these units. Now, we would like to study the simplest quantum system, the hydrogen atom, where our "particle" is the atom's electron, and the force field is due solely to the atom's nucleus, a single proton, located at the origin. Therefore, we switch to spherical coordinates, and the potential function is just $V = -\frac{e^2}{\rho}$, where e is the electric charge of a proton (and, of course, negative the charge of the electron). Again, we choose units so that $e = 1$, and we have the simplified Schrödinger's equation

$$i\psi_t = -\frac{1}{2}\nabla^2\psi - \frac{1}{\rho}\psi.$$

a) Separate out *time*, that is, let $u = T(t)v$ where v is a function of the space variables. Show that the space part of the equation becomes

$$\nabla^2 v + \left(\lambda + \frac{2}{\rho}\right)v = 0,$$

where λ is the separation constant.

b) If we write this equation in spherical coordinates, as in Exercise 2, we get

$$\frac{1}{\rho^2}(\rho^2 v_\rho)_\rho + \frac{1}{\rho^2 \sin\phi}(v_\phi \sin\phi)_\phi$$

$$+ \frac{1}{\rho^2 \sin^2\phi}v_{\theta\theta} + \left(\lambda + \frac{2}{\rho}\right)v = 0.$$

Writing $v(\rho, \theta, \phi) = R(\rho)\Theta(\theta)\Phi(\phi)$, show that this equation separates into

$$\rho^2 R'' + 2\rho R' + (\lambda \rho^2 + 2\rho - \mu)R = 0$$
$$\Theta'' + \gamma \Theta = 0$$
$$\Phi'' \sin^2 \phi + \Phi' \sin \phi \cos \phi + \mu \Phi \sin^2 \phi - \gamma \Phi = 0.$$

c) Conclude that we must have $\gamma = k^2$, where $k = 0, 1, 2, \dots$.

d) In the Φ-equation, change variables via $x = \cos \phi$ and show that the resulting equation is

$$[(1 - x^2)y']' + \left(\mu - \frac{k^2}{1 - x^2}\right) y = 0.$$

When $k \neq 0$, this is *almost* Legendre's equation and is called the **associated Legendre equation of order k**, about which we'll say more in Exercise 6.

e) Now for the R-equation: As we'll see in Chapter 9, it turns out that we must have $\mu = \ell(\ell + 1)$, $\ell = 0, 1, 2, \dots$, and that $\lambda < 0$. So we write $\lambda = -\beta^2$, and we have

$$R'' + \frac{2}{\rho}R' + \left(-\beta^2 + \frac{2}{\rho} - \frac{\ell(\ell + 1)}{\rho^2}\right)R = 0.$$

Make the change of variable $x = 2\beta\rho$, and show that the resulting equation is

$$R'' + \frac{2}{x}R' + \left(-\frac{1}{4} + \frac{1}{\beta x} - \frac{\ell(\ell + 1)}{x^2}\right)R = 0$$

(where, now, $R' = \frac{dR}{dx}$).

f) Next, make the change of dependent variable

$$R(x) = x^a e^{bx} w(x),$$

and determine all values of a and b so that the equation becomes

$$xw'' + [(2\ell + 1) + 1 - x]w' + \left(\frac{1}{\beta} - \ell - 1\right) w = 0.$$

We look at this equation in the following exercise.

5. **Laguerre's ODE, cont.:** Laguerre's equation is the ODE

$$xy'' + (1 - x)y' + \lambda y = 0.$$

a) Show that if y is a solution of Laguerre's equation, then $y^{(m)} = \frac{d^m y}{dx^m}$ is a solution of the **associated Laguerre's equation**

$$xu'' + (1 + m - x)u' + (\lambda - m)u = 0.$$

As we'll see, the polynomial solutions of

$$xy'' + (1 - x)y' + ny = 0$$

are the Laguerre polynomials $L_n(x)$, $n = 0, 1, 2, \ldots$. Conclude that the **associated Laguerre polynomial**

$$L_n^m(x) = \frac{d^m}{dx^m} L_n(x)$$

is a polynomial solution of the corresponding associated Laguerre's equation.

Note: Similarly, our solutions here must be bounded as $x \to 0^+$, and, at the other end, it turns out that we need

$$\lim_{x \to \infty} \sqrt{x} \, e^{-x/2} y(x) = 0.$$

Once more, it turns out that the only such solutions are polynomials.

b) Show that the w-equation in Exercise 4e is, indeed, the associated Laguerre's equation.

6. **Associated Legendre's ODE:** The three-dimensional wave equation is, as you might guess,

$$u_{tt} = c^2 \nabla^2 u = c^2(u_{xx} + u_{yy} + u_{zz}).$$

Therefore, if we wish to study the vibration of a ball, we should look at this equation in spherical coordinates. As above, we have

$$\frac{1}{c^2} u_{tt} = (\rho^2 u_\rho)_\rho + \frac{1}{\rho^2 \sin \phi}(u_\phi \sin \phi)_\phi + \frac{1}{\rho^2 \sin^2 \phi} u_{\theta\theta}.$$

a) First "separate out" t, then ρ, and show that we have

$$T'' + c^2 \lambda T = 0, \quad R'' + \frac{2}{\rho}R' + \left(\lambda - \frac{\gamma}{\rho^2}\right)R = 0,$$

$$\frac{1}{\sin \phi}(v_\phi \sin \phi)_\phi + \frac{1}{\sin^2 \phi} v_{\theta\theta} + \gamma v = 0,$$

where $v = v(\theta, \phi)$ and λ and γ are separation constants.

b) Show that the change of dependent variable $w = r^{1/2}R$ turns the R-equation into

$$w'' + \frac{1}{r}w' + \left(\lambda - \frac{\gamma + \frac{1}{4}}{r^2}\right)w = 0.$$

What equation is this, essentially?

c) Now separate $v(\theta, \phi) = \Theta(\theta)\Phi(\phi)$, and show that the result is

$$\Theta'' + \beta\Theta = 0, \quad (\Phi' \sin \phi)' \sin \phi + (\gamma \sin^2 \Phi - \beta)\Phi = 0.$$

d) Similar to what we did in Exercise 2f, use the change of variable $x = \cos \phi$ to transform the ϕ-equation into

$$[(1 - x^2)y']' + \left(\gamma - \frac{\beta}{1 - x^2}\right)y = 0.$$

Here, again, is the **associated Legendre's equation:** Again, note that the change to sphericals makes it necessary that we have y bounded at the poles, that is, at $x = \pm 1$.

e) Explain why we must have $\beta = k^2$, where k is an integer, so that we write the equation as

$$[(1 - x^2)y']' + \left(\gamma - \frac{k^2}{1 - x^2}\right)y = 0.$$

7. **Chebyshev Polynomials and ODEs**

a) Use Euler's formula

$$e^{i\theta} = \cos \theta + i \sin \theta$$

to prove De Moivre's Theorem

$$\cos n\theta + i \sin n\theta = (\cos \theta + i \sin \theta)^n.$$

b) Use De Moivre's Theorem and the fact that $a + bi = c + di$ if and only if $a = c$ and $b = d$ to prove the trig identities

$$\cos 2\theta = 2 \cos^2 \theta - 1, \qquad \sin 2\theta = 2 \sin \theta \cos \theta,$$
$$\cos 3\theta = 4 \cos^3 \theta - 3 \cos \theta, \quad \sin 3\theta = \sin \theta (4 \cos^2 \theta - 1)$$

and

$$\cos 4\theta = 8 \cos^4 \theta - 8 \cos^2 \theta + 1,$$
$$\sin 4\theta = \sin \theta (8 \cos^3 \theta - 4 \cos \theta).$$

In other words, for the polynomials

$$T_2(x) = 2x^2 - 1, \qquad S_1(x) = 2x$$
$$T_3(x) = 4x^3 - 3x, \qquad S_2(x) = 4x^2 - 1$$
$$T_4(x) = 8x^4 - 8x^2 + 1, \quad S_3(x) = 8x^3 - 4x,$$

we have

$$\cos nx = T_n(\cos\theta) \quad \text{and} \quad \sin nx = (\sin\theta)S_{n-1}(\cos\theta).$$

c) Use the binomial theorem,

$$(a+b)^n = \sum_{k=0}^{n} \binom{n}{k} a^{n-k}b^k,$$

to prove that, for $n = 0, 1, 2, \ldots$, there exist polynomials $T_n(x)$ and $S_n(x)$ such that

$$\cos n\theta = T_n(\cos\theta) \quad \text{and} \quad \sin(n+1)\theta = S_n(\cos\theta)\sin\theta.$$

These are called the **Chebyshev polynomials of the first and second kind**, respectively. Further, show that if n is even, then T_n and S_n are even functions, while if n is odd, then T_n and S_n are odd. (See Exercise 1d, Section 7.6.)

d) Since $\cos n\theta = T_n(\cos\theta)$ is a solution of the ODE

$$y'' + n^2 y = 0,$$

use the substitution $x = \cos\theta$ to show that $T_n(x)$ satisfies **Chebyshev's equation of the first kind**

$$(1 - x^2)y'' - xy' + n^2 y = 0.$$

e) Although not so obvious, it can be shown that the function $f_n(\theta) = \frac{\sin(n+1)\theta}{\sin\theta}$ satisfies the ODE

$$(\sin\theta)y'' + 2(\cos\theta)y' + n(n+2)\sin\theta y = 0.$$

Use the substitution $x = \cos\theta$ to show that $S_n(x)$ satisfies **Chebyshev's equation of the second kind**

$$(1 - x^2)y'' - 3xy' + n(n+2)y = 0.$$

Note: As with Legendre's equation, we must have y bounded as $x \to 1^-$ and $x \to -1^+$.

7.2 Ordinary Points and Power Series Solutions; Chebyshev, Hermite and Legendre Polynomials

All of the ODEs derived in the previous section have variable coefficients. You may remember that the method of power series is, in general, used to solve such equations. So, we look for solutions of the form

$$y = \sum_{n=0}^{\infty} a_n (x - x_0)^n,$$

where $x = x_0$ is the point "about which we expand the series" (and, usually, an important point, for example, the point where the initial conditions are given) and the a_n are unknown constants to be determined. The resulting power series is actually the Taylor series of a solution. And we know, from the theory of ODEs, that if $x = x_0$ is an *ordinary point* of the equation, we can expect to find two linearly independent solutions, while if $x = x_0$ is a *singular point*, this need not be the case.

Definition 7.1 *Given the ODE*

$$y'' + P(x)y' + Q(x)y = 0$$

we say that $x = x_0$ is an **ordinary point** *of the equation if $P(x)$ and $Q(x)$ are analytic at $x = x_0$. Otherwise, we say that $x = x_0$ is a* **singular point***.*

(Remember that $f(x)$ is analytic at $x = x_0$ if the Taylor series of f,

$$\sum_{n=0}^{\infty} \frac{f^{(n)}(x_0)}{n!} (x - x_0)^n,$$

converges to f on an interval $x_0 - r < x < x_0 + r$, $r > 0$. In practice, polynomials are analytic everywhere, since a polynomial is *its own* Taylor series. Also, the functions e^x, $\sin x$ and $\cos x$ are analytic everywhere. One obvious way for a function *not* to be analytic at $x = x_0$ is if $P(x)$ or $Q(x)$ has the factor $x - x_0$ in its denominator.)

Example 1 Hermite's equation: Hermite's ODE $y'' - 2xy' + \lambda y = 0$ has no singular points.

Example 2 Legendre's and Chebyshev's equations: Legendre's equation can be written as

$$y'' - \frac{2x}{1 - x^2} y' + \frac{\lambda}{1 - x^2} y = 0;$$

similarly, we may rewrite Chebyshev's equations:

$$y'' - \frac{x}{1-x^2}y' + \frac{\lambda}{1-x^2}y = 0$$

and

$$y'' - \frac{3x}{1-x^2}y' + \frac{\lambda}{1-x^2}y = 0.$$

We are interested in solving these equations only for $-1 \leq x \leq 1$. Note that all three are singular at $x = \pm 1$, while every x in $-1 < x < 1$ is an ordinary point of each.

Example 3 Bessel's and Laguerre's equations: We rewrite Bessel's ODE as

$$y'' + \frac{1}{x}y' + \frac{x^2 - \alpha^2}{x^2}y = 0$$

and Laguerre's equation as

$$y'' + \frac{1-x}{x}y' + \frac{\lambda}{x}y = 0.$$

Clearly, each is singular at $x = 0$ and nowhere else.

For various reasons, we would like to solve each of these equations at $x = x_0 = 0$. We may solve the equations from Examples 1 and 2 using standard power series solutions. Further, from the theory of ODEs, the radius of convergence of each solution will be $r = |x_0|$, where x_0 is the singular point nearest the origin.

What about using power series for Bessel and Laguerre? Let's first look at a simpler example.

Example 4 Solve the Cauchy–Euler equation

$$2x^2 y'' + 3xy' - 2y = 0.$$

We let $y = x^r$ and find that our two independent solutions are

$$y_1 = \sqrt{|x|} \quad \text{and} \quad y_2 = \frac{1}{x^2}.$$

Neither of these functions is analytic at $x = 0$; therefore, no power series of the form

$$\sum_{n=0}^{\infty} a_n x^n$$

will lead to either solution.

Although occasionally we *will* find power series solutions at a singular point, we cannot, in general, expect to do so. Therefore, we need a more general method to deal with these situations—for this, we wait for the next section.

For now, let's solve Legendre's equation, leaving Chebyshev's and Hermite's equations for the exercises.

LEGENDRE'S EQUATION

$$(1 - x^2)y'' - 2xy' + \lambda y = 0, \quad -1 < x < 1, y \text{ bounded as } x \to \pm 1.$$

We let $y = \sum\limits_{i=0}^{\infty} a_i x^i$ and plug into the ODE. The result is

$$2a_2 + (\lambda - 2)a_1 + \lambda a_0 + \sum_{i=2}^{\infty} [(i+2)(i+1)a_{i+2} + (\lambda - i^2 - i)a_i]x^n = 0.$$

So we have a_0, a_1 arbitrary,

$$a_2 = \frac{(2 - \lambda)a_1 - \lambda a_0}{2},$$

and the recurrence relation is

$$a_{i+2} = \frac{i(i+1) - \lambda}{(i+2)(i+1)}a_i, \qquad i = 2, 3, \dots .$$

Now, we want solutions that are bounded at $x = \pm 1$. However, if we look at

$$\lim_{i \to \infty} \frac{a_{i+2}}{a_i} = \lim_{i \to \infty} \frac{i(i+1) - \lambda}{(i+2)(i+1)} = 1,$$

we see that if $a_0 \neq 0$, we get an infinite series of even powers of x, which behaves like the geometric series $\sum\limits_{i=0}^{\infty} x^{2i}$. Similarly, $a_1 \neq 0$ gives us a series which behaves like $\sum\limits_{i=0}^{\infty} x^{2i+1}$. Each of these series diverges at $x = \pm 1$ (and is unbounded at $x = 1$). (This unrigorous treatment can be made precise, of course.)

So the only way that we can have a bounded solution is if the series terminates, that is, if it is a polynomial. When will this happen?

Suppose $\lambda = n(n+1)$, where n is a positive integer. Then

$$a_{n+2} = \frac{n(n+1) - n(n+1)}{(n+2)(n+1)}a_n = 0.$$

In this case, we'll have

$$a_{n+2} = a_{n+4} = \dots = 0,$$

i.e., if n is odd, the series of odd powers will be a polynomial; similarly for n even and the even powers. In each case, the *other* half of the series will still be infinite—the only way to eliminate it will be to choose $a_0 = 0$ or $a_1 = 0$, respectively.

Essentially, we have found that the numbers

$$\lambda_n = n(n+1), \qquad n = 0, 1, 2, \ldots,$$

are the *eigenvalues* of the given Legendre *boundary-value problem*, while the eigenfunctions are the corresponding polynomial solutions. Since any constant multiple of an eigenfunction also is an eigenfunction, we define the nth **degree Legendre polynomial** $P_n(x)$ to be the polynomial solution for which $P_n(1) = 1$.

In order to compute some Legendre polynomials, we remember that, for each n, we have

$$a_2 = \frac{(2-\lambda)a_1 - \lambda a_0}{2} = \frac{[2 - n(n+1)]a_1 - n(n+1)a_0}{2}$$

along with the recurrence formula

$$a_{i+2} = \frac{i(i+1) - n(n+1)}{(i+2)(i+1)} a_i, \qquad i = 2, 3, \ldots.$$

So, we have, for the first few Legendre polynomials:

$\lambda_0 = 0$:

Choose $a_1 = 0 \Rightarrow a_3 = a_5 = a_7 = \ldots = 0$. Then,

$$a_2 = 2a_1 - 0a_0 = 0$$
$$\Rightarrow a_4 = a_6 = a_8 = \ldots = 0.$$

So, choosing $a_0 = 1$ gives us

$$P_0(x) = 1.$$

$\lambda_1 = 2$:

Choose $a_0 = 0 \Rightarrow a_2 = a_4 = a_6 = \ldots = 0$. The recurrence formula here is

$$a_{i+2} = \frac{i(i+1) - 2}{(i+2)(i+1)} a_i,$$

so $a_3 = 0$, implying that $a_5 = a_7 = a_9 = \ldots = 0$. Thus,

$$P_1(x) = x.$$

$\lambda_2 = 6$:

Choose $a_1 = a_3 = a_5 = \ldots = 0$. Letting $a_0 = 1$, we have $a_2 = \frac{0a_1 - 6a_0}{12} = -3$. The recurrence formula is

$$a_{i+2} = \frac{i(i+1) - 6}{(i+2)(i+1)} a_i,$$

so $a_4 = a_6 = \ldots = 0$. So an eigenfunction is $f_2(x) = -3x^2 + 1$. Dividing by $f_2(1) = -2$, we have

$$P_2(x) = \frac{3}{2}x^2 - \frac{1}{2}.$$

We may continue computing polynomials in this manner. The next few turn out to be

$$P_3(x) = \frac{1}{2}(5x^3 - 3x)$$

$$P_4(x) = \frac{1}{8}(35x^4 - 30x^2 + 3)$$

$$P_5(x) = \frac{1}{8}(63x^5 - 70x^3 + 15x)$$

(see Exercise 2, below, and Exercise 1 of Section 7.6). We plot P_0 through P_5 in Figure 7.1. Notice that if n is even, then P_n is even and, if n is odd, then P_n is odd. We prove this in Exercise 1b of Section 7.6.

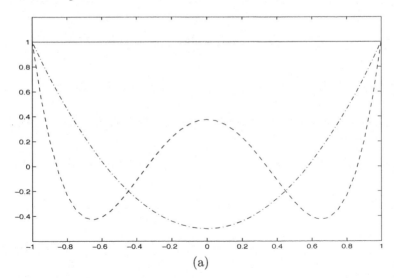

(a)

FIGURE 7.1
MATLAB graphs of the first six Legendre polynomials: (a) P_0, P_2 and P_4 and (b) P_1, P_3 and P_5. (In each case, solid, dash-dotted and dashed, respectively.)

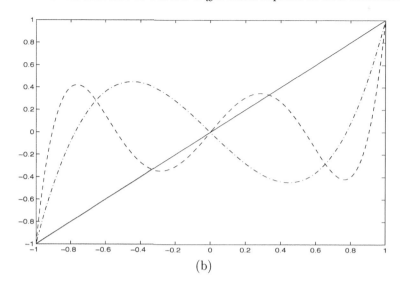

(b)

FIGURE 7.1 continued.

One can show (with a *lot* of work) that the Legendre polynomials themselves satisfy a recurrence relation. In fact, as we'll see, Hermite's equation, Laguerre's equation and both Chebyshev's equations have polynomial eigenfunctions and, in each case, these polynomials satisfy a recurrence relation.

Each such set of polynomials is called a set of **orthogonal polynomials**, for reasons given in Chapter 8, and referred to briefly in Section 7.6. There, we'll list the recurrence formulas and other properties of the orthogonal polynomials.

In the exercises, we derive the Chebyshev and the Hermite polynomials.

Exercises 7.2

1. For each equation, classify each point on the x-axis as an ordinary point or a singular point.

   a) $(x + 2)y'' - y' + y = 0$

   b) $x(x + 1)y'' + 3xy' + (x + 1)y = 0$

   c) $(x^2 + 1)y'' + 2xy' + 5y = 0$

2. Derive the Legendre polynomials P_3, P_4 and P_5.

3. Proceed as we did in the case of Legendre's equation to find the eigenvalues and eigenfunctions for each of the following:

   a) Chebyshev's equation of the first kind

   $$(1 - x^2)y'' - xy' + \lambda y = 0, \quad -1 < x < 1, y \text{ bounded as } x \to \pm 1.$$

Calculate the first four **Chebyshev polynomials of the first kind**, T_0, T_1, T_2 and T_3, where, in each case, we *normalize* the polynomial so that $T_n(1) = 1$.

b) Chebyshev's equation of the second kind

$$(1 - x^2)y'' - 3xy' + \lambda y = 0, \quad -1 < x < 1, y \text{ bounded as } x \to \pm 1.$$

Calculate the first four **Chebyshev polynomials of the second kind**. So, calculate S_0, S_1, S_2 and S_3, where we normalize by choosing $S_n(1) = n + 1$.

c) Hermite's equation

$$y'' - 2xy' + \lambda y = 0, \quad -\infty < x < \infty, e^{-x^2} y \to 0 \text{ as } x \to \pm\infty.$$

That is, calculate the first four **Hermite polynomials**, H_0, H_1, H_2 and H_3. In each case, choose the polynomial so that its leading term, that is, the term with the highest power, is of the form $2^n x^n$.

4. **MATLAB**

a) Referring to Exercise 3 of Section 7.1, plot the graphs of the functions

$$y_n = e^{x^2/2} H_n(x), \quad n = 0, 1, 2, 3,$$

where H_n is the nth-degree Hermite polynomial from the previous exercise.

b) Use BVP4C to solve the problem

$$y'' + (\lambda - x^2)y = 0, \quad -L \le x \le L,$$
$$y(-L) = y(L) = 0,$$

for various large values of L, and compare the first four eigenfunctions with the functions in part (a).

5. The associated Legendre equation, again, is

$$[(1 - x^2)y']' + \left(\gamma - \frac{k^2}{1 - x^2}\right) y = 0, \quad -1 < x < 1,$$

y bounded as $x \to \pm 1, k$ a positive integer.

We may solve this as we solved Legendre's equation, but there is a slicker way of dealing with it.

a) If y satisfies Legendre's equation,

$$(1 - x^2)y'' - 2xy' + \lambda y = 0,$$

differentiate this equation k times to show that $y^{(k)}$ satisfies the equation

$$(1 - x^2)y^{(k+2)} - 2(k + 1)xy^{(k+1)} + [\lambda - k(k + 1)]y = 0.$$

b) Next, make the change of dependent variable

$$y(x) = (1 - x^2)^{k/2} z(x)$$

in the associated Legendre equation.

c) Conclude that if $f_\gamma(x)$ is a **Legendre function**, that is, a solution of Legendre's equation

$$(1 - x^2)y'' - 2xy' + \gamma y = 0,$$

then

$$g_\gamma^k(x) = (1 - x^2)^{k/2} f_\gamma^{(k)}(x)$$

is a solution of the associated Legendre equation. In particular, show that the **associated Legendre function of degree n and order k**

$$P_n^k(x) = (1 - x^2)^{k/2} P_n^{(k)}(x)$$

is a solution of

$$(1 - x^2)y'' - 2xy' + \left[n(n+1) - \frac{k^2}{1 - x^2} \right] y = 0.$$

(It turns out that, as with Legendre's equation, the associated Legendre's equation has bounded solutions if and only if $\gamma = n(n+1)$, $n = 0, 1, 2, \ldots$. These bounded solutions are the associated Legendre functions.)

7.3   The Method of Frobenius; Laguerre Polynomials

We saw that both Laguerre's equation and Bessel's equation are singular at $x = 0$. Actually, there are two types of singular points, one "good" and the other "not so good."

Definition 7.2 *Suppose the ODE*

$$y'' + P(x)y' + Q(x)y = 0$$

has a singular point at $x = x_0$. Let's rewrite the equation in the form

$$(x - x_0)^2 y'' + (x - x_0)p(x)y' + q(x)y = 0.$$

*If p and q are analytic at $x = x_0$, we say that x_0 is a **regular** singular point of the ODE; otherwise, it is an **irregular** singular point.*

Example 1 The equation

$$y'' + \frac{1}{(x+1)^2(x-2)}y' - \frac{1}{(x-2)^2}y = 0$$

is singular at $x = -1$ and $x = 2$. If we multiply by $(x-2)^2$, the equation becomes

$$(x-2)^2 y'' + \frac{x-2}{(x+1)^2}y' - y = 0.$$

Since $p(x) = \frac{1}{(x+1)^2}$ and $q(x) = -1$ are analytic at $x = 2$, this point is a *regular* singular point.

If, instead, we multiply the ODE by $(x+1)^2$, we get

$$(x+1)^2 y'' + (x+1)\frac{x-2}{x+1}y' - (x+1)^2 y = 0.$$

Then, $q(x) = (x+1)^2$ is analytic at $x = -1$, but $p(x) = \frac{x-2}{x+1}$ is not. Therefore, $x = -1$ is an *irregular* singular point.

Example 2 Bessel's equation

$$x^2 y'' + xy' + (x^2 - \alpha^2)y = 0$$

and Laguerre's equation

$$x^2 y' + x(1-x)y' + xy = 0$$

both have a *regular* singular point at $x = 0$. Now, it turns out that if an ODE has a *regular* singular point at $x = x_0$, then it always possesses one solution of the form

$$y = (x-x_0)^r \sum_{n=0}^{\infty} a_n(x-x_0)^n$$

(and, sometimes, two). The procedure for finding such solutions is called the **method of Frobenius**. Let's begin with an example.

Example 3 Try to find Frobenius series solutions for

$$2x^2 y'' + xy' - (x+1)y = 0$$

about the regular singular point $x = 0$. We let

$$y = x^r \sum_{n=0}^{\infty} a_n x^n = \sum_{n=0}^{\infty} a_n x^{n+r},$$

and we proceed as we did with power series solutions. So,

$$y' = \sum_{n=0}^{\infty} (n+r)a_n x^{n+r-1}$$

and

$$y'' = \sum_{n=0}^{\infty}(n+r)(n+r-1)a_n x^{n+r-2}.$$

Substituting into the ODE gives us

$$\sum_{n=0}^{\infty}2(n+r)(n+r-1)a_n x^{n+r} + \sum_{n=0}^{\infty}(n+r)a_n x^{n+r}$$

$$- \sum_{n=0}^{\infty}a_n x^{n+r+1} + \sum_{n=0}^{\infty}a_n x^{n+r} = 0$$

or, after changing the index in the third series and pulling out the first term of each of the other three series,

$$[2r(r-1)+r-1]a_0$$

$$+ \sum_{n=1}^{\infty}\left\{[2(n+r)(n+r-1)+(n+r)-1]a_n - a_{n-1}\right\}x^n = 0.$$

Now we can determine the values of r for which a solution is possible. The first term must be zero and, since we don't want $a_0 = 0$ (else we'll have the 0 solution!), r must satisfy the **indicial equation**

$$2r(r-1)+r-1 = 0.$$

Therefore, the only viable values of r are

$$r = -\frac{1}{2}, \ 1.$$

We take each in turn.

Case 1: $r = 1$

In this case, the recurrence formula

$$[2(n+r)(n+r-1)+(n+r)-1]a_n - a_{n-1} = 0$$

becomes

$$a_n = \frac{1}{2(n+1)^2 - (n+1) - 1}a_{n-1}$$

$$= \frac{1}{n(2n+3)}a_{n-1}, \qquad n = 1, 2, \ldots$$

(note the similarity between the denominator and the indicial equation). Therefore, the first few coefficients are

$$a_1 = \frac{1}{5}a_0, \ a_2 = \frac{1}{14}a_1 = \frac{1}{70}a_0, \ a_3 = \frac{1}{27}a_2 = \frac{1}{1890}a_0, \cdots$$

and, letting $a_0 = 1$, the corresponding solution is

$$y = x\left[1 + \frac{1}{5}x + \frac{1}{70}x^2 + \frac{1}{1890}x^3 + \cdots\right].$$

Case 2: $r = -\frac{1}{2}$

Here, the recurrence formula becomes

$$\left[2\left(n - \frac{1}{2}\right)\left(n - \frac{3}{2}\right) + \left(n - \frac{1}{2}\right) - 1\right]a_n - a_{n-1} = 0$$

or

$$a_n = \frac{1}{n(2n-3)}a_{n-1}, \qquad n = 1, 2, \ldots .$$

So the first few coefficients are

$$a_1 = -a_0, \ a_2 = \frac{1}{2}a_1 = -\frac{1}{2}a_0, \ a_3 = \frac{1}{9}a_2 = -\frac{1}{18}a_0, \cdots$$

and the corresponding Frobenius solution is

$$y = x^{-1/2}\left[1 - x - \frac{1}{2}x^2 - \frac{1}{18}x^3 + \cdots\right].$$

(See Exercise 3.)

So, in this case, we were fortunate to be able to find two linearly independent solutions. It's natural to ask what could have "gone wrong," that is, under what circumstances might we *not* be able to find a second solution this way? Obviously, if the indicial equation has only one (double) root, we're stuck, at least for now. Are there any other such situations?

Take a look at the second recurrence formula in the previous example—certainly it never gave us any trouble for $n = 1, 2, \ldots$. However, it's easy enough to imagine that there may be values of r for which the corresponding denominator *does* become zero for a positive integer n. Again, we look at some examples; however, let us mention that, even if only one Frobenius solution to a problem exists, it is *always* possible to derive from this solution a second, linearly independent solution.

Example 4 Let's look for Frobenius solutions of

$$xy'' + 2y' - y = 0.$$

Plugging in $y = \sum\limits_{n=0}^{\infty} a_n x^{n+r}$, we arrive at

$$r(r+1)a_0 + \sum_{n=1}^{\infty}[(n+r)(n+r+1)a_n - a_{n-1}]x^n = 0.$$

The indicial equation $r(r+1) = 0$ has roots $r = 0, -1$. Also, the recurrence formula is

$$a_n = \frac{1}{(n+r)(n+r+1)}a_{n-1}, \qquad n = 1, 2, \ldots .$$

For $r = 0$, we have

$$a_n = \frac{1}{n(n+1)}a_{n-1}, \qquad n = 1, 2, \ldots,$$

which leads to a solution (see Exercise 2). As for $r = -1$, the recurrence formula becomes

$$a_n = \frac{1}{(n-1)n}a_{n-1}$$

and, for $n = 1$, we get $a_1 = \frac{1}{0}a_0$, which is undefined.

Now, what caused this to happen, and is there any way around it? As we'll see in Exercise 9, this occurs whenever the two indicial roots differ by an integer! As for getting around it, we usually can't. Here, though, let's go back and rewrite the troublesome recurrence formula in its original form:

$$(n-1)na_n = a_{n-1}.$$

Again, $n = 1$ gives us $0a_1 = a_0$. Of course, we could choose $a_0 = 0$ but, in that case, we'll wind up with the same solution we got from $r = 0$ (try it!). However, as the next example shows, there *are* cases where the right side already is zero, and the difficulty is circumvented.

Example 5 Find all Frobenius solutions of

$$x^2 y'' - x^2 y' + (x^2 - 2)y = 0.$$

Proceeding as usual, we arrive at the equation

$$(r+1)(r-2)a_0 + [(r-1)(r+2)a_1 - ra_0]x$$
$$+ \sum_{n=2}^{\infty}[((n+r)(n+r-1) - 2)a_n - (n+r-1)a_{n-1} + a_{n-2}]x^n = 0.$$

The indicial roots are $r = -1, 2$. For $r = 2$, we have no problems (see Exercise 3). As for $r = -1$, we get

$$-2a_1 + a_0 = 0$$

and

$$n(n-3)a_n = (n-2)a_{n-1} - a_{n-2}, \qquad n = 2, 3, \ldots .$$

So we have $a_1 = \frac{1}{2}a_0$ and, from $n = 2$, $a_2 = \frac{1}{2}a_0$. What happens when $n = 3$? We get

$$0a_3 = a_2 - a_1$$
$$= \frac{1}{2}a_0 - \frac{1}{2}a_0$$
$$= 0,$$

that is, the recurrence formula is true for $n = 3$ *regardless of the choice of a_3*. So a_3 is arbitrary! Therefore, we may choose $a_3 = 0$ (although we don't have to).

The question still remains as to how to find a second linearly independent solution in those cases where there is only one Frobenius solution. It turns out that, in all cases, there is a so-called *logarithmic* solution. If y_1 is the Frobenius solution then, for the case $r_1 = r_2$, the second solution takes the form

$$y_2 = y_1 \ln x + x^{r_1} \sum_{n=0}^{\infty} c_n x^n$$

while, for the case $r_1 - r_2 = N$, an integer, we have

$$y_2 = y_1 \ln x + x^{r_2} \sum_{n=0}^{\infty} d_n x^n.$$

There are various ways to derive these solutions—for example, via reduction of order—and they can be found in most ODE texts. We choose a different tact when solving Bessel's equation in Section 7.5, although the resulting y_2 will be equivalent to a linear combination of the y_1 and y_2 above.

Let's finish this section with a derivation of the Laguerre polynomials.

Example 6 Laguerre's equation and the Laguerre polynomials: We wish to find all polynomial solutions of

$$xy'' + (1-x)y' + \lambda y = 0, \qquad 0 < x < \infty,$$

i.e., we wish to find the *eigenvalues* λ which have polynomial *eigenfunctions*. We let

$$y = x^r \sum_{n=0}^{\infty} a_n x^n,$$

and the ODE implies that we must have

$$r^2 a_0 + \sum_{n=0}^{\infty} [(n+r+1)^2 a_{n+1} + (\lambda - n - r)a_n]x^n = 0.$$

Then the indicial equation $r^2 = 0$ has double root $r = 0$. There will be one Frobenius solution and a logarithmic solution. The latter, of course, cannot be a polynomial.

So with $r = 0$, the Frobenius solution has the recurrence formula

$$a_{n+1} = \frac{n - \lambda}{(n+1)^2} a_n, \qquad n = 0, 1, 2, \dots .$$

Thus, we will have a polynomial solution if and only if λ is a nonnegative integer. Let's compute the first few of these **Laguerre polynomials**. Setting $a_0 = 1$, we have

$\lambda = 0$: $a_{n+1} = \frac{n}{(n+1)^2} a_n, n = 0, 1, 2, \dots$

$\Rightarrow a_1 = 0 = a_2 = a_3 = \dots$ and $L_0(x) = 1$.

$\lambda = 1$: $a_{n+1} = \frac{n-1}{(n+1)^2} a_n, n = 0, 1, 2, \dots$

$\Rightarrow a_1 = -1, \quad a_2 = 0 = a_3 = a_4 = \dots$ and $L_1(x) = 1 - x$.

$\lambda = 2$: $a_{n+1} = \frac{n-2}{(n+1)^2} a_n, n = 0, 1, 2, \dots$

$\Rightarrow a_1 = -2, \quad a_2 = -\frac{1}{4}a_1 = \frac{1}{2}, \quad a_3 = 0 = a_4 = a_5 = 0 \dots$

and $L_2(x) = 1 - 2x + \frac{1}{2}x^2$.

More generally, if $\lambda = N$, we find $L_N(x)$ from the recurrence formula

$$
\begin{aligned}
a_{n+1} &= \frac{n - N}{(n+1)^2} a_n \\
&= \frac{(n - N)}{(n+1)^2} \frac{(n - 1 - N)}{n^2} a_{n-1} \\
&= \cdots \\
&= \frac{(n - N)}{(n+1)^2} \frac{(n - 1 - N)}{n^2} \cdots \frac{(1 - N)}{1^2}.
\end{aligned}
$$

Ultimately (see Exercise 5), we have

$$L_n(x) = \sum_{k=0}^{n} \frac{(-1)^k n!}{(k!)^2 (n-k)!} x^k$$

as the polynomial solution of the equation

$$xy'' + (1 - x)y' + ny = 0.$$

Exercises 7.3

1. For each ODE, classify all singular points as either regular or irregular.

   a) $x^2(x-1)y'' - y' + 2y = 0$

   b) $x^3 y'' + xy' - y = 0$

   c) $y'' + y' + \frac{1}{(x+2)^4 x^2} y = 0$

2. Finish finding the Frobenius solution for Example 4, and show that it can be written as

$$y = \sum_{n=0}^{\infty} \frac{1}{n!(n+1)!} x^n.$$

3. Write the solutions from Example 3 in summation notation.

4. Use the method of Frobenius to show that the given ODE has the given general solution:

   a) $xy'' + 2y' + xy = 0, y = \frac{1}{x}(c_1 \cos x + c_2 \sin x)$

   b) $xy'' - y' + 4x^3 y = 0, y = c_1 \cos(x^2) + c_2 \sin(x^2)$

5. Show that the formula given in Example 5 for the Laguerre polynomial $L_n(x)$ is correct.

6. Each of the following ODEs has an indicial equation with a double root $r = r_1 = r_2$. Use the formula

$$y_2 = y_1 \ln x + x^{r_1} \sum_{n=0}^{\infty} a'_n(r_1) x^n$$

   to construct a second solution. Is it clear that your two solutions are linearly independent?

   a) $(x^2 - x^3)y'' - 3xy' + 4y = 0$

   b) $(x^2 + x^3)y'' - (x + x^2)y' + y = 0$

7. Use the method of Frobenius to solve Bessel's equation

$$x^2 y'' + xy' + (x^2 - \alpha^2)y = 0$$

   for the given value of α:

   a) $\alpha = \pi$

   b) $\alpha = \frac{1}{2}$

   c) $\alpha = \frac{3}{2}$

8.   a) Use the substitution $u = \sqrt{x}\, y$ to turn Bessel's equation into

$$u'' + \left(1 + \frac{1 - 4\alpha^2}{4x^2}\right) u = 0.$$

   b) Use part (a) to show that the solution for Bessel's equation with $\alpha = \frac{1}{2}$ also can be written as

$$y = \frac{1}{\sqrt{x}}(c_1 \cos x + c_2 \sin x).$$

9. Given the ODE
$$x^2 y'' + x p(x) y' + q(x) y = 0,$$

where p and q are polynomials, show that if the Frobenius indicial equation is $f(r) = 0$, then the recurrence formula will be

$$f(n + r)a_n = (\text{term(s) involving } a_k, \text{ with } k < n).$$

Suppose now that the roots of the indicial equation differ by a integer, that is, suppose that the roots are $r = r_1$ and $r = r_1 + N$, where N is a natural number. Show that, in the recurrence formula corresponding to $r = r_1$, we have

$$0 \cdot a_N = \ldots .$$

7.4   Interlude: The Gamma Function

As you may have seen in Exercise 6 of the previous section, when solving Bessel's equations we run into expressions that look very much like factorials, except that the individual terms are not integers. The extension of the factorial to these kinds of expressions is called the **gamma function** (only because its symbol is the letter gamma)

$$\Gamma(x) = \int_0^\infty t^{x-1} e^{-t}\, dt.$$

The integral is improper, of course—possibly at both ends, depending on the value of x. It's not hard to show that the right end causes no problems, as $e^{-t} \to 0$ much more rapidly than any power of t. At the left end, it turns out that $\Gamma(x)$ behaves like

$$\int_0 t^{x-1}\, dt,$$

which converges for $x > 0$.

Now, what about its relation to the factorial? First, we have

$$\Gamma(1) = \int_0^\infty e^{-t}\, dt = 1.$$

Next, let's relate $\Gamma(x+1)$ and $\Gamma(x)$, using our old standby, integration by parts:

$$\begin{aligned}
\Gamma(x+1) &= \int_0^\infty t^x e^{-t}\, dt \\
&= -t^x e^{-t}\Big|_0^\infty + x \int_0^\infty t^{x-1} e^{-t}\, dt \\
&= x\Gamma(x)
\end{aligned}$$

(see Exercise 1). So, for x an integer, we have

$$\begin{aligned}
\Gamma(2) &= 1\Gamma(1) = 1 \\
\Gamma(3) &= 2\Gamma(2) = 2 \cdot 1 \\
\Gamma(4) &= 4\Gamma(3) = 3 \cdot 2 \cdot 1,
\end{aligned}$$

and it should be clear (and we prove it in Exercise 1) that

$$\Gamma(n+1) = n!$$

More generally, it's easy to see that

$$\begin{aligned}
\Gamma(\alpha+n) &= (\alpha+n-1)\Gamma(\alpha+n-1) \\
&= \cdots \\
&= (\alpha+n-1)(\alpha+n-2)\cdots\alpha\Gamma(\alpha).
\end{aligned}$$

(Again, see Exercise 1.) We may also show that

$$\Gamma\left(\frac{1}{2}\right) = \sqrt{\pi},$$

in terms of which we may then find $\Gamma\left(\frac{n}{2}\right)$, where n is an odd integer.

It turns out that the domain $-1 < x < \infty$ is not good enough and that we need to extend $\Gamma(x)$ to the left of -1. Of course, we would like the extension to satisfy the property $\Gamma(x+1) = x\Gamma(x)$, so we actually use this equation to *define* the extension! Since $\Gamma(x+1)$ has domain $-2 < x < \infty$, we define

$$\Gamma(x) = \frac{\Gamma(x+1)}{x}, \qquad -2 < x < -1.$$

Now, we've defined $\Gamma(x)$ on $-2 < x < -1$, so we do it again!

$$\Gamma(x) = \frac{\Gamma(x+1)}{x}, \qquad -3 < x < -2.$$

Of course, we continue this process indefinitely. (Note the use of $<$ and not $\leq$. This is due to the fact that there's no way to define $\Gamma(0)$, hence no way to extend the definition as above to the negative integers.)†

See Figure 7.2 for the graph of the gamma function.

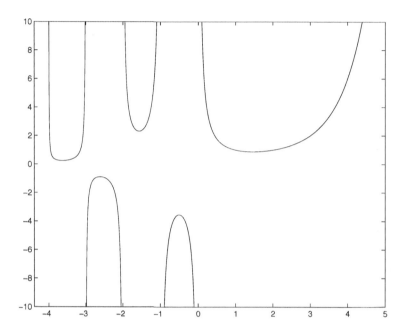

FIGURE 7.2
MATLAB graph of the *gamma function*, $y = \Gamma(x)$.

Another function which shows up often is

$$\psi(x) = \frac{d}{dx} \ln \Gamma(x+1), \qquad x > 0.$$

What does it look like? Well,

$$\Gamma'(x) = \int_0^\infty t^{x-1} e^{-t} \ln t \, dt \quad (\text{why?}),$$

---

†It turns out that $\displaystyle\lim_{x \to n} \frac{1}{\Gamma(x)} = 0$ for $n = 0, -1, -2, \ldots$. Therefore, we often define $\frac{1}{\Gamma(n)} = 0$ for these values. Then it can be shown that the function $\frac{1}{\Gamma(x)}$ is everywhere continuous! See Exercise 4.

where we have interchanged the order of integration and differentiates (it turns out that we can, here). Then

$$
\begin{aligned}
\psi(x) &= \frac{\Gamma'(x+1)}{\Gamma(x+1)} \\
&= \frac{x\Gamma'(x) + \Gamma(x)}{x\Gamma(x)} \\
&= \frac{\Gamma'(x)}{\Gamma(x)} + \frac{1}{x} \\
&= \psi(x-1) + \frac{1}{x}.
\end{aligned}
$$

If $x = n$, a positive integer, then

$$
\begin{aligned}
\psi(n) &= \psi(n-1) + \frac{1}{n} \\
&= \psi(n-2) + \frac{1}{n} + \frac{1}{n-1} \\
&= \cdots \\
&= \psi(0) + \sum_{k=1}^{n} \frac{1}{k}.
\end{aligned}
$$

We write $\phi(n) = \sum_{k=1}^{n} \frac{1}{k}$; as for the other term,

$$
\psi(0) = \Gamma'(1) = \int_{0}^{\infty} e^{-t} \ln t \, dt.
$$

Although it's not obvious, this integral converges, to a famous irrational number

$$
-\gamma = -.57721\ldots .
$$

The positive number γ is known as **Euler's**[‡] **constant.** $\Big($It can be shown, in fact, that

$$
\gamma = \lim_{n \to \infty} \left(1 + \frac{1}{2} + \cdots + \frac{1}{n} - \ln n \right). \Big)
$$

So we may write

$$
\psi(n) = -\gamma + \phi(n).
$$

[‡]Sometimes referred to as the Euler–Mascheroni constant.

Exercises 7.4

1. Use mathematical induction to prove that

    a) $\Gamma(n+1) = n!$ for $n = 1, 2, \ldots$.

    b) $\Gamma(\alpha + n) = (\alpha + n - 1)(\alpha + n - 2) \cdots \alpha \Gamma(\alpha)$ for $n = 1, 2, \ldots$.

2.   a) Use the substitution $t = u^2$ to show that $\Gamma\left(\frac{1}{2}\right) = \sqrt{\pi}$.

    b) Compute $\Gamma\left(\frac{3}{2}\right), \Gamma\left(\frac{5}{2}\right)$ and $\Gamma\left(-\frac{1}{2}\right)$.

    c) Generalize part (a) and show that

$$\int_0^\infty e^{-u^n}\, du = \frac{1}{n}\Gamma\left(\frac{1}{n}\right).$$

3. Use mathematical induction to prove that

    a) $\psi(n) = \psi(0) + \sum\limits_{k=1}^{n} \frac{1}{k}$, for $n = 1, 2, \ldots$.

    b) $\psi(x + n) = \psi(x) + \sum\limits_{k=1}^{n} \frac{1}{x+k}$, for $n = 1, 2, \ldots$.

4. **MATLAB:** Plot the function $f(x) = \frac{1}{\Gamma(x)}$.

5. The **beta function** is defined as

$$B(x, y) = \int_0^1 t^{x-1}(1-t)^{y-1}dt, \qquad x > 0, y > 0.$$

We prove that

$$B(x, y) = \frac{\Gamma(x)\Gamma(y)}{\Gamma(x+y)}.$$

    a) Proceed as in Exercise 2a to write

$$\Gamma(x)\Gamma(y) = 4\int_0^\infty e^{-u^2} u^{2x-1}\, du \int_0^\infty e^{-v^2} v^{2y-1}\, dv.$$

    b) Combine this into a double integral, change to polar coordinates and then separate the resulting double integral into the product of two integrals.

    c) Make the substitution $t = \sin^2\theta$ in the definition of $B(x, y)$.

    d) Let $x = y$ to obtain the **Legendre duplication formula**

$$\frac{2^{2x-1}}{\sqrt{\pi}}\Gamma(x)\Gamma\left(x + \frac{1}{2}\right) = \Gamma(2x).$$

    e) **MATLAB:** Plot the graph of $z = B(x, y)$.

7.5   Bessel Functions

Now let's bring the method of Frobenius to bear on Bessel's equation

$$x^2 y'' + xy' + (x^2 - \alpha^2)y = 0.$$

Again, we let $y = x^r \sum_{n=0}^{\infty} a_n x^n$, and the ODE implies that we must have

$$(r^2 - \alpha^2)a_0 + [(r+1)^2 - \alpha^2]a_1 x$$

$$+ \sum_{n=2}^{\infty} \{[(n+r)^2 - \alpha^2]a_n + a_{n-2}\}x^{n+r} = 0.$$

So the indicial equation $r^2 - \alpha^2 = 0$ has two roots $r = \pm\alpha$ (so long as $\alpha \neq 0$). For $r = \alpha > 0$, we have

$$[(\alpha+1)^2 - \alpha^2]a_1 = 0 \Rightarrow a_1 = 0$$

and

$$[(n+\alpha)^2 - \alpha^2]a_n + a_{n-2} = 0, \qquad n = 2, 3, \ldots$$

or

$$n(n+2\alpha)a_n = -a_{n-2}, \qquad n = 2, 3, \ldots .$$

So all of the odd-numbered coefficients are zero:

$$a_1 = a_3 = a_5 = \ldots = 0.$$

As for the evens, we have

$$a_n = -\frac{1}{n(n+2\alpha)}a_{n-2}, \qquad n = 2, 4, 6, \ldots$$

or, letting $n = 2k$,

$$a_{2k}(\alpha) = -\frac{1}{2k(2k+2\alpha)}a_{2k-2} = -\frac{1}{2'k2'(k+\alpha)}a_{2k-2}$$

$$= \frac{1}{2k(2k+2\alpha)}\frac{1}{(2k-2)(2k-2+2\alpha)}a_{2k-4}$$

$$= \frac{1}{2^2 k(k-1)2^2(k+\alpha)(k-1+\alpha)}a_{2k-4}$$

$$= \cdots$$

$$= \frac{(-1)^k}{2^k k! 2^k (k+\alpha)(k-1+\alpha)\cdots(1+\alpha)}a_0$$

$$= \frac{(-1)^k \Gamma(1+\alpha)}{2^{2k} k! \Gamma(k+\alpha+1)}a_0, \qquad k = 1, 2, \ldots .$$

(See Exercise 3.) So we have the Frobenius solution

$$y_1 = a_0 x^\alpha \sum_{k=0}^{\infty} \frac{(-1)^k \Gamma(1+\alpha)}{2^{2k} k! \Gamma(k+\alpha+1)} x^{2k}.$$

For various reasons, it's traditional to choose

$$a_0 = \frac{1}{2^\alpha \Gamma(\alpha+1)},$$

giving us the solution

$$J_\alpha(x) = \frac{x^\alpha}{2^\alpha} \sum_{k=0}^{\infty} \frac{(-1)^k}{2^{2k} k! \Gamma(k+\alpha+1)} x^{2k}$$

$$= \sum_{k=0}^{\infty} \frac{(-1)^k}{k! \Gamma(k+\alpha+1)} \left(\frac{x}{2}\right)^{2k+\alpha}.$$

This solution is called the **Bessel function of the first kind of order α.**
It's a valid solution of Bessel's equation of order α for any $\alpha \geq 0$. If n is an
integer $(n \geq 0)$, we have

$$J_n(x) = \sum_{k=0}^{\infty} \frac{(-1)^k}{k!(k+n)!} \left(\frac{x}{2}\right)^{2k+n}.$$

The graphs of J_0, J_1 and J_2 can be seen in Figure 7.3.

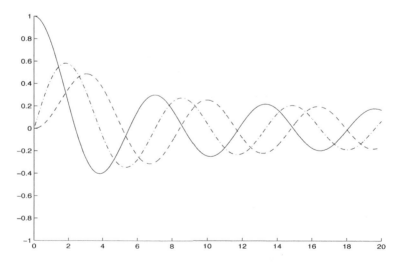

FIGURE 7.3
**MATLAB graphs of the Bessel functions of the first kind, J_0 (solid),
J_1 (dash-dotted) and J_2 (dashed).**

For $r = -\alpha$, we have

$$[(-\alpha + 1)^2 - \alpha^2]a_1 = 0$$

and

$$[(n - \alpha)^2 - \alpha^2]a_n = -a_{n-2}, \qquad n = 2, 3, \ldots$$

or

$$(1 - 2\alpha)a_1 = 0$$

and

$$n(n - 2\alpha)a_n = -a_{n-2}, \qquad n = 2, 3, \ldots .$$

We suspect there may be a problem if 2α is an integer (that is, if $r_1 - r_2 = \alpha - (-\alpha)$ is an integer!). So, for now, suppose this is not the case. Then, proceeding as above, we wind up with

$$a_1 = a_3 = a_5 = \ldots = 0$$

and

$$a_{2k} = \frac{(-1)^k \Gamma(1 - \alpha)}{2^{2k} k! \Gamma(k - \alpha + 1)} a_0,$$

and, choosing $a_0 = \frac{1}{2^\alpha \Gamma(1-\alpha)}$, we have the solution

$$J_{-\alpha}(x) = \sum_{k=0}^{\infty} \frac{(-1)^k}{k! \Gamma(k - \alpha + 1)} \left(\frac{x}{2}\right)^{2k-\alpha},$$

which is, of course, the **Bessel function of the first kind of order** $-\alpha$. Note that J_α is bounded at $x = 0$, while $J_{-\alpha}$ is not. Therefore, they are linearly independent and, at least for $x > 0$, the general solution of Bessel's equation is

$$y = c_1 J_\alpha(x) + c_2 J_{-\alpha}(x).$$

What happens when α is an integer or a half-integer? In Exercise 7 of Section 7.3 we saw that, for $\alpha = \frac{1}{2}$ and $\alpha = \frac{3}{2}$, there was a second Frobenius solution. In general, suppose $\alpha = \frac{2m+1}{2}$, $m = 0, 1, 2, \ldots$. Then, for $r = -\alpha = -\frac{2m+1}{2}$, the recurrence formulas become

$$-2ma_1 = 0$$

and

$$n(n - 2m - 1)a_n = -a_{n-2}, \qquad n = 2, 3, 4, \ldots .$$

In the case $\alpha = \frac{1}{2}$, that is, $m = 0$, we have a_1 arbitrary, and no problems thereafter. So we may take $a_1 = 0$, and $J_{-1/2}(x)$ is just given by the $J_{-\alpha}(x)$ formula, above. Similarly, for $\alpha = \frac{3}{2}, \frac{5}{2}, \frac{7}{2}, \ldots$, that is, for $m = 1, 2, 3, \ldots$, we have $a_1 = a_3 = a_5 = \ldots = 0$. Then, the left side of the recurrence formula is zero only when n is odd, so there is no problem here, either, and $J_{-\alpha}(x)$ is, again, as above.

Now, when $\alpha = m = 1, 2, 3, \ldots$, the formulas become

$$(1 - 2m)a_1 = 0 \Rightarrow a_1 = 0$$

and

$$n(n - 2m)a_n = -a_{n-2}, \qquad n = 2, 3, \ldots .$$

So $a_1 = a_3 = a_5 = \ldots = 0$. However, if $a_0 \neq 0$, then all of the even $a_2, a_4, \ldots, a_{2n-2}$ are nonzero, while, for $n = 2m$, we get the contradiction

$$0 \cdot a_{2m} = -a_{2m-2}.$$

Therefore, we will not have a second Frobenius solution. (As in Example 4 of the previous section, we could *choose* $a_0 = a_2 = \ldots = a_{2m-2} = 0$, in which case a_{2m} is arbitrary and we can start from there. But it turns out that we get the solution $(-1)^n J_n(x)$. See Exercise 3.)

So we must manufacture a second linearly independent solution. There are various ways to do this but, for Bessel's equations, the standard approach is as follows. First, when α is *not* an integer, we define the function

$$Y_\alpha(x) = \frac{(\cos \pi \alpha) J_\alpha(x) - J_{-\alpha}(x)}{\sin \pi \alpha}.$$

This function is (a) well defined for all nonintegral α and (b) a solution of Bessel's equation of order α (why?). Further, (c) J_α and Y_α are linearly independent (again, see Exercise 3). $Y_\alpha(x)$ is called the **Bessel function of the second kind of order α**, or the **Weber function of order α**.

Next, we extend the idea to $\alpha = n = 0, 1, 2, \ldots$ by looking at the above as a function of α. As $\alpha \to n$, $Y_\alpha(x) \to$

$$\frac{\cos n\pi J_n(x) - J_{-n}(x)}{\sin n\pi}.$$

Here, $J_{-n}(x)$ is interpreted to be the second Frobenius solution for $\alpha = n$. We mentioned above that $J_{-n}(x) = (-1)^n J_n(x)$. Therefore,

$$\frac{\cos n\pi J_n(x) - J_{-n}(x)}{\sin n\pi} = \frac{(-1)^n J_n(x) - (-1)^n J_n(x)}{\sin n\pi} = \frac{0}{0},$$

so we may treat α as a variable and use L'Hôpital's rule! Therefore, we define

$$Y_n(x) = \lim_{\alpha \to n} Y_\alpha(x)$$

$$= \frac{1}{\pi} \lim_{\alpha \to n} \left[\frac{\partial}{\partial \alpha} J_\alpha(x) - \pi \tan \pi \alpha J_\alpha(x) - \frac{\frac{\partial}{\partial \alpha} J_{-\alpha}(x)}{\cos \pi \alpha} \right];$$

so, realizing that this all can be done rigorously, we write

$$Y_n(x) = \frac{1}{\pi} \left[\frac{\partial}{\partial \alpha} J_\alpha(x) - (-1)^n \frac{\partial}{\partial \alpha} J_{-\alpha}(x) \right] \Bigg|_{\alpha=n}.$$

For $\alpha = 0$, it turns out that we get

$$Y_0(x) = \frac{2}{\pi} \sum_{k=0}^{\infty} \frac{(-1)^k}{k! \Gamma(k+1)} \left[\ln \frac{x}{2} - \psi(k) \right] \left(\frac{x}{2} \right)^{2k}$$

$$= \frac{2}{\pi} \left[J_0(x) \left(\gamma + \ln \frac{x}{2} \right) - \sum_{k=0}^{\infty} \frac{(-1)^k \phi(k)}{(k!)^2} \left(\frac{x}{2} \right)^{2k} \right].$$

(See Exercise 12.)

More generally, it can be shown that $Y_n(x)$ is of the form

$$J_n(x)(A \ln x + B) + x^{-n} \sum_{k=0}^{\infty} a_k x^k, \qquad n = 1, 2, 3, \ldots,$$

where the constants A and a_0 are nonzero. Hence, $Y_n(x)$, $n = 0, 1, 2, \ldots$, is unbounded as $x \to 0^+$ and, in general, J_n and Y_n are linearly independent. We plot Y_0, Y_1 and Y_2 in Figure 7.4.

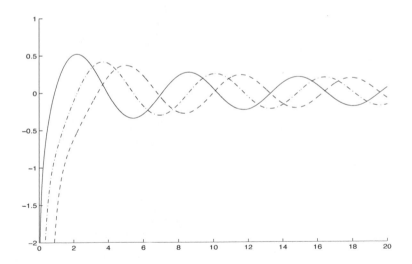

FIGURE 7.4
MATLAB graphs of the Bessel functions of the second kind, Y_0 (solid), Y_1 (dash-dotted) and Y_2 (dashed).

Bessel functions exhibit a number of important properties, some of which we explore in the exercises. One that we'll need in order to solve Bessel's eigenvalue problem below is the fact that each function $J_\alpha(x)$, $\alpha \geq 0$, has infinitely many positive roots. Although we don't prove it here,[§] we can see why it might be true by looking at what we did in Exercise 8 of Section 7.3. There we showed that the substitution $u = \sqrt{x}\, y$ turns Bessel's equation into

$$u'' + \left(1 + \frac{1 - 4\alpha^2}{4x^2}\right) u = 0.$$

For large values of x, this equation is "close to" the equation

$$u'' + u = 0,$$

which, of course, leads to the solution given in Exercise 8b of Section 7.3 namely,

$$y = \frac{1}{\sqrt{x}}(c_1 \cos x + c_2 \sin x).$$

This solution has infinitely many zeros $x_1, x_2, \ldots \to \infty$ (why?), and so it's not hard to believe that the same is true of the Bessel functions. Further, it turns out that $x_{n+1} - x_n \to \pi$ as $n \to \infty$.

BESSEL'S EIGENVALUE PROBLEM

Now, what about Bessel's eigenvalue problem from Exercise 1b in Section 7.1? Again, it is the R-equation arising from the application of separation of variables to the wave (and heat) equation in polar coordinates. Generally, then, we are asked to find the eigenvalues and eigenfunctions of the problem

$$x^2 y'' + xy' + (\lambda x^2 - n^2)y = 0, \qquad 0 < x < L,$$
$$y \text{ bounded as } x \to 0^+, y(L) = 0,$$

where n is a nonnegative integer. We proceed as usual:

Case 1: $\lambda = 0$
$$x^2 y'' + xy' - n^2 y = 0$$

is a Cauchy–Euler equation. If $n = 0$, we have general solution

$$y = c_1 + c_2 \ln x,$$

while for $n > 0$ we have

$$y = c_1 x^n + c_2 x^{-n}.$$

In either case, the boundary condition gives $c_1 = c_2 = 0$ (why?), and $\lambda = 0$ is not an eigenvalue.

[§] See, e.g., Georgi Tolstov's excellent book *Fourier Series*.

Case 2: $\lambda < 0, \lambda = -k^2$

This case is treated in Exercise 9, where we see that there are no negative eigenvalues.

Case 3: $\lambda > 0, \lambda = k^2$

Here, of course, we have

$$x^2 y'' + xy' + (k^2 x^2 - n^2)y = 0.$$

Proceeding as in Exercise 1d of Section 7.1, we see that the general solution is

$$y = c_1 J_n(kx) + c_2 Y_n(kx).$$

As Y_α is unbounded as $x \to 0^+$, we must have $c_2 = 0$. Then, the right end boundary condition gives us $c_1 = 0$ *unless* k is such that

$$J_n(kL) = 0.$$

As mentioned above, the Bessel function J_n has an infinite sequence of positive zeros, $x_{n,1}, x_{n,2}, \ldots \to \infty$, so we have an infinite sequence of eigenvalues

$$k_{n,m} = \frac{x_{n,m}}{L}, \qquad m = 1, 2, 3, \ldots,$$

with corresponding eigenfunctions

$$y_{n,m} = J_n\left(\frac{x_{n,m}}{L} x\right), \qquad m = 1, 2, 3, \ldots.$$

ZEROS OF $J_n(x)$; STURM COMPARISON THEOREM, REVISITED

In Chapter 8, we'll need to know the zeros of the Bessel functions $J_n(x)$, $n = 0, 1, 2, \ldots$. We list the first 20 positive zeros of J_0, J_1 and J_2 in Table 7.1.¶ We note here that the table bears out the results of the Sturm Comparison Theorem, mentioned in Exercise 20, Section 1.7. In this setting, the theorem implies that if $\alpha_1 < \alpha_2$, then between any two zeros of J_{α_2} there lies a zero of J_{α_1}. Notice, too, that the difference between zeros seems to be tending toward approximately 3.14, as expected.

Exercises 7.5

1. Write down the general solution of each equation, on $0 < x < \infty$.

    a) $x^2 y'' + xy' + (x^2 - 5)y = 0$

    b) $x^2 y'' + xy' + (x^2 - 9)y = 0$

    c) $xy'' + y' + xy = 0$

¶From the *Handbook of Mathematical Functions* by Abramowitz and Stegun.

	J_0	J_1	J_2
1)	2.40483	3.83171	5.13562
2)	5.52008	7.01559	8.41724
3)	8.65373	10.17347	11.61984
4)	11.79153	13.32369	14.79595
5)	14.93092	16.47063	17.95982
6)	18.07106	19.61586	21.11700
7)	21.21164	22.76008	24.27011
8)	24.35247	25.90367	27.42057
9)	27.49348	29.04683	30.56920
10)	30.63461	32.18968	33.71652
11)	33.77852	35.33231	36.86286
12)	36.91710	38.47477	40.00845
13)	40.05843	41.61709	43.15345
14)	43.19979	44.75932	46.29800
15)	46.34119	47.90146	49.44216
16)	49.48261	51.04354	52.58602
17)	52.62405	54.18555	55.72963
18)	55.76551	57.32753	58.87302
19)	58.90698	60.46946	62.01622
20)	62.04847	63.61136	65.15927

TABLE 7.1
The first 20 positive zeros of J_0, J_1 and J_2.

2. Verify the calculations in deriving $a_{2k}(\alpha)$ for the first solution of Bessel's equation.

3. Show that $J_{-n}(x) = (-1)^n J_n(x)$ (where $J_{-n}(x)$ is constructed by choosing $a_0 = a_2 = \ldots = a_{2n-2} = 0$).

4. Establish the following properties of Bessel functions.

   a) $J_0(0) = 1$
   b) $J_\alpha(0) = 0$ for $\alpha > 0$

    c) $J_0'(x) = -J_1(x)$

    d) $\frac{d}{dx}[xJ_1(x)] = xJ_0(x)$

5.   a) Show that $\frac{d}{dx}[x^\alpha J_\alpha(x)] = x^\alpha J_{\alpha-1}(x)$.

    b) Show that $\frac{d}{dx}[x^{-\alpha} J_\alpha(x)] = -x^{-\alpha} J_{\alpha+1}(x)$.

    c) Use the results of parts (a) and (b) to show that

$$J_\alpha'(x) = \frac{1}{\alpha}[J_{\alpha-1}(x) - J_{\alpha+1}(x)].$$

    d) Use the results of parts (a) and (b) to show that

$$J_{\alpha+1}(x) = \frac{2\alpha}{x}J_\alpha(x) - J_{\alpha-1}(x).$$

    e) Express $J_2(x)$ and $J_3(x)$ in terms of $J_0(x)$ and $J_1(x)$.

6. Show that the functions $Y_\alpha(x)$ also satisfy the relations in the previous exercise.

7.   a) Use Rolle's Theorem to show that if $f(x)$ has infinitely many zeros on $0 < x < \infty$, then so does $f'(x)$.

    b) Solve the Bessel eigenvalue problem

$$x^2 y'' + xy' + (\lambda x^2 - n^2)y = 0, \qquad 0 < x < L,$$
$$y \text{ bounded as } x \to 0^+, y'(L) = 0.$$

8. **Hankel Functions or Bessel Functions of the Third Kind:** The Hankel functions of order α, of the first and second kinds, are defined by, respectively,

$$H_\alpha^{(1)}(x) = J_\alpha(x) + iY_\alpha(x) \quad \text{and}$$
$$H_\alpha^{(2)}(x) = J_\alpha(x) - iY_\alpha(x),$$

where i is the imaginary number satisfying $i^2 = -1$. Show that each of these functions satisfies Bessel's equation of order α.

9. **Modified Bessel Functions:** The ODE

$$x^2 y'' + xy' - (x^2 + \alpha^2)y = 0$$

is called the **modified Bessel's equation of order α.**

    a) Use the method of Frobenius to show that

$$I_\alpha(x) = \sum_{k=0}^{\infty} \frac{1}{k!\Gamma(k+\alpha+1)} \left(\frac{x}{2}\right)^{2k+\alpha}$$

is a solution for $\alpha \geq 0$ and that

$$I_{-\alpha}(x) = \sum_{k=0}^{\infty} \frac{1}{k!(k - \alpha + 1)} \left(\frac{x}{2}\right)^{2k-\alpha}$$

is a solution for $\alpha > 0$, α not an integer. Each of these is a **modified Bessel function of the first kind of order α or $-\alpha$,** respectively.

b) Show that, if $\alpha > 0$ is not an integer, the function

$$K_{\alpha}(x) = \frac{\pi}{2} \frac{I_{-\alpha}(x) - I_{\alpha}(x)}{\sin \pi \alpha}$$

also is a solution. This is the **modified Bessel function of the second kind of order α.** We may define $K_n(x)$, $n = 0, 1, 2, \ldots$, as we defined $Y_n(x)$, by taking $\lim_{\alpha \to n} K_{\alpha}(x)$.

c) Verify that $I_{\alpha}(x) > 0$ for all x, for any $\alpha \geq 0$. Also, show that $I_0(0) = 1$ and $I_{\alpha}(0) = 0$ for $\alpha > 0$.

d) It turns out that the functions $K_n(x)$, $n = 0, 1, 2, \ldots$, are unbounded as $x \to 0^+$. Use this, and the results of part (c), to show that, for each $k = 1, 2, 3, \ldots$, the problem

$$x^2 y'' + xy' - (k^2 x^2 + n^2)y = 0,$$
$$y \text{ bounded as } x \to 0^+, y(L) = 0,$$

has only the trivial solution.

e) Show formally that $I_{\alpha}(ix) = i^{\alpha} J_{\alpha}(x)$.

f) Use part (e) to show that

$$I_0'(x) = I_1(x),$$

then do the same using the series in part (a). (One also can show that $K_0'(x) = K_1(x)$.)

We provide the graphs of I_0, I_1, I_2 and K_0, K_1, K_2 in Figures 7.5 and 7.6, respectively.

10. **Bessel's equation in disguise:** There are many ODEs which can be turned into Bessel's equation via an appropriate change of variables. Actually, let's go the other way—let's *start* with Bessel's equation and try to generalize it.

a) Let $x = at^b$, for constants a and b, to turn Bessel's equation into

$$t^2 y'' + ty' + (a^2 b^2 t^{2b} - b^2 \alpha^2)y = 0,$$

where $y' = \frac{dy}{dt}$.

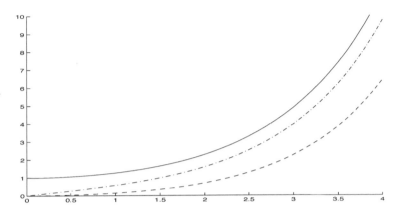

FIGURE 7.5
MATLAB graphs of I_0 (solid), I_1 (dash-dotted) and I_2 (dashed), modified Bessel functions of the first kind.

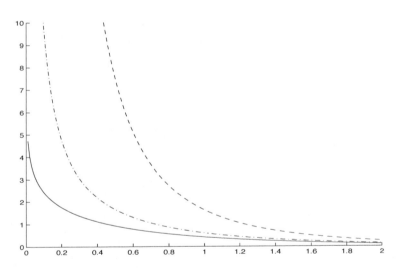

FIGURE 7.6
MATLAB graphs of K_0 (solid), K_1 (dash-dotted) and K_2 (dashed), modified Bessel functions of the second kind.

b) Next, change dependent variables via $z(t) = t^c y(t)$, for c constant, to get

$$t^2 z'' + (1 - 2c)tz' + (a^2 b^2 t^{2b} + c^2 - b^2 \alpha^2)y = 0.$$

(Compare with Exercise 8a, Section 7.3.)

c) Show that the general solution of this last equation is

$$z = t^2[c_1 J_\alpha(at^b) + c_2 Y_\alpha(at^b)].$$

d) Use these results to find the general solution of **Airy's equation**, $y'' + xy = 0$.

11. From ODEs, you may remember **Ricatti's equation**,

$$y' + by^2 = cx^m.$$

a) If $m = -2$, show that the substitution $v(x) = xy(x)$ turns Ricatti's equation into a separable equation.

b) More generally, show that the substitution

$$y = \frac{1}{bu}u', \text{ where } u = u(x),$$

transforms Ricatti's equation into

$$u'' - bcx^m u = 0.$$

c) Use the above, and Exercise 9, to solve the Ricatti equation

$$y' - y^2 = x^2.$$

12. Derive the expression given for $Y_0(x)$.

13. **Integral form for Bessel functions:** It turns out that

$$J_n(x) = \frac{1}{2\pi i^n} \int_{-\pi}^{\pi} e^{i(x\cos\theta - n\theta)} \, d\theta, \qquad n = 0, 1, 2, \dots .$$

We do so only for the case $n = 0$, as follows:

a) Use the Maclaurin series for e^x to expand $e^{ix\cos\theta}$, in order to show that

$$\int_{-\pi}^{\pi} e^{ix\cos\theta} \, d\theta = \sum_{m=0}^{\infty} \frac{(ix)^m}{m!} \int_{-\pi}^{\pi} \cos^m \theta \, d\theta.$$

(You may assume that we may integrate term-by-term.)

b) Use the *binomial theorem* and *Euler's formula* to show that

$$\int_{-\pi}^{\pi} \cos^m \theta \, d\theta = \begin{cases} 0, & \text{if } m \text{ is odd}, \\ \binom{2k}{k}/2^{2k}, & \text{if } m = 2k \text{ is even}. \end{cases}$$

(Remember that $\binom{n}{k}$ is just "n-choose k," $\binom{n}{k} = \frac{n!}{k!(n-k)!}$.)

c) Conclude that $J_0(x) = \frac{1}{2\pi} \int_{-\pi}^{\pi} e^{ix\cos\theta}\, d\theta$.

d) If f is a *real* function with $|f(\theta)| \leq 1$ on $-\pi \leq \theta \leq \pi$, explain why

$$-1 \leq \frac{1}{2\pi} \int_{-\pi}^{\pi} f(\theta)d\theta \leq 1.$$

(One can make a similar statement for complex functions and, thus, conclude that $|J_n(x)| \leq 1$ for any x, for any $n = 0, 1, 2, \ldots$.)

7.6   Recap: A List of Properties of Bessel Functions and Orthogonal Polynomials

In this final section, we list a number of the important properties of the functions we have dealt with in this chapter. We prove some of these properties in the exercises; others were proven earlier, and some we state without proof. The various *Fourier* series are listed here for convenience, but are dealt with in the following chapter.

BESSEL FUNCTIONS

OF THE FIRST KIND

$J_\alpha(x), \alpha \geq 0,$ and $J_{-\alpha}(x), \alpha > 0$ and α not an integer.

OF THE SECOND KIND   (Weber functions)

$$Y_\alpha(x) = \frac{\cos\pi\alpha \cdot J_\alpha(x) - J_{-\alpha}(x)}{\sin\pi\alpha}, \quad \alpha > 0 \text{ and } \alpha \text{ not an integer,}$$

$Y_n(x) = \lim_{\alpha \to n} Y_\alpha(x), n$ an integer.

OF THE THIRD KIND   (Hankel functions)

$$H_\alpha^{(1)}(x) = J_\alpha(x) + iY_\alpha(x),$$
$$H_\alpha^{(2)}(x) = J_\alpha(x) - iY_\alpha(x).$$

All are solutions of the ODE

$$x^2 y'' + xy' + (x^2 - \alpha^2)y = 0$$

or

$$(xy')' + \frac{x^2 - \alpha^2}{x}y = 0, \qquad 0 < x < \infty.$$

Recurrence relation:

$$f_{\alpha+1}(x) = \frac{\alpha}{x} f_\alpha(x) - f_{\alpha-1}(x).$$

Orthogonality relation:

$$\int_0^1 x J_n(k_i x) J_n(k_j x) dx = \begin{cases} 0, & \text{if } i \neq j, \\ \frac{1}{2} J_{n+1}^2(k_i), & \text{if } i = j \end{cases}$$

for each $n = 0, 1, 2, \ldots$. (Here, the numbers k_i are the roots of J_n.)

Series representation:

$$J_\alpha(x) = \sum_{k=0}^\infty \frac{(-1)^k}{k! \Gamma(k+\alpha+1)} \left(\frac{x}{2}\right)^{2k+\alpha}, \qquad \alpha \geq 0.$$

Fourier–Bessel series:

$$f(x) \sim \sum_{n=1}^\infty c_n J_\alpha(k_n x) \quad (\alpha \text{ fixed}), \qquad 0 \leq x \leq 1,$$

$$\text{where } c_n = \frac{2}{J_{\alpha+1}^2(k_n)} \int_0^1 x f(x) J_\alpha(k_n x) dx$$

and $k_n = n^{\text{th}}$ positive root of J_α.

LEGENDRE POLYNOMIALS $P_n(x), n = 0, 1, 2, \ldots$

Solutions of ODE

$$(1 - x^2) y'' - 2xy' + n(n+1)y = 0$$

or

$$[(1 - x^2)y']' + n(n+1)y = 0, \qquad -1 < x < 1.$$

$P_0(x) = 1, P_1(x) = x$

Recurrence relation:

$$n P_n(x) = (2n - 1)x P_{n-1}(x) - (n - 1) P_{n-2}(x).$$

Orthogonality relation:

$$\int_{-1}^1 P_n(x) P_m(x) dx = \begin{cases} \dfrac{2}{2n+1}, & \text{if } n = m, \\ 0, & \text{if } n \neq m. \end{cases}$$

$$P_n(\pm 1) = (\pm 1)^n$$

Series representation:

$$P_n(x) = \sum_{k=0}^{\left[\frac{n}{2}\right]} \frac{(-1)^k (2n-2k)!}{2^n k!(n-k)!(n-2k)!} x^{n-2k}. \; \|$$

Rodrigues's formula:

$$P_n(x) = \frac{1}{2^n n!} \frac{d^n}{dx^n} [(x^2-1)^n].$$

Fourier–Legendre series:

$$f(x) \sim \sum_{n=1}^{\infty} c_n P_n(x), \qquad -1 \leq x \leq 1,$$

$$\text{where } c_n = \frac{2n+1}{2} \int_{-1}^{1} f(x) P_n(x) dx.$$

CHEBYSHEV POLYNOMIALS

OF THE FIRST KIND $\; T_n(x), n = 0, 1, \ldots$

Solutions of ODE

$$(1-x^2)y'' - xy' + n^2 y = 0$$

or

$$[\sqrt{1-x^2}\, y']' + n^2 \sqrt{1-x^2}\, y = 0, \qquad -1 < x < 1.$$

$$T_0(x) = 1, T_1(x) = x$$

Recurrence relation:

$$T_n(x) = 2x T_{n-1}(x) - T_{n-2}(x).$$

Orthogonality relation:

$$\int_{-1}^{1} \frac{T_n(x) T_m(x)}{\sqrt{1-x^2}} dx = \begin{cases} \pi, & \text{if } n = m = 0, \\ \dfrac{\pi}{2}, & \text{if } n = m > 0, \\ 0, & \text{if } n \neq m. \end{cases}$$

$\|$ Here and following, $\left[\frac{n}{2}\right]$ equals the greatest integer that is $\leq \frac{n}{2}$ (remember the *greatest integer function* $f(x) = [x]$).

$$T_n(\pm 1) = (\pm 1)^n$$

Series representation:

$$T_n(x) = \sum_{k=0}^{\left[\frac{n}{2}\right]} \binom{n}{2k} x^{n-2k}(x^2 - 1)^k,$$

$$\text{where } \binom{n}{m} = \frac{n!}{m!(n-m)!} = \text{``}n\text{-choose-}m.\text{''}$$

$$T_n(\cos\theta) = \cos n\theta$$

Fourier–Chebyshev series (first kind):

$$f(x) \sim \sum_{n=1}^{\infty} c_n T_n(x), \qquad -1 \le x \le 1,$$

$$\text{where } c_n = \frac{2}{\pi} \int_{-1}^{1} f(x) T_n(x) dx.$$

OF THE SECOND KIND $S_n(x), n = 0, 1, \ldots$

Solutions of ODE

$$(1 - x^2)y'' - 3xy' + n(n+2)y = 0$$

or

$$[(1 - x^2)^{3/2} y']' + n(n+2)\sqrt{1 - x^2}\, y = 0, \qquad -1 < x < 1.$$

$$S_0(x) = 1, S_1(x) = 2x$$

Recurrence relation:

$$S_n(x) = 2x S_{n-1}(x) - S_{n-2}(x).$$

Orthogonality relation:

$$\int_{-1}^{1} S_n(x) S_m(x) \sqrt{1 - x^2}\, dx = \begin{cases} \dfrac{\pi}{2}, & \text{if } n = m, \\ 0, & \text{if } n \ne m. \end{cases}$$

$$S_n(\pm 1) = (\pm 1)^n (n + 1)$$

Series representation:

$$S_n(x) = \sum_{k=0}^{\left[\frac{n}{2}\right]} \binom{n+1}{2k+1} x^{n-2k}(x^2-1)^k.$$

$S_n(\cos\theta) = \frac{\sin(n+1)\theta}{\sin\theta}$ (where we use L'Hôpital's rule if $\theta = 0, \pm\pi, \ldots$)

Fourier–Chebyshev series (second kind):

$$f(x) \sim \sum_{n=1}^{\infty} c_n S_n(x), \qquad -1 \le x \le 1,$$

$$\text{where } c_n = \frac{2}{\pi} \int_{-1}^{1} f(x)S_n(x)dx.$$

LAGUERRE POLYNOMIALS $\quad L_n(x), n = 0,1,2,\ldots$

Solutions of ODE

$$xy'' + (1-x)y' + ny = 0$$
$$\text{or}$$
$$(xe^{-x}y')' + ne^{-x}y = 0, \qquad 0 < x < \infty.$$

$L_0(x) = 1, L_1(x) = 1 - x$

Recurrence relation:

$$nL_n(x) = (2n - 1 - x)L_{n-1}(x) - (n-1)L_{n-2}(x).$$

Orthogonality relation:

$$\int_0^{\infty} L_n(x)L_m(x)e^{-x}\,dx = \begin{cases} 1, \text{ if } n = m, \\ 0, \text{ if } n \ne m. \end{cases}$$

Series representation:

$$L_n(x) = \sum_{k=0}^{n} \binom{n}{k} \frac{(-1)^k}{k!} x^k.$$

Rodrigues's formula:

$$L_n(x) = \frac{1}{n!} e^x \frac{d^n}{dx^n}(x^n e^{-x}).$$

Fourier–Laguerre series:

$$f(x) \sim \sum_{n=0}^{\infty} c_n L_n(x), \qquad -1 \leq x < 1,$$

$$\text{where } c_n = \int_0^{\infty} f(x) L_n(x) e^{-x} \, dx.$$

HERMITE POLYNOMIALS $H_n(x), n = 0, 1, 2, \ldots$

Solutions of ODE

$$y'' - 2xy' + 2ny = 0$$

or

$$[e^{-x^2} y']' + 2n e^{-x^2} y = 0, \qquad -\infty < x < \infty.$$

$H_0(x) = 1, H_1(x) = 2x$

Recurrence relation:

$$H_n(x) = 2x H_{n-1}(x) - 2(n-1) H_{n-2}(x).$$

Orthogonality relation:

$$\int_{-\infty}^{\infty} H_n(x) H_m(x) e^{-x^2} \, dx = \begin{cases} 2^n n! \sqrt{\pi}, & \text{if } n = m, \\ 0, & \text{if } n \neq m. \end{cases}$$

Series representation:

$$H_n(x) = \sum_{k=0}^{\left[\frac{n}{2}\right]} \frac{(-1)^k n!}{k!(n-2k)!} (2x)^{n-2k}.$$

Rodrigues's formula:

$$H_n(x) = (-1)^n e^{x^2} \frac{d^n}{dx^n} (e^{-x^2}).$$

Fourier–Hermite series:

$$f(x) \sim \sum_{n=0}^{\infty} c_n H_n(x), \qquad -\infty < x < \infty,$$

$$\text{where } c_n = \frac{1}{2^n n! \sqrt{\pi}} \int_{-\infty}^{\infty} f(x) H_n(x) e^{-x^2} \, dx.$$

Exercises 7.6

1. a) For each set of orthogonal polynomials, write down the third through fifth polynomials, and make sure that these match what we got in Section 7.2. (See Section 7.3 for L_0, L_1 and L_2.)

    b) For each set of orthogonal polynomials $f_n(x)$ except Laguerre (why not Laguerre?), use the *recurrence relation* to show that

    $$n \text{ even} \Rightarrow f_n \text{ is an even function,}$$
    $$n \text{ odd} \Rightarrow f_n \text{ is an odd function.}$$

    c) **MATLAB:** For each set of orthogonal polynomials, use the recurrence relation to write a MATLAB program which generates the first N polynomials.

    d) **MATLAB:** Plot the graphs of the first six of each set of orthogonal polynomials, as we did for the Legendre polynomials in Figure 7.1.

2. For each set of functions in this section, we gave two differential equations. Show that the two equations are equivalent in each case. In each case, the second equation is said to be in **self-adjoint form**. This idea is treated in detail in Chapter 8.

3. Here, we prove the so-called *orthogonality relations* for Legendre polynomials, for $n \neq m$.

    a) P_n and P_m satisfy, respectively,

    $$[(1 - x^2)P_n']' + n(n + 1)P_n = 0$$

    and

    $$[(1 - x^2)P_m']' + m(m + 1)P_m = 0.$$

    Multiply the top equation by P_m and the bottom by P_n, and subtract.

    b) Integrate the resulting equation from $x = -1$ to $x = 1$. You'll need to use integration by parts.

4. a) Proceed as in Exercise 3, for the Laguerre polynomials.

    b) Do the same for the Hermite polynomials.

5. a) Use the fact that $T_n(\cos\theta) = \cos n\theta$ to prove the orthogonality relation for the Chebyshev polynomials of the first kind (both for $n \neq m$ and for $n = m$).

    b) Similarly, use the property $S_n(\cos\theta) = \frac{\sin(n+1)\theta}{\sin\theta}$ to prove the orthogonality relation for the Chebyshev polynomials of the second kind (both for $n \neq m$ and for $n = m$).

6.   a) Use Rodrigues's formula for the Legendre polynomials to show that we write the *associated Legendre functions* as

$$P_m^n(x) = \frac{1}{2^m m!}(1 - x^2)^{n/2}\frac{d^{m+n}}{dx^{m+n}}[(x^2 - 1)^m]$$

(*m* and *n* are integers, of course).

b) For which values of *n* and *m* will $P_m^n(x)$ be a polynomial?

c) Show that

$$P_m^n(\cos\theta)$$

is a polynomial in $\sin\theta$ and/or $\cos\theta$, for *any* choice of *n* and *m*.

7. Here we prove the orthogonality relation for the functions $J_\alpha(k_i x)$, $i = 1, 2, 3, \ldots$, where $\alpha > 0$ and the numbers k_i are the positive zeros of the Bessel function $J_\alpha(x)$.

a) Show that $J_\alpha(k_i x)$ is a solution of the problem

$$x^2 y'' + xy' + (k_i^2 x^2 - \alpha^2)y = 0,$$
$$y \text{ bounded as } x \to 0^+, y(1) = 0.$$

b) Letting $y_i = J_\alpha(k_i x)$ and $y_j = J_\alpha(k_j x)$, $i \neq j$, we have

$$x^2 y_i'' + xy_i' + (k_i^2 x^2 - \alpha^2)y_i = 0$$
$$x^2 y_j'' + xy_j' + (k_j^2 x^2 - \alpha^2)y_j = 0.$$

Multiply the first by y_j and the second by y_i, and subtract. Then, integrate the resulting equation from 0 to 1 and conclude that we have

$$\int_0^1 x J_\alpha(k_i x) J_\alpha(k_j x)dx = 0.$$

c) More generally, for any $L > 0$, show that

$$\int_0^L x J_\alpha\left(\frac{k_i x}{L}\right) J_\alpha\left(\frac{k_j x}{L}\right) dx = 0.$$

8. What about when $i = j$ in Exercise 6?

a) Multiply the differential equation in Exercise 6a by y' and, using integration by parts, show that it follows that

$$[y'(1)]^2 + (k_i^2 - \alpha^2)[y(1)]^2 = 2k_i^2 \int_0^1 xy^2 \, dx.$$

b) Conclude that

$$\int_0^1 x J_\alpha^2(k_i x)\,dx = \frac{1}{2}[J_\alpha'(k_i)]^2.$$

c) Use the result of Exercise 5b from the previous section to show that we can rewrite this as

$$\int_0^1 x J_\alpha^2(k_i x)\,dx = \frac{1}{2}J_{\alpha+1}^2(k_i)$$

and, more generally, as

$$\int_0^L x J_\alpha^2\left(\frac{k_i x}{L}\right)\,dx = \frac{L^2}{2}J_{\alpha+1}^2(k_i).$$

9. Proceed as in Exercises 7 and 8, and show that if k_i, $i = 1,2,3,\dots$ are the positive zeros of the *derivative* J_α', then we still have

$$\int_0^1 x J_\alpha(k_i x) J_\alpha(k_j x)\,dx = 0 \quad \text{for} \quad i \neq j$$

and, more generally, that

$$\int_0^L x J_\alpha\left(\frac{k_i x}{L}\right) J_\alpha\left(\frac{k_j x}{L}\right)\,dx = 0 \quad \text{for} \quad i \neq j.$$

Show that, in this case, for $i = j$ we have

$$\int_0^1 x J_\alpha^2(k_i x)\,dx = \frac{k_i^2 - \alpha^2}{2k_i^2}J_\alpha^2(k_i).$$

Prelude to Chapter 8

In the 1830s, Charles Sturm (1803–1852) and Joseph Liouville* (1809–1882), both in Paris, embarked on a study of second-order boundary-value problems, resulting in the consolidation and generalization of the ideas from the previous chapter. Here, we introduce these *Sturm–Liouville problems*, and we find that their eigenvalues and eigenfunctions share the same important properties which characterize the eigenvalues and eigenfunctions of Section 3.7. Indeed, the sets of trigonometric functions comprising the various Fourier series, as well as the sets of special functions, all shake out as particular cases in the *Sturm–Liouville theory*.

The theory can be extended to higher order equations, where we touch upon the ideas of *adjoint* and *self-adjoint* problems. (This terminology, though, didn't arise until the early 1900s, in the work of David Hilbert (1862–1943) and others.) For our purposes, the most important result is the fact that the special functions are complete on their defining intervals, in the same sense as are the trigonometric functions. Sturm and Liouville saw that this was the case and established *Bessel's inequality* and *Parseval's*[†] *equality* in this setting of *generalized Fourier series*. (Bessel and Parseval had derived them for the trigonometric case.) However, they were not in a position to state and prove a general theorem on completeness, due to the problem that Fourier series converge pointwise only at points of continuity. All of this eventually was "made nice" following 1907, when Ernst Fischer (1875–1959) introduced a different kind of convergence, the so-called *mean-square convergence*, which entails looking at integrals of the functions involved and, thus, circumvents the difficulties associated with points of discontinuity. (Actually, the integrals were not Riemann integrals, but the new *Lebesgue integrals*, devised by Henri Lebesgue (1875–1941) around the turn of the century, and in some sense the culmination of the work of the great analysts—Dirichlet, Cauchy, Georg Friedrich Riemann (1826–1866), Liouville, Weierstrass, et al.—whose investigations had been spurred, initially, by the work of Fourier.)

---

*Liouville was very productive in many areas of mathematics, but he is also known as the person who resurrected the work of Evariste Galois (1811–1832). If you're not familiar with Galois's story, you really should check it out.
[†]Marc-Antoine Parseval (d. 1836).

8

Sturm–Liouville Theory and Generalized Fourier Series

8.1 Sturm–Liouville Problems

Before moving on to equations in higher dimensions, we pause to look again at Fourier series, with any eye toward seeing if there are other sets of functions which behave like the functions discussed in Section 3.7. It may seem strange that our approach will hinge upon looking again at ODE eigenvalue problems. However, you will remember that, in Chapters 1 and 3, when dealing with the separated eigenvalue problems

$$y'' + \lambda y = 0, \qquad 0 \leq x \leq L,$$

$$y(0) = 0 \text{ or } y'(0) = 0,$$

$$y(L) = 0 \text{ or } y'(L) = 0,$$

we found that the eigenvalues and eigenfunctions had a number of important properties—we list these and, in parentheses, we compare them with similar eigenproperties of matrices.

1) *The eigenvalues are real.* (A matrix need not have real eigenvalues.)

2) *There are infinitely many eigenvalues.*[*] ($n \times n$ matrices have exactly n eigenvalues, if we include multiplicities.)

3) *Each eigenvalue has multiplicity one*, i.e., if y_1 is an eigenfunction corresponding to λ_1, then the only eigenfunctions corresponding to λ_1 are of the form cy_1. (This was not the case with matrices and eigenvectors.)

4) *If y_1 and y_2 are eigenfunctions corresponding to different eigenvalues λ_1 and λ_2, then y_1 and y_2 are* **orthogonal** *on $[0, L]$*, i.e.,

$$\langle y_1, y_2 \rangle = \int_0^L y_1(x)y_2(x)dx = 0.$$

---

[*]Actually, they are *countably* infinite, meaning that they can be put into a 1-1 correspondence with the natural numbers.

(Eigenvectors corresponding to different eigenvalues need not be perpendicular.)

5) *The eigenfunctions form a* **complete** *set in the space of piecewise smooth functions on* $[0, L]$, i.e., for any such f, there exist constants $c_1, c_2, \ldots,$ such that

$$f(x) = \sum_{n=1}^{\infty} c_n y_n(x), \qquad 0 \le x \le L,$$

where y_n is the eigenfunction corresponding to λ_n. (For matrices, we *sometimes* found that the eigenvectors of an $n \times n$ matrix spanned $\mathbb{R}^n$.)

Let's start by rewriting the ODE $y'' + \lambda y = 0$ as

$$y'' = -\lambda y$$

so that it looks like a matrix eigenvalue problem (modulo the minus sign). More generally, then, we'll be looking at eigenvalue ODEs of the form

$$L[y] = -\lambda y,$$

where L is a linear differential operator; so, above, we have $L[y] = y''$. Further, we'll restrict ourselves to second-order ODEs, as these form the vast majority of the problems we have encountered which arise from separation of variables. So we consider eigenvalue ODEs of the form

$$L[y] = a_0(x)y'' + a_1(x)y' + a_2(x)y = -\lambda y. \tag{8.1}$$

For starters, we may put this equation into a sort of standard form. Remember that, when solving the first-order ODE

$$a_0(x)y' + a_1(x)y = b(x),$$

we divided by a_0 and multiplied by the integrating factor

$$r(x) = e^{\int \frac{a_1(x)}{a_0(x)} dx}$$

to rewrite the equation as

$$(ry')' = r(x)b(x).$$

Well, we may do the same with the first two terms of *any* linear ODE. So, we rewrite equation (8.1) as

$$y'' + \frac{a_1(x)}{a_0(x)}y' + \frac{a_2(x)}{a_0(x)}y = -\lambda \frac{1}{a_0(x)}y$$

and then multiply by the integrating factor $r(x) = e^{\int \frac{a_1(x)}{a_0(x)} dx}$ to arrive at

$$(r(x)y(x))' + q(x)y(x) = -\lambda w(x)y(x).$$

This particular eigenvalue ODE is called a **Sturm–Liouville equation** and any operator

$$L[y] = (ry')' + qy$$

is called a **Sturm–Liouville operator**, or to have been put into **Sturm–Liouville form**. Finally, any eigenvalue problem of the form

$$L[y] = (ry')' + qy = -\lambda wy, \qquad a \le x \le b,$$
$$a_1 y(a) + a_2 y'(a) + a_3 y(b) + a_4 y'(b) = 0,$$
$$b_1 y(a) + b_2 y'(a) + b_3 y(b) + b_4 y'(b) = 0,$$

where $a_1, \ldots, a_4, b_1, \ldots, b_4$ are constants and the two boundary conditions are independent (i.e., neither implies the other), is called a **Sturm–Liouville problem**. We're interested in a few specific types of Sturm–Liouville problems, as they arise frequently in applications, often via separation of variables. Specifically, most problems that we run into will be of the form

$$L[y] = (ry')' + qy = -\lambda wy, \qquad a < x < b,$$
$$a_1 y(a) + a_2 y'(a) = 0,$$
$$b_1 y(b) + b_2 y'(b) = 0,$$

where

1) $w(x), q(x), r(x)$ and $r'(x)$ are continuous on $a < x < b$.

2) $w(x) > 0$ and $r(x) > 0$ on $a < x < b$.

The boundary conditions—one at $x=a$, the other at $x=b$—are called **separated boundary conditions**.

Definition 8.1 *If, in addition, the Sturm–Liouville problem above also satisfies*

1) w, q, r and r' are continuous on $a \le x \le b$,

2) $w(x) > 0$ and $r(x) > 0$ for $x = a$ and $x = b$

we say that it is a **regular Sturm–Liouville problem***.*

If, instead, the problem satisfies at least one of the following:

1) $w(a) = 0$ or $w(b) = 0$ or $r(a) = 0$ or $r(b) = 0$

2) any of w, q or r becomes infinite at $x = a$ or $x = b$

3) $a = -\infty$ or $b = \infty$,

we call the problem a **singular Sturm–Liouville problem***. (Of course, there are many other ways a problem can become singular, for example, if one of the functions involved becomes infinite on the interval $a < x < b$.)*

Example 1 The system

$$y'' = (y')' = -\lambda y, \qquad 0 \le x \le L,$$
$$y(0) = 0 \text{ or } y'(0) = 0,$$
$$y(L) = 0 \text{ or } y'(L) = 0$$

is a regular Sturm–Liouville problem, with $r(x) \equiv w(x) \equiv 1$, $q(x) \equiv 0$.

Example 2 The Cauchy–Euler equation

$$x^2 y'' + axy' + by = -\lambda y, \qquad a, b \text{ constant}$$

can be rewritten

$$(x^a y')' + bx^{a-1}y = -\lambda x^{a-1}y.$$

So, e.g.,

$$(x^a y')' + bx^{a-1}y = -\lambda x^{a-1}y$$
$$y(1) = y(e) = 0$$

is a regular Sturm–Liouville problem, while, with boundary conditions

$$y(0) = y(e) = 0,$$

it would be a *singular* Sturm–Liouville problem.

Example 3 We often need to solve *Legendre's equation*

$$(1 - x^2)y'' - 2xy' + \lambda y = 0$$

on the interval $-1 \le x \le 1$. We can rewrite the equation as

$$[(1 - x^2)y']' + \lambda y = 0,$$

and we see that the equation, subject to boundary conditions at $x = \pm 1$, is a *singular* Sturm–Liouville problem.

Analogous versions of properties 1–5 hold for the general regular Sturm–Liouville problem, as we'll show in the following section. A few of the exercises below will serve as preparation or motivation for these proofs.

Exercises 8.1

1. Write each ODE in Sturm–Liouville form:

   a) $y'' + 3y' - 2y = 0$

   b) $y'' - xy = 0$ (Airy's equation)

   c) $x^2 y'' + xy' - 6y = 0$

d) $xy'' + (1 - x)y' + ny = 0$ (Laguerre's equation of order n)

e) $y'' - 2xy' + 2ny = 0$ (Hermite's equation of order n)

f) $x^2 y'' + xy' + (x^2 - a^2)y = 0$ (Bessel's equation of order a)

g) $(1 - x^2)y'' - 3xy' + \lambda y = 0$ (Chebyshev's equation of the second kind)

2. Write the ODE in Sturm–Liouville form, then determine if the Sturm–Liouville problem is regular or singular.

a) $y'' + \lambda y = 0$
$y(-1) + 3y'(-1) = 5y(2) - 7y'(2) = 0$

b) $y'' + 2y' - 5\lambda y = 0$
$y(0) = y'(3) = 0$

c) $(1 - x^2)y'' - xy' + \lambda y = 0$ (Chebyshev's equation of the first kind)
$y(-1) = y(1) = 0$

d) $x^2 y'' + xy' + \lambda y = 0$
$y(0) = y'(1) = 0$

3. Suppose that y_1 and y_2 are linearly independent solutions of

$$(ry')' + qy = 0$$

on a given interval (possibly of infinite extent). If W is the Wronskian

$$W[y_1, y_2] = y_1 y_2' - y_2 y_1',$$

show that

$$(rW)' = 0$$

and, then, that rW is constant on the given interval. (Assume that r and q are well-enough behaved.)

4. (For the proof of Theorem 8.4 in the following section.) Remember that every complex number z can be written as $z = a + bi$, where a and b are real numbers. We define the complex conjugate of z to be the complex number

$$\bar{z} = a - bi.$$

Prove the following:

a) For any complex number z, $z \cdot \bar{z}$ is real.

b) For any complex numbers z_1 and z_2,

    i) $\overline{z_1 + z_2} = \bar{z}_1 + \bar{z}_2$

    ii) $\overline{z_1 z_2} = \bar{z}_1\, \bar{z}_2$

c) If z is real, then $\bar{z} = z$.

It follows from the above that any complex function $f(x)$, x real, can be written as $f(x) = g(x) + ih(x)$, where g and h are real.

  d) Show that $\overline{f'(x)} = \overline{f(x)}'$.

Finally, putting everything together, and using the definition of the equality of two numbers

$$a + bi = c + di \quad \text{if and only if} \quad a = c \text{ and } b = d,$$

prove:

  e) If y_1 is an eigenfunction, corresponding to the eigenvalue λ_1, of the problem

$$a_0(x)y'' + a_1(x)y' + a_2(x)y = -\lambda y$$
$$A_1 y(a) + A_2 y'(a) = B_1 y(b) + B_2 y'(b),$$

    then $\bar{y}_1$ also is an eigenfunction, corresponding to the eigenvalue $\bar{\lambda}_1$.

5. (**Perpendicularity/orthogonality of eigenvectors:**   compare to proof of Theorem 8.3 in the following section.) Remember that the inner product of functions is analogous to the dot product of vectors, and the orthogonality of functions is analogous to the perpendicularity of vectors. Here, we'd like to prove that the eigenvectors of a symmetric matrix are orthogonal. Specifically, suppose that the real matrix A is $n \times n$ and symmetric, that is, $A^T = A$, and suppose that v and w are real eigenvectors, corresponding to real eigenvalues λ_1 and λ_2, respectively, with $\lambda_1 \neq \lambda_2$. Let's agree that all vectors are column vectors, so that the *dot* product is the same as a *matrix* product, that is,

$$v \cdot w = v^T w.$$

  a) Show that $vAw - wAv = 0$. (Hints: $(AB)^T = B^T A^T$, and $v_1 \cdot v_2 = v_2 \cdot v_1$.)

  b) Show that $vAw - wAv = (\lambda_2 - \lambda_1)v \cdot w$.

  c) Conclude that v and w are perpendicular.

In general, if A is a real $n \times n$ matrix, then A^T is called the **Hermitian adjoint** of A, and A is called **self-adjoint** if $A^T = A$. What if the entries are allowed to be complex? First, when we talked about the inner product of complex *functions* (Exercise 15, Section 3.3), we saw that it had to be defined as

$$\langle f, g \rangle = \int_a^b \overline{f(x)} g(x) dx,$$

or else it would not possess the properties that we would like an inner product to have. Similarly, the inner product of two complex vectors must be defined as

$$\langle \boldsymbol{v}, \boldsymbol{w} \rangle = \bar{v}^T w,$$

where

$$\bar{v} = \begin{bmatrix} \bar{v}_1 \\ \vdots \\ \bar{v}_n \end{bmatrix}.$$

(Refer here and below to Exercise 4.) In this setting, the **Hermitian adjoint** of a complex matrix A is defined to be

$$A^* = \bar{A}^T,$$

where $\bar{A}$ results from taking the complex conjugate of every element of A. A is said to be **self-adjoint** if

$$A^* = A.$$

d) Show that if A is self-adjoint, then its eigenvectors are orthogonal.

e) Show that if A is *real* and self-adjoint, then its eigenvalues are real.

(Actually, it can be shown that the eigenvalues of *any* complex self-adjoint matrix must be real.)

6. (For the proof of Theorem 8.3 in the following section.) Generalize **Green's first and second identities** (Exercise 23, Section 1.7) by showing that the Sturm–Liouville operator

$$L[y] = (ry')' + qy$$

satisfies

a) $\int_a^b y_1 L[y_2] dx = r(x) y_1(x) y_2'(x)\big|_a^b - \int_a^b r y_1' y_2' \, dx + \int_a^b q y_1 y_2 \, dx$

b) $\int_a^b (y_1 L[y_2] - y_2 L[y_1]) dx = r(x)[y_1(x) y_2'(x) - y_2(x) y_1'(x)]\big|_a^b$

(so long as everything is well-enough behaved). The latter is called **Green's second identity** or **Green's formula** for L. When written in differential form

$$y_1 L[y_2] - y_2 L[y_1] = \frac{d}{dx}[r(y_1 y_2' - y_2 y_1')]$$

it is known as **Lagrange's identity** for L.

c) Show that we can*not* do anything similar if the operator is first-order, that is, of the form

$$L[y] = a_0(x)y' + a_1(x)y.$$

d) However, show that we *can* simplify

$$\int_a^b (y_1 L[y_2] - y_2 L^*[y_1])dx$$

if we define

$$L[y] = a_0(x)y' + a_1(x)y,$$
$$L^*[y] = -a_0(x)y' + a_1(x)y.$$

(We will call L^* the **adjoint** of the operator L.)

7. The Sturm–Liouville equation $(ry')' + qy = -\lambda wy$, $a \le x \le b$, may be put into a simpler form, as follows. (We assume $r > 0$ and $p > 0$ on $a \le x \le b$.)

a) Let

$$t = \int_a^x f(z)dz, \quad f(z) > 0 \text{ on } a \le z \le b.$$

Show that the equation becomes

$$Y'' + \frac{(rf)'}{rf}Y' + \frac{q}{f^2 r}Y = -\lambda \frac{w}{f^2 r}Y,$$

where $Y(t) = y(x)$ and the prime represents differentiation by t.

b) Show that the specific choice

$$f(x) = \sqrt{\frac{w(x)}{r(x)}}$$

leads to the equation

$$Y'' + \frac{1}{2}\frac{(rw)'}{rw}Y' + \frac{q}{r}Y = -\lambda Y.$$

c) Now show that the change of *dependent* variable $Y(t) = g(t)z(t)$ leads to the equation

$$gz'' + \left[2g' + \frac{1}{2}\frac{(rw)'}{rw}g\right]z' + \left[\frac{q}{r}g + \frac{1}{2}\frac{(rw)'}{rw}g' + g''\right]z = -\lambda z.$$

d) Finally, show that, by choosing

$$g = \frac{1}{\sqrt[4]{rw}},$$

the equation becomes

$$z'' + Q(t)z = -\lambda z,$$

where

$$Q = -\frac{7[(rw)']^2}{16(rw)^2} - \frac{(rw)''}{4rw}.$$

This is called the **Liouville normal form** of the original equation.

8.2   Regular and Periodic Sturm–Liouville Problems

It turns out that statements 1–5 of the previous section are "mostly" true for the regular Sturm–Liouville problem

$$L[y] = (ry')' + qy = -\lambda wy, \qquad a < x < b,$$
$$a_1 y(a) + a_2 y'(a) = b_1 y(b) + b_2 y'(b) = 0. \tag{8.2}$$

We'll state here without proof two of the theorems which are too difficult to prove at this level. Then we'll prove the remaining three, using some of the results from the exercises of the previous section.

Theorem 8.1 *The eigenvalues of the regular Sturm–Liouville problem (8.2) form an infinite sequence*

$$\lambda_1 < \lambda_2 < \lambda_3 < \dots$$

with

$$\lim_{n \to \infty} \lambda_n = \infty.$$

Theorem 8.2 *The eigenfunctions of the regular Sturm–Liouville problem (8.2) form a complete set in the space of piecewise smooth functions on $a \le x \le b$ (complete in the sense that functions which differ at finitely many points are represented by the same series of eigenfunctions).*

For proofs see, e.g., the classic text *Theory of Ordinary Differential Equations* by Coddington and Levinson.

Now we'd like to prove the remaining three results from the previous section for the regular Sturm–Liouville problem (8.2):

$$L[y] = (ry')' + qy = -\lambda w y,$$
$$a_1 y(a) + a_2 y'(a) = b_1 y(b) + b_2 y'(b) = 0.$$

First, what can we say with respect to orthogonality of the eigenfunctions? Suppose y_1 and y_2 are eigenfunctions corresponding to eigenvalues λ_1 and λ_2, respectively, with $\lambda_1 \neq \lambda_2$. In Exercise 6 of the previous section, we showed that we always have

$$\int_a^b (y_1 L[y_2] - y_2 L[y_1]) dx = r(x)[y_1(x)y_2'(x) - y_2(x)y_1'(x)]\Big|_a^b.$$

But, from the ODE, we also have, for eigenfunctions y_1 and y_2,

$$\int_a^b (y_1 L[y_2] - y_2 L[y_1]) dx = (\lambda_1 - \lambda_2) \int_a^b w(x) y_1(x) y_2(x) dx.$$

Let's show that the boundary terms disappear. They become

$$r(b)[y_1(b)y_2'(b) - y_2(b)y_1'(b)] - r(a)[y_1(a)y_2'(a) - y_2(a)y_1'(a)].$$

If we can show that the first bracketed term is zero, then we're done (why?). Now, since we do *not* have $b_1 = b_2 = 0$, let's first assume that $b_1 \neq 0$. Then, from the second boundary condition, we get

$$y_1(b) = -\frac{b_2}{b_1} y_1'(b)$$
$$y_2(b) = -\frac{b_2}{b_1} y_2'(b),$$

so that

$$y_1(b)y_2'(b) - y_2(b)y_1'(b) = -\frac{b_2}{b_1} y_1'(b)y_2'(b) + \frac{b_2}{b_1} y_1'(b)y_2'(b) = 0.$$

If, instead, $b_2 \neq 0$, then

$$y_1'(b) = -\frac{b_1}{b_2} y_1(b)$$
$$y_2'(b) = -\frac{b_1}{b_2} y_2(b),$$

from which it again follows that

$$y_1(b)y_2'(b) - y_2(b)y_1'(b) = 0.$$

Therefore, we have proved the following.

Theorem 8.3 *Suppose that $\lambda_1 \neq \lambda_2$ are eigenvalues, with eigenfunctions y_1 and y_2, respectively, of the regular Sturm–Liouville problem (8.2). Then,*

$$\int_a^b w(x)y_1(x)y_2(x)dx = 0.$$

So y_1 and y_2 are not necessarily orthogonal unless $w(x) \equiv 1$. What we say is that y_1 and y_2 are

*orthogonal with respect to the **weight function** w,*

and Theorem 8.3 says that the eigenfunctions of the regular Sturm–Liouville problem (8.2) are pairwise orthogonal with respect to w. Often, we'll just say that the functions are orthogonal and, in the case $w(x) \equiv 1$, we say that they are **simply orthogonal**.

Next, using the results of Exercise 5 of Section 8.1, it is easy to show that the eigenvalues must be real.

Theorem 8.4 *The eigenvalues of a regular Sturm–Liouville problem are real.*

PROOF Suppose that $\lambda = c + di$ is an eigenvalue, with eigenfunction $y(x) = u(x) + iv(x)$. Then Exercise 5 tells us that $\bar{\lambda} = c - di$ is an eigenvalue, with eigenfunction $u(x) - iv(x)$.

From the proof of Theorem 8.3, we know that

$$(\lambda_1 - \lambda_2) \int_a^b w(x)y_1(x)y_2(x)dx = 0$$

$$= [(c+di) - (c-di)] \int_a^b w(x)[u(x) + iv(x)][u(x) - iv(x)]dx$$

$$= 2di \int_a^b w(x)[u(x)^2 + v(x)^2]dx = 0.$$

The integral is positive (why?), forcing $d = 0$. Therefore, $\lambda = c$ is real. ∎

Note that this theorem justifies our considering only the cases $\lambda > 0$, $\lambda = 0$ and $\lambda < 0$ back in Section 1.7.

Now, what about the multiplicity of the eigenvalues?

Theorem 8.5 *If y_1 and y_2 are eigenfunctions of (8.2) corresponding to the same eigenvalue λ, then $y_2 = cy_1$ for some constant c.*

PROOF We're given that

$$L[y_1] = (ry_1')' + (q + \lambda w)y_1 = 0$$
$$L[y_2] = (ry_2')' + (q + \lambda w)y_2 = 0.$$

Then,
$$y_1 L[y_2] - y_2 L[y_1] = y_1(ry_2')' - y_2(ry_1')' = 0.$$

Integrating by parts, we get (see Exercise 2)

$$\int_a^x [y_1(ry_2')' - y_2(ry_1')']dx = r(x)[y_1(x)y_2'(x) - y_2(x)y_1'(x)]$$
$$- r(a)[y_1(a)y_2'(a) - y_2(a)y_1'(a)] = 0.$$

Since we already showed, in proving Theorem 8.3, that the last term is zero, and since $r(x) > 0$ on $a \leq x \leq b$, we must have

$$y_1(x)y_2'(x) - y_2(x)y_1'(x) \equiv 0 \quad \text{on} \quad a \leq x \leq b.$$

What does this do for us? Well, it sure looks like a determinant; in fact, it's the Wronskian, $W(y_1, y_2; x)$! And remember from ODEs that $W \equiv 0$ implies that y_1 and y_2 are linearly dependent, i.e., that $y_2 = cy_1$. ∎

So far we have considered only problems with separated boundary conditions. These certainly are not the most general types of boundary conditions. Let's look at the following important example.

Example 1 Find all eigenvalues and eigenfunctions of the eigenvalue problem

$$y'' + \lambda y = 0,$$
$$y(-\pi) = y(\pi),$$
$$y'(-\pi) = y'(\pi).$$

Proceeding as in Chapter 1, we find that there are no negative eigenvalues (see Exercise 2a, below). We do get $\lambda_0 = 0$, with $y_0 = 1$, and, interestingly, we get

$$y_n = n^2, \quad y_n = c_n \cos nx + d_n \sin nx$$

for any choice of the constants c_n and d_n. The eigenfunctions here, of course, are the functions comprising the Fourier series on $-\pi \leq x \leq \pi$. So, we have a countably infinite number of real eigenvalues, and the eigenfunctions are orthogonal and complete. However, *each positive eigenvalue has multiplicity 2.*

Definition 8.2 *The Sturm–Liouville problem*

$$L[y] = (ry')' + qy = -\lambda wy, \quad a < x < b,$$
$$y(a) = y(b), \quad\quad\quad\quad\quad\quad\quad\quad (8.3)$$
$$y'(a) = y'(b),$$

where

1) $w(x), q(x), r(x)$ and $r'(x)$ are continuous on $a \le x \le b$,

2) $w(x) > 0$ and $r(x) > 0$ on $a \le x \le b$,

3) $r(a) = r(b)$,

is called a **periodic Sturm–Liouville problem,** and the boundary conditions are **periodic boundary conditions.**

One may prove analogous versions of Theorems 8.1, 8.2, 8.3 (see Exercise 5) and 8.4 for these problems. Although we do not have uniqueness of eigenfunctions, we do have the following theorem.

Theorem 8.6 *The eigenvalues of a periodic Sturm–Liouville problem form a sequence*

$$\lambda_0 < \lambda_1 \le \lambda_2 < \lambda_3 \le \lambda_4 < \cdots,$$

where, whenever we have "=," *that particular eigenvalue has multiplicity 2; otherwise, each eigenvalue has multiplicity one.*[†]

Exercises 8.2

1. Solve the regular Sturm–Liouville problem, then verify directly, by integration, the orthogonality of the eigenfunctions. (Note: Make sure to use the correct weight function!)

   a) $y'' + \lambda y = 0, y(0) + y'(0) = y'(L) = 0$
   b) $x^2 y'' - xy' + (\lambda + 1)y = 0, y(1) = y(2) = 0$

2. Find all eigenvalues and eigenfunctions of the periodic Sturm–Liouville problem, and show by direct integration that the eigenfunctions are simply orthogonal.

   a) $y'' + \lambda y = 0,$
      $y(-L) = y(L),$
      $y'(-L) = y'(L).$ (Compare with Exercise 9, Section 1.7.)
   b) $y'' + (\lambda - 2)y = 0,$
      $y(0) = y(L),$
      $y'(0) = y'(L).$

3. If the functions $\phi_1, \phi_2, \ldots$ are orthogonal with respect to w on a given interval, what can we say about the functions $\sqrt{w}\,\phi_1, \sqrt{w}\,\phi_2, \ldots$?

4. Show that $\int_a^x [y_1(ry_2')' - y_2(ry_1')']dx = r(y_1 y_2' - y_2 y_1')\big|_a^x.$

[†] Again, for a proof, see Coddington and Levinson's *Theory of Ordinary Differential Equations.*

5. Prove that the eigenfunctions of the periodic Sturm–Liouville problem (8.3) are orthogonal with respect to the weight function w, and that its eigenvalues are real.

6. Why can there be no *more* than two linearly independent eigenfunctions associated with an eigenvalue of the periodic Sturm–Liouville problem (8.3)?

7. Find all eigenvalues of the following problems. Do the results contradict Theorem 8.1?

   a) $y'' + \lambda y = 0$,
      $2y(0) - y(1) = 2y'(0) + y'(1) = 0$

   b) $y'' + \lambda y = 0$,
      $y(0) - y(1) = y'(0) + y'(1) = 0$

8. **Rayleigh quotient, revisited:** Here we generalize Exercise 26 of Section 1.7, using Green's first identity from Exercise 5 of the previous section.

   a) Suppose that we are given the regular Sturm–Liouville problem

   $$(ry')' + qy = -\lambda wy, \qquad a < x < b,$$
   $$a_1 y(a) + a_2 y'(a) = b_1 y(b) + b_2 y'(b) = 0,$$

   where $a_1 a_2 \leq 0$ and $b_1 b_2 \geq 0$ (and $a_1^2 + a_2^2 > 0$ and $b_1^2 + b_2^2 > 0$, that is, we don't have $a_1 = a_2 = 0$ or $b_1 = b_2 = 0$). If, in addition, $q(x) \leq 0$ on $a \leq x \leq b$, show that the problem has no negative eigenvalues. When will zero be an eigenvalue?

   b) Show that the same is true for the periodic Sturm–Liouville problem

   $$(ry')' + qy = -\lambda wy, \qquad a < x < b,$$
   $$y(a) = y(b),$$
   $$y'(a) = y'(b),$$

   where, again, we have $q(x) \leq 0$ on $a \leq x \leq b$. When will zero be an eigenvalue?

9. Suppose that y_1 and y_2 are solutions of the problem

   $$(ry')' + qy = -\lambda_0 wy$$

   for which $y_1(a) = 1, y_1'(a) = 0, y_2(a) = 0, y_2'(a) = 1$. (Assume that p, q, r and r' are continuous, $r > 0$ and $p > 0$, or any intervals $a \leq x \leq b$ in this problem.)

a) Why are y_1 and y_2 unique?

b) Show that the periodic Sturm–Liouville problem (8.3) has two linearly independent eigenfunctions corresponding to λ_0 if and only if $y_1'(b) = y_2(b) = 0$ and $y_1(b) = y_2'(b) = 1$.

10. In this problem, we consider regular Sturm–Liouville problems where, in addition, functions y which satisfy the boundary conditions also satisfy $y(b)y'(b) - y(a)y'(a) \leq 0$.

   a) Show that any function y which satisfies a Dirichlet or a Neumann condition at each end also satisfies

   $$y(b)y'(b) - y(a)y'(a) = 0.$$

   b) What conditions must a_1, a_2, b_1 and b_2 satisfy so that we're guaranteed that if y satisfies the Robin conditions

   $$a_1 y(a) + a_2 y'(a) = b_1 y(b) + b_2 y'(b) = 0,$$

   then y also satisfies

   $$y(b)y'(b) - y(a)y'(a) \leq 0?$$

   c) Use Green's first identity (Exercise 22, Section 1.7 and Exercise 6 of the previous section) to prove that if an eigenfunction of the regular Sturm–Liouville problem

   $$y'' + \lambda y = 0$$
   $$a_1 y(a) + a_2 y'(a) = b_1 y(b) + b_2 y'(b) = 0$$

   also satisfies

   $$y(b)y'(b) - y(a)y'(a) \leq 0,$$

   then its corresponding eigenvalue is nonnegative. (Thus, if all functions which satisfy the boundary conditions also satisfy the last equation, then the problem has only nonnegative eigenvalues.)

   d) In part (c), suppose that $\lambda = 0$ is an eigenvalue. What is its corresponding eigenfunction? Further, what can be said about which boundary conditions we actually must have started with?

11. a) Show that

   $$y_1 y_2^{(4)} - y_2 y_1^{(4)} = \frac{d}{dx}(y_1 y_2''' - y_2 y_1''' - y_1' y_2'' + y_2' y_1'').$$

   b) Use the above to show that the eigenfunctions of the problem

   $$y^{(4)} + \lambda y = 0, \quad y(0) = y'(0) = y(L) = y'(L) = 0$$

   are simply orthogonal.

c) List all combinations of simple homogeneous boundary conditions (of the form $y''(0) = 0$, $y'''(L) = 0$, etc.), two at each end, for which the eigenfunctions of the problem are simply orthogonal.

d) Investigate the above situation, but with the operator y''' instead of $y^{(4)}$ (you may assume that you have the "best possible" boundary conditions).

12. Suppose that r is continuous and that $r(x) > 0$ on $0 \leq x \leq 1$. Show that the eigenfunctions are simply orthogonal for the problem

$$(ry^{(n)})^{(n)} + \lambda y = 0,$$

$$y(0) = y'(0) = \cdots = y^{(n-1)}(0) = y(1) = y'(1) = \cdots = y^{(n-1)}(1),$$

for any positive integer n.

13. Using the properties

$$T_n(\cos \theta) = \cos n\theta$$

$$S_n(\cos \theta) = \frac{\sin(n+1)\theta}{\sin \theta},$$

show that

a) The Chebyshev polynomials of the first kind, $T_n(x)$, are orthogonal with respect to the weight function $w(x) = \frac{1}{\sqrt{1-x^2}}$ on $1- \leq x \leq 1$.

b) The Chebyshev polynomials of the second kind, $S_n(x)$, are orthogonal with respect to the weight function $w(x) = \sqrt{1 - x^2}$ on $-1 \leq x \leq 1$.

14. Analogous to the adjoint of a matrix operator, we would like to define the adjoint of a linear differential operator L. Then, the analog of a Hermitian matrix will be what is called a self-adjoint linear differential operator (although the term Hermitian still is occasionally used).

Suppose we have the operator

$$L[y] = a_0(x)y'' + a_1(x)y' + a_2(x)y.$$

a) Use integration by parts to show that

$$\int_a^b y_1 L[y_2]dx = [y_1 a_0 y_2' - (y_1 a_0)' y_2 + y_1 a_1 y_2]\Big|_a^b$$

$$+ \int_a^b y_2[(y_1 a_0)'' - (y_1 a_1)' + y_1 a_2]dx.$$

In practice, we hope that the boundary conditions force the non-integral terms on the right to be zero. Then, the **adjoint** of L is the operator in the last integral:

$$L^*[y] = (a_0 y)'' - (a_1 y)' + a_2 y.$$

b) Conclude that $y_1 L[y_2] - y_2 L^*[y_1]$ is an *exact derivative*, that is, that

$$y_1 L[y_2] - y_2 L^*[y_1] = \frac{d}{dx} F(x; y_1, y_2, y_1', y_2')$$

for some function F.

c) L is **self-adjoint** if $L^* = L$. Show that L is self-adjoint if and only if $a_1 = a_0'$.

d) Show that the Sturm–Liouville operator

$$L[y] = (ry')' + qy$$

is always self-adjoint. What about

$$L[y] = ry' + q?$$

e) In general, what is $(L^*)^* = L^{**}$?

8.3 Singular Sturm–Liouville Problems; Self-Adjoint Problems

There are many important applications which involve Sturm–Liouville problems which fail to be regular. Specifically, we'll look at problems

$$L[y] = (ry')' + qy = -\lambda wy, \qquad a < x < b,$$

for which at least one of the following is true:

1) $r = 0$ at an endpoint.

2) w or q becomes infinite at an endpoint.

3) $a = -\infty$ or $b = \infty$.

In each case we have, as mentioned before, a *singular* Sturm–Liouville problem. Important examples include the following.

Example 1 Bessel's equation of order α,

$$(xy')' - \frac{\alpha^2}{x} y = -\lambda x^2 y, \qquad 0 < x < L,$$

is singular at $x = 0$.

Example 2 Legendre's equation,

$$[(1 - x^2)y']' = -\lambda y, \qquad -1 < x < 1,$$

is singular at $x = \pm 1$.

Example 3 Chebyshev's equations,

$$[\sqrt{1 - x^2}\, y']' = -\lambda \frac{y}{\sqrt{1 - x^2}}, \qquad -1 < x < 1,$$

and

$$[(1 - x^2)^{3/2}y']' = -\lambda\sqrt{1 - x^2}\, y, \qquad -1 < x < 1,$$

are singular at $x = \pm 1$.

Example 4 Hermite's equation,

$$(e^{-x^2}y')' = -\lambda e^{-x^2}y, \qquad -\infty < x < \infty,$$

is singular at $x = \pm\infty$.

Example 5 Laguerre's equation,

$$(xe^{-x}y')' = -\lambda e^{-x}y, \qquad 0 < x < \infty,$$

is singular at $x = 0$ and at $x = \infty$.

Note that we haven't specified boundary conditions in any of these examples. The reason, of course, is that solutions often fail to exist at singular points, so we certainly cannot expect a solution to attain a specified value at such a point.

In Chapter 7, we saw that certain natural boundary conditions often arise at singular points, due to the nature of the physical problem being solved. However, we would like to do something a little more satisfying, mathematically. To that extent, we put the Sturm–Liouville theory into the more general setting of adjoint operators.

ADJOINT OPERATORS AND SELF-ADJOINT EIGENVALUE PROBLEMS

Motivated by Exercise 14 of Section 8.2, we make the following definition.

Definition 8.3 *Given the operator*

$$L[y] = a_0(x)y'' + a_1(x)y' + a_2(x)y,$$

the **adjoint** *of L is the operator L^* defined by*

$$L^*[y] = (a_0 y)'' - (a_1 y)' + a_2 y.$$

(The adjoint can be generalized and defined for any linear differential operator. See Exercise 7.)

Example 6 Many of our eigenvalue problems entail solving $L[y] = y'' = -\lambda y$. The adjoint of L is

$$L^*[y] = y'',$$

that is, L is its own adjoint, whence the following definition.

Definition 8.4 *If $L^* = L$, we say that L is **self-adjoint**.*

Example 7 The operator $L[y] = ay'' + by' + cy$, where a, b and c are constants, has adjoint

$$L^*[y] = ay'' - by' + cy.$$

Hence, L is self-adjoint if and only if $b = 0$.

Example 8 The Cauchy–Euler operator $L[y] = ax^2 y'' + bxy' + cy$, where a, b and c are constant, has adjoint

$$L^*[y] = ax^2 y'' + (4a - b)xy' + (2a - b + c)y.$$

So L is self-adjoint if and only if $2a = b$.

More generally, in Exercise 14 of the previous section we found necessary and sufficient conditions for L to be self-adjoint.

Theorem 8.7 *The operator $L[y] = a_0(x)y'' + a_1(x)y' + a_2(x)y$ is self-adjoint if and only if $a_0' = a_1$.*

Therefore, there are many second-order operators which are *not* self-adjoint. So, what about Sturm–Liouville operators? We have

$$L[y] = (ry')' + qy = ry'' + r'y' + qy;$$

therefore,

> **a second-order linear differential operator is self-adjoint if and only if it is a Sturm–Liouville operator, and any second-order linear differential operator can be put into self-adjoint form.**

(There is no equivalent statement for higher-order operators.) Of course, $\boldsymbol{L[y]}$ **is self-adjoint if and only if** $\boldsymbol{L[y] + \lambda wy}$ **is self-adjoint** (why?).

Now, we also showed in Exercise 14 of the previous section that L and L^* satisfy the following.

Theorem 8.8 *For the operators L and L^* given in Definition 8.3,*

$$y_1 L[y_2] - y_2 L^*[y_1] = \frac{d}{dx}[a_0 y_1 y_2' - a_0 y_1' y_2 + a_1 y_1 y_1 - a_0' y_1 y_2]$$

(Lagrange's identity)

and

$$\int_a^b (y_1 L[y_2] - y_2 L^*[y_1]) dx = (a_0 y_1 y_2' - a_0 y_1' y_2 + a_1 y_1 y_2 - a_0' y_1 y_2)\big|_a^b.$$

(Green's formula)

It's easy to show that, for L self-adjoint, we have

$$a_0 y_1 y_2' - a_0 y_1' y_2 + a_1 y_1 y_2 - a_0' y_1 y_2 = a_0 (y_1 y_2' - y_2 y_1'),$$

and it was this expression's equaling zero at $x = a$ and $x = b$ that was key to our proving the theorems in Section 8.2 for regular and periodic Sturm–Liouville problems. As for more general Sturm–Liouville problems, it's quite possible that the above expression does *not* disappear at the boundaries. Therefore, we have the following definition.

Definition 8.5 *The general Sturm–Liouville problem*

$$L[y] = (ry')' + qy = -\lambda py,$$
$$a_1 y(a) + a_2 y'(a) + a_3 y(b) + a_4 y'(b) = 0,$$
$$b_1 y(a) + b_2 y'(a) + b_3 y(b) + b_4 y'(b) = 0$$

*is a **self-adjoint problem** if all well-behaved y_1 and y_2 which satisfy the boundary conditions also satisfy*

$$r(y_1 y_2' - y_2 y_1')\big|_a^b = 0.$$

(More generally, an nth-order eigenvalue problem is self-adjoint if

1) The operator L is self-adjoint

2) $\int_a^b (y_1 L[y_2] - y_2 L[y_1]) dx = 0$

for all y_1, y_2 satisfying the boundary conditions.)

Example 9 All regular and periodic Sturm–Liouville problems are self-adjoint.

Example 10 Consider the problem

$$y'' + qy = -\lambda wy,$$
$$y(0) - y'(1) = y'(0) + y(1) = 0.$$

It has boundary conditions which are neither separated nor periodic—so the problem is neither regular nor periodic. Nevertheless, we have

$$r(y_1 y_2' - y_2 y_1')\big|_0^1 = y_1(1) y_2'(1) - y_2(1) y_1'(1) - y_1(0) y_2'(0) + y_2(0) y_1'(0),$$

and this expression is zero if y_1 and y_2 satisfy the boundary conditions (why?). Thus, the problem *is* self-adjoint.

Now, back to singular Sturm–Liouville problems and their boundary conditions. Although it seems like we're cheating, we try to give conditions at the singular points which will make the problem self-adjoint. Actually, though, hindsight is 20-20, and these problems were solved before any talk of self-adjointness—and the conditions which *do* make the problems self-adjoint give us these same solutions. Further, these boundary conditions are consistent with those derived physically and mentioned in Chapter 7.

So, for Legendre's equation, the associated Legendre's equation and both Chebyshev's equations, we have singularities of the form $r(\pm 1) = 0$; thus, these problems are self-adjoint provided we have

$$y, y' \text{ bounded as } x \to 1^- \text{ and } x \to -1^+$$

(see Exercise 3c). The Legendre and Chebyshev polynomials satisfy these properties, of course, and it turns out that they are the only solutions to do so.[‡]

Bessel's equation is singular both because $r(0) = 0$ *and* because $q(0+) = \infty$. To make a long story short, essentially we need to stipulate that

$$y, y' \text{ be bounded as } x \to 0^+$$

and that they be such that any improper integrals involving $q(x) = -\frac{\alpha^2}{x}$ converge. Again, the solutions end up being the Bessel functions.

To be complete, although it is not done in practice, we may do something similar for equations on unbounded intervals. However, here the situation is delicate and, for a rigorous discussion, one should consult the references cited in Section 7.1. That said, and guided by the physics of the problems, Laguerre becomes a self-adjoint problem if we stipulate the conditions

$$y, y' \text{ bounded as } x \to 0+,$$
$$\sqrt{x}\, e^{-\frac{x}{2}} y, \sqrt{x}\, e^{-\frac{x}{2}} y' \to 0 \text{ as } x \to \infty.$$

Similarly for Hermite, if we have y satisfy

$$e^{-\frac{x^2}{2}} y, e^{-\frac{x^2}{2}} y' \to 0 \text{ as } x \to \pm\infty.$$

Note that it was the self-adjointness of regular and periodic Sturm–Liouville problems that allowed us to prove that their eigenvalues are real and their eigenfunctions orthogonal. So it should be no surprise if the same were true for self-adjoint problems in general and, indeed, this turns out to be so.

ADJOINT BOUNDARY CONDITIONS

To be more precise, for a given boundary-value problem we may define what we call its **adjoint boundary-value problem**. To do so, we look at an example.

[‡]See, e.g., Sagan's *Boundary and Eigenvalue Problems in Mathematical Physics*.

Example 11 Given the boundary-value problem

$$L[y_2] = y_2'' + 2y_2' - 3y_2 = 0,$$
$$y_2(0) = y_2'(1) = 0,$$

we wish to find boundary conditions for y_1 so that (from *Green's second identity*)

$$\int_0^1 [y_1 L[y_2] - y_2 L^*[y_1]) dx$$

$$= [a_0(y_1 y_2' - y_1' y_2) + (a_1 - a_0')y_1 y_2]\big|_0^1$$

$$= (y_1 y_2' - y_1' y_2 + 2y_1 y_2)\big|_0^1 = 0.$$

Applying the y_2 conditions, we have

$$y_1(1)y_2'(1) - y_1'(1)y_2(1) + 2y_1(1)y_2(1)$$
$$- y_1(0)y_2'(0) - y_2(0)y_1'(0) - 2y_1(0)y_2(0)$$
$$= y_2(1)[y_1(1) - y_1'(0)] - y_2'(0)y_1(0).$$

Since there are no restrictions on $y_2(1)$ and $y_2'(0)$, we need to have

$$y_1(0) = y_1(1) - y_1'(1) = 0.$$

We call these **adjoint boundary conditions**, and "the" adjoint boundary-value problem is

$$L^*[y] = y'' - 2y' - 3y = 0,$$
$$y(0) = y(1) - y'(1) = 0.$$

We say "the" because the adjoint boundary conditions are not unique (why?).

Finally, we say that a boundary-value problem is self-adjoint if

1) $L^* = L$.

2) The adjoint boundary conditions are equivalent to the given boundary conditions (in the sense that any function satisfies the latter if and only if it satisfies the former).

Exercises 8.3

1. Find all eigenvalues and eigenfunctions of the singular Sturm–Liouville problem

   a) $x^2 y'' + xy' + \lambda y = 0$; y, y' bounded as $x \to 0^+$; $y(1) = 0$

   b) $y'' + \lambda y = 0$; $y(0) + y'(0) = 0$; $y, y' \to 0$ as $x \to \infty$

2. In each case, supposing the boundary conditions are such that the eigen-functions are orthogonal with respect to the weight functions w, find w.

   a) $xy'' + \lambda y = 0$

   b) $y'' + 2y' + \lambda y = 0$

   c) $y'' + 2xy' - \lambda xy = 0$

3. a) Suppose that the problem

$$(ry')' + qy = -\lambda wy,$$
$$y, y' \text{ bounded as } x \to a^+,$$
$$b_1 y(b) + b_2 y'(b) = 0$$

   has a singularity of the form $r(a) = 0$. Show that

$$\int_a^b (y_1 L[y_2] - y_2 L[y_1]) dx = 0$$

   for all y_1, y_2 which satisfy the boundary conditions.

   b) Do the same for the problem

$$(ry')' + qy = -\lambda wy,$$
$$a_1 y(a) + a_2 y'(a) = 0,$$
$$y, y' \to 0 \text{ as } x \to \infty.$$

   c) Show that Legendre's and both Chebyshev's equations, subject to the boundary condition

$$y, y' \text{ bounded as } x \to 1^- \text{ and } x \to -1^+,$$

   form self-adjoint systems.

4. Decide if the problem, as it stands, is self-adjoint, and prove your result.

   a) $y'' + \lambda y = 0, y(0) = y'(0) - y(1) = 0$

   b) $x^2 y'' + 2xy' + (\lambda - x^3)y = 0, y(2) = y'(3) = 0$

   c) $y'' + 2y' + \lambda y = 0, y(0) = y'(1) = 0$

   d) $y'' + \lambda y = 0, y(0) + y'(0) + 2y'(1) = y'(0) - y(1) = 0$

5. What conditions must a and b satisfy if

$$y'' + \lambda y = 0, \quad y(0) + ay'(1) = y'(0) + by(1) = 0$$

   is to be self-adjoint?

6. Here we generalize Exercise 11 of Section 8.2. Remember, the PDE for the *Euler–Bernoulli beam* is

$$w_{tt} + \alpha^4 w_{xxxx} = 0,$$

where α^4 is constant. Letting $\alpha^4 = 1$ and separating variables, $w = T(t)Y(x)$, we arrive at the x-ODE

$$y^{(4)} + \lambda y = 0.$$

Now, the four standard *energy-conserving* boundary conditions for the beam are:

Clamped:	$w = w_x = 0$ at $x = x_0$;	
Pinned:	$w = w_{xx} = 0$ at $x = x_0$;	
Roller-supported:	$w_x = w_{xxx} = 0$ at $x = x_0$;	
Free:	$w_{xx} = w_{xxx} = 0$ at $x = x_0$.	

(It's easy to see that the first set of conditions is quite reasonable, from a physical/geometric standpoint. If one digs a little deeper or, still better, if one knows the physics of the problem, then the other three make perfect sense, as well. See Appendix D.)

It's easy enough to separate the boundary conditions, as well. So, prove that the problem

$$y^{(4)} + \lambda y = 0, \qquad 0 < x < 1,$$

subject to any of the above boundary conditions at $x = 0$ and $x = 1$, is self-adjoint.

7. A different way of looking at what we did in Exercise 12 of Section 8.2 is as follows. Given the ODE

$$L[y] = a_0(x)y'' + a_1(x)y' + a_2(x)y = 0,$$

we would like to extend the idea of an *exact equation* to this equation. Remember, a first-order equation is exact if there is a function $F(x, y)$ such that the equation can be written in the form $\frac{d}{dx} F(x, y) = 0$. We try to do the same for the above equation, realizing that the best we can expect here is that $F = F(x, y, y')$. So, we wish to find F so that

$$\frac{d}{dx} F(x, y, y') = a_0(x)y'' + a_1(x)y' + a_2(x)y.$$

However,

$$\frac{d}{dx} F(x, y, z) = F_x \frac{dx}{dx} + F_y \frac{dy}{dx} + F_z \frac{dz}{dx} = F_x + F_y y' + F_z z'.$$

Letting y' play the role of z, we have

$$F_x + F_y y' + F_{y'} y'' = a_0(x)y'' + a_1(x)y' + a_2(x)y. \qquad (8.4)$$

a) Why can we conclude that $F = a_0(x)y' + g(x, y)$, for some function g?

b) Substitute the expression from part (a) back into equation (8.4) to conclude that g must satisfy

$$g_x = a_0(x)y \quad \text{and} \quad g_y = a_1(x) - a_0'(x).$$

c) Conclude that $g = (a_1 - a_0')y + h(x)$, for some function h, and thus that we must have

$$a_1' - a_0'' = a_2 \quad \text{and} \quad h(x) = \text{constant}.$$

Therefore, there is such an F only if $a_1' - a_0'' = a_2$, and, in this case, we have $F = a_0 y' + (a_1 - a_0')y$, which reduces the original ODE to

$$a_0 y' + (a_1 - a_0')y = c, \quad c = \text{arbitrary constant}.$$

d) Conversely, show that if $a_1' - a_0'' = a_2$, then $F = a_0 y' + (a_1 - a_0')y$ satisfies (8.4) and, therefore, the original ODE is exact.

Now, suppose that the original ODE is *not* exact, and suppose we'd like to find an integrating factor, that is, a function $v(x)$ such that

$$v a_0 y'' + v a_1 y' + v a_2 y = 0$$

is exact.

e) Show that this new equation is exact if and only if v satisfies

$$a_0 v'' + (2a_0' - a_1)v' + (a_0'' - a_1' + a_2)v = 0.$$

In other words, v is an integrating factor for the equation $L[y] = 0$ if and only if v satisfies *the adjoint equation* $L^*[v] = 0$.

f) If $L^*[v] = 0$ is not exact, and $u(x)$ is an integrating factor, what ODE must u satisfy?

g) Use the above to find the general solution of the equation

$$xy'' + (3x + 2)y' + (2x + 3)y = 0, \qquad x > 0.$$

8. More generally, we define the adjoint of the nth-order linear differential operator

$$L[y] = a_0(x)y^{(n)} + a_1(x)y^{(n-1)} + \ldots + a_n(x)y$$

to be the operator

$$L^*[y] = (-1)^n (a_0 y)^{(n)} - (-1)^{n-1} (a_1 y)^{(n-1)} \pm \ldots + a_n y.$$

In fact, one can compute this operator in a manner similar to that of the previous problem.

a) Determine if the operator is self-adjoint:

    i) $L[y] = y^{(6)} + 3y'' - 5xy$

    ii) $L[y] = y^{(4)} - xy''$

    iii) $L[y] = y^{(2m)}, m = 1, 2, \ldots$

    iv) $L[y] = y^{(2m+1)}, m = 1, 2, \ldots$ (compare with Exercise 11c of Section 8.2)

b) Show that any operator of the form

$$L[y] = [a_0(x)y'']'' + [a_1(x)y']' + a_2(x)y$$

is self-adjoint.

c) Show that any operator of the form

$$L[y] = [a_0(x)y^{(n)}]^{(n)} + [a_1(x)y^{(n-1)}]^{(n-1)}$$
$$+ \ldots + [a_{n-1}(x)y']' + a_n(x)y$$

is self-adjoint.

9. Find the adjoint boundary-value problem for the given problem:

a) $L[y] = a_0 y'' + a_1 y' + a_2 y = 0$, a_0, a_1, a_2 constant, $y(0) - y'(0) = y(1) + y'(1) = 0$

b) $L[y] = a_0(x)y'' + a_1(x)y' + a_2(x)y = 0$, $y'(0) = y(1) = 0$

In each case, what condition on the coefficients will make the problem self-adjoint? (By the way, it turns out that things are more complicated for operators of higher order.)

8.4 The Mean-Square or L^2 Norm and Convergence in the Mean

We have seen that many Sturm–Liouville problems have eigenfunctions which are complete on the interval in question. So, suppose we have such a problem on the interval $a \leq x \leq b$, with eigenfunctions $\phi_1, \phi_2, \ldots$. We can expand "any" function $f(x)$ into a series

$$f(x) \sim \sum_{n=1}^{\infty} c_n \phi_n(x)$$

which, as it turns out, will behave just like the various trigonometric Fourier series, i.e., it will converge to f pointwise except at finitely many points. So, below, we'll generalize the idea of Fourier series to include *basis functions* $\phi_1, \phi_2, \ldots$ from any complete, orthogonal set.

However, before doing so, we change the mathematical setting. This will allow us to consider more than just the piecewise smooth functions. In addition, the new setting will be more satisfying mathematically, in that we no longer will have to include the codicil "except at finitely many points."

We begin by generalizing the idea of inner product.

Definition 8.6 *Given the* weight function $w(x)$ *with the properties* $w(x) > 0$ *and w continuous on $a < x < b$, the* **inner product, with respect to w,** *of the functions f and g is defined to be*

$$\langle f, g \rangle = \int_a^b f(x) g(x) w(x) dx$$

(as long as the integral exists).

Of course, if $w(x) \equiv 1$, this is the inner product defined in Section 3.1. Also, being analogous to the dot product of vectors, the inner product satisfies the following properties (see Exercise 1):

$$\langle f, g \rangle = \langle g, f \rangle$$
$$\langle f, g + h \rangle = \langle f, g \rangle + \langle f, h \rangle$$
$$\langle cf, g \rangle = c \langle f, g \rangle.$$

As defined earlier, if $\langle f, g \rangle = 0$, we say that f and g are **orthogonal with respect to w**.

Continuing the analogy with the dot product, remember that the length of a vector is given by

$$\|\boldsymbol{v}\| = \sqrt{\boldsymbol{v} \cdot \boldsymbol{v}}.$$

Vector length is an example of what is called a *norm*; two of the important properties of a norm—obviously satisfied by vectors—are

$$\|\boldsymbol{v}\| \geq 0$$
$$\|\boldsymbol{v}\| = 0 \text{ if and only if } \boldsymbol{v} = \boldsymbol{0}.$$

The norm automatically leads to the "distance between two vectors" (actually, the distance between their endpoints, in standard position),

$$\|\boldsymbol{v} - \boldsymbol{w}\| = \sqrt{\langle \boldsymbol{v} - \boldsymbol{w}, \boldsymbol{v} - \boldsymbol{w} \rangle}.$$

Of course, $\|\boldsymbol{v} - \boldsymbol{w}\| = 0$ if and only if $\boldsymbol{v} = \boldsymbol{w}$. (A distance formula is a special case of what is called a *metric* on the given vector space; this particular kind of metric is said to be *induced by* the corresponding norm.)

We wish to extend these ideas (along with one more vector property, dealt with in Exercise 6) to inner products of functions. So we define the norm of a function.

Definition 8.7 *With the same conditions as in Definition 8.6, we define the* **mean-square** *or* $\boldsymbol{L^2}$ *("L-2")* **norm** *of a function f on $a \leq x \leq b$ to be*

$$\|f\| = \sqrt{\langle f, f \rangle} = \left\{ \int_a^b [f(x)]^2 w(x) dx \right\}^{1/2}.$$

Now there is a slight problem—there are *many* functions f which satisfy $\|f\| = 0$ (give some examples); so, on the surface, it seems that we cannot extend the first property of the vector norms to functions. (This, of course, is intimately related to our "except at finitely many points" problem.) Our way around this is to alter our definition of "=." Taking our cue from the vector case, we agree to say that f is the *zero-function* if and only if $\|f\| = 0$ and that $f = g$ if and only if $\|f - g\| = 0$. (Actually, we say that we *identify g with f*.) This is a common mathematical practice and, although it seems that we're cheating, we're justified because all of our operations are based on integrals. Also, as we've seen, this identification seems to make sense physically. In fact, we could almost as easily have done it back in Chapter 3, although we would have had to give up on the beautiful idea of pointwise convergence of Fourier series.

When dealing with vectors, we found that it was convenient if our basis vectors were of length one. The same is true here.

Definition 8.8 *The set of functions* $\psi_1, \psi_2, \ldots$ *is* **orthonormal** *(with respect to w, on $a \leq x \leq b$) if*

1) They are orthogonal.

2) $\|\psi_1\| = \|\psi_2\| = \cdots = 1.$

Remember, we can turn a vector into a unit vector by dividing by its own length. Here we do the same with functions.

Theorem 8.9 *Suppose* $\phi_1, \phi_2, \ldots$ *are orthogonal with respect to w on $a \leq x \leq b$. Then the functions*

$$\psi_1 = \frac{\phi_1}{\|\phi_1\|}, \quad \psi_2 = \frac{\phi_2}{\|\phi_2\|}, \ldots$$

form an orthonormal set with respect to w, on $a < x < b$. Here, we say that we have **normalized** the set of functions $\phi_1, \phi_2, \ldots$.

PROOF   If $n = m$, then

$$\langle \psi_n, \psi_m \rangle = \int_a^b \frac{\phi_n}{\|\phi_n\|} \frac{\phi_n}{\|\phi_n\|} w \; dx$$

$$= \frac{1}{\|\phi_n\|^2} \int_a^b \phi_n^2 w \; dx$$

$$= \frac{1}{\|\phi_n\|^2} \|\phi_n\|^2 = 1.$$

If $n \neq m$, then

$$\langle \psi_n, \psi_m \rangle = \frac{1}{\|\phi_n\| \|\phi_m\|} \int_a^b \phi_n \phi_m w \; dx = 0 \quad \text{(why?)}.$$

∎

Example 1 The functions $\phi_n = \sin \frac{n\pi x}{L}$, $n = 1, 2, \ldots$, are simply orthogonal on $0 \leq x \leq L$. Therefore, since

$$\|\phi_n\|^2 = \int_0^L \sin^2 \frac{n\pi x}{L} \; dx = \frac{L}{2},$$

the functions $\psi_n = \sqrt{\frac{2}{L}} \sin \frac{n\pi x}{L}$, $n = 1, 2, \ldots$, are (simply) orthonormal on $0 \leq x \leq L$.

Example 2 The Legendre polynomials P_n, $n = 0, 1, 2, \ldots$, are simply orthogonal on $-1 \leq x \leq 1$. Since

$$\|P_n\| = \sqrt{\frac{2n+1}{2}},$$

the polynomials $\psi_n = \sqrt{\frac{2n+1}{2}} P_n$ are (simply) orthonormal on $-1 \leq x \leq 1$.

Example 3 The Chebyshev polynomials of the first kind, T_n, $n = 0, 1, 2, \ldots$, are orthogonal with respect to the $w(x) = \frac{1}{\sqrt{1-x^2}}$ on $-1 \leq x \leq 1$. Since

$$\|T_n\| = \begin{cases} \sqrt{\pi}, & \text{if } n = 0, \\ \sqrt{\dfrac{\pi}{2}}, & \text{if } n \geq 1, \end{cases}$$

the polynomials

$$\psi_0 = \frac{1}{\sqrt{\pi}}, \quad \psi_n = \sqrt{\frac{2}{\pi}} T_n, \quad n = 1, 2, \ldots,$$

are orthonormal with respect to $w(x) = \frac{1}{\sqrt{1-x^2}}$ on $-1 \le x \le 1$.

Now, back in Chapter 3, when considering pointwise convergence of a sequence of functions, we wanted

$$\lim_{n \to \infty} |g_n(x) - g(x)| = 0$$

for each value of x. The only distance formula or metric at our disposal was the absolute value function and, in a sense, we had to look at the above limit for each x, separately. However, we now have a new metric, induced by the norm of a function, which looks at functions in a *global*, as opposed to a pointwise, sense.

Definition 8.9 *Given a sequence of functions* $g_1, g_2, g_3, \ldots$, *we say that the sequence* **converges in the mean** *to g if*

$$\lim_{n \to \infty} \|g_n - g\| = 0.$$

Actually, to be more precise, we call this convergence **mean-square convergence** *or* L^2 **convergence** *to distinguish it from other types of convergence in the mean (e.g., mean-cubed convergence or L^3 convergence).*

Can we say that either pointwise or mean-square convergence is *stronger* than the other, that is, that any sequence of functions which converges pointwise also must converge in the mean, or vice versa? Surprisingly (well, not if you've done a fair amount of analysis), the answer is no, as we'll see in the exercises.

Exercises 8.4

1. Prove the following properties of the inner product with respect to a weight function, w:

   a) $\langle f, g \rangle = \langle g, f \rangle$

   b) $\langle f, g + h \rangle = \langle f, g \rangle + \langle f, h \rangle = \langle g + h, f \rangle$

   c) $\langle cf, g \rangle = \langle f, cg \rangle = c \langle f, g \rangle$, for any constant c

2. a) The functions 1, $\cos \frac{\pi x}{L}$, $\cos \frac{2\pi x}{L}, \ldots$ are simply orthogonal on $0 \le x \le L$. Use them to construct a set which is (simply) orthonormal on $0 \le x \le L$.

   b) Construct a set of functions which is (simply) orthonormal on $-L \le x \le L$.

3. Construct a set of polynomials which is orthonormal with respect to the weight function w, on the given interval.

   a) $w(x) = \sqrt{1 - x^2}$, on $-1 \leq x \leq 1$

   b) $w(x) = e^{-x}$, on $0 \leq x < \infty$

   c) $w(x) = e^{-x^2}$, on $-\infty < x < \infty$

4. Given the set of functions $g_n(x) = x^n$, $n = 1, 2 \ldots$, on $0 \leq x \leq 1$:

   a) Show that the sequence converges in the mean to $g(x) = 0$ (with respect to the weight function $w(x) = 1$).

   b) Show that the sequence does *not* converge pointwise to the function $g(x) = 0$.

5. Given the set of functions

$$f_n(x) = \begin{cases} n, \text{ if } 0 < x < \dfrac{1}{n}, \\ 0, \text{ otherwise,} \end{cases} \quad \text{on} \quad 0 \leq x \leq 1,$$

   a) Show that the sequence converges pointwise.

   b) Show that the sequence does *not* converge in the mean (with respect to the weight function $w(x) = 1$).

6.   a) Show that $\|f - g\| = 0$ implies that $\|f\| = \|g\|$.

   b) Remember that the vector dot product satisfies $\boldsymbol{v} \cdot \boldsymbol{w} = \|\boldsymbol{v}\| \|\boldsymbol{w}\| \cos\theta$, where θ is the angle between the vectors. Therefore, we always have $|\boldsymbol{v} \cdot \boldsymbol{w}| \leq \|\boldsymbol{v}\| \|\boldsymbol{w}\|$. Prove that the inner product of functions, with respect to the weight function in (x), satisfies

$$|\langle f, g \rangle| \leq \|f\| \|g\|.$$

   This is the **Schwarz inequality**. (Hint: $F(\lambda) = \|f + \lambda g\|^2 = \langle f + \lambda g, f + \lambda g \rangle \geq 0$; therefore, F is a quadratic polynomial in λ with only one root.)

   c) Using the Schwarz inequality, and comparing $\|f + g\|^2$ and $(\|f\| + \|g\|)^2$, prove the **triangle inequality**

$$\|f + g\| \leq \|f\| + \|g\|.$$

7. Give a different proof of the Schwarz inequality by using the fact that

$$\int_a^b \int_a^b [f(x)g(y) - f(y)g(x)]^2 w(x)w(y)\,dx\,dy \geq 0.$$

8. If we generalize the vector statement $|\boldsymbol{v} \cdot \boldsymbol{w}|^2 \leq \|\boldsymbol{v}\|^2 \|\boldsymbol{w}\|^2$, we get **Cauchy's inequality**

$$\left(\sum_{i=1}^{n} a_i b_i \right)^2 \leq \sum_{i=1}^{n} a_i^2 \sum_{i=1}^{n} b_i^2.$$

Use mathematical induction to prove this inequality.

9. Suppose we start with the $\mathbb{R}^3$ vector $\boldsymbol{v} = a\hat{\boldsymbol{\imath}} + b\hat{\boldsymbol{\jmath}} + c\hat{\boldsymbol{k}}$.

a) Show that the vector in the x-y plane which most closely approximates $\boldsymbol{v}$ is the vector $\boldsymbol{v}_1 = a\hat{\boldsymbol{\imath}} + b\hat{\boldsymbol{\jmath}}$, i.e., show that the quantity $\|\boldsymbol{v} - (A\hat{\boldsymbol{\imath}} + B\hat{\boldsymbol{\jmath}})\|$ is minimized by choosing $A = a$ and $B = b$. ($\boldsymbol{v}_1$, of course, is the projection of $\boldsymbol{v}$ onto the x-y plane.)

b) Conclude that $\|\boldsymbol{v}_1\| \leq \|\boldsymbol{v}\|$. When are they equal?

c) We generalize the above result as follows. Suppose $\boldsymbol{v} = a\hat{\boldsymbol{u}}_1 + b\hat{\boldsymbol{u}}_2 + c\hat{\boldsymbol{u}}_3$ for a set $\hat{\boldsymbol{u}}_1, \hat{\boldsymbol{u}}_2, \hat{\boldsymbol{u}}_3$ of orthonormal (i.e., perpendicular and of length one) vectors. Show that the quantity

$$\|\boldsymbol{v} - (A\hat{\boldsymbol{u}}_1 + B\hat{\boldsymbol{u}}_2)\|$$

is minimized when $A = \boldsymbol{v} \cdot \hat{\boldsymbol{u}}_1$ and $B = \boldsymbol{v} \cdot \hat{\boldsymbol{u}}_2$.

d) Conclude that

$$\|(\boldsymbol{v} \cdot \hat{\boldsymbol{u}}_1)\hat{\boldsymbol{u}}_1 + (\boldsymbol{v} \cdot \hat{\boldsymbol{u}}_2)\hat{\boldsymbol{u}}_2\| \leq \|\boldsymbol{v}\|.$$

Then conclude that we have "=" if and only if $\boldsymbol{v}$ is perpendicular to $\hat{\boldsymbol{u}}_3$.

e) Conclude that

$$(\boldsymbol{v} \cdot \hat{\boldsymbol{u}}_1)^2 + (\boldsymbol{v} \cdot \hat{\boldsymbol{u}}_2)^2 \leq \|\boldsymbol{v}\|^2.$$

When we extend this idea to functions in the following section, this inequality will be called **Bessel's inequality**.

f) Conclude that we have "=" in the above statement if and only if $\boldsymbol{v}$ is perpendicular to $\hat{\boldsymbol{u}}_3$. The equation

$$(\boldsymbol{v} \cdot \boldsymbol{u}_1)^2 + (\boldsymbol{v} \cdot \boldsymbol{u}_2)^2 = \|\boldsymbol{v}\|^2$$

will be generalized, in the case of functions, to what will be *Parseval's equality*. Actually, though, this equation, here, is a very famous theorem, in disguise. Which theorem?

10. Given the function $f(x)$ on $0 \leq x \leq \pi$, using weight function $w(x) = 1$,

a) Find the value of c_1 which minimizes

$$\|f - c_1 \sin x\|^2.$$

What's the significance of this number?

b) Find the values of c_1 and c_2 which minimize

$$\|f - c_1 \sin x - c_2 \sin 2x\|^2$$

two different ways:

    i) Using calculus of two variables, with c_1 and c_2 the independent variables;

    ii) Using algebra, by completing the square in each variable, c_1 and c_2.

Again, what is the significance of these numbers?

In each case, we are trying to find the linear combination which most closely approximates f in the mean-square or L^2 sense. The quantity being minimized is called the **mean-square error**.

11. Do the same as in Exercise 10, but for the given function f on $-1 \le x \le 1$, and where P_0, P_1 and P_2 are the first three Legendre polynomials.

    a) $f(x) = x^4$

    b) $f(x) = \begin{cases} 0, \text{ if } -1 \le x < 0, \\ x, \text{ if } 0 \le x \le 1 \end{cases}$

    c) $f(x) = x^2$

In each case, what is the mean-square error?

8.5   Generalized Fourier Series; Parseval's Equality and Completeness

Now we are ready to look at generalized Fourier series in the setting of mean-square convergence. We begin with an orthogonal set of functions, with respect to the weight function $w(x)$, on an interval $a \le x \le b$, where we may have $a = -\infty$ or $b = \infty$. Then, given any function f, we wish to see if we can expand f into an infinite series of the functions $\phi_1, \phi_2, \ldots$. Remembering that an infinite series is defined in terms of the limit of its partial sums, we first try to determine, for each N, the values of $c_1, c_2, \ldots, c_N$ which minimize

$$\left\| f - \sum_{n=1}^{N} c_n \phi_n \right\|.$$

We can do this in a number of ways (see Exercise 10b in the previous section), and we choose to do so algebraically.

Since the above expression is nonnegative, we may look at

$$\left\| f - \sum_{n=1}^{N} c_n \phi_n \right\|^2 = \left\langle f - \sum_{n=1}^{N} c_n \phi_n, f - \sum_{n=1}^{N} c_n \phi_n \right\rangle$$

$$= \langle f, f \rangle - 2 \left\langle f, \sum_{n=1}^{N} c_n \phi_n \right\rangle + \left\langle \sum_{n=1}^{N} c_n \phi_n, \sum_{m=1}^{M} c_m \phi_m \right\rangle$$

$$\text{(why } m\text{?)}$$

$$= \langle f, f \rangle - 2 \sum_{n=1}^{N} c_n \langle f, \phi_n \rangle + \sum_{n=1}^{N} \sum_{m=1}^{M} c_n c_m \langle \phi_n, \phi_m \rangle.$$

However, $\langle \phi_n, \phi_m \rangle = 0$ unless $n = m$, so we have

$$= \langle f, f \rangle - 2 \sum_{n=1}^{N} c_n \langle f, \phi_n \rangle + \sum_{n=1}^{N} c_n^2 \langle \phi_n, \phi_n \rangle$$

$$= \sum_{n=1}^{N} [c_n^2 \|\phi_n\|^2 - 2c_n \langle f, \phi_n \rangle] + \langle f, f \rangle.$$

Now, we complete the square inside the summation:

$$= \sum_{n=1}^{N} \|\phi_n\|^2 \left[c_n^2 - \frac{2\langle f, \phi_n \rangle}{\|\phi_n\|^2} + \frac{\langle f, \phi_n \rangle^2}{\|\phi_n\|^4} - \frac{\langle f, \phi_n \rangle^2}{\|\phi_n\|^4} \right]$$

$$+ \langle f, f \rangle$$

$$= \sum_{n=1}^{N} \|\phi_n\|^2 \left[c_n - \frac{\langle f, \phi_n \rangle}{\|\phi_n\|^2} \right]^2 - \sum_{n=1}^{N} \frac{\langle f, \phi_n \rangle^2}{\|\phi_n\|^2} + \langle f, f \rangle.$$

Finally, since the variables c_n appear only in the squared term of the first sum, it should be clear that the expression is minimized when each term in this sum is zero. Therefore, the coefficients we're after are

$$c_n = \frac{\langle f, \phi_n \rangle}{\|\phi_n\|^2}.$$

Definition 8.10 *Given an orthogonal set of functions $\phi_1, \phi_2, \ldots$ on an interval $a \leq x \leq b$, and any function f defined on this interval, the constants*

$$c_n = \frac{\langle f, \phi_n \rangle}{\|\phi_n\|^2}, \qquad n = 1, 2, \ldots,$$

are called the **Fourier coefficients** of f with respect to the orthogonal set $\phi_1, \phi_2, \ldots$ (as long as they are defined). The sum

$$\sum_{n=1}^{\infty} c_n \phi_n$$

is called the **Fourier series** of f with respect to these functions and, as before, we write

$$f \sim \sum_{n=1}^{\infty} c_n \phi_n.$$

Of course, if the functions ϕ_n are orthonormal, the Fourier coefficients are just $c_n = \langle f, \phi_n \rangle$.

Example 1 Given the simply orthogonal set $\left\{ \sin \frac{n\pi x}{L} \right\}_{n=1}^{\infty}$ on $0 \leq x \leq L$, the Fourier coefficients of any function f are

$$
\begin{aligned}
c_n &= \frac{\langle f, \sin \frac{n\pi x}{L} \rangle}{\left\| \sin \frac{n\pi x}{L} \right\|^2} \\
&= \frac{2}{L} \int_0^L f(x) \sin \frac{n\pi x}{L} \, dx,
\end{aligned}
$$

which, of course, are just the Fourier sine coefficients derived in Chapter 3. So the Fourier series is just the old Fourier sine series.

In fact, our derivation of the Fourier coefficients here really is just a "cleaner" and more general version of that used in Section 3.3 to derive the trigonometric Fourier coefficients.

Example 2 With respect to the simply orthogonal set of Legendre polynomials P_n, $n = 0, 1, 2, \ldots$, on $-1 \leq x \leq 1$, the Fourier coefficients of any f are

$$
\begin{aligned}
c_n &= \frac{\langle f, P_n \rangle}{\|P_n\|^2}, \qquad n = 0, 1, 2, \ldots, \\
&= \frac{2n+1}{2} \int_{-1}^{1} f(x) P_n(x) dx,
\end{aligned}
$$

and the Fourier–Legendre series for f is

$$f \sim \sum_{n=0}^{\infty} c_n P_n.$$

The big question, of course, is, "When does the generalized Fourier series of f actually converge in the mean to f?" Obviously, it does so if and only if

$$\lim_{N \to \infty} \left\| f - \sum_{n=1}^{N} c_n \phi_n \right\| = 0, \qquad c_n = \frac{\langle f, \phi_n \rangle}{\|\phi_n\|^2},$$

but under what conditions on f and the c_n will this happen? Going back to where we minimized the quantity inside the limit, we have, from letting $c_n = \frac{\langle f, \phi_n \rangle}{\|\phi_n\|^2}$,

$$0 \leq \left\| f - \sum_{n=1}^{N} c_n \phi_n \right\|^2 = \|f\|^2 - \sum_{n=1}^{N} \|\phi_n\|^2 \left[\frac{\langle f, \phi_n \rangle}{\|\phi_n\|^2} \right]^2$$

$$= \|f\|^2 - \sum_{n=1}^{N} c_n^2 \|\phi_n\|^2.$$

This makes sense so long as $\|f\|$ is finite, in which case we rewrite it as

$$\sum_{n=1}^{N} c_n^2 \|\phi_n\|^2 \leq \|f\|^2, \text{ for any } N = 1, 2, 3, \dots .$$

Now, the left side is the Nth partial sum of an infinite series of nonnegative terms, and these are *bounded above* (by $\|f\|^2$). Therefore, the infinite series must converge and must satisfy the famous **Bessel's inequality:**

$$\sum_{n=1}^{\infty} c_n^2 \|\phi_n\|^2 \leq \|f\|^2.$$

For convergence to f, we need

$$\left\| f - \sum_{n=1}^{N} c_n \phi_n \right\|^2 = \|f\|^2 - \sum_{n=1}^{N} c_n^2 \|\phi_n\|^2 \to 0 \text{ as } N \to \infty,$$

that is, we need to have "=" in Bessel's inequality. This is the equally famous **Parseval's equality:**

$$\sum_{n=1}^{\infty} c_n^2 \|\phi_n\|^2 = \|f\|^2,$$

and we have proved the following theorem.

Theorem 8.10 *Suppose that $\|f\| < \infty$. Then the Fourier series for f, in terms of the orthogonal set $\{\phi_n\}_{n=1}^{\infty}$, converges to f if and only if Parseval's equality holds, that is, if and only if*

$$\sum_{n=1}^{\infty} c_n^2 \|\phi_n\|^2 = \|f\|^2, \text{ where } c_n = \frac{\langle f, \phi_n \rangle}{\|\phi_n\|^2}.$$

Now we may talk about completeness in the mean-square or L^2 sense.

Definition 8.11 *Given the set* $\phi_1, \phi_2, \ldots$, *orthogonal with respect to w on* $a \le x \le b$, *we say that this set is* **complete in the mean-square sense** *if, for every function f satisfying $\|f\| < \infty$, the Fourier series for f converges in the mean to f. In other words,* **the set is complete if and only if, for every f satisfying $\|f\| < \infty$, Parseval's equality holds.**

Unfortunately, there is no general method for proving that a given set of orthogonal functions is complete. However, it has been proven, for example, that the eigenfunctions of any regular or periodic Sturm–Liouville problem form a complete set, and the same has been proven for various sets of orthogonal polynomials. **In particular, the trigonometric functions of Chapter 3, and the Legendre polynomials, associated Legendre functions,[§] Chebyshev (both kinds), Laguerre and Hermite polynomials all form complete sets in the mean-square or L^2 sense.[¶] So, too, do the functions** $\{J_n(\frac{x_m x}{a})\}_{m=1}^{\infty}$, **for any $n = 0, 1, 2, \ldots$, where $\{x_m\}_{m=1}^{\infty}$ is the set of the positive roots of J_n.**

There *is* a fairly easy way to show that a set $\phi_1, \phi_2, \ldots$ is *not* complete. Let's first think about vectors; given a set of k perpendicular vectors in $\mathbb{R}^n$, how do we know if they span $\mathbb{R}^n$? Easy—if $k < n$, it doesn't span the space. However, we can't use the same argument in these *function spaces* because they are *infinite dimensional*. But, going back to vectors, if $k < n$, then there's at least one dimension unaccounted for—we can produce a nonzero vector which is perpendicular to the given vectors. The same idea holds here.

Theorem 8.11 *If the orthogonal set $\phi_1, \phi_2, \ldots$ is complete, and if f is orthogonal to each of the ϕ_n, then we must have $f \equiv 0$ (in the mean-square sense).*

PROOF   Since $\phi_1, \phi_2, \ldots$ form a complete set, we have

$$f = \sum_{n=1}^{\infty} c_n \phi_n, \text{ where } c_n = \frac{\langle f, \phi_n \rangle}{\|\phi_n\|^2}.$$

But f is orthogonal to each ϕ_n, so $\langle f, \phi_n \rangle = 0$ for each n. It follows that $f \equiv 0$.   ∎

Example 3 Do the functions $\cos x, \cos 2x, \ldots$ form a complete set on $0 \le x \le \pi$? Take $f(x) = 1$. Then

$$\langle f, \cos nx \rangle = \int_0^{\pi} \cos nx \, dx = 0 \quad \text{for each} \quad n = 1, 2, \ldots .$$

[§] P_n^m, for *fixed* m, for $n = m, m+1, \ldots$.
[¶] In fact, they all actually converge pointwise in the same way as the trigonometric series.

Therefore, f is orthogonal to each function $\cos nx$; thus, the set of functions is not complete.

Before getting to the exercises, let's look at a few examples involving Parseval's equality.

Example 4 *Parseval's equality for the trigonometric Fourier series:* Given $f(x)$ or $-\pi \leq x \leq \pi$, with $\|f\| < \infty$, we have

$$f(x) = \frac{a_0}{2} + \sum_{n=1}^{\infty}(a_n \cos nx + b_n \sin nx).$$

Then, $\|1\|^2 = 2\pi$, $\|\cos nx\|^2 = \|\sin nx\|^2 = \pi$ for $n = 1, 2, \ldots$. So Parseval's equality becomes

$$\|f\|^2 = \left(\frac{a_0}{2}\right)^2 \cdot 2\pi + \pi \sum_{n=1}^{\infty}(a_n^2 + b_n^2)$$

or

$$\frac{1}{\pi}\|f\|^2 = \frac{a_0^2}{2} + \sum_{n=1}^{\infty}(a_n^2 + b_n^2).$$

We extend this idea in Exercise 6.

Example 5 Use the above version of Parseval's theorem, and the function $f(x) = x$, to rederive Euler's series

$$\frac{\pi^2}{6} = \sum_{n=1}^{\infty}\frac{1}{n^2}$$

(see Exercise 19, Section 3.4).

First,

$$\|f\|^2 = \int_{-\pi}^{\pi} x^2 \, dx = \frac{2\pi^3}{3}.$$

Also, the Fourier coefficients were computed in the above-mentioned exercise:

$$a_0 = a_n = 0, \qquad n = 1, 2, \ldots,$$

$$b_n = \frac{1}{\pi}\int_{-\pi}^{\pi} x \sin nx \, dx = \frac{2}{n}(-1)^{n+1}.$$

Then, Parseval's equality becomes

$$\frac{1}{\pi}\frac{2\pi^3}{3} = 4\sum_{n=1}^{\infty}\frac{1}{n^2},$$

which, after a little algebra, gives us what we want.

Exercises 8.5

1. Show that the Fourier coefficients for both the trigonometric Fourier series and the Fourier cosine series are the same as what we get using $c_n = \frac{\langle f, \phi_n \rangle}{\|\phi_n\|^2}$.

2. Calculate the first four terms of the Fourier–Legendre series (in parts a and b) for

   a) $f(x) = \begin{cases} 0, \text{ if } -1 \le x < 0, \\ x, \text{ if } 0 \le x \le 1. \end{cases}$

   b) $f(x) = \cos \pi x$.

   c) Calculate the complete Fourier–Legendre series for $f(x) = x^2$.

3. **Generalized Fourier Series:** In Example 2 we showed that if

$$f(x) = \sum_{n=0}^{\infty} c_n P_n(x), \qquad -1 \le x \le 1,$$

then the **Fourier–Legendre coefficients** of f are

$$c_n = \frac{2n+1}{2} \int_{-1}^{1} f(x) P_n(x) dx, \qquad n = 0, 1, 2, \dots.$$

   a) For a function $f(x)$ on $0 \le x < \infty$, if

$$f(x) = \sum_{n=0}^{\infty} c_n L_n(x),$$

   show that the **Fourier–Laguerre coefficients** of f are

$$c_n = \int_{0}^{\infty} f(x) L_n(x) e^{-x} dx, \qquad n = 0, 1, 2, \dots.$$

   b) For a function $f(x)$ on $-\infty < x < \infty$, if

$$f(x) = \sum_{n=0}^{\infty} c_n H_n(x),$$

   show that the **Fourier–Hermite coefficients** of f are

$$c_n = \frac{1}{2^n n! \sqrt{\pi}} \int_{-\infty}^{\infty} f(x) H_n(x) e^{-x^2} dx, \qquad n = 0, 1, 2, \dots.$$

c) For a function $f(x)$ on $0 \le x \le 1$, if

$$f(x) = \sum_{i=1}^{\infty} c_i J_\alpha(k_i x),$$

where the numbers k_i are the positive roots of the Bessel function J_α, show that the **Fourier–Bessel coefficients** of f are

$$c_i = \frac{2}{J_{\alpha+1}^2(k_i)} \int_0^1 x f(x) J_\alpha(k_i x) dx.$$

(Refer to Exercises 7 and 8, Section 7.5.)

d) Use the properties established in Exercise 4 of Section 7.5 to derive the following Fourier–Bessel series:

$$1 \sim 2 \sum_{n=1}^{\infty} \frac{1}{k_n J_1(k_n)} J_0(k_n x), \qquad 0 < x < 1,$$

$$x^2 \sim \sum_{n=1}^{\infty} \frac{k_n^2 - 4}{k_n^3 J_1(k_n)} J_0(k_n x), \qquad 0 < x < 1,$$

$$x^m \sim 2 \sum_{n=1}^{\infty} \frac{1}{k_n J_{m+1}(k_n)} J_m(k_n x), \qquad 0 < x < 1, \quad m = 0, 1, 2, \dots .$$

The numbers k_n are the positive zeros of J_0 in the first two, and of J_m in the last.

e) **MATLAB:** Plot the graphs of

$$2 \sum_{n=1}^{N} \frac{1}{k_n J_1(k_n)} J_0(k_n x) \quad \text{and} \quad \sum_{n=1}^{N} \frac{k_n^2 - 4}{k_n^3 J_1(k_n)} J_0(k_n x)$$

on $-1 \le x \le 3$, for various values of N. Refer to Table 7.1, Section 7.5.

f) **MATLAB:** Do the same as in part (e) for

$$2 \sum_{n=1}^{N} \frac{1}{k_n J_2(k_n)} J_1(k_n x) \quad \text{and} \quad 2 \sum_{n=1}^{N} \frac{1}{k_n J_3(k_n)} J_2(k_n x),$$

where, in the first sum, the k_n are the positive roots of J_1, while in the second, they're the positive roots of J_2. Again, refer to Table 7.1, Section 7.5.

g) Suppose, instead, that the numbers k_i, $i = 1, 2, 3, \dots$, are the roots of J_α'. Show that the Fourier–Bessel series for $f(x)$, $0 \le x \le 1$, in this case is

$$f(x) \sim \sum_{i=1}^{\infty} c_i J_\alpha(k_i x)$$

with

$$c_i = \frac{2k_i^2}{(k_i^2 - \alpha^2)J_\alpha^2(k_i)} \int_0^1 x f(x) J_\alpha(k_i x) dx.^{\|}$$

(Refer to Exercise 9, Section 7.5.)

4. One may derive the **Fourier–Chebyshev coefficients** as in the previous exercise. However, we proceed as follows:

a) Suppose we have

$$f(x) = \frac{c_0}{2} + \sum_{n=1}^{\infty} c_n T_n(x), \qquad -1 \leq x \leq 1.$$

Let $x = \cos\theta$, $0 \leq \theta \leq \pi$, and use the more familiar looking result to conclude that

$$c_n = \frac{2}{\pi} \int_{-1}^1 f(x) T_n(x) dx, \qquad n = 0, 1, 2, \dots .$$

b) Similarly, supposing that

$$f(x) = \sum_{n=1}^{\infty} c_n S_n(x), \qquad -1 \leq x \leq 1,$$

use the same substitution to show that we have

$$c_n = \frac{2}{\pi} \int_{-1}^1 f(x) S_n(x) dx.$$

5. Find the first three terms of the Fourier–Laguerre series for $f(x) = e^{-2x}$ (don't forget the weight function).

6. a) Explain why every piecewise continuous function $f(x)$ on $a \leq x \leq b$ also satisfies $\|f\|^2 = \int_a^b f^2(x) w(x) dx < \infty$ for any weight function w.

    b) Give an example of a class of functions $f(x)$ on $0 \leq x \leq 1$ such that $\int_0^1 |f(x)| dx$ is infinite (and, therefore, f is not piecewise continuous), but $\int_0^1 f^2(x) dx < \infty$.

7. a) Show that Parseval's equality for the trigonometric Fourier series of a function f on $-L \leq x \leq L$ is

$$\frac{1}{L} \int_{-L}^{L} f^2(x) dx = \frac{a_0^2}{2} + \sum_{n=1}^{\infty} (a_n^2 + b_n^2).$$

$^{\|}$It turns out here that, for $\alpha = 0$, we must include $k_1 = 0$. One may then use l'Hôpital's rule in the "variable" k_1, if need be. (For a more general treatment of this situation, see Pinsky's *Partial Differential Equations and Boundary-Value Problems with Applications*.)

b) Show that Parseval's inequality, for the Fourier sine and Fourier cosine series of f on $0 \le x \le L$, gives us

$$\frac{2}{L} \int_0^L f^2(x)dx = \sum_{n=1}^{\infty} b_n^2 = \frac{a_0^2}{2} + \sum_{n=1}^{\infty} a_n^2.$$

8. In Exercises 19 and 20 of Section 3.4, we talked about some of the wonderful results involving infinite series that Euler had derived, without the benefit of Fourier analysis. We saw that, using Fourier series, we could duplicate some of these results rather easily. Now, with Parseval's equality, we add yet another weapon to our arsenal.

   a) For a warmup, use the Fourier sine series for $f(x) = 1$ on $0 \le x \le \pi$ to rederive the sum

   $$\frac{\pi^2}{8} = \sum_{n=1}^{\infty} \frac{1}{(2n-1)^2}.$$

   b) Now use the Fourier cosine series for $f(x) = x^2$, on $0 \le x \le \pi$, to show that

   $$\frac{\pi^4}{90} = \sum_{n=1}^{\infty} \frac{1}{n^4}.$$

   c) Next, use the Fourier cosine series for $f(x) = x$, on $0 \le x \le 1$, to derive

   $$\frac{\pi^4}{96} = \sum_{n=1}^{\infty} \frac{1}{(2n-1)^4}.$$

   d) Use the results of parts (b) and (c) to find the sum

   $$\frac{1}{2^4} + \frac{1}{4^4} + \frac{1}{6^4} + \frac{1}{8^4} + \cdots .$$

9. Remember that, in Exercise 3, Section 8.2, we showed that if $\phi_1, \phi_2, \ldots$ are orthogonal with respect to w on an interval, then $\sqrt{w}\,\phi_1, \sqrt{w}\,\phi_2, \ldots$ are simply orthogonal on the same interval. Now, supposing that $\phi_1, \phi_2, \ldots$ are complete with respect to w, show that the functions $\sqrt{w}\,\phi_1, \sqrt{w}\,\phi_2, \ldots$ are (simply) complete on the same interval.

10. If $\phi_1, \phi_2, \ldots$ is an orthogonal set on a given interval, and if $\langle f, \phi_n \rangle = \langle g, \phi_n \rangle$, $n = 1, 2, \ldots$, must the functions f and g be identical, in the mean-square sense, i.e., must $\|f - g\| = 0$? Why or why not?

11. Generalize Parseval's equality and show that if

$$f(x) = \sum_{n=1}^{\infty} c_n \phi_n(x) \quad \text{and} \quad g(x) = \sum_{n=1}^{\infty} d_n \phi_n(x)$$

on a given interval, where $\{\phi_n\}_{n=1}^\infty$ is a complete orthogonal set on the interval, then

$$\langle f, g \rangle = \sum_{n=1}^\infty c_n d_n \|\phi_n\|^2.$$

12. **The Hanging Chain:** Bessel functions first arose in the study of the hanging chain (by Daniel Bernoulli in 1732). So, suppose we have a chain of length L, attached at the top and able to *swing*. Letting

$$u(x,t) = \text{location of point } x \text{ at time } t$$

as in Figure 8.1, it turns out that u satisfies the boundary-value problem

$$u_{tt} = gxu_{xx} + gu_x, \qquad 0 < x < L, t > 0,$$
$$u(x,0) = f(x),$$
$$u_t(x,0) = h(x),$$
$$u(0,t) \text{ bounded}, \ u(L,t) = 0.$$

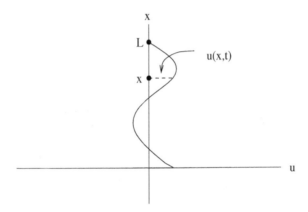

FIGURE 8.1
The hanging chain.

Here, g is the constant gravitational acceleration at the earth's surface.

a) Solve this problem.

b) **MATLAB:** Letting $L = 1$, graph the first five vibration modes.

c) What is the vibration frequency of the nth mode?

Prelude to Chapter 9

Armed with the special functions, we're now in a position to solve the Big Three PDEs in two and three spatial dimensions. We look mostly at problems on bounded domains with simple geometry—rectangular, spherical and the like—and we'll find that, again, we can use the Fourier method to solve them.

Fourier, in fact, after deriving the two- and three-dimensional heat equations, proceeded to solve them as he had solved the one-dimensional version. Poisson followed with solutions of heat problems in polar and spherical coordinates and, ultimately, with his 1835 treatise *Théorie Mathematique de la Chaleur (Mathematical Theory of Heat)*. As for the wave equation, Euler had dealt with the vibrating drumhead much earlier and gave the product solutions for the rectangular and circular drumhead in 1759, solving the latter which eventually would be known as the Bessel functions of the first kind. It remained only to look at infinite linear combinations of these solutions.

In that same year, Euler and Lagrange independently provided cylindrical and spherical wave solutions of the wave equation on all of three-dimensional space. And in the early 19th century, Poisson derived the three-dimensional version of d'Alembert's solution, from which it's easy to see that solutions satisfy *Huygens's Principle* (which Christiaan Huygens (1629–1695) had shown is satisfied by light waves, based on his wave theory of light).

Euler and Lagrange both had written Laplace's equation in polar and spherical coordinates. It was then Legendre, while studying gravitational attraction, who solved the spherical version in the 1780s, with some help from Laplace. And it was here that he encountered the polynomials which now bear his name and which are a special case of the *spherical harmonics* which form part of the solution to the spherical heat and wave equations.

As we've seen, Poisson was responsible for showing that the gravitational potential must satisfy the *non*homogeneous Laplace's equation—that is, *Poisson's equation*—in regions where mass is present. It was while studying these problems that he also provided his elegant closed form solution to Laplace's equation in polar coordinates, a solution now referred to as *Poisson's integral formula*. Also, around 1813, Poisson was the first to apply Laplace's and Poisson's equations to the study of electricity.

9

PDEs in Higher Dimensions

9.1 PDEs in Higher Dimensions: Examples and Derivations

THE HEAT/DIFFUSION EQUATION IN THREE DIMENSIONS

We now derive the heat equation in three space dimensions. Suppose we have a solid three-dimensional piece of material, with constant mass density ρ, specific heat σ and thermal conductivity k (all as defined in Section 2.2). Suppose also that there is a heat source/sink throughout the material, given by

$$f(x, y, z, t) = \text{rate at which heat is added/removed, per unit volume,}$$
$$\text{at point } (x, y, z) \text{ at time } t.$$

We wish, then, to determine the temperature function

$$u(x, y, z, t) = \text{temperature at point } (x, y, z), \text{ at time } t,$$

by computing in two different ways, the rate at which heat enters a typical differential element. Our element, in this case, is the rectangular "box" shown in Figure 9.1.

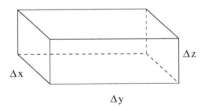

FIGURE 9.1
Three-dimensional differential element.

Proceeding as in Section 2.2, the time rate of change of the heat content of the box, at time t, is

$$\sigma \rho \Delta x \Delta y \Delta z u_t(x, y, z, t).$$

(Actually, we probably should, as in (2.3), use $u_t\left(x + \frac{\Delta x}{2}, y + \frac{\Delta y}{2}, z + \frac{\Delta z}{2}, t\right)$ here. However, it really doesn't matter (why?). Also, as then, we may approach things more rigorously, which we do in Exercise 1.)

Now we need the general statement of Fourier's Law, which states that the heat flux across any differential element of *area* at the point (x, y, z) is

$$\Phi(x, y, z, t) = -k\frac{du}{dn}(x, y, z, t),$$

where $\frac{du}{dn}$, of course, is the directional derivative of u in the direction normal to the area element (realizing that there are two such directions). So,

$$\Phi(x, y, z, t) = -k\nabla u \cdot \hat{n}.^*$$

Here, $\hat{n}$ is the unit normal, and ∇u is the **temperature gradient** (hence the use of this terminology in the one-dimensional case). Figure 9.2 illustrates the flux for various orientations of the differential area and ∇u, letting $k = 1$. (Actually, we graph the vector $\Phi\hat{n}$, making clear the direction.)

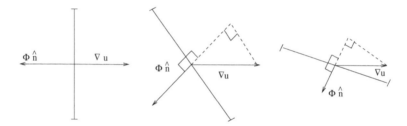

FIGURE 9.2
Flux, including direction, for various orientations of the face of the differential element.

Now we go back to Figure 9.1 and compute the inward flow across the front and the back, that is, across the two faces $x = $ constant. For the front, the inward flow is

$$\Phi(x + \Delta x, y, z, t)\Delta y\Delta z = -k\nabla u \cdot (-\hat{\imath})\Delta y\Delta z = ku_x(x + \Delta x, y, z, t)\Delta y\Delta z;$$

for the back, it is

$$-k\nabla u \cdot \hat{\imath}\Delta y\Delta z = -ku_x(x, y, z, t)\Delta y\Delta z.$$

The total contribution is

$$k\Delta y\Delta z[u_x(x + \Delta x, y, z, t) - u_z(x, y, z, t)],$$

---

*In all cases, ∇ involves only the *space* variables, x, y and z.

which, of course, is essentially the same as (2.5). In fact, it should be fairly obvious that each component behaves the same way.

Finally, the rate at which heat is added to the box is

$$f(x, y, z, t)\Delta x \Delta y \Delta z.$$

Putting everything together, we have

$$\sigma \rho u_t(x, y, z, t) = k \left[\frac{u(x + \Delta x, y, z, t) - u(x, y, z, t)}{\Delta x} \right.$$
$$+ \frac{u(x, y + \Delta y, z, t) - u(x, y, z, t)}{\Delta y}$$
$$\left. + \frac{u(x, y, z + \Delta z, t) - u(x, y, z, t)}{\Delta z} \right]$$
$$+ f(x, y, z, t)$$

and, letting $\Delta x \to 0, \Delta y \to 0$ and $\Delta z \to 0$, we have the heat equation

$$u_t = \alpha^2 \nabla^2 u + f,$$

where the thermal diffusivity, as before, is $\alpha^2 = \frac{k}{\sigma \rho}$ and $\nabla^2 u$ is the Laplacian

$$\nabla^2 u = u_{xx} + u_{yy} + u_{zz}.$$

Of course, the heat equation in two dimensions can be derived in the same manner.

OTHER APPLICATIONS OF THE HEAT/DIFFUSION EQUATION

Diffusion in three dimensions

As with the one-dimensional heat equation, if we replace "temperature" by "concentration," then Fourier's Law is known as Fick's Law and the derivation of the diffusion equation for the substance proceeds almost exactly as above.

Diffusion-convection/mathematical biology

Suppose we are looking at algae on the surface of the ocean. Let

$$u(x, y, t) = \text{concentration of algae at point } (x, y), \text{ at time } t.$$

The algae certainly are carried along by ocean currents and, in addition, they are particle-like and undergo diffusion. Thus, they satisfy the two-dimensional *diffusion-convection equation*, which turns out to be

$$u_t = \alpha^2 \nabla^2 u - v_1 u_x - v_2 u_y,$$

where $v(x, y, t) = v_1(x, y, t)\hat{i} + v_2(x, y, t)\hat{j}$ is the velocity of the current at point (x, y), at time t. See Exercise 3.

THE WAVE EQUATION IN TWO DIMENSIONS

Here we derive the equation for the vibrations of a membrane (or drumhead). We now have what, essentially, is a two-dimensional problem, with the membrane being the two-dimensional analog of the string. As such, we make the same assumption, that is, if we let

$$u(x, y, t) = \text{height of membrane at point } (x, y), \text{ at time } t,$$

we assume that each point (x, y) of the membrane possesses only vertical motion.

As always, we consider a differential rectangle of the membrane, as in Figure 9.3a. The only forces acting on this element are those due to the rest of the membrane pulling on it, that is, due to the *tension* along the four edges. Assuming that the tension per unit length τ is constant, we can "add these up" along each side. The resultants are, by symmetry, at the center of each edge. See Figure 9.3b.

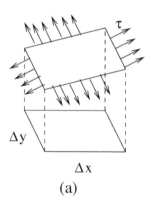

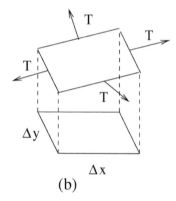

(a)   (b)

FIGURE 9.3
The (a) forces per unit length and (b) resultants of those forces acting along the edges of a differential element.

As in the case of the string, the horizontal components of these forces must sum to zero. As for the vertical components, we treat each direction separately, as in Section 2.3 and Figure 2.5. Thus, the sum of the vertical forces is

$$T\{\Delta y[u_x(x + \Delta x, y, t) - u_x(x, y, t)] + \Delta x[u_y(x, y + \Delta y, t) - u_y(x, y, t)]\}.$$

Using *Newton's 2nd Law*, we must equate this to

$$\text{mass} \cdot \text{acceleration} = \rho \Delta x \Delta y \cdot u_{tt}(x, y, t).$$

Here, ρ is the mass per unit area, and, again, we approximate u_{tt} at the corner, rather than the midpoint, of the rectangle. So the result is

$$\rho u_{tt} = T\left[\frac{u_x(x+\Delta x, y, t) - u_x(x, y, t)}{\Delta x} + \frac{u_y(x, y+\Delta y, t) - u_y(x, y, t)}{\Delta y}\right]$$

and, letting $\Delta x \to 0$ and $\Delta y \to 0$, we arrive at the wave equation

$$u_{tt} = c^2\nabla^2 u,$$

where $c^2 = \sqrt{T/\rho}$ is the wave speed and $\nabla^2 u$ is the Laplacian

$$\nabla^2 u = u_{xx} + u_{yy}.$$

As with the one-dimensional wave equation, we may include the effect of a *load*

$$f(x, y, t) = \text{force per unit area at point } (x, y), \text{ at time } t,$$

resulting in the PDE

$$u_{tt} = c^2\nabla^2 u + \frac{1}{\rho}f(x, y, t).$$

OTHER APPLICATIONS OF THE WAVE EQUATION

One might expect that the propagation of waves in three-dimensional media is described by the three-dimensional wave equation, and this certainly is the case.

Compression waves in liquids and gases

Given a fluid with negligible viscosity, let

$$p(x, y, z, t) = P(x, y, z, t) - P_\ell(x, y, z, t).$$

Here, P is the hydrostatic pressure, P_ℓ is the equilibrium hydrostatic pressure (in the absence of motion) and p is called the incremental pressure. Then it can be shown that, in certain circumstances, p satisfies the wave equation

$$p_{tt} = c^2\nabla^2 p,$$

where c is the wave velocity. In particular, *the propagation of sound waves* is governed by the wave equation. See Exercise 6.

Elastic waves in solids

In the study of the vibrational motion of an elastic solid, we let

$$\boldsymbol{R}(x, y, z, t) = \text{displacement of point initially at } (x, y, z), \text{ at time } t.$$

It can be shown that any vector field can be resolved into the gradient of a scalar and the curl of a zero-divergence vector, that is, that there exist *potentials* ϕ and $\boldsymbol{H}$ such that

$$\boldsymbol{R} = \nabla\phi + \nabla \times \boldsymbol{H}, \text{ with } \nabla \cdot \boldsymbol{H} = 0.$$

If no body force is present, it can be shown that ϕ and $\boldsymbol{H}$ satisfy

$$\phi_{tt} = \frac{\lambda + 2u}{\rho}\nabla^2\phi, \quad \boldsymbol{H}_{tt} = \frac{u}{\rho}\nabla^2\boldsymbol{H},$$

where λ and u are the so-called **Lamé constants** for the material and ρ is the constant mass density. See Exercise 7.

Electromagnetic waves

The general form of Maxwell's equations in a vacuum is given in Appendix D:

$$\nabla \cdot \boldsymbol{E} = 4\pi\rho \qquad\qquad \nabla \cdot \boldsymbol{B} = 0$$
$$\boldsymbol{E}_t = c\nabla \times \boldsymbol{B} - 4\pi\boldsymbol{J} \qquad\qquad \boldsymbol{B}_t = -c\nabla \times \boldsymbol{E},$$

where c is the speed of light. It can be shown that

$$\boldsymbol{E}_{tt} = c^2\nabla^2\boldsymbol{E} - 4\pi\boldsymbol{J}_t - 4\pi c^2\nabla\rho$$

and

$$\boldsymbol{B}_{tt} = c^2\nabla^2\boldsymbol{B} + 4\pi c\nabla \times \boldsymbol{J}.$$

In particular, if $\rho \equiv 0$ and $\boldsymbol{J} \equiv 0$, we have

$$\boldsymbol{E}_{tt} = c^2\nabla^2\boldsymbol{E} \quad \text{and} \quad \boldsymbol{B}_{tt} = c^2\nabla^2\boldsymbol{B}.$$

See Exercise 8. (For a vector function $\boldsymbol{F} = F_1\hat{\imath} + F_2\hat{\jmath} + F_3\hat{k}$ and $\nabla^2\boldsymbol{F} = \nabla^2 F_1\hat{\imath} + \nabla^2 F_2\hat{\jmath} + \nabla^2 F_3\hat{k}$.)

THE LAPLACE/POTENTIAL EQUATION AND POISSON'S EQUATION

Back in Section 2.5 we derived Laplace's and Poisson's equations in electro-statics. If the magnetic field does not change with respect to time, so that $\boldsymbol{B}_t = 0$, two of Maxwell's equations become

$$\nabla \cdot \boldsymbol{E} = 4\pi\rho \quad \text{and} \quad \nabla \times \boldsymbol{E} = \boldsymbol{0}.$$

Again, ρ is the change density. The latter equation implies that $\boldsymbol{E}$ has a potential $-\phi$:

$$\boldsymbol{E} = -\nabla\phi.^\dagger$$

---

† Again, so long as the domain is simply-connected.

Then the former gives us Poisson's equation

$$\nabla^2 \phi = -4\pi\rho.$$

OTHER APPLICATIONS OF LAPLACE'S AND POISSON'S EQUATIONS

Steady state problems

Of course, when looking at steady state solutions of *any* of the homogeneous heat and wave equation examples, we see that they must satisfy Laplace's equation. Similarly, steady state solutions of the *non*homogeneous heat and wave equations, with time-independent source terms, will satisfy Poisson's equation.

Magnetostatics

From Appendix D, the static Maxwell's equations are

$$\nabla \cdot \boldsymbol{E} = 4\pi\rho \qquad \nabla \cdot \boldsymbol{B} = 0$$
$$\nabla \times \boldsymbol{B} = 0 \qquad \nabla \times \boldsymbol{E} = 0.$$

We already used the first and last to show that there is an electric potential ϕ such that $\nabla^2 \phi = -4\pi\rho$. Similarly, it is easy to show that $\boldsymbol{B}$ has a magnetic potential, $\boldsymbol{B} = -\nabla\psi$, and that ψ satisfies Laplace's equation $\nabla^2\psi = 0$.

Newtonian gravity

It can be shown that the Newtonian gravitational field

$$\boldsymbol{F}(x,y,z) = \text{force per unit mass at } (x,y,z),$$

due to a distribution of mass, has a gravitational potential $\psi = \psi(x,y,z)$, so that

$$\boldsymbol{F} = \nabla\psi,$$

and that ψ must satisfy Laplace's equation $\nabla^2\psi = 0$ in empty space, while it satisfies $\nabla^2\psi = -4\pi\rho$ in regions where the density of matter is $\rho = \rho(x,y,z)$.

Velocity of an incompressible and irrotational fluid (i.e., of a perfect fluid)

If a fluid is incompressible, its velocity satisfies $\nabla \cdot \boldsymbol{v} = 0$; similarly, if irrotational, its velocity satisfies $\nabla \times \boldsymbol{v} = 0$. Thus, $\boldsymbol{v}$ has a velocity potential ϕ which satisfies $\nabla^2\phi = 0$. See Exercises 4 and 5.

EQUATIONS OF CONTINUITY

We may generalize to two and three dimensions the continuity equations discussed in Section 2.2.

Fluid flow

If there is no source, the equation of continuity is

$$\rho_t + \nabla \cdot (\rho \boldsymbol{v}) = 0,$$

where, again, ρ is the fluid's density and $\boldsymbol{v}$ its velocity. See Exercise 4.

Electric current

In general, the equation of continuity relating charge density ρ with current density J, again, with no source term, is

$$\rho_t + \nabla \cdot (\rho \boldsymbol{J}) = 0.$$

(The derivation is similar to that of the fluid flow continuity equation.)

Exercises 9.1

1. Here we provide a more rigorous, and more general, derivation of the heat equation in three dimensions. Suppose we have a simply-connected piece of material with mass density $\rho = \rho(x, y, z)$, specific heat $\sigma = \sigma(x, y, z)$ and thermal conductivity $k = k(x, y, z)$. Let V be an arbitrary subset of the material, with boundary S.

   a) Conclude that the rate at which heat enters V is

   $$\iiint\limits_{V} \sigma \rho u_t \, dv$$

   (you may assume that $\frac{d}{dt} \iiint f(x, y, z, t)dv = \iiint f_t(x, y, z, t)dv$).

   b) Show that this must equal

   $$\iint\limits_{S} k\nabla u \cdot \hat{\boldsymbol{n}} \, ds.$$

   c) Use the Divergence Theorem on the result from part (b), and the arbitrariness of V, to conclude that we must have

   $$u_t = \frac{1}{\sigma \rho} \nabla \cdot (k\nabla u).$$

2. Here, we do the same as in Exercise 1, but for the two-dimensional wave equation. Let D be any subset of a two-dimensional vibrating membrane, with boundary curve C. Let $\rho = \rho(x, y)$ be the mass density (per unit area) of the membrane. We apply $F = ma$ to D.

a) Conclude that we have

$$ma = \iint_D \rho u_{tt} \, dA.$$

b) Show that the vertical force at each point along C is $T\frac{du}{dn}$‡ and, thus, that the total vertical force acting on the piece of membrane is

$$\int_C T\nabla u \cdot \hat{\boldsymbol{n}} \, ds.$$

c) Use Green's Theorem in part (b), and the arbitrariness of D, to conclude that we must have

$$T\nabla^2 u = \rho u_{tt}.$$

(Note: T will be constant—again, an approximation—as in the one-dimensional case.)

3. **Two-dimensional diffusion-convection equation:** We'd like to derive the contribution of convection to the diffusion-convection equation, and we proceed very much as in Section 5.1. So, suppose we have a differential rectangle of size $\Delta x \times \Delta y$, and suppose the velocity of the current is the *constant*

$$v = v_1 \hat{\imath} + v_2 \hat{\jmath}.$$

Show that the net inflow into the element is approximately

$$
\begin{aligned}
u_t(x,y,t)\Delta x \Delta y = &-v_1 \Delta t[u(x+\Delta x,y,t) - u(x,y,t)]\Delta y \\
&- v_2 \Delta t[u(x,y+\Delta y,t) - u(x,y,t)]\Delta x \\
&+ \text{terms of order } (\Delta t)^2 \Delta x \Delta y
\end{aligned}
$$

and, thus, upon dividing by $\Delta x \Delta y \Delta t$ and letting each term go to zero, we have

$$u_t = -\vec{v} \cdot \nabla u = -v_1 u_x - v_2 u_y.$$

4. Derive the equation of continuity

$$\rho_t + \nabla \cdot (\rho \boldsymbol{v}) = 0$$

for a fluid with density $\rho = \rho(x,y,z,t)$ and velocity $\boldsymbol{v} = \boldsymbol{v}(x,y,z,t)$. Show that if the fluid is *incompressible*, that is, if its density doesn't change, then we have

$$\nabla \cdot \boldsymbol{v} = 0.$$

(See Exercise 13, Chapter 2, Section 2.2.)

---

‡Since we assume the membrane to be perfectly flexible.

5. a) Suppose we have a fluid with two-dimensional velocity field $\boldsymbol{v} = \boldsymbol{v}(x, y, t) = v_1 \hat{\boldsymbol{i}} + v_2 \boldsymbol{j}$. Use Green's Theorem to show that if

$$\oint_C \boldsymbol{v} \cdot d\boldsymbol{r} = \oint_C v_1 \, dx + v_2 \, dy = 0$$

around any simple closed curve C, then we must have

$$v_{1y} = v_{2x}.$$

b) Now use Stokes's Theorem to show that if the three-dimensional velocity field satisfies

$$\oint_C \boldsymbol{v} \cdot d\boldsymbol{r} = 0$$

around any simple closed curve C, then we must have

$$\nabla x \boldsymbol{v} = \boldsymbol{0}.$$

Of course, part (a) is a special case of part (b); in each case we say that the fluid is *irrotational*. Explain why this term makes sense.

6. **Compression waves in fluids and gases:** Suppose we'd like to consider waves in a nonviscous fluid, for example, sound waves in air. We let

$$P(x, y, z, t) = \text{hydrostatic pressure at point } (x, y, z), \text{ at time } t$$

and

$$P_\ell(x, y, z) = \text{equilibrium pressure in the absence of motion.}$$

We look at the **incremental pressure**

$$p(x, y, z, t) = P(x, y, z, t) - P_\ell(x, y, z).$$

Now, if

$$\boldsymbol{R}(x, y, z, t) = \text{displacement of fluid initially at } (x, y, z), \text{ at time } t,$$

Hooke's Law *for a fluid* says that

$$p = -B\nabla \cdot \boldsymbol{R}$$

(as long as $\nabla \cdot \boldsymbol{R}$ is small), where the proportionality constant B is called the **bulk modulus**

a) Consider the incremental pressure forces acting on the differential element of size $\Delta x \times \Delta y \times \Delta z$ in Figure 9.1. Realizing that pressure is essentially a *negative tension,* show that the x-direction component of force is

$$-p_x(x, y, z, t)\Delta x \Delta y \Delta z$$

and, therefore, that the total force due to p on the element is

$$-\nabla p \Delta x \Delta y \Delta z.$$

We call ∇p the **pressure gradient**, for obvious reasons.

b) Use Newton's 2nd Law to show that we must have

$$-\nabla p = \rho \boldsymbol{R}_{tt},$$

when $\rho = \rho(x, y, z, t)$ is the density of the fluid.

c) Finally, eliminate $\boldsymbol{R}$ and show that the result is that p satisfies the wave equation

$$p_{tt} = c^2 \nabla^2 p, \S$$

where $c^2 = \frac{B}{\rho}$.

7. **Elastic waves in solids:** Supposing we have a homogeneous and isotropic solid, the normal stresses σ_x, σ_y and σ_z and the shearing stresses τ_{yz}, τ_{zx} and τ_{xy} satisfy

$$\sigma_x = \lambda \nabla \cdot \boldsymbol{R} + 2\mu u_x, \qquad \tau_{yz} = \mu(w_y + v_z),$$
$$\sigma_y = \lambda \nabla \cdot \boldsymbol{R} + 2\mu v_y, \qquad \tau_{zx} = \mu(u_z + w_x),$$
$$\sigma_z = \lambda \nabla \cdot \boldsymbol{R} + 2\mu w_z, \qquad \tau_{xy} = \mu(v_x + u_y).$$

Here, $\boldsymbol{R}(x, y, z, t) = u(x, y, z, t)\hat{\boldsymbol{i}} + v(x, y, z, t)\hat{\boldsymbol{j}} + w(x, y, z, t)\hat{\boldsymbol{k}}$ is the displacement of the point initially at (x, y, z), and λ and μ are the so-called *Lamé constants* for the material. If $\boldsymbol{f}(x, y, z, t) = f_1\hat{\boldsymbol{i}} + f_2\hat{\boldsymbol{j}} + f_3\hat{\boldsymbol{k}}$ is the *load* or *body force,* then it can be shown that the total force at each point is given by

$$x\text{-direction:} \ (\sigma_x)_x + (\tau_{xy})_y + (\tau_{zx})_z + \rho f_1,$$
$$y\text{-direction:} \ (\sigma_y)_y + (\tau_{xy})_x + (\tau_{yz})_z + \rho f_2,$$
$$z\text{-direction:} \ (\sigma_z)_z + (\tau_{zx})_x + (\tau_{yz})_y + \rho f_3.$$

a) Show that the equation of motion for the solid is

$$(\lambda + \mu)\nabla(\nabla \cdot \boldsymbol{R}) + u\nabla^2 \boldsymbol{R} + \rho \boldsymbol{f} = \rho \boldsymbol{R}_{tt}.$$

---

§You may interchange the order of differentiation wherever needed.

b) Show that $\nabla^2 \boldsymbol{R} = \nabla(\nabla \cdot \boldsymbol{R}) - \nabla \times (\nabla \times \boldsymbol{R})$ and, thus, that the equation of motion can be rewritten as

$$(\lambda + 2u)\nabla(\nabla \cdot \boldsymbol{R}) - u\nabla \times (\nabla \times \boldsymbol{R}) + \rho f = \rho \boldsymbol{R}_{tt}.$$

c) It turns out¶ that we may find *potentials* ϕ and $\boldsymbol{H}$ and ψ and $\boldsymbol{K}$ such that

$$\boldsymbol{R} = \nabla\phi + \nabla \times \boldsymbol{H} \quad \text{and} \quad f = \nabla\psi + \nabla x \boldsymbol{K}.$$

Show that the equation of motion can be rewritten as

$$\nabla[(\lambda + 2u)\nabla^2\phi + \rho\psi - \rho\phi_{tt}] + \nabla x[u\nabla^2\boldsymbol{H} + \rho\boldsymbol{K} - \rho\boldsymbol{H}_{tt}] = 0$$

and, thus, will be satisfied if $\phi, \boldsymbol{H}, \psi$ and $\boldsymbol{K}$ satisfy

$$\phi_{tt} = \frac{\lambda + 2u}{\rho}\nabla^2\phi + \psi, \qquad \boldsymbol{H}_{tt} = \frac{u}{\rho}\nabla^2\boldsymbol{H} + \boldsymbol{K}.$$

8. **Electromagnetic waves:** Show that Maxwell's equations in a vacuum imply that

$$\boldsymbol{E}_{tt} = c^2\nabla^2\boldsymbol{E} - 4\pi\boldsymbol{J}_t - 4\pi c^2\nabla\rho$$

and

$$\boldsymbol{B}_{tt} = c^2\nabla^2\boldsymbol{B} + 4\pi c\nabla \times \boldsymbol{J}.$$

(You'll need the identity established in Exercise 7b.)

9.2   The Heat and Wave Equations on a Rectangle; Multiple Fourier Series

Now we solve the heat, wave and Laplace equations on two- and three-dimensional domains with rectangular and circular/spherical boundaries. This may seem much too restrictive. Although it is natural to solve one-dimensional problems on intervals, the situation in higher dimensions is much more complicated, with there being infinitely many possible shapes. It turns out, however, that there are so-called **conformal mappings**‖ (from *complex analysis*) which may be used to transform more complicated domains into these simpler regions, or combinations thereof.

We begin by considering the heat and wave equations on a rectangle in this section.

¶See, e.g., *Methods of Theoretical Physics* by Morse and Feshbach.
‖In fact, the **Riemann Mapping Theorem** guarantees that any *simply connected* region in the plane, of finite extent, can be mapped *conformally* onto a circle, where the problem can be solved and then mapped back to the original region.

THE TWO-DIMENSIONAL HEAT EQUATION

Let's start by solving the two-dimensional heat equation on a finite rectangle, with the temperature held at zero along the edges. It will proceed almost exactly as did the one-dimensional case—separation of variables, boundary-value problems, superposition of solutions—and, at the end, we'll find the need for an extended version of the Fourier series.

So, holding off on the initial condition, we begin with

$$u_t = \alpha^2 (u_{xx} + u_{yy}) = \alpha^2 \nabla^2 u, \qquad 0 < x < a, 0 < y < b, t > 0,$$
$$u(0, y, t) = u(a, y, t) = u(x, 0, t) = u(x, b, t) = 0, \quad 0 < x < a, 0 < y < b, t > 0.$$

First, separate variables by letting $u(x, y, t) = X(x)Y(y)T(t)$. As in Exercise 18, Section 1.6, the PDE leads us to

$$\frac{T'}{\alpha^2 T} = \frac{X''}{X} + \frac{Y''}{Y} = -\lambda$$
$$\Rightarrow \quad T' + \alpha^2 \lambda T = 0, \qquad \frac{X''}{X} = -\frac{Y''}{Y} - \lambda = -\gamma$$
$$\Rightarrow \quad T' + \alpha^2 \lambda T = 0, \quad X'' + \gamma X = 0, \quad Y'' + (\gamma - \lambda)Y = 0.$$

Similarly, we separate the boundary conditions:

$$X(0) = X(a) = Y(0) = Y(b) = 0.$$

So we're led to the two eigenvalue problems

$$X'' + \gamma X = 0 \qquad\qquad Y'' + (\lambda - \gamma)Y = 0$$
$$X(0) = X(a) = 0 \qquad Y(0) = Y(b) = 0,$$

which we've solved many times. The X-boundary-value problem has eigenvalues and eigenfunctions

$$\gamma_n = \frac{n^2 \pi^2}{a^2}, \quad X_n(x) = \sin \frac{n\pi x}{a}, \qquad n = 1, 2, 3, \dots .$$

Then, for each such γ, the Y-ODE is

$$Y'' + \left(\lambda - \frac{n^2 \pi^2}{a^2} \right) Y = 0, \qquad n = 1, 2, 3, \dots,$$

and, for each n, we must have

$$\lambda - \frac{n^2 \pi^2}{a^2} = \frac{m^2 \pi^2}{b^2}, \qquad m = 1, 2, 3, \dots$$

that is, for each pair n, m, we have the eigenvalue

$$\lambda_{n,m} = \frac{n^2 \pi^2}{a^2} + \frac{m^2 \pi^2}{b^2}, \qquad n = 1, 2, 3, \dots, m = 1, 2, 3, \dots$$

with corresponding eigenfunction

$$Y(y) = Y_m(y) = \sin \frac{m\pi y}{b}.$$

Now, for each pair n, m, the solution to the T-equation is

$$T_{n,m}(t) = e^{-\alpha^2 \lambda_{n,m} t} = e^{-\alpha^2 \pi^2 \left(\frac{n^2}{a^2} + \frac{m^2}{b^2}\right)t},$$

giving us the product solutions

$$\begin{aligned} u_{n,m}(x, y, t) &= T_{n,m}(t)X_n(x)Y_m(y) \\ &= e^{-\alpha^2 \pi^2 \left(\frac{n^2}{a^2} + \frac{m^2}{b^2}\right)t} \sin \frac{n\pi x}{a} \sin \frac{m\pi y}{b}, \\ & n = 1, 2, 3, \ldots; m = 1, 2, 3, \ldots . \end{aligned}$$

Finally, any linear combination of these solutions is a solution of the PDE and the boundary conditions, so we have the *general solution*

$$\begin{aligned} u(x, y, t) &= \sum_{n=1}^{\infty} \sum_{m=1}^{\infty} c_{n,m} u_{n,m}(x, y, t) \\ &= \sum_{n=1}^{\infty} \sum_{m=1}^{\infty} c_{n,m} e^{-\alpha^2 \pi^2 \left(\frac{n^2}{a^2} + \frac{m^2}{b^2}\right)t} \sin \frac{n\pi x}{a} \sin \frac{m\pi y}{b}, \end{aligned}$$

where the numbers $c_{n,m}$ are, of course, arbitrary constants.

Now, how about the initial condition? As in Section 2.6, for certain special initial conditions, we can solve the problem immediately.

Example 1 Solve the initial-boundary-value problem

$$u_t = \nabla^2 u, \qquad 0 < x < 1, 0 < y < 2, t > 0,$$
$$u(0, y, t) = u(1, y, t) = u(x, 0, t) = u(x, 2, t) = 0,$$
$$u(x, y, 0) = 3 \sin 6\pi x \sin 2\pi y + 7 \sin \pi x \sin \frac{3\pi y}{2}.$$

The general solution is

$$u(x, y, t) = \sum_{n=1}^{\infty} \sum_{m=1}^{\infty} c_{n,m} e^{-\left(n^2 + \frac{m^2}{2}\right)\pi^2 t} \sin n\pi x \sin \frac{m\pi y}{2}.$$

Then,

$$\begin{aligned} u(x, y, 0) &= \sum_{n=1}^{\infty} \sum_{m=1}^{\infty} c_{n,m} \sin n\pi x \sin \frac{m\pi y}{2} \\ &= 3 \sin 6\pi x \sin 2\pi y - 7 \sin \pi x \sin \frac{3\pi y}{2}. \end{aligned}$$

Thus, we have that each $c_{n,m} = 0$ except for two cases:

$$n = 6, m = 4 \Rightarrow c_{6,4} = 3; \quad n = 1, m = 3 \Rightarrow c_{1,3} = -7.$$

So our final solution is

$$u(x, y, t) = 3e^{-44\pi^2 t} \sin 6\pi x \sin 2\pi y - 7e^{-\frac{11}{2}\pi^2 t} \sin \pi x \sin \frac{3\pi y}{2}.$$

Of course, the big question is, what happens when the initial condition is *not* so amenable, that is, what happens for general initial condition

$$u(x, y, 0) = f(x, y)?$$

Here we need to extend the concept of Fourier series to functions of several variables. We will *not* give a detailed treatment here. Suffice it to say that multiple Fourier series behave quite like the one-dimensional kind with similar convergence properties. But how do we calculate the coefficients?

We *could* proceed as we did before, this time by considering functions $\sin \frac{n\pi x}{a} \sin \frac{m\pi y}{b}$, $n = 1, 2, 3, \ldots; m = 1, 2, 3, \ldots$, on the rectangle $0 \leq x \leq a$, $0 \leq y \leq b$. Specifically, we can show that these functions are orthogonal on the rectangle, etc. (see Exercise 8).

However, our nonrigorous derivation follows directly from the one-variable case. So, given $f(x, y)$ on the rectangle $0 \leq x \leq a$, $0 \leq y \leq b$, for each (fixed) y we can expand in a Fourier sine series in x:

$$f(x, y) = \sum_{n=1}^{\infty} d_n(y) \sin \frac{n\pi x}{a},$$

where, for each y,

$$d_n(y) = \frac{2}{a} \int_0^a f(x, y) \sin \frac{n\pi x}{a} dx, \qquad n = 1, 2, \ldots .$$

Then, for each n, $d_n(y)$ also can be expanded as a Fourier sine series:

$$d_n(y) = \sum_{m=1}^{\infty} c_{n,m} \sin \frac{n\pi y}{b},$$

with

$$c_{n,m} = \frac{2}{b} \int_0^b d_n(y) \sin \frac{m\pi x}{b}$$
$$= \frac{2}{b} \int_0^b \left[\frac{2}{a} \int_0^a f(x, y) \sin \frac{n\pi x}{a} dx \right] \sin \frac{m\pi y}{b} dy,$$
$$n = 1, 2, 3, \ldots; m = 1, 2, 3, \ldots .$$

Thus, putting everything together, we have the **double Fourier sine series** for f on the rectangle $0 \leq x \leq a$, $0 \leq y \leq b$,

$$f(x,y) \sim \sum_{n=1}^{\infty} \sum_{m=1}^{\infty} c_{n,m} \sin \frac{n\pi x}{a} \sin \frac{m\pi y}{b},$$

where

$$c_{n,m} = \frac{4}{ab} \int_0^a \int_0^b f(x,y) \sin \frac{n\pi x}{a} \sin \frac{m\pi y}{b} \, dy dx.$$

It follows that the solution of the initial-boundary-value problem

$$u_t = \alpha^2 (u_{xx} + u_{yy}),$$
$$u(x,y,0) = f(x,y),$$
$$u(0,y,t) = u(a,y,t) = u(x,0,t) = u(x,b,t) = 0$$

is

$$u = \sum_{n=1}^{\infty} \sum_{m=1}^{\infty} c_{n,m} e^{-\alpha^2 \pi^2 \left(\frac{n^2}{a^2} + \frac{m^2}{b^2} \right) t} \sin \frac{n\pi x}{a} \sin \frac{n\pi y}{b},$$

with the above values for the constants $c_{n,m}$.

THE TWO-DIMENSIONAL WAVE EQUATION

The two-dimensional wave equation, modeling the vibrations of a rectangular membrane, is solved analogously. So, suppose we're given

$$u_{tt} = c^2 (u_{xx} + u_{yy}) = c^2 \nabla^2 u,$$
$$u(x,y,0) = f(x,y),$$
$$u_t(x,y,0) = g(x,y),$$
$$u(0,y,t) = u(a,y,t) = u(x,0,t) = u(x,b,t) = 0.$$

Separating the PDE and boundary conditions leads to the ODEs

$$T'' + c^2 \lambda T = 0, \quad X'' + \gamma X = 0, \quad Y'' + (\lambda - \gamma) Y = 0$$

and boundary conditions

$$X(0) = X(a) = Y(0) = Y(b) = 0.$$

Proceeding as above (see Exercise 2), we arrive at the general solution

$$u(x,y,t) = \sum_{n=1}^{\infty} \sum_{m=1}^{\infty} \sin \frac{n\pi x}{a} \sin \frac{m\pi y}{b} [c_{n,m} \cos c\sqrt{\lambda_{n,m}} \, t + d_{n,m} \sin c\sqrt{\lambda_{n,m}} \, t],$$

where, as above,

$$\lambda_{n,m} = \pi^2 \left(\frac{n^2}{a^2} + \frac{m^2}{b^2} \right)$$

and

$$c_{n,m} = \frac{4}{ab} \int_0^a \int_0^b f(x,y) \sin \frac{n\pi x}{a} \sin \frac{m\pi y}{b} \, dy dx,$$

$$d_{n,m} = \frac{4}{ab\sqrt{\lambda_{n,m}}} \int_0^a \int_0^b g(x,y) \sin \frac{n\pi x}{a} \sin \frac{m\pi y}{b} \, dy dx,$$

$$n = 1, 2, 3, \ldots; m = 1, 2, 3, \ldots .$$

It's interesting to look at the *X-Y* eigenfunctions

$$v_{n,m}(x,y) = \sin \frac{n\pi x}{a} \sin \frac{m\pi y}{b}$$

for each of these problems and, in particular, to look at the role they play in the case of the vibrating membrane. Remember that the solution of the *one*-dimensional wave equation is

$$u(x,t) = \sum_{n=1}^{\infty} \sin \frac{n\pi x}{L} \left(c_n \cos \frac{n\pi ct}{L} + d_n \sin \frac{n\pi ct}{L} \right).$$

There, the *X*-eigenfunction

$$X_n(x) = \sin \frac{n\pi x}{L}, \qquad n = 1, 2, 3, \ldots,$$

is the nth *mode of vibration*, corresponding to the nth *vibration frequency*

$$\nu_n = \frac{nc}{2L}.$$

For the present case, $v_{n,m}(x,y)$ is called the $(n,m)^{\text{th}}$ **mode of vibration** of the membrane, corresponding to the $(n,m)^{\text{th}}$ **frequency**

$$\nu_{n,m} = \frac{c\sqrt{\lambda_{n,m}}}{2\pi} = \frac{c\sqrt{n^2 b^2 + m^2 a^2}}{2ab}$$

(why?). In Table 9.1 we list the frequencies and modes for $n = 1, 2, 3$ and $m = 1, 2, 3$ for a 2×3 membrane, with $c = 1$. Figure 9.4 shows these nine vibration modes.

$v_{1,1} = \sin \frac{\pi x}{2} \sin \frac{\pi y}{3}$	$v_{1,2} = \sin \frac{\pi x}{2} \sin \frac{2\pi y}{3}$	$v_{1,3} = \sin \frac{\pi x}{2} \sin \pi y$
$\nu_{1,1} = \frac{\sqrt{13}}{12}$	$\nu_{1,2} = \frac{5}{12}$	$\nu_{1,3} = \frac{\sqrt{45}}{12}$
$v_{2,1} = \sin \pi x \sin \frac{\pi y}{3}$	$v_{2,2} = \sin \pi x \sin \frac{2\pi y}{3}$	$v_{2,3} = \sin \pi x \sin \pi y$
$\nu_{2,1} = \frac{\sqrt{20}}{12}$	$\nu_{2,2} = \frac{\sqrt{52}}{12}$	$\nu_{2,3} = \frac{\sqrt{72}}{12}$
$v_{3,1} = \sin \frac{3\pi x}{2} \sin \frac{\pi y}{3}$	$v_{3,2} = \sin \frac{3\pi x}{2} \sin \frac{2\pi y}{3}$	$v_{3,3} = \sin \frac{3\pi x}{2} \sin \pi y$
$\nu_{3,1} = \frac{\sqrt{82}}{12}$	$\nu_{3,2} = \frac{\sqrt{97}}{12}$	$\nu_{3,3} = \frac{\sqrt{117}}{12}$

TABLE 9.1
The vibration modes, and corresponding vibration frequencies, for a 2×3 membrane, for $n = 1, 2, 3$ and $m = 1, 2, 3$.

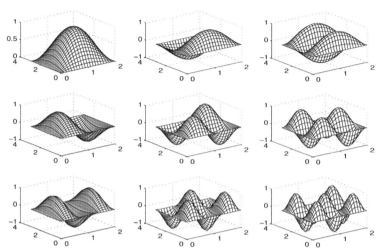

FIGURE 9.4
MATLAB graphs of the vibration modes given in Table 9.1.

Now, in the one-dimensional case, you'll remember that, for each *mode*, there are points along the string which remain fixed—the *nodes* for that particular *mode*. As can be seen in Figure 9.4, the modes in the two-dimensional case possess *curves* which have this property; these are the **nodal lines** (or **nodal curves**) corresponding to each mode. Figure 9.5 shows the nodal lines for each of the modes in Figure 9.4.

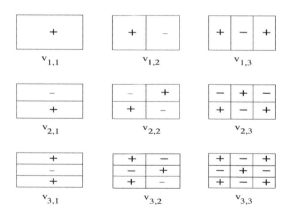

FIGURE 9.5
The nodal lines for the modes in Figure 9.4. The +/− signs give the sign of the mode throughout each cell.

The "+" and "−" signs represent those regions where $v_{n,m} > 0$ and $v_{n,m} < 0$, respectively.

An interesting special case is that of the square membrane. Specifically, we take $a = b = 1$, along with $c = 1$, and we plot the $(1, 2)^{\text{th}}$ and $(2, 1)^{\text{th}}$ modes in Figure 9.6, along with the corresponding nodal lines in Figure 9.7.

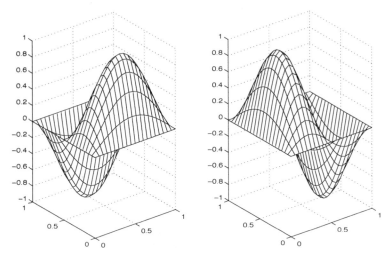

FIGURE 9.6
MATLAB graphs of the modes $v_{1,2} = \sin \pi x \sin 2\pi y$ and $v_{2,1} = \sin 2\pi x \sin \pi y$ for a 1×1 membrane.

FIGURE 9.7
Nodal lines and signs for the modes in Figure 9.6.

Here, it's immediately obvious that $\nu_{1,2} = \nu_{2,1}$ and that the corresponding modes are symmetric. What is the significance of this? In the one-dimensional case, to each frequency there corresponds a unique mode. In other words, each eigenvalue has multiplicity one. Here, however, we have $\lambda_{1,2} = \lambda_{2,1}$, so that this number is an eigenvalue of multiplicity two (at least!). What this means is that any linear combination of the eigenfunctions $v_{1,2}$ and $v_{2,1}$ also is an eigenfunction corresponding to this particular frequency. So, instead of $v_{1,2}$ and $v_{2,1}$, we could use, for example, the linearly independent eigenfunctions

$$w_1 = v_{1,2} + v_{2,1}, \qquad w_2 = v_{1,2} - v_{2,1}$$

(see Exercise 17). These functions** are shown in Figure 9.8; their nodal lines are in Figure 9.9.

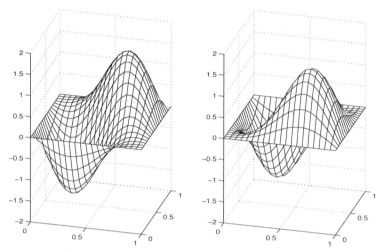

FIGURE 9.8
MATLAB graphs of the linear combinations $w_1 = -v_{1,2} - v_{2,1}$ and $w_2 = v_{1,2} - v_{2,1}$.

FIGURE 9.9
Nodal lines and signs for the modes in Figure 9.8.

In general, of course, determining the multiplicity of the $(n,m)^{\text{th}}$ frequency requires finding all pairs (n_1, m_1) satisfying $n^2 b^2 + m^2 a^2 = n_1^2 b^2 + m_1^2 a^2$. It can be shown that *no* frequency has multiplicity greater than one if the ratio b/a is irrational. Conversely, if b/a is rational, then one may use elementary number theory to determine multiplicities.[††]

One may use double Fourier series to solve Laplace's equation on a rectangular solid, as well. See Exercise 10.

[**]Actually, we graph $w_1 = -v_{1,2} - v_{2,1}$ to get a better picture.
[††]For a nice treatment, see *Partial Differential Equations and Boundary-Value Problems* by Mark A. Pinsky.

EIGENVALUES AND EIGENFUNCTIONS OF THE LAPLACE OPERATOR

Note that, in both the heat and wave equations, if we began by separating only time from the space variables, that is, if we had let

$$u(x, y, t) = T(t)\Phi(x, y)$$

at the start, then we would be led to the *Helmholtz* PDE, with Dirichlet boundary conditions,

$$\nabla^2\Phi + \lambda\Phi = 0, \qquad 0 < x < a, 0 < y < b,$$
$$\Phi(0, y) = \Phi(a, y) = \Phi(x, 0) = \Phi(x, b) = 0.$$

Then, in solving this system, we actually found that *the eigenvalues and eigenfunctions of the Laplace operator, subject to the Dirichlet condition, on the rectangle $0 < x < a, 0 < y < b$, are*

$$\lambda_{n,m} = \frac{n^2\pi^2}{a^2} + \frac{m^2\pi^2}{b^2}, \; \Phi_{n,m}(x, y) = \sin\frac{n\pi x}{a}\sin\frac{m\pi y}{b}.$$

We'll say much more in Section 9.6.

Exercises 9.2

1. **Helmholtz equation:** When solving the two- (or three-) dimensional heat and wave equations, we *may* choose to begin by separating time from the space variables. Given the three-dimensional wave equation

   $$u_{tt} = c^2\nabla^2 u,$$

   show that letting $u(x, y, z, t) = \Phi(x, y, z)T(t)$ leads to the **Helmholtz equation**

   $$\nabla^2\Phi + \tilde{\lambda}\Phi = 0,$$

   where $\tilde{\lambda} = \lambda/c$ and λ is the separation constant.

2. Work through the derivation of the *general solution* of the two-dimensional wave equation.

3. Find the general solution of the two-dimensional heat equation $u_t = \nabla^2 u$ subject to the boundary conditions

   a) $u(0, y, t) = u(a, y, t) = u_y(x, 0, t) = u_y(x, b, t) = 0$

   b) $u_x(0, y, t) = u_x(a, y, t) = u(x, 0, t) = u(x, b, t) = 0$

   c) $u_x(0, y, t) = u_x(a, y, t) = u_y(x, 0, t) = u_y(x, b, t) = 0$

   d) $u(0, y, t) = u_x(a, y, t) = u(x, 0, t) = u(x, b, t) = 0$

4. Find the general solution of the two-dimensional wave equation $u_{tt} = \nabla^2 u$ subject to the boundary conditions

    a) $u(0, y, t) = u(a, y, t) = u_y(x, 0, t) = u(x, b, t) = 0$

    b) $u_x(0, y, t) = u_x(a, y, t) = u(x, 0, t) = u(x, b, t) = 0$

5.   a) Solve the two-dimensional heat IBVP

$$u_t = \nabla^2 u,$$
$$u(x, y, 0) = 2 + 5\cos \pi x \cos \pi y,$$
$$u_x(0, y, t) = u_x(1, y, t) = u_y(x, 0, t) = u_y(x, 2, t) = 0.$$

    What's the *steady state* solution?

    b) **MATLAB:** Plot the solution of part (a) for various values of t.

    c) Solve the two-dimensional wave IBVP

$$u_{tt} = \nabla^2 u,$$
$$u(x, 0) = 3\sin 4y - 5\cos 2x \sin y,$$
$$u_t(x, 0) = 7\cos x \sin 3y,$$
$$u_x(0, y, t) = u_x(\pi, y, t) = u(x, 0, t) = u(x, \pi, t) = 0.$$

    d) **MATLAB:** Plot the solution of part (b) for various values of t.

6. Compute the double Fourier sine series for $f(x, y)$ on the rectangle $0 \le x \le \pi$, $0 \le y \le \pi$

    a) $f(x, y) = 1$

    b) $f(x, y) = \begin{cases} 0, \text{ if } x < y \\ 1, \text{ if } x \ge y \end{cases}$

    c) $f(x, y) = xy$

    d) **MATLAB:** Plot the truncated double Fourier series $\left(\sum\limits_{n=1}^{N} \sum\limits_{m=1}^{M} \right)$ for each function above, on the rectangle $-\pi \le x \le 3\pi$, $-\pi \le y \le 3\pi$, for various values of N and M.

7. Show that any well-enough behaved function $f(x, y)$ on $0 \le x \le \pi$, $0 \le y \le \pi$ can be expanded into the following double Fourier series:

    a) $f(x, y) \sim \sum\limits_{m=1}^{\infty} \frac{c_{0,m}}{2} \sin my + \sum\limits_{n=1}^{\infty} \sum\limits_{m=1}^{\infty} c_{n,m} \cos nx \sin my$, where

$$c_{n,m} = \frac{4}{\pi^2} \int_0^\pi \int_0^\pi f(x, y) \cos nx \sin my \, dy dx,$$
$$n = 0, 1, 2, 3, \ldots; m = 1, 2, 3, \ldots .$$

b) $f(x,y) \sim \sum\limits_{n=1}^{\infty} \frac{c_{n,0}}{2} \sin nx + \sum\limits_{n=1}^{\infty} \sum\limits_{m=1}^{\infty} c_{n,m} \sin nx \cos my$, where

$$c_{n,m} = \frac{4}{\pi^2} \int_0^{\pi} \int_0^{\pi} f(x,y) \sin nx \cos my \, dydx,$$

$$n = 1,2,3,\ldots; m = 0,1,2,3,\ldots.$$

c) $f(x,y) \sim \frac{c_{00}}{4} + \sum\limits_{m=1}^{\infty} \frac{c_{0,m}}{2} \cos my + \sum\limits_{n=1}^{\infty} \frac{c_{n,0}}{2} \cos nx$

$+ \sum\limits_{n=1}^{\infty} \sum\limits_{m=1}^{\infty} c_{n,m} \cos nx \cos my$, where

$$c_{n,m} = \frac{4}{\pi^2} \int_0^{\pi} \int_0^{\pi} f(x,y) \cos nx \cos my \, dydx,$$

$$n = 0,1,2,\ldots; m = 0,1,2,\ldots.$$

8. Show that the functions $v_{n,m}(x,y) = \sin\frac{n\pi x}{a} \sin\frac{m\pi y}{b}$, $n = 1,2,3,\ldots$; $m = 1,2,3,\ldots$ are orthogonal on the rectangle $0 \le x \le a$, $0 \le y \le b$. Use this orthogonality to derive the formula for $c_{n,m}$, the $(n,m)^{\text{th}}$ coefficient in the double Fourier sine series, for a function $f(x,y)$.

9. Use the results of Exercise 6 to solve the following IBVPs.

a) $u_t = \nabla^2 u$,
$u(x,y,0) = 1$,
$u(0,y,t) = u(\pi,y,t) = u(x,0,t) = u(x,\pi,t) = 0$.

b) $u_{tt} = \nabla^2 u$,
$u(x,y,0) = xy$,
$u_t(x,y,0) = 1$,
$u(0,y,t) = u(\pi,y,t) = u(x,0,t) = u(x,\pi,t) = 0$.

c) **MATLAB:** Plot each (truncated) solution for various values of t.

10. **Laplace's equation on a rectangular solid:** Here we solve the three-dimensional Laplace equation on a rectangular solid.

a) First, we solve the *Dirichlet problem* where $u \equiv 0$ on the boundary, except on the two faces $z = $ constant. Specifically, show that the solution of the problem

$$\nabla^2 u = u_{xx} + u_{yy} + u_{zz} = 0, \qquad 0 < x < a, 0 < y < b, 0 < z < c,$$
$$u(x,y,0) = f(x,y), u(x,y,c) = g(x,y),$$
$$u(x,0,z) = u(x,b,z) = u(0,y,z) = u(a,y,z) = 0$$

is

$$u(x,y,z) = \sum\limits_{n=1}^{\infty} \sum\limits_{m=1}^{\infty} \sin\frac{n\pi x}{a} \sin\frac{m\pi y}{b} [c_{n,m} \cosh \pi\alpha_{n,m} z + d_{n,m} \sinh \pi\alpha_{n,m} z],$$

where

$$c_{n,m} = \frac{4}{ab} \int_0^a \int_0^b f(x,y) \sin \frac{n\pi x}{a} \sin \frac{m\pi y}{b} \, dy dx$$

and

$$d_{n,m} = -c_{n,m} \coth \pi \alpha_{n,m} c$$

$$+ \frac{4}{ab \sinh \pi \alpha_{n,m} c} \int_0^a \int_0^b g(x,y) \sin \frac{n\pi x}{a} \sin \frac{m\pi y}{b} \, dy dx,$$

$$n = 1,2,3,\ldots; m = 1,2,3,\ldots.$$

Here, $\alpha_{n,m} = \sqrt{\frac{n^2}{a^2} + \frac{m^2}{b^2}}$, and coth is the hyperbolic cotangent, $\coth x = \frac{\cosh x}{\sinh x}$.

b) *Without doing too much work*, write down the solution to

$$\nabla^2 u = u_{xx} + u_{yy} + u_{zz} = 0,$$
$$u(x,0,z) = f(x,z), u(x,b,z) = g(x,z),$$
$$u(0,y,z) = u(a,y,z) = u(x,y,0) = u(x,y,c) = 0.$$

c) Solve the *Dirichlet problem*

$$\nabla^2 u = u_{xx} + u_{yy} + u_{zz} = 0,$$
$$u(x,y,0) = 3\sin \pi x \sin y, u(x,y,2\pi) = 0,$$
$$u(x,0,z) = 2\sin 4\pi x \sin z, u(x,\pi,z) = 0,$$
$$u(0,y,z) = 0, \quad u(1,y,z) = \sin 3y \sin 3z.$$

11. a) Calculate the first (lowest) nine frequencies, and draw the corresponding nodal patterns, for the vibrating membrane modeled by the PDE $u_{tt} = u_{xx} + u_{yy}$, subject to the boundary conditions

$$u(0,y,t) = u(\pi,y,t) = u_y(x,0,t) = u_y(x,\pi,t) = 0.$$

b) **MATLAB:** Plot the corresponding modes from part (a).

c) In general, what seems to be the relationship between the pair (n,m) and the corresponding number of nodal lines? (To be consistent, if $u \equiv 0$ along an edge, we call the edge a nodal line.)

12. Solve the diffusion-convection IBVP

$$u_t = u_{xx} + u_{yy} + 2\alpha u_x + 2\beta u_y - ku,$$
$$u(x, y, 0) = f(x, y),$$
$$u(0, y, t) = u(a, y, t) = u(x, 0, t) = u(x, b, t) = 0.$$

Here, α, β and k are nonnegative constants.

13. a) Show formally that any well-enough behaved function $f(x, y, z)$, on the rectangular solid $0 \le x \le \pi, 0 \le y \le \pi, 0 \le z \le \pi$, can be expanded into the triple Fourier series

$$\sum_{n=1}^{\infty} \sum_{m=1}^{\infty} \sum_{p=1}^{\infty} c_{n,m,p} \sin nx \sin my \sin pz,$$

where

$$c_{n,m,p} = \frac{8}{\pi^3} \int_0^\pi \int_0^\pi \int_0^\pi f(x, y, z) \sin nx \sin my \sin pz \; dz\,dy\,dx,$$
$$n = 1, 2, 3, \ldots; m = 1, 2, 3, \ldots; p = 1, 2, 3, \ldots .$$

b) Solve the three-dimensional heat IBVP

$$u_t = u_{xx} + u_{yy} + u_{zz}, \qquad 0 < x < \pi, 0 < y < \pi, 0 < z < \pi,$$
$$u(x, y, z, 0) = f(x, y, z),$$
$$u(x, y, 0) = u(x, y, \pi) = u(x, 0, z) = u(x, \pi, z)$$
$$= u(0, y, z) = u(\pi, y, z) = 0.$$

14. a) Given $f(x, y)$ on $0 \le x \le a, 0 \le y \le b$, perform a change to new variables ξ, η so that

$$F(\xi, \eta) = f(x(\xi), y(\eta))$$

on the square $0 \le \xi \le \pi, 0 \le \eta \le \pi$.

b) Compute the double Fourier sine series for $F(\xi, \eta)$, then change back to the variables x, y and show that we have the same coefficients $c_{n,m}$ that were derived in this section.

c) Use the same idea on each of the double Fourier series in Exercise 7 to calculate each kind of double Fourier series for a function on the more general domain $0 \le x \le a, 0 \le y \le b$. Thus, without loss of generality, we may assume in any of these double Fourier series problems that we have a function with domain $0 \le x \le \pi$, $0 \le y \le \pi$ (similar to the one-variable case).

15. Given $f(x, y)$ on $0 \le x \le \pi, 0 \le y \le \pi$,

a) Consider the function

$$g_1(x,y) = \begin{cases} f(x,y), & \text{if } 0 \le x \le \pi, 0 \le y \le \pi, \\ f(-x,-y), & \text{if } -\pi \le x < 0, -\pi \le y < 0, \\ -f(-x,y), & \text{if } -\pi \le x < 0, 0 \le y \le \pi, \\ -f(x,-y), & \text{if } 0 \le x \le \pi, -\pi \le y < 0. \end{cases}$$

i) What does the graph of $z = g_1(x,y)$ look like?

ii) Show that if we have

$$f(x,y) \sim \sum_{n=1}^{\infty}\sum_{m=1}^{\infty} c_{n,m} \sin nx \sin my,$$

then this same series converges to $g_1(x,y)$ (except, possibly, on the boundaries and at discontinuities). Compute the coefficients $c_{n,m}$ in terms of the function g_1.

This is the two-variable analog of the statement that the Fourier series of an odd function is a pure sine series.

b) Similarly, construct a function $g_2(x,y)$ on $-\pi \le x \le \pi, -\pi \le y \le \pi$, for the series in Exercise 7a. Again, what are the coefficients, in terms of g_2?

c) Do the same, but for a function g_3, for the series in Exercise 7b.

d) Do the same, but for a function g_4, for the series in Exercise 7c.

Again, each of these is a special case of what is the general double Fourier series for a function on $-a \le x \le a, -b \le y \le b$.

16. **General double Fourier series:** Given a function $f(x,y)$ on $-\pi \le x \le \pi, -\pi \le y \le \pi$,

a) Find functions g_1, g_2, g_3 and g_4 which exhibit the symmetries in Exercises 15a, b, c and d, respectively, and for which we have

$$f(x,y) = g_1(x,y) + g_2(x,y) + g_3(x,y) + g_4(x,y).$$

b) Use the series in Exercise 14 to show that we have

$$f(x,y) \sim \frac{a_{0,0}}{4} + \frac{1}{2}\sum_{n=1}^{\infty}(a_{n,0}\cos nx + a_{0,n}\cos ny$$
$$+ b_{0,n}\sin ny + c_{n,0}\sin nx)$$
$$+ \sum_{n=1}^{\infty}\sum_{m=1}^{\infty}(a_{n,m}\cos nx \cos my + b_{n,m}\cos nx \sin my$$
$$+ c_{n,m}\sin nx \cos my + d_{n,m}\sin nx \sin my),$$

where

$$a_{n,m} = \frac{4}{\pi^2} \int_0^\pi \int_0^\pi f(x,y) \cos nx \cos my \, dydx,$$
$$n = 0,1,2,\ldots; m = 0,1,2,\ldots;$$

$$b_{n,m} = \frac{4}{\pi^2} \int_0^\pi \int_0^\pi f(x,y) \cos nx \sin my \, dydx,$$
$$n = 0,1,2,\ldots; m = 1,2,\ldots;$$

$$c_{n,m} = \frac{4}{\pi^2} \int_0^\pi \int_0^\pi f(x,y) \sin nx \cos my \, dydx,$$
$$n = 1,2,3,\ldots; m = 0,1,2,\ldots;$$

and

$$d_{n,m} = \frac{4}{\pi^2} \int_0^\pi \int_0^\pi f(x,y) \sin nx \sin my \, dydx,$$
$$m = 1,2,3,\ldots; m = 1,2,\ldots.$$

17. Prove that if $f_1(x,y)$ and $f_2(x,y)$ are linearly independent on a region, then so are

$$g_1(x,y) = f_1(x,y) + f_2(x,y),$$
$$g_2(x,y) = f_1(x,y) - f_2(x,y).$$

More generally, for which choices of constants a, b, c and d will

$$h_1(x,y) = af_1(x,y) + bf_2(x,y),$$
$$h_2(x,y) = cf_1(x,y) + df_2(x,y)$$

be linearly independent?

18. **Nonhomogeneous equations:**

a) Proceed as in Section 4.4 to solve the heat equation problem with source term

$$u_t = \nabla^2 u + \sin 2x \sin 3y, \qquad 0 < x < \pi, 0 < y < \pi, t > 0,$$
$$u(x,y,0) = \sin 4x \sin 7y,$$
$$u(x,0,t) = u(x,\pi,t) = u(0,y,t) = u(\pi,y,t) = 0.$$

b) **MATLAB:** Plot the solution of part (a) for various values of t.

c) More generally, solve the problem

$$u_t = \nabla^2 u + F(x,y,t),$$
$$u(x,y,0) = 0,$$
$$u(x,0,t) = u(x,\pi,t) = u(0,y,t) = u(\pi,y,t) = 0.$$

You may assume that F can be expanded in a double Fourier series

$$F(x, y, t) = \sum_{n=1}^{\infty} \sum_{m=1}^{\infty} F_{n,m}(t) \sin nx \sin my.$$

9.3 Laplace's Equation in Polar Coordinates: Poisson's Integral Formula

In order to solve these problems on domains which have *circular* boundaries, we must resort to polar coordinates. Let us then compute the Laplacian in polar coordinates (which, in fact, we already did in Exercise 7 of Section 2.5). So, given

$$\nabla^2 u = u_{xx} + u_{yy},$$

we let

$$x = x(r, \theta) = r \cos \theta,$$
$$y = y(r, \theta) = r \sin \theta.$$

It turns out to be easier to do this *backwards*, that is, write

$$u_r = u_x x_r + u_y y_r$$
$$= u_x \cos \theta + uy \sin \theta$$

and

$$u_\theta = u_x x_\theta + u_y y_\theta$$
$$= -u_x r \sin \theta + u_y r \cos \theta.$$

Then, solving for u_x and u_y, we have

$$u_x = u_r \cos \theta - \frac{1}{r} u_\theta \sin \theta,$$
$$u_y = \frac{1}{r} u_\theta \cos \theta + u_r \sin \theta.$$

It follows that

$$u_{xx} = \frac{\partial}{\partial x}(u_x) = \frac{\partial}{\partial x}\left(u_r \cos\theta - \frac{1}{r}u_\theta \sin\theta\right)$$

$$= \left(u_r \cos\theta - \frac{1}{r}u_\theta \sin\theta\right)_r \cos\theta$$

$$- \frac{1}{r}\left(u_r \cos\theta - \frac{1}{r}u_\theta \sin\theta\right)_\theta \sin\theta$$

$$= u_{rr}\cos^2\theta + \frac{1}{r^2}u_{\theta\theta}\sin^2\theta - \frac{2}{r}u_{r\theta}\sin\theta\cos\theta$$

$$+ \frac{1}{r}u_r \sin^2\theta + \frac{2}{r^2}u_\theta \sin\theta\cos\theta$$

and, similarly,

$$u_{yy} = u_{rr}\sin^2\theta + \frac{1}{r^2}u_{\theta\theta}\cos^2\theta + \frac{2}{r}u_{r\theta}\sin\theta\cos\theta$$

$$+ \frac{1}{r}u_r \cos^2\theta - \frac{2}{r^2}u_\theta \sin\theta\cos\theta.$$

Adding, we have

$$u_{xx} + u_{yy} = u_{rr} + \frac{1}{r}u_r + \frac{1}{r^2}u_{\theta\theta}.$$

Now we may solve the two-dimensional heat, wave and Laplace equations on a disk. We begin with Laplace, the least involved.

LAPLACE'S EQUATION ON A DISK

Here we solve the Laplace equation, on the disk $0 \le r \le a$, with a Dirichlet boundary condition—the so-called **interior Dirichlet problem**—leaving other types of boundary conditions (as well as somewhat more complicated geometries) for the exercises. So we must solve

$$u_{rr} + \frac{1}{r}u_r + \frac{1}{r^2}u_{\theta\theta} = 0, \qquad 0 < r < a, -\infty < \theta < \infty,$$

$$u(a,\theta) = f(\theta), \qquad -\infty < \theta < \infty.$$

However, the change to polars necessitates further restrictions. First, since the point (r,θ) is the same as the point $(r,\theta + 2\pi)$, we must require

$$u(r,\theta + 2\pi) = u(r,\theta)$$

for each θ, and each r in $0 < r < a$. Also, why have we been avoiding $r = 0$? From our experience with polar coordinates, we see that it's possible to have equations with solutions that are unbounded at the origin. These solutions certainly are not continuous, so we must require

$$\lim_{r \to 0} u(r,\theta) = L < \infty, \qquad -\infty < \theta < \infty,$$

or, in short,

$$u(0, \theta) < \infty.$$

Much of the separation of variables work already was done in Exercises 11 and 29, Section 1.6. There, we separated the ODEs and got

$$r^2 R'' + rR' - \lambda R = 0, \qquad \Theta'' + \lambda \Theta = 0.$$

Here, we also need to separate the auxiliary conditions:

$$u(r, \theta + 2\pi) = u(r, \theta) \Rightarrow R(r)\Theta(\theta + 2\pi) = R(r)\Theta(\theta)$$
$$\Rightarrow \Theta(\theta + 2\pi) = \Theta(\theta), \qquad -\infty < \theta < \infty,$$

and

$$\lim_{r \to 0} u(r, \theta) = \lim_{r \to 0} R(r)\Theta(\theta) = \Theta(\theta) \lim_{r \to 0} R(r) \quad \text{(why?)}$$
$$< \infty \Rightarrow \lim_{r \to 0} R(r) < \infty \quad \text{(again, why?)}.$$

Again, in short, we write $R(0) < \infty$.

We also showed that the eigenvalues and eigenfunctions of the Θ-problem

$$\Theta'' + \lambda \Theta = 0,$$
$$\Theta(\theta + 2\pi) = \Theta(\theta), \text{ for all } \theta,$$

are

$$\lambda_n = n^2, n = 0, 1, 2, \ldots; \quad \Theta_0(\theta) = c_0, \Theta_n(\theta) = c_n \cos n\theta + d_n \sin n\theta,$$
$$n = 1, 2, 3, \ldots .$$

Then, from our earlier solution of the R-equation, we have

$$\lambda_0 = 0, r^2 R'' + rR' = 0 \Rightarrow R_0(r) = a_0 + b_0 \ln r,$$
$$\lambda_n = n^2, r^2 R'' + rR' - n^2 R = 0 \Rightarrow R_n(r) = a_n r^n + b_n r^{-n},$$
$$n = 1, 2, 3, \ldots .$$

However, the condition $\lim_{r \to 0^+} R(r) = 0$ forces $b_n = 0$, $n = 0, 1, 2, \ldots$. So our surviving product solutions are

$$u_0(r, \theta) = c_0,$$
$$u_n(r, \theta) = r^n (c_n \cos n\theta + d_n \sin n\theta), \qquad n = 1, 2, 3, \ldots,$$

and, thus, our *general solution* is

$$u(r, \theta) = c_0 + \sum_{n=1}^{\infty} r^n (c_n \cos n\theta + d_n \sin n\theta).$$

Finally, we apply the Dirichlet condition:

$$u(a, \theta) = f(\theta) = c_0 + \sum_{n=1}^{\infty} a^n (c_n \cos n\theta + d_n \sin n\theta).$$

First, it follows that we must have started with an f of period 2π, or the original problem would not have been well-posed (why?). Then, we may restrict ourselves to the interval $-\pi \le \theta \le \pi$, in which case it is clear that the series must be the Fourier series of f on $-\pi \le \theta \le \pi$, that is, that we must have

$$c_0 = \frac{a_0}{2}, \quad a^n c_n = a_n \quad \text{and} \quad a^n d_n = b_n, \qquad n = 1, 2, 3, \ldots,$$

where

$$a_n = \frac{1}{\pi} \int_0^{2\pi} f(\theta) \cos n\theta \ d\theta, ^{\ddagger\ddagger} \qquad n = 0, 1, 2, \ldots,$$

and

$$b_n = \frac{1}{\pi} \int_0^{2\pi} f(\theta) \sin n\theta \ d\theta, \qquad n = 1, 2, 3, \ldots .$$

Our solution, then, is

$$u(r, \theta) = \frac{a_0}{2} + \sum_{n=1}^{\infty} \left(\frac{r}{a}\right)^n (a_n \cos n\theta + b_n \sin n\theta).$$

(By the way, we *could,* instead, have used the equivalence of our·θ-problem with the periodic Sturm–Liouville problem

$$\Theta'' + \lambda\Theta = 0, \qquad -\pi < \theta < \pi$$
$$\Theta(-\pi) = \Theta(\pi)$$
$$\Theta'(-\pi) = \Theta'(\pi)$$

—see Exercise 19, Section 1.7 and Example 1, Section 8.2—in order to set the problem on $-\pi \le \theta \le \pi$ from the start.)

We may, instead, have a Neumann or a Robin boundary condition (or any combination of boundary conditions, of course!). We leave these for the exercises, except that we must mention that the Neumann problem here, as when we solved it in Chapter 4, requires a *consistency condition.* Specifically, if our boundary condition is of the form

$$u_r(a, \theta) = f(\theta),$$

---

‡‡Why may we integrate on $0 \le \theta \le 2\pi$ instead of $-\pi \le \theta \le \pi$? Actually, we need to be careful—see Exercise 19h.

then we will find that we must have

$$\int_0^{2\pi} f(\theta)d\theta = \int_{-\pi}^{\pi} f(\theta)d\theta = 0.$$

As before, all this says is that there is *no net flux across the boundary*. More generally, as we saw in Exercise 14c, Section 4.3, we must have

$$\oint_C \frac{\partial u}{\partial n}ds = 0,$$

where C is the boundary curve of the region in question. This sometimes is called the *theorem of the vanishing flux*.

Interestingly, it turns out that we may rewrite the solution to the interior Dirichlet problem as an integral. Specifically, the solution is given by the famous **Poisson's integral formula**

$$u(r,\theta) = \frac{a^2 - r^2}{2\pi} \int_0^{2\pi} \frac{f(\phi)}{a^2 - 2ar\cos(\theta - \phi) + r^2}d\phi,$$

which we prove, after a few remarks.

Poisson's formula tells us that the value of u at any point in the interior of the disk is a weighted average of its values on the boundary; the **Poisson kernel**

$$\frac{a^2 - r^2}{a^2 - 2ar\cos(\theta - \phi) + r^2}$$

tells us how to *weight* each of the boundary values. In fact, the denominator is just the square of the distance between the point (r, θ) and each point (a, ϕ) on the boundary (see Figure 9.10 and Exercise 13). It's the law of cosines again, of course.

Also, if we let $r = 0$ in Poisson's integral formula, we have the result that

$$u(0,\theta) = \frac{1}{2\pi} \int_0^{2\pi} f(\theta)d\theta,$$

that is, that the value of u at the center is just the average of its values along the boundary. It is not hard to *translate* this to *any* disk in the x-y plane. In other words, we have that *the value of any harmonic function at the center of a disk is equal to the average of its values on the boundary*. This is the **Mean Value Property** for harmonic functions.

We can say even more. Suppose u is a nonconstant harmonic function on any well-enough behaved closed region R, with nonempty interior in the x-y plane.* Then, from the Extreme Value Theorem, u attains a maximum value at some point in the region.

---

*To be precise, u is harmonic on the region's interior.

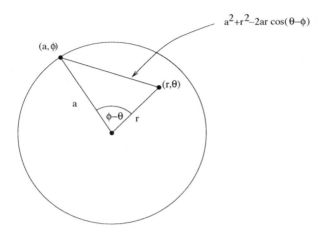

FIGURE 9.10
The geometry of Poisson's integral formula.

Where? Let $P \in \text{int}(R)$. Then we may draw a circle, centered at P, which lies within R; thus, $u(P)$ = average of u along this circle. Since u is not constant, it follows that there must be a point Q on the circle such that $u(Q) > u(P)$ (why?). In other words, we have shown that the maximum value of u *cannot* occur at any point in the interval of R. This is the famous **Maximum Principle for harmonic functions**, that the maximum must occur on the boundary of R. (For a detailed proof, see Theorem C.5 in Appendix C.)

PROOF of *Poisson's Integral Formula*:

We need to prove that

$$\frac{a_0}{2} + \sum_{n=1}^{\infty} \left(\frac{r}{a}\right)^n (a_n \cos n\theta + b_n \sin n\theta)$$

$$= \frac{a^2 - r^2}{2\pi} \int_0^{2\pi} \frac{f(\theta)}{a^2 - 2ar \cos(\theta - \phi) + r^2} d\phi,$$

where

$$a_n = \frac{1}{\pi} \int_0^{2\pi} f(\theta) \cos n\theta \, d\theta, \qquad n = 0, 1, 2, \ldots,$$

$$b_n = \frac{1}{\pi} \int_0^{2\pi} f(\theta) \sin n\theta \, d\theta, \qquad n = 1, 2, 3, \ldots,$$

and we proceed much as we did in Section 3.5. As then, we begin by writing

$$\frac{a_0}{2} + \sum_{n=1}^{\infty} \left(\frac{r}{a}\right)^n (a_n \cos n\theta + b_n \sin n\theta)$$

$$= \frac{1}{2\pi} \int_0^{2\pi} \left[1 + 2\sum_{n=1}^{\infty} \left(\frac{r}{a}\right)^n (\cos n\theta \cos n\phi + \sin n\theta \sin n\phi)\right] f(\phi)d\phi$$

$$= \frac{1}{2\pi} \int_0^{2\pi} \left[1 + 2\sum_{n=1}^{\infty} \left(\frac{r}{a}\right)^n \cos n(\theta - \phi)\right] f(\phi)d\phi$$

$$= \frac{1}{2\pi} \int_0^{2\pi} \left[1 + \sum_{n=1}^{\infty} \left(\frac{r}{a}\right)^n \left(e^{in(\theta-\phi)} + e^{-in(\theta-\phi)}\right)\right] f(\phi)d\phi$$

$$= \frac{1}{2\pi} \int_0^{2\pi} \left[1 + \frac{re^{i(\theta-\phi)}}{a - re^{i(\theta-\phi)}} + \frac{re^{-in(\theta-\phi)}}{a - re^{-i(\theta-\phi)}}\right] f(\phi)d\phi$$

$$= \frac{1}{2\pi} \int_0^{2\pi} \frac{a^2 - r^2}{a^2 - 2ar\cos(\theta - \phi) + r^2} f(\phi)d\phi$$

(see Exercise 15a).

Exercises 9.3

1. Evaluate in both Cartesian coordinates and polar coordinates—make sure you get the same answer both ways.

    a) $\nabla^2(xy)$
    b) $\nabla^2(2x^3 + 3x^2y - y^2)$

2. Evaluate

    a) $\nabla^2(y^2(x^2 + y^2)^4)$
    b) $\nabla^2(r^n), n = 1, 2, 3, \ldots$
    c) $\nabla^2(\cos n\theta), n = 1, 2, 3, \ldots$

3. a) Use symmetry to solve the interior Dirichlet problem

$$\nabla^2 u = 0, \quad 0 < r < 2, \quad -\infty < \theta < \infty,$$
$$u(2, \theta) = 4.$$

   b) Use symmetry to solve the Dirichlet problem on an *annulus*,

$$\nabla^2 u = 0, \quad 2 < r < 5, -\infty < \theta < \infty,$$
$$u(2, \theta) = 4,$$
$$u(5, \theta) = -1.$$

c) Use symmetry to solve the **exterior Dirichlet problem**

$$\nabla^2 u = 0, \quad r > 2, -\infty < \theta < \infty,$$
$$u(2, \theta) = 4.$$

d) Show that the function $u = 4 + \frac{r}{2}\cos\theta - \frac{2}{r}\cos\theta$ *also* satisfies the problem in part (c). Therefore, we must be very careful when solving exterior problems. Really, it turns out that we need only specify the additional requirement that u be bounded as $r \to \infty$. However, note that there is *no* solution if we stipulate the condition that $u \to 0$ as $r \to \infty$!

4. More generally, solve the **exterior Dirichlet problem**

$$\nabla^2 u = 0, \quad r > a, -\infty < \theta < \infty,$$
$$u(a, \theta) = f(\theta),$$
$$u \text{ bounded as } r \to \infty.$$

5. More generally, solve the Dirichlet problem on an *annulus*,

$$\nabla^2 u = 0, \quad a < r < b, -\infty < \theta < \infty,$$
$$u(a, \theta) = f_1(\theta),$$
$$u(b, \theta) = f_2(\theta).$$

6. Solve the **interior Neumann problem**

$$\nabla^2 u = 0, \quad 0 < r < a, -\infty < \theta < \infty,$$
$$u_r(a, \theta) = g(\theta).$$

What consistency condition must g satisfy, and where in the solution of the problem does the need for this condition arise?

7. Solve the **interior Robin problem**

$$\nabla^2 u = 0, \quad 0 < r < a, -\infty < \theta < \infty,$$
$$u_r(a, \theta) + hu(a, \theta) = f(\theta), h \text{ constant.}$$

Need f satisfy a consistency condition? If so, what is it?

8. a) Solve the general interior Dirichlet problem on a wedge,

$$\nabla^2 u = 0, \quad 0 < r < a, \quad 0 < \theta < \alpha < 2\pi,$$
$$u(a, \theta) = f(\theta),$$
$$u(r, 0) = u(r, \alpha) = 0.$$

b) Solve the general interior Neumann problem on a wedge,

$$\nabla^2 u = 0, \quad 0 < r < a, 0 < \theta < \alpha < 2\pi,$$
$$u_r(a, \theta) = g(\theta),$$
$$u(r, 0) = u(r, \alpha) = 0.$$

Need g satisfy a consistency condition? If so, what is it?

9. **Helmholtz equation:** Solve the Helmholtz boundary-value problem

$$\nabla^2 u + k^2 u = 0, \quad 0 < r < a, k \text{ constant},$$
$$u(a, \theta) = f(\theta).$$

10. Explain why the problem

$$\nabla^2 u = 0, \quad 0 < r < a, -\infty < \theta < \infty,$$
$$u_r(a, \theta) = 1$$

doesn't make sense, physically.

11. If u is continuous on $r \leq a$ and harmonic on $r < a$, and $u(a, \theta) = 2 + 4\sin\theta$,

a) Find the maximum and minimum values of u on the disk $r \leq a$.

b) Find the value of u at the origin.

12. a) Why does the fact that $\nabla^2 u = 0$ suggest that a harmonic function u cannot attain a maximum or minimum on the interior of a region?

b) Assuming that the *maximum principle for harmonic functions* is true, state and prove the **Minimum Principle for harmonic functions**.

c) **Liouville's Theorem:** Prove that a function $u \not\equiv constant$ which is harmonic on the x-y plane cannot be bounded above or below.

13. a) Refer to Figure 9.10, where L is the distance between the point (r, θ) and any boundary point (a, ϕ). Show that L^2 is the denominator of the Poisson kernel.

b) Show that Poisson's integral formula can be written in vector form as

$$u(\boldsymbol{r}) = \frac{a^2 - |\boldsymbol{r}|^2}{2\pi a} \oint_C \frac{u(\boldsymbol{r'})}{|\boldsymbol{r} - \boldsymbol{r'}|^2} ds,$$

where C, of course, is the circle $r = a$.

14. a) Show formally that Poisson's integral formula *does* satisfy Laplace's equation.

b) Show that

$$\frac{a^2 - r^2}{2\pi} \int_0^{2\pi} \frac{1}{a^2 - 2ar\cos(\theta - \phi) + r^2}d\phi = 1$$

for any choice of r, θ and a.

15. a) Justify the steps in the derivation of Poisson's integral formula.

b) Show that the solution of the *exterior* Dirichlet problem (Exercise 4) can be written

$$u(r, \theta) = \frac{r^2 - a^2}{2\pi} \int_0^{2\pi} \frac{f(\phi)}{a^2 - 2ar\cos(\theta - \phi) + r^2}d\phi,$$

and, thus, show that if $u(r, \theta, a)$ is the solution of the *interior* Dirichlet problem on the disk $r < a$, then the solution of the *exterior* Dirichlet, with the same boundary condition on $r = a$, is $u(a, \theta, r)$.

16. **Poisson's integral formula for the interior Neumann problem:**

a) From our proof of Poisson's integral formula, we know that

$$\frac{1}{2} + \sum_{n=1}^{\infty} \left(\frac{r}{a}\right)^n \cos n\alpha = \frac{1}{2}\frac{a^2 - r^2}{a^2 - 2ar\cos(\theta - \phi) + r^2}.$$

Use this to show that

$$\sum_{n=1}^{\infty} \frac{r^n}{n}\cos n\alpha = -\frac{1}{2}\ln(1 + r^2 - 2r\cos\alpha).$$

b) Generalizing part (a), show that the solution of the interior Neumann problem (Exercise 6) can be written as

$$u(r, \theta) = c - \frac{a}{2\pi}\int_0^{2\pi} f(\phi)\ln[a^2 - 2ar\cos(\theta - \phi) + r^2]d\phi,$$

where c is an arbitrary constant.

17. **Electrostatics: z-independent Dirichlet problem in a cylinder.** Suppose we have an infinite cylinder of radius a with the potential distributed uniformly along its length, so that $u(a, \theta, z) = f(\theta)$. If the resulting potential is $u(r, \theta, z)$, determine the electric field $\mathbf{E} = -\nabla u$ at any point, whether inside or outside the cylinder.

18. **Gravitation: Radially symmetric Dirichlet problem in spherical coordinates.** We showed in Exercise 2, Section 7.1 (and we show again in Exercise 1, Section 9.5) that the Laplacian in spherical coordinates is

$$\nabla^2 u = u_{\rho\rho} + \frac{2}{\rho}u_\rho + \frac{1}{\rho}(u_{\phi\phi} + u_\phi \cot\phi + u_{\theta\theta} \cos^2\phi)$$

$$= \frac{1}{\rho^2}\left[(\rho^2 u_\rho)_\rho + \frac{1}{\sin\phi}(u_\phi \sin\phi)_\phi + \frac{1}{\sin^2\phi}u_{\theta\theta}\right].$$

Now, suppose that matter is uniformly distributed on the surface of the sphere $\rho = a$, so that $u(a, \theta, \phi) = $ constant.

   a) Show that the gravitational field $\boldsymbol{F} = \nabla u = \boldsymbol{0}$ inside the sphere.

   b) Show that, for points outside the sphere, the radius of the sphere is irrelevant (and, in fact, it turns out that the gravitational field outside the sphere is the same as if all of the matter were concentrated at the origin).

19. **Equipotential curves and surfaces:** Curves (in two dimensions) and surfaces (in three dimensions) along which a potential function is constant are called, not surprisingly, **equipotential curves** and **surfaces**. In the case of steady state heat problems, these curves and surfaces are referred to as **isotherms** (as you may have seen on weather maps).

   In Parts a–c, describe the equipotential curves for a harmonic function on the unit disk subject to the given boundary condition.

   a) $u(1, \theta) = 3\cos\theta$

   b) $u(1, \theta) = \sin 2\theta$

   c) $u(1, \theta) = 2 - \cos 2\theta$

   We were fortunate that the initial conditions were so easy to deal with, of course. What if things are more complicated? There are identities that we can establish, the use of which will enable us to write many of our series solutions in closed form. To begin with,

   d) Show that the Maclaurin series for $f(x) = \ln(1 + x)$ is

$$\ln(1 + x) = \sum_{n=1}^{\infty} \frac{(-1)^{n+1}x^n}{n}.$$

   What's its interval of convergence?

   Next, the *complex logarithm* function is defined so that it's consistent with the real function $\ln x$. In particular, it satisfies

$$\log(1 + z) = \sum_{n=1}^{\infty} \frac{(-1)^{n+1}z^n}{n} \quad \text{for} \quad |z| < 1.$$

(Remember that the *modulus* of $z = x + iy$ is defined to be $|x + iy| = \sqrt{x^2 + y^2}$.)

We need now to define the *polar form* of a complex number; however, this form entails nothing more than writing the number in polar coordinates. Thus, we have

$$z = x + iy = r\cos\theta + ir\sin\theta,$$

where $r = |z|$ and $\theta = \tan^{-1}\frac{y}{x}$ (and we must take care when handling the $\tan^{-1}$, of course). Thus,

$$z = re^{i\theta} \quad (\text{why?}).$$

Finally, the complex logarithm also satisfies the other rules of logs (to be precise, we *define* it so that it does). So,

$$
\begin{aligned}
\log z = \log re^{i\theta} &= \log r + i\theta \log e \\
&= \ln r + i\theta.
\end{aligned}
$$

We would like to substitute $z = re^{i\theta}$ into the Maclaurin series above to see what "pops out."

e) Show that

$$\log(1 + r^{i\theta}) = \frac{1}{2}\ln(1 + 2r\cos\theta + r^2) + i\tan^{-1}\left(\frac{r\sin\theta}{1 + r\cos\theta}\right).$$

f) Now substitute $z = re^{i\theta}$ into the Maclaurin series and, taking real and imaginary parts, show that

$$\sum_{n=1}^{\infty}\frac{(-1)^{n+1}}{n}r^n\cos n\theta = \frac{1}{2}\ln(1 + 2r\cos\theta + r^2),$$

$$\sum_{n=1}^{\infty}\frac{(-1)^{n+1}}{n}r^n\sin n\theta = \tan^{-1}\left(\frac{r\sin\theta}{1 + r\cos\theta}\right).$$

g) Show, further, that

$$\sum_{n=1}^{\infty}\frac{1}{n}r^n\cos n\theta = \frac{1}{2}\ln(1 - 2r\cos\theta + r^2)$$

$$\sum_{n=1}^{\infty}\frac{1}{n}r^n\sin n\theta = \tan^{-1}\left(\frac{r\sin\theta}{1 - r\cos\theta}\right).$$

After all that, solve, and describe the equipotential curves for the problem $\nabla^2 u = 0$, $0 < r < 1$, subject to the boundary conditions in parts (h) and (i).

h) $u(1, \theta) = 0, -\pi < \theta \leq \pi$

i) $u(1, \theta) = \begin{cases} 1 & 0 \leq \theta < \pi, \\ 0, & \pi \leq \theta < 2\pi \end{cases}$

j) Finally, show that the function

$$f(r, \theta) = \frac{1 - r^2}{2\pi} \int_0^{2\pi} \frac{\sin \phi}{1 - 2r \cos(\theta - \phi) + r^2} d\phi, \quad |r| < 1,$$

is constant along each line $y = $ constant, while

$$f(r, \theta) = \frac{1 - r^2}{2\pi} \int_0^{2\pi} \frac{\cos \phi}{1 - 2r \cos(\theta - \phi) + r^2} d\phi, \quad |r| < 1,$$

is constant along each line $x = $ constant.

9.4   The Wave and Heat Equations in Polar Coordinates

Θ-INDEPENDENT WAVE EQUATION ON A DISK

The method of solution for the heat equation in polar coordinates is almost identical to that of the wave equation, so we leave the former for the exercises. In each case, the actual solution is fairly complicated, so we begin by considering the special case *of radially symmetric*, or θ-independent, vibrations of a circular drumhead. Our problem, then, is

$$u_{tt} = \nabla^2 u, \qquad 0 < r < 1, -\infty < \theta < \infty, t > 0$$

with initial shape and velocity

$$u(r, \theta, 0) = f(r),$$
$$u_t(r, \theta, 0) = 0,$$

and boundary condition

$$u(1, \theta, t) = 0.$$

As usual, we have the auxiliary condition

$$\lim_{r \to 0^+} u(r, \theta, t) < \infty,$$

and it should be clear from the initial and boundary condition that our solutions will not depend on θ.

So we begin by writing

$$u_{tt} = u_{rr} + \frac{1}{r}u_r + \frac{1}{r^2}u_{\theta\theta}$$

$$= u_{rr} + \frac{1}{r}u_r,$$

and, separating variables, we arrive at

$$rR'' + R' + \lambda r R = 0, \quad T'' + \lambda T = 0$$
$$R(1) = 0, \quad R(0) < \infty.$$

We solve the R eigenvalue problem as always.

Case 1: $\lambda = 0$

Here we have the Cauchy–Euler equation

$$rR'' + R' = 0$$

with general solution

$$R(r) = c_1 + c_2 \ln r.$$

Then, the boundary conditions imply that $c_1 = c_2 = 0$ (why?). So $\lambda = 0$ *is not an eigenvalue.*

Case 2: $\lambda < 0, \lambda = -k^2 (k > 0)$

Now we solve

$$rR'' + R' - k^2 r R = 0.$$

We *could* solve this equation using the method of Frobenius but, instead, we proceed as in Exercise 9, Section 7.5. So, first we multiply by r, then make the substitution $x = kr$, to arrive at the *modified Bessel's equation of order 0,*

$$x^2 R'' + x R' - x^2 R = 0.$$

Thus, the general solution is

$$R(r) = c_1 I_0(kr) + c_2 K_0(kr),$$

where, of course, I_0 and K_0 are the *modified Bessel functions of order 0*. Then, since K_0 is unbounded at the origin, we must have $c_2 = 0$ and, since $I_0(x) > 0$ for $x > 0$, we also have $c_1 = 0$. Then, *there are no negative eigenvalues.*

Case 3: $\lambda > 0, \lambda = k^2 (k > 0)$

Finally, we have the ODE

$$rR'' + R' + k^2 r R = 0.$$

Again, we multiply by r and let $x = kr$. This time, we get Bessel's equation of order 0,

$$x^2 y'' + x y' + x^2 y = 0,$$

so our general solution is

$$R(r) = c_1 J_0(kr) + c_2 Y_0(kr).$$

Here, J_0 and Y_0 are, of course, Bessel functions of order 0. Since Y_0 is unbounded as $r \to 0^+$, we have $c_2 = 0$. Then, the boundary condition at the edge gives us

$$c_1 J_0(k) = 0.$$

This implies that $c_1 = 0$ *unless k is a zero of J_0*. Remembering that J_0 has infinitely many positive roots $k_1, k_2, \ldots \to \infty$, we see that the eigenvalues are

$$\lambda_n = k_n^2, \qquad n = 1, 2, 3, \ldots,$$

with corresponding eigenfunctions

$$R_n(r) = J_0(k_n r), \qquad n = 1, 2, 3, \ldots .$$

Back to the T-equation, we have

$$T'' + k_n^2 T = 0$$

with general solution

$$T_n(t) = c_n \cos k_n t + d_n \sin k_n t.$$

Therefore, we have the product solutions

$$
\begin{aligned}
u_n(r, \theta, t) &= R_n(r) T_n(t) \\
&= J_0(k_n r)(c_n \cos k_n t + d_n \sin k_n t)
\end{aligned}
$$

and, thus, the *general solution*

$$
\begin{aligned}
u(r, \theta, t) &= \sum_{n=1}^{\infty} u_n(r, \theta, t) \\
&= \sum_{n=1}^{\infty} J_0(k_n r)(c_n \cos k_n t + d_n \sin k_n t).
\end{aligned}
$$

Finally, we apply the initial conditions:

$$u(r, \theta, 0) = f(r) = \sum_{n=1}^{\infty} c_n J_0(k_n r),$$

and this just says that the series is a Fourier–Bessel series for $f(r)$ on $0 \le r \le 1$ (see Exercise 3c in Section 8.5). Similarly,

$$u_t(r, \theta, 0) = 0 = \sum_{n=1}^{\infty} d_n k_n J_0(k_n r),$$

so we have $d_n = 0$, $n = 1, 2, 3, \ldots$, and our final solution is

$$u(r, \theta, t) = \sum_{n=1}^{\infty} c_n J_0(k_n r) \cos k_n t,$$

where

$$c_n = \frac{2}{J_1^2(k_n)} \int_0^1 r f(r) J_0(k_n r) dr.$$

As with the vibrating string and rectangular membrane, we may look at the individual *modes* of vibration

$$R_n(r) = J_0(k_n r), \qquad n = 1, 2, 3, \ldots,$$

as well as their *nodal patterns*. The first columns of Figures 9.11 and 9.12 show, respectively, the first three vibration modes and corresponding nodal lines.

Before moving on, it is interesting to note that, although each product solution u_n is periodic, the infinite series u is *not* periodic! (Why is this the case?)

GENERAL WAVE EQUATION ON A DISK

Okay, now that we're finished with our warmup problem, let's solve the problem of the vibrating circular drumhead in its full generality. So we have

$$u_{tt} = c^2 \nabla^2 u = c^2 \left(u_{rr} + \frac{1}{r} u_r + \frac{1}{r^2} u_{\theta\theta} \right),$$

$$0 < r < a, -\infty < \theta < \infty, t > 0,$$

$$u(r, \theta, 0) = f(r, \theta),$$

$$u_t(r, \theta, 0) = g(r, \theta),$$

$$u(a, \theta, t) = 0.$$

As with Laplace, we have the auxiliary condition

$$u(\theta + 2\pi) = u(\theta), \qquad \lim_{r \to 0^+} u(r, \theta) < \infty.$$

Separating variables, we have

$$r^2 R'' + r R' + (\lambda r^2 - \gamma) R = 0, \qquad \Theta'' + \gamma \Theta = 0, \qquad T'' + c^2 \lambda T = 0,$$

along with the conditions

$$R(a) = 0, \qquad R(0) < \infty \quad \text{and} \quad \Theta(\theta + 2\pi) = \Theta(\theta).$$

As before, the Θ-equation implies that we have

$$\gamma_n = n^2, \qquad \Theta_0(0) = c_0, \Theta_n(\theta) = c_n \cos n\theta + d_n \sin n\theta, \qquad n = 0, 1, 2, \ldots .$$

Then, the R-equation becomes

$$r^2 R'' + r R' + (\lambda r^2 - n^2) R = 0,$$
$$R(a) = 0, \quad R(0) < \infty.$$

Again, we solve the R eigenvalue problem.

Case 1: $\lambda = 0$
   Here we have the Cauchy–Euler equation

$$r^2 R'' + r R' - n^2 R = 0$$

for each $n = 0, 1, 2, \ldots$. For $n = 0$, the general solution is

$$R(r) = c_1 + c_2 \ln r,$$

and the boundary conditions imply that $c_1 = c_2 = 0$.
   For $n \neq 0$,

$$R(r) = c_1 r^n + c_2 r^{-n}.$$

Again, the boundary conditions lead to $c_1 = c_2 = 0$. Thus, $\lambda = 0$ *is not an eigenvalue, for any choice of* n.

Case 2: $\lambda < 0, \lambda = -k^2 (k > 0)$
   We solve here

$$r^2 R'' + r R' - (k^2 r^2 + n^2) R = 0.$$

As before, the change of variable $x = kr$ turns this into the *modified Bessel's equation* of order n,

$$x^2 y'' + x y' - (x^2 + n^2) y = 0.$$

Thus, the general solution is

$$R(r) = c_1 I_n(kr) + c_2 K_n(kr),$$

where I_n and K_n are the modified Bessel functions of order n. Since K_n is unbounded as $r \to 0$, we must have $c_2 = 0$ and, since $I_n(x) > 0$ for $x > 0$, we must also have $c_1 = 0$. Thus, *there are no negative eigenvalues for any choice of* n.

Case 3: $\lambda > 0, \lambda = k^2 (k > 0)$
   Now we have the ODE

$$r^2 R'' + r R' + (k^2 r^2 - n^2) R = 0.$$

Again, we perform the change of variable $x = kr$, the result being *Bessel's equation of order* n,

$$x^2 y'' + x y' + (x^2 - n^2) y = 0.$$

The general solution, of course, is

$$R(r) = c_1 J_n(kr) + c_2 Y_n(kr),$$

where J_n and Y_n are Bessel functions of order n. Since Y_n is unbounded as $r \to 0$, we have $c_0 = 0$. Then, the boundary condition at $r = a$ gives us

$$R(a) = c_1 J_n(ka) = 0.$$

This implies that $c_1 = 0$ *unless ka is a zero of the Bessel function J_n.* Remembering that each Bessel function J_n has infinitely many positive zeros $x_{n,m}$, $m = 1, 2, 3, \ldots$ (with $x_{n,m} \to \infty$), we see that the eigenvalues are

$$\lambda_{n,m} = (k_{n,m})^2 = \left(\frac{x_{n,m}}{a}\right)^2, \qquad n = 0, 1, 2, \ldots; m = 1, 2, 3, \ldots,$$

with corresponding eigenfunctions

$$R_{n,m}(r) = J_n\left(\frac{x_{n,m}r}{a}\right), \qquad n = 0, 1, 2, \ldots; m = 1, 2, 3, \ldots .$$

Finally, we solve the T-equation

$$T'' + \frac{c^2 x_{n,m}^2}{a^2} T = 0$$

to get

$$T_{n,m}(t) = c_{n,m} \cos\left(\frac{cx_{n,m}t}{a}\right) + d_{n,m} \sin\left(\frac{cx_{n,m}t}{a}\right).$$

Putting everything together, we have, for each pair n, m, the product solution

$$u_{n,m}(r, \theta, t) = J_n\left(\frac{x_{n,m}r}{a}\right)(C_n \cos n\theta + D_n \sin n\theta)$$

$$\cdot \left[c_{n,m} \cos\left(\frac{cx_{n,m}t}{a}\right) + d_{n,m} \sin\left(\frac{cx_{n,m}t}{a}\right)\right]$$

and, thus, our *general solution is*

$$u(r, \theta, t) = \sum_{n=0}^{\infty} \sum_{m=1}^{\infty} u_{n,m}(r, \theta, t).$$

To determine the constants we must, of course, apply the initial conditions. First,

$$u(r, \theta, 0) = f(r, \theta) = \sum_{n=0}^{\infty} \sum_{m=1}^{\infty} c_{n,m} J_n\left(\frac{x_{n,m}r}{a}\right)(C_n \cos n\theta + D_n \sin n\theta),$$

$$0 \leq r \leq a, -\pi \leq \theta \leq \pi.$$

This means that the series is just a *double Fourier series* for f (where one of the series is actually a *Fourier–Bessel* series—again, see Section 8.5). In fact, proceeding nonrigorously, we can rewrite the equation as

$$f(r,\theta) = \sum_{n=1}^{\infty} C_0 c_{0,m} J_0\left(\frac{x_{0,m}r}{a}\right) + \sum_{n=1}^{\infty}\left\{\left[C_n \sum_{n=1}^{\infty} c_{n,m} J_n\left(\frac{x_{n,m}r}{a}\right)\right]\cos n\theta\right.$$
$$\left.+ \left[D_n \sum_{m=1}^{\infty} c_{n,m} J_n\left(\frac{x_{n,m}r}{a}\right)\right]\sin n\theta\right\},$$

and we see that the series is just the Fourier series in θ, for each (fixed) r, for $f(r,\theta)$ on $-\pi \le \theta \le \pi$. Therefore, for the Fourier coefficients a_n, b_n, we have

$$\frac{a_0}{2} = C_0 \sum_{m=1}^{\infty} c_{0,m} J_0\left(\frac{x_{0,m}r}{a}\right) = \frac{1}{2\pi}\int_{-\pi}^{\pi} f(r,\theta)d\theta,$$

$$a_n = C_n \sum_{m=1}^{\infty} c_{n,m} J_n\left(\frac{x_{n,m}r}{a}\right) = \frac{1}{\pi}\int_{-\pi}^{\pi} f(r,\theta)\cos n\theta\, d\theta,$$

$$b_n = D_n \sum_{n=1}^{\infty} c_{n,m} J_n\left(\frac{x_{n,m}r}{a}\right) = \frac{1}{\pi}\int_{-\pi}^{\pi} f(r,\theta)\sin n\theta\, d\theta, 0 \le r \le a.$$

In turn, each series is a Fourier–Bessel series for the corresponding function of r on the right.

As for the initial velocity, we have, again,

$$u_t(r,\theta,0) = g(r,\theta),$$

which leads to

$$cC_0 \sum_{m=1}^{\infty} d_{0,m} X_{0,m} J_0\left(\frac{x_{0,m}r}{a}\right) = \frac{1}{2\pi}\int_{-\pi}^{\pi} g(r,\theta)d\theta,$$

$$cC_n \sum_{m=1}^{\infty} d_{n,m} x_{n,m} J_n\left(\frac{x_{n,m}r}{a}\right) = \frac{1}{\pi}\int_{-\pi}^{\pi} g(r,\theta)\cos n\theta\, d\theta,$$

$$cD_n \sum_{m=1}^{\infty} d_{n,m} x_{n,m} J_n\left(\frac{x_{n,m}r}{a}\right) = \frac{1}{\pi}\int_{-\pi}^{\pi} g(r,\theta)\sin n\theta\, d\theta$$

(see Exercises 3 and 5).

Phew! Now, remembering that we can write

$$C_n \cos n\theta + D_n \sin n\theta = C\cos n(\theta - \theta_1),$$

for constants C and θ_1, we look at the vibration modes

$$v_{n,m}(r,\theta) = J_n(x_{n,m},r)\cos n\theta,$$

where we have taken $a = 1$, each corresponding to the vibration frequency

$$\nu_{n,m} = \frac{cx_{n,m}}{2\pi a} = \frac{cx_{n,m}}{2\pi}.$$

Figure 9.11 shows the modes for $n = 0, 1, 2$; $m = 1, 2, 3$, while Figure 9.12 gives the corresponding nodal patterns. (Again, $c = 1$.)

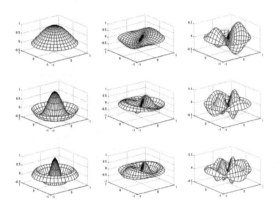

FIGURE 9.11
MATLAB graphs of the modes $V_{n,m} = J_n(x_{n,m}r)\cos n\theta$ for $n = 0, 1, 2$ and $m = 1, 2, 3$. Again, $x_{n,m}$ is the mth positive zero of J_n.[*]

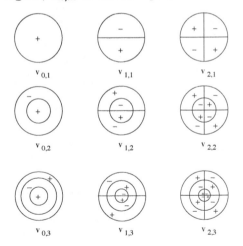

FIGURE 9.12
Nodal lines and signs for the modes in Figure 9.11.

---

[*]Thanks to Udaak Z. George, University of Sussex, UK, for pointing out that this figure was incorrect in the first edition, and for providing us with the correct MATLAB code.

Exercises 9.4

1. Use the results of Exercise 3d, Section 8.5, to solve the vibrating drum problem

$$u_{tt} = \nabla^2 u, \qquad 0 < r < 1, -\infty < \theta < \infty, t > 0,$$
$$u(1, r, \theta) = 0,$$

subject to the initial conditions

a) $u(r, \theta, 0) = 1 - 2r^2,$
   $u_t(r, \theta, 0) = 0$

b) $u(r, \theta, 0) = 0,$
   $u_t(r, \theta, 0) = 1$

c) $u(r, \theta, 0) = 2J_0(k_3 r),$
   $u_t(r, \theta, 0) = J_0(k_1 r) - J_0(k_2 r)$

where k_n is the nth positive zero of J_0.

d) **MATLAB:** Plot the solutions (truncated, if necessary) in r-θ-u space for various times t.

2. a) Solve the θ-independent *heat equation* on a disk,

$$u_t = \alpha^2 \nabla^2 u, \qquad 0 < r < 1, -\infty < \theta < \infty, t > 0,$$
$$u(r, \theta, 0) = f(r),$$
$$u(1, \theta, t) = 0.$$

What is the steady state temperature?

b) Do the same, but, instead, with the *Neumann* boundary condition

$$u_r(1, \theta, t) = 0.$$

(You'll need to look at Exercise 9, Section 7.6.) What is the steady state temperature? Explain, physically, why

$$h(t) = \int_0^{2\pi} \int_0^1 u(r, \theta, t) r \, dr \, d\theta$$

must be constant, and find its value, in terms of f.

3. a) Justify the expressions derived from the initial velocity $g(r, \theta)$ at the end of this section.

b) Write down integrals representing the values of the products C_n $c_{n,m}$; $C_n d_{n,m}$; $D_n c_{n,m}$; and $D_n d_{n,m}$ in the solution for the general vibrating circular membrane problem, for the case $a = 1$.

4. Solve the membrane problem

$$u_{tt} = c^2 \nabla^2 u, \qquad 0 < r < 1, -\infty < \theta < \infty, t > 0,$$
$$u(1, \theta, t) = 0,$$

subject to initial conditions

a) $u(r, \theta, 0) = 5 J_4(x_{4,1} r) \cos 4\theta - J_2(x_{2,3} r) \sin 2\theta,$
   $u_t(r, \theta, 0) = 0$

b) $u(r, \theta, 0) = J_0(x_{0,1} r),$
   $u_t(r, \theta, 0) = J_2(x_{2,1} r) \sin 2\theta$

c) **MATLAB:** Plot both solutions for various times t.

5. **Heat equation on a circular wedge:** Solve the heat equation on a wedge

$$u_t = \nabla^2 u, \qquad 0 < r < 1, 0 < \theta < \alpha < 2\pi,$$
$$u(r, \theta, 0) = f(\theta),$$
$$u(1, \theta, t) = u(r, 0, t) = u(r, \alpha, t) = 0.$$

(The wave equation is solved similarly, of course.)

6. **Electrostatics: θ-independent Dirichlet problem in a cylinder.** In Exercise 17 of the previous section, we considered z-independent solutions of Laplace's equation on an *infinite* cylinder. Now, instead, we look at a *finite* cylinder, with θ-independent boundary conditions. Refer to Figure 9.13 for Exercises 6 and 7.

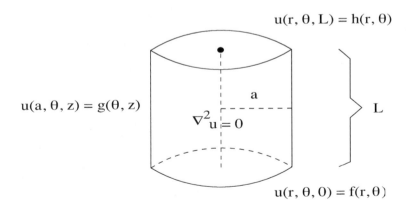

FIGURE 9.13
General Dirichlet problem in a cylinder.

a) Solve

$$\nabla^2 u = 0, \qquad 0 < r < 1, 0 < z < L,$$
$$u(r, \theta, 0) = 0,$$
$$u(r, \theta, L) = h(r),$$
$$u(1, \theta, z) = 0.$$

b) Solve

$$\nabla^2 u = 0, \qquad 0 < r < 1, 0 < z < L,$$
$$u(r, \theta, 0) = 0,$$
$$u(r, \theta, L) = 0,$$
$$u(1, \theta, z) = g(z).$$

(See Exercise 9, Section 7.5.)

c) Solve

$$\nabla^2 u = 0, \qquad 0 < r < 1, 0 < z < L,$$
$$u(r, \theta, 0) = f(r),$$
$$u(r, \theta, L) = 0,$$
$$u(1, \theta, z) = 0.$$

d) Solve

$$\nabla^2 u = 0, \qquad 0 < r < 1, 0 < z < L,$$
$$u(r, \theta, 0) = f(r),$$
$$u(r, \theta, L) = h(r),$$
$$u(1, \theta, z) = g(z).$$

7. **General Dirichlet problem in a cylinder:** Continuing the previous exercise, we now solve Laplace's equation in cylindrical coordinates with general boundary conditions. (See Figure 9.13.)

a) Solve

$$\nabla^2 u = 0, \qquad 0 < r < 1, 0 < z < L,$$
$$u(r, \theta, 0) = f(r, \theta),$$
$$u(r, \theta, L) = 0,$$
$$u(a, \theta, z) = 0.$$

b) Solve

$$\nabla^2 u = 0, \qquad 0 < r < 1, 0 < z < L,$$
$$u(r, \theta, 0) = 0,$$
$$u(r, \theta, L) = 0,$$
$$u(a, \theta, z) = g(\theta, z).$$

(Hint: same as in Exercise 6b)

8. **Radial vibrations of a ball: spherical coordinates.** As we'll see in the following section, if a uniform solid sphere of radius 1 undergoes vibrations in the radial direction, and if its boundary sphere is held in place, then, if $u = u(\rho, \theta, \phi, t)$ in spherical coordinates, u must satisfy

$$u_{tt} = c^2 \nabla^2 u = c^2 \left(u_{\rho\rho} + \frac{2}{\rho} u_\rho \right), \qquad 0 < \rho < 1, t > 0,$$

$$u(\rho, \theta, \phi, 0) = f(\rho),$$
$$u_t(\rho, \theta, \phi, 0) = g(\rho),$$
$$u(1, \theta, \phi, t) = 0.$$

a) Show that the PDE separates, via $u = T(t)R(\rho)$, into the ODEs

$$R'' + \frac{2}{\rho} R' + \lambda R = 0, \qquad T'' + \lambda T = 0,$$

with boundary conditions

$$R(1) = 0, \qquad R(0) < \infty.$$

b) As in Exercise 6b, Section 7.1, show that the substitution $R(\rho) = \rho^{-1/2} y(\rho)$ turns the R-equation into

$$\rho^2 y'' + \rho y' + \left(\lambda \rho^2 - \frac{1}{4} \right) y = 0,$$

$$y(1) = 0, \qquad \lim_{\rho \to 0^+} \rho^{-1/2} y(\rho) < \infty.$$

c) Solve the problem.

9.5 Problems in Spherical Coordinates

Of course, when solving three-dimensional problems with cylindrical or spherical boundaries, it only makes sense to switch to the corresponding type of coordinates. Since we've dealt with cylindrical problems in the exercises, here we concentrate on sphericals. (See Figure 9.14.)

Back in Exercise 2a of Section 7.1, we derived an expression for the Laplacian in spherical coordinates:

$$\nabla^2 u = \frac{1}{\rho^2} \left[(\rho^2 u_\rho)_\rho + \frac{1}{\sin\phi} (u_\phi \sin\phi)_\phi + \frac{1}{\sin^2\phi} u_{\theta\theta} \right].^*$$

****Warning:** Again, we use θ for the *polar* angle (*longitude*) and ϕ for the angle "down" from the vertical (*latitude*).

(We provide a slicker derivation in Exercise 1.) So let's begin by finding the potential inside a sphere, given the potential on the boundary—the *interior Dirichlet problem for a ball*. See Figure 9.15.

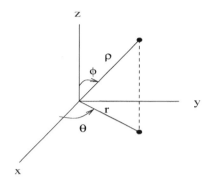

FIGURE 9.14
Spherical coordinates.

Θ-INDEPENDENT DIRICHLET PROBLEM ON A BALL

For a warmup, we first solve the θ-independent problem

$$\nabla^2 u = 0, \qquad 0 < \rho < 1,$$
$$u(1, \theta, \phi) = f(\phi).$$

So we have

$$(\rho^2 u_\rho)_\rho + \frac{1}{\sin \phi}(u_\phi \sin \phi)_\phi = 0,$$

which, upon separation via $u = R(\rho)\Phi(\phi)$, becomes

$$\frac{[\rho^2 R'(\rho)]'}{R(\rho)} = -\frac{[\Phi'(\phi) \sin \phi]'}{\Phi(\phi) \sin \phi} = -\lambda$$

or

$$[\rho^2 R'(\rho)]' - \lambda R(\rho) = 0$$

and

$$[\Phi'(\phi) \sin \phi]' + \lambda \Phi(\phi) \sin \phi = 0.$$

The change to sphericals also makes it necessary to stipulate that Φ be continuous at the poles, that is, at $\phi = 0$ and $\phi = \pi$; so, certainly, Φ must be bounded as $\phi \to 0^+$ and $\phi \to \pi^-$. Now, the Φ-equation looks familiar—in Exercise 2e, Section 7.1, we saw that the substitution $x = \cos \phi$ gives us the problem

$$[(1 - x^2)\Phi']' + \lambda \Phi = 0, \qquad -1 < x < 1,$$
$$\Phi(x) \text{ bounded as } x \to -1^+ \text{ and as } x \to 1^-,$$

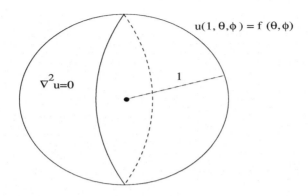

FIGURE 9.15
General Dirichlet problem on a ball of radius 1.

which is just Legendre's equation. In Section 7.2, we saw that the eigenvalues and eigenfunctions are

$$\lambda_n = n(n+1), \qquad n = 0, 1, 2, \ldots$$

and

$$\Phi_n(x) = P_n(x), \qquad n = 0, 1, 2, \ldots,$$

where P_n is the nth degree Legendre polynomial. Thus (being loose with our notation),

$$\Phi_n(\phi) = P_n(\cos\phi), \qquad n = 0, 1, 2, \ldots$$

and, in addition, the R-equation becomes

$$\rho^2 R'' + 2\rho R' - n(n+1)R = 0, \qquad 0 < \rho < 1, n = 0, 1, 2, \ldots,$$
$$R \text{ bounded as } \rho \to 0^+.$$

This, of course, is a Cauchy–Euler equation with general solution

$$R(\rho) = c_1 \rho^n + c_2 \rho^{-1-n}.$$

Since the second solution becomes infinite as $\rho \to 0^+$, we have the solution

$$R_n(\rho) = \rho^n, \qquad n = 0, 1, 2, \ldots.$$

Thus, the *general solution* is

$$u(\rho, \theta, \phi) = \sum_{n=0}^{\infty} c_n \rho^n P_n(\cos\phi).$$

Finally, the boundary condition gives us

$$u(1, \theta, \phi) = f(\phi) = \sum_{n=0}^{\infty} c_n P_n(\cos \phi)$$

and, again letting $x = \cos \theta$, we have

$$f(\cos^{-1} x) = \sum_{n=0}^{\infty} c_n P_n(x).$$

Thus, the c_n are just the Fourier–Legendre coefficients of the function $f(\cos^{-1} x)$,

$$\begin{aligned}
c_n &= \frac{2n+1}{2} \int_{-1}^{1} f(\cos^{-1} x) P_n(x) dx \\
&= \frac{2n+1}{2} \int_{0}^{\pi} f(\phi) P_n(\cos \phi) \sin \phi \; d\phi
\end{aligned}$$

(see Example 2, Section 8.5).

GENERAL DIRICHLET PROBLEM ON A BALL

Okay, now let's generalize to the case where the boundary condition involves a function of both ϕ *and* θ. We'd like to solve

$$\begin{aligned}
\nabla^2 u &= 0, \qquad 0 < \rho < 1, \\
u(1, \theta, \phi) &= f(\theta, \phi).
\end{aligned}$$

Now we have

$$(\rho^2 u_\rho)_\rho + \frac{1}{\sin \phi}(u_\phi \sin \phi)_\phi + \frac{1}{\sin^2 \phi} u_{\theta\theta} = 0.$$

Letting $u(\rho, \theta, \phi) = R(\rho)\Theta(\theta)\Phi(\phi)$ leads to

$$-\frac{[\rho^2 R'(\rho)]'}{R(\rho)} = \frac{1}{\sin \phi} \frac{[\Phi'(\phi) \sin \phi]'}{\Phi(\phi)} + \frac{1}{\sin^2 \phi} \frac{\Theta''(\theta)}{\Theta(\theta)} = -\mu$$

or

$$\rho^2 R'' + 2\rho R' - \mu R = 0$$

and

$$\frac{[\Phi' \sin \phi]' \sin \phi}{\Phi} + \mu \sin^2 \phi = -\frac{\Theta''}{\Theta} = \lambda.$$

Finally, the ODEs for Θ and Φ are

$$\Theta'' + \lambda\Theta = 0, \quad [\Phi' \sin \phi]' \sin \phi + (\mu \sin^2 \phi - \lambda)\Phi = 0,$$

with the usual conditions that Φ be bounded at $\phi = 0$ and $\phi = \pi$, and

$$\Theta(0) = \Theta(2\pi), \quad \Theta'(0) = \Theta'(2\pi).$$

Again, we must have $\lambda_n = n^2$ and

$$\Theta_n(\theta) = c_n \cos n\theta + d_n \sin n\theta, \qquad n = 0, 1, 2, \dots .$$

Then, as in the θ-independent case, we let $x = \cos \phi$ in the Φ-equation, resulting in (see Exercise 11)

$$(1 - x^2)\Phi''(x) - 2x\Phi'(x) + \left(\mu - \frac{n^2}{1 - x^2} \right) \Phi = 0, \qquad -1 < x < 1,$$

Φ bounded as $x \to -1^+$ and $x \to 1^-$.

This, of course, is the **associated Legendre's equation of order n** (see Exercise 4, Section 8.1, and Exercise 4, Section 8.2). It has bounded solutions if and only if $\mu = m(m+1)$, $m = 0, 1, 2, \dots$, and these bounded solutions are the *associated Legendre functions*

$$\Phi(x) = P_m^n(x) = (1 - x^2)^{n/2} P_m^{(n)}(x).$$

So our Φ-solutions are

$$\Phi_{n,m}(\phi) = P_m^n(\cos \phi) = \sin^m(\phi) P_m^{(n)}(\cos \phi) \quad \text{(why?)},$$
$$n = 0, 1, 2, \dots; m = 0, 1, 2, \dots .$$

The R-equation again becomes

$$\rho^2 R'' + 2\rho R' - m(m+1)R = 0, \qquad m = 0, 1, 2, \dots,$$

with bounded solutions

$$R_m(\rho) = \rho^m, \qquad m = 0, 1, 2, \dots .$$

Putting it all together, the general solution to our problem is

$$u(\rho, \theta, \phi) = \sum_{n=0}^{\infty} \sum_{m=n}^{\infty} \rho^m P_m^n(\cos \phi)(c_{n,m} \cos n\theta + d_{n,m} \sin n\theta)^\dagger$$

(see Exercise 7).

Finally, we apply the boundary conditions

$$u(1, \theta, \phi) = f(\theta, \phi) = \sum_{n=0}^{\infty} \sum_{m=n}^{\infty} P_m^n(\cos \phi)(c_{n,m} \cos n\theta + d_{n,m} \sin n\theta),$$

---

†Why does the second summation start at n and not at zero?

and we may solve (unrigorously!) for the constants as in our treatment of the wave equation in the previous section. (See Exercise 5.)

Completeness of the functions P_m^n (for each fixed n) on $[-1, 1]$ and of the set $\{1, \cos nx, \sin nx\}_{n=1}^{\infty}$ on $[-\pi, \pi]$ means that the functions

$$C_m^0(\theta, \phi) = P_m(\cos \phi), \qquad m = 0, 1, 2, \ldots;$$
$$C_m^n(\theta, \phi) = P_m^n(\cos \phi) \cos n\theta, \qquad m = 0, 1, 2, \ldots; n = 1, 2, 3, \ldots;$$
$$S_m^n(\theta, \phi) = P_m^n(\cos \phi) \sin n\theta, \qquad m = 0, 1, 2 \ldots; n = 1, 2, 3, \ldots$$

are *complete on the sphere*. These are the so-called **spherical harmonics**, and we'll discuss them further at the end of this section, after we solve the heat equation in spherical coordinates (the solution of which *also* involves these spherical harmonics). Actually, it turns out that the functions in our solution—the so-called **solid harmonics**—can be written in Cartesian coordinates as polynomials in x, y and z! See Exercise 12.

DIFFUSION OF HEAT IN A BALL

Suppose we have a ball of radius 1, made of a homogeneous material with constant mass density and with the temperature of the boundary held at $0°$. Then the temperature distribution satisfies the problem (assuming that the thermal diffusivity is $\alpha^2 = 1$)

$$u_t = \nabla^2 u, \qquad 0 < \rho < 1,$$
$$u(\rho, \theta, \phi, 0) = f(\rho, \theta, \phi) \qquad \text{(initial temperature distribution)},$$
$$u(1, \theta, \phi, t) = 0.$$

We begin to solve this problem by separating variables, as in Exercise 6, Section 8.1. Letting

$$u = T(t)v(\rho, \theta, \phi)$$

leads to the equation

$$T' + \gamma T = 0$$

and the Helmholtz equation

$$\nabla^2 v + \gamma v = 0.$$

Separating the latter via $v(\rho, \theta, \phi) = R(\rho)H(\theta, \phi)$, we get

$$-\frac{[\rho^2 R'(\rho)]'}{R(\rho)} - \gamma \rho^2 = \frac{\frac{1}{\sin \phi}(H_\phi \sin \phi)_\phi + \frac{1}{\sin^2 \phi}H_{\theta\theta}}{H} = -\mu,$$

which, apart from the $-\gamma \rho^2$ term, is exactly what we had above, for Laplace's equation. In particular, we must have $\mu = m(m+1)$, $m = 0, 1, 2, \ldots$, and the functions are, again, the *spherical harmonics*

$$H_m^n(\theta, \phi) = P_m^n(\cos\phi)(c_{n,m}\cos n\theta + d_{n,m}\sin n\theta),$$
$$n = 0, 1, 2, \ldots; m = 0, 1, 2, \ldots$$

(although, as above, we need only look at $m = n, n+1, \ldots$).
   Then, the R-equation is

$$\rho^2 R'' + 2\rho R' + [\gamma\rho^2 - m(m+1)]R = 0$$

with boundary conditions

$$R(0) < \infty, \quad R(1) = 0.$$

This equation, which we met in Exercise 4, Section 7.1, is *almost* Bessel's equation and is called the **spherical Bessel's equation**. The substitution

$$w(\rho) = \rho^{1/2}R(\rho)$$

turns it into

$$\rho^2 w'' + \rho w' + (\gamma\rho^2 - \alpha^2)w = 0,$$

where $\alpha^2 = m(m+1) + \frac{1}{4} = \left(m + \frac{1}{2}\right)^2$. The new boundary conditions are

$$w(0) = 0 \quad (\text{why?}), \quad w(1) = 0.$$

This w-equation was solved earlier (where?). We must have $\gamma = y_{m,\ell}^2$, and the resulting solutions are

$$w_{m,\ell}(\rho) = J_{m+\frac{1}{2}}(y_{m,\ell}\rho), \qquad m = 0, 1, 2, \ldots; \ell = 1, 2, 3, \ldots,$$

where

$$y_{m,\ell} = x_{m+\frac{1}{2},\ell} = \text{the } \ell^{\text{th}} \text{ positive zero of } J_{m+\frac{1}{2}}.$$

It follows that

$$R_{m,\ell}(\rho) = \frac{1}{\sqrt{y_{m,\ell}\rho}} J_{m+\frac{1}{2}}(y_{m,\ell}\rho).$$

Finally, the T-equation is

$$T' + \gamma T = T' + \ell^2 T = 0$$

with solution

$$T_k(t) = e^{-\ell^2 t}.$$

Putting it all together, we have the product solutions

$$u_{n,m,\ell} = e^{-\ell^2 t} H_m^n(\theta, \phi) \cdot \frac{1}{\sqrt{\rho}} J_{m+\frac{1}{2}}(y_{m,\ell}\rho)$$

$$\ell = 1, 2, 3, \ldots, n = 0, 1, 2, \ldots, m = n, n+1, n+2, \ldots$$

(what happened to the $\sqrt{y_{m,k}}$ term?), and we form the general solution and then apply the initial condition as always (i.e., so long as we have all the right orthogonality relationships. See Exercise 16).

The new special functions which were solutions of the spherical Bessel's equation are called, not surprisingly, the **spherical Bessel functions of the first kind**, and we write

$$j_n(x) = \sqrt{\frac{\pi}{2x}} J_{n+\frac{1}{2}}(x), \qquad n = 0, 1, 2, \ldots .$$

(As usual, the $\sqrt{\frac{\pi}{2}}$ is a *normalization factor*, to make $j_0(0) = 1$.) We *almost* met $j_0(x)$ back in Exercise 8b, Section 7.3, where we showed that the general solution to Bessel's equation of order $1/2$ is

$$y = \frac{1}{\sqrt{x}}(c_1 \cos x + c_2 \sin x).$$

In fact, we can show that

$$j_0(x) = \frac{\sin x}{x}$$

and, using recurrence formulas, we can write each $j_n(x)$ in a closed form which involves only $\sin x$, $\cos x$ and powers of x (see Exercises 11–13). We graph the first four spherical Bessels in Figure 9.16.

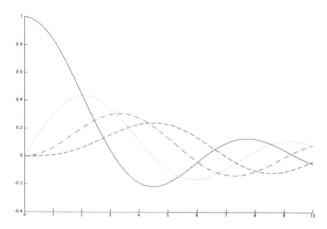

FIGURE 9.16
MATLAB graphs of the spherical Bessel functions j_0, j_1, j_2 and j_3 (solid, dotted, dash-dotted and dashed, respectively).

Now, back to those *other* important special functions, the spherical harmonics. First, we already have the usual *harmonic functions*, the solutions of Laplace's equation. So we have the so-called

Rectangular harmonics: $\sin nx \sinh ny$, etc.

Harmonics on a disk: $r^n \cos n\theta$, $r^n \sin n\theta$

Solid harmonics (in a ball): $\rho^n C_m^n(\theta, \phi)$, $\rho^n S_m^n(\theta, \phi)$.

Now, the spherical harmonics are *not* harmonic functions. So what are they? They are what we might call *boundary harmonics*, in the following sense. When solving Laplace's equation on the unit square, say, we arrive at the ODE $X'' + n^2 X = 0$ (or $Y'' + n^2 Y = 0$). This, of course, is just the one-dimensional Helmholtz equation. The solutions $\{\sin nx\}_{n=1}^{\infty}$ or $\{\cos nx\}_{n=0}^{\infty}$ are complete on the interval $0 \le x \le 1$, i.e., in a sense, on the *x-boundary* of the square. We might call these functions *linear harmonics* (although we don't).

When solving Laplace's equation on the *disk*, we separated variables to get $\Theta'' + n^2\Theta = 0$, again, the Helmholtz equation. The solutions $\{\cos n\theta, \sin n\theta\}_{n=0}^{\infty}$ form a complete set along the boundary of the disk (and, actually, along *any* circle centered at the origin). We could call them *circular harmonics*.

Finally, the spherical harmonics are just solutions of $\nabla^2 v(\theta, \phi) + n^2 v(\theta\phi) = 0$, once more the Helmholtz equation. And the solutions $S_m^n(\theta, \phi)$ are complete *on the boundary sphere* (and, in fact, on any sphere centered at the origin).

One final note: the spherical harmonics corresponding to $n = 0$ are, again, just the θ-independent **zonal harmonics**

$$C_m^0(\theta, \phi) = P_m(\cos \phi).$$

(Why "zonal"?)

Exercises 9.5

1. **Laplacian in spherical coordinates:** Again, the Laplacian in cylindricals is

$$\nabla^2 u = u_{xx} + u_{yy} + u_{zz} = u_{rr} + \frac{1}{r}u_r + \frac{1}{r^2}u_{\theta\theta} + u_{zz}.$$

Since θ is the same in cylindricals and sphericals, we must deal only with the r and z terms, as follows (see Figure 9.14).

a) Conclude from the figure that

$$z = \rho \cos \phi$$

and

$$r = \rho \sin \phi.$$

Thus, transforming from (x, y) to (r, θ) should involve the same process as transforming from (z, r) to (ρ, ϕ).

b) Conclude that

$$u_{zz} + u_{rr} = u_{\rho\rho} + \frac{1}{\rho}u_\rho + \frac{1}{\rho^2}u_{\phi\phi}.$$

c) Show also that

$$u_r = \frac{r}{\rho}u_\rho + \frac{\cos\phi}{\rho}u_\phi$$

and, thus, that

$$\nabla^2 u = u_{\rho\rho} + \frac{1}{\rho}u_\rho + \frac{1}{\rho^2}(u_{\phi\phi} + u_\phi \cot\phi + u_{\theta\theta}\cos^2\phi).$$

d) Show that this is equivalent to

$$\nabla^2 u = \frac{1}{\rho^2}\left[(\rho^2 u_\rho)_\rho + \frac{1}{\sin\phi}(u_\phi \sin\phi)_\phi + \frac{1}{\sin^2\phi}u_{\theta\theta}\right].$$

2. Solve the ODE $\rho^2 R'' + 2\rho R' - n(n+1)R = 0$, $n = 0, 1, 2, \ldots$.

3. Solve the Laplace equation inside the sphere $\rho = 1$ subject to the boundary condition

   a) $u(1,\theta,\phi) = 3P_5(\cos\phi) - 7P_2(\cos\phi)$
   b) $u(1,\theta,\phi) = \cos^2\phi$
   c) $u(1,\theta,\phi) = -\sin\theta + 2P_7^4(\cos\phi)\cos 4\theta$

4. a) Solve, formally, the *exterior* Dirichlet problem

$$\nabla^2 u = 0, \quad \rho > 1,$$
$$u(1,\theta,\phi) = f(\phi).$$

   b) Solve the interior and exterior Dirichlet problems

$$\nabla^2 u = 0, \quad 0 < \rho < a,$$
$$u(a,\theta,\phi) = f(\phi)$$

   and

$$\nabla^2 u = 0, \quad \rho > a,$$
$$u(a,\theta,\phi) = f(\phi).$$

5. It can be shown that the orthogonality relationship of the associated Legendre functions is

$$\int_{-1}^{1} P_m^n(x)P_k^n(x)dx = \begin{cases} 0, & \text{if } m \neq k, \\ \dfrac{2(n+m)!}{(2m+1)(m-n)!}, & \text{if } m = k. \end{cases}$$

Use this to determine the constants $c_{n,m}$ and $d_{n,m}$ in the solution of Laplace's equation,

$$u(\rho,\theta,\phi) = \sum_{n=0}^{\infty}\sum_{m=n}^{\infty} \rho^m P_m^n(\cos\phi)(c_{n,m}\cos n\theta + d_{n,m}\sin n\theta)$$

(as integrals involving $u(1,\theta,\phi) = f(\theta,\phi)$).

6. Reverse the order of summation in the series

$$\sum_{n=0}^{\infty}\sum_{m=n}^{\infty} a_{n,m}$$

and use mathematical induction to prove your result.

7. Find the general solution of the *exterior* Dirichlet problem

$$\nabla^2 u = 0, \quad 1 < \rho < \infty,$$
$$u(1,\theta,\phi) = f(\theta,\phi).$$

8. Write down the general solution of the three-dimensional vibration problem

$$u_{tt} = \nabla^2 u, \quad 0 < \rho < 1,$$

subject to the boundary condition

$$u(1,\theta,\phi,t) = 0.$$

9. Show that the substitution $x = \cos\phi$ changes the ϕ-problem

$$[\Phi'\sin\phi]'\sin\phi + (\mu\sin^2\phi - \lambda)\Phi = 0, \quad 0 < \phi < \pi,$$
$$\Phi \text{ bounded as } \phi \to 0^+, \pi^-$$

to the x-problem

$$(1-x^2)\Phi'' - 2x\Phi' + \left(\mu - \frac{\lambda}{1-x^2}\right)\Phi = 0, \quad -1 < x < 1,$$
$$\Phi \text{ bounded as } x \to -1^+, 1^-.$$

10. In Exercise 6 of Section 7.6, we showed that we may write the *associated Legendre functions* in the form

$$P_m^n(x) = \frac{1}{2^m m!}(1-x^2)^{n/2}\frac{d^{m+n}}{dx^{m+n}}[(x^2-1)^m].$$

Use this expression (and Exercise 3, Section 7.2, if necessary) to answer the following:

a) Write $\rho^2 P_2^1(\cos\phi)\cos\theta$ as a polynomial in x, y and z.

b) Do the same for $\rho^3 P_3^2(\cos\phi)\sin 2\theta$.

c) Do the same for $\rho P_1^3(\cos\phi)\cos 3\theta$.

d) Do the same for $\rho^3 P_3^3(\cos\phi)\sin 3\theta$.

e) What is the degree, in x, y and z, of the polynomial

$$\rho^m P_m^n(\cos\phi)\cos n\theta?$$

f) What about

$$\rho^m P_m^n(\cos\phi)\sin n\theta?$$

g) What is the degree of the z-part of each of these polynomials?

11. Show that the *spherical Bessel function of the first kind of order* 0 can be written

$$j_0(x) = \frac{\sin x}{x}$$

a) By using the fact that the general solution of Bessel's equation of order $1/2$ is

$$y = \frac{1}{\sqrt{x}}(c_1 \cos x + c_2 \sin x)$$

b) By using the Frobenius series solution for $J_{1/2}(x)$ and the result of Exercise 4d, Section 7.5.

12. a) From Exercise 5, Section 7.5, conclude that

$$x J_\alpha'(x) = x J_{\alpha-1}(x) - \alpha J_\alpha(x).$$

b) Use part (a) to show that

$$j_1(x) = \frac{1}{x^2}\sin x - \frac{1}{x}\cos x.$$

c) Now use the recurrence formulas in Exercise 5d of that same section to compute $j_2(x)$ and $j_3(x)$.

13. Continuing our discussion of the spherical Bessel functions,

a) Rewrite the identity in Exercise 12a as

$$x^{-(\alpha+1)} J_{\alpha+1}(x) = -\frac{1}{x}\frac{d}{dx}[x^{-\alpha}J_\alpha(x)],$$

and use mathematical induction to show that we must then have

$$x^{-(\alpha+n)} J_{\alpha+n}(x) = \left(-\frac{1}{x}\right)^n \frac{d^n}{dx^n}[x^{-\alpha}J_\alpha(x)], \quad n = 1, 2, 3, \ldots .$$

b) Conclude that

$$x^{-\left(\frac{1}{2}+n\right)} J_{n+\frac{1}{2}}(x) = \sqrt{\frac{2}{\pi}} \left(-\frac{1}{x}\right)^n \frac{d^n}{dx^n} \left[\frac{\sin x}{x}\right].$$

c) Use the relationship in part (b) to compute $j_1(x)$, $j_2(x)$ and $j_3(x)$ (and make sure your answers match those in the previous exercise).

14.   a) Show that the spherical Bessel functions satisfy the orthogonality relationship

$$\int_0^1 x^2 j_n(y_{n,k}x) j_n(y_{n,j}x)dx = \begin{cases} 0, & \text{if } k \neq j, \\ \dfrac{1}{2} j_{n+1}^2(y_{n,k}), & \text{if } k = j, \end{cases}$$

and, more generally, that

$$\int_0^L x^2 j_n(y_{n,k}x) j_n(y_{n,j}x)dx = \begin{cases} 0, & \text{if } k \neq j, \\ \dfrac{L^3}{2} j_{n+1}^2(y_{n,k}), & \text{if } k = j. \end{cases}$$

b) Show that the spherical harmonics are simply orthogonal on the sphere, that is, show that

$$\iint_S C_{m_1}^{n_1}(\theta, \phi) C_{m_2}^{n_2}(\theta, \phi)dS = 0 \text{ if } m_1 \neq m_2 \text{ or } n_1 \neq n_2,$$

$$\iint_S S_{m_1}^{n_1}(\theta, \phi) S_{m_2}^{n_2}(\theta, \phi)dS = 0, \text{ if } m_1 \neq m_2 \text{ or } n_1 \neq n_2,$$

$$\iint_S C_{m_1}^{n_1}(\theta, \phi) S_{m_2}^{n_2}(\theta, \phi)dS = 0 \text{ for all } m_1, m_2, n_1 \text{ and } n_2,$$

and

$$\iint_S [C_m^n(\theta, \phi)]^2 dS = \iint_S [S_m^n(\theta, \phi)]^2 dS$$

$$= \begin{cases} \dfrac{4\pi}{2m+1}, & \text{if } n = 0, \\ \dfrac{2\pi(m+n)!}{(2m+1)(m-n)!} & \text{if } n > 0, \end{cases} \quad 0 \leq n \leq m.$$

Here, S is *any* sphere centered at the origin. (Note: One often sees the spherical harmonics in complex form,

$$Y_m^n(\theta, \phi) = P_m^n(\cos \phi)e^{in\theta}, \qquad n = \ldots, -2, -1, 0, 1, 2, \ldots,$$

where we define $P_m^n = P_m^{-n}$ for $n < 0$.)

c) Show that the eigenfunctions of the spherical heat and wave equations are simply orthogonal on the ball of radius 1, that is, show that

$$\iiint_V w_{k_1 m_1 n_1}(\rho, \theta, \phi) w_{k_2 m_2 n_2}(\rho, \theta, \phi) dV = 0$$

if $k_1 \neq k_2$ or $m_1 \neq m_2$ or $n_1 \neq n_2$, where V is the unit ball and

$$w_{kmn}(\rho, \theta, \phi) = S_m^n(\theta, \phi) j_m(y_{m,k}\rho).$$

(Again, $Y_{m,k}$ is the kth positive root of $J_{m+\frac{1}{2}}$.)

Use the results of parts (a) and (b) of the previous exercise.

15. **Quantum mechanics and the hydrogen atom, revisited:** Back in Exercise 4, Section 7.1, we saw that the wave function ψ for the electron in a hydrogen atom satisfies Schrödinger's equation,

$$i\hbar\psi_t = -\frac{\hbar^2}{2m}\nabla^2\psi - \frac{e^2}{\rho}\psi.$$

Again, let's set $\hbar = m - \ell = 1$, so that we must solve

$$i\psi_t = -\frac{1}{2}\nabla^2\psi - \frac{1}{\rho}\psi.$$

In addition, we must have

$$\lim_{\rho\to\infty}\psi = 0.$$

(To be more precise, we really need $\iiint \psi \, dV < \infty$, where the integral is taken over all of three-dimensional space, for each t.)

a) Separate the variables by setting

$$\psi(\rho, \theta, \phi, t) = T(t)R(\rho)u(\theta, \phi).$$

Show that the result can be written as

$$T' - \frac{i\lambda}{2}T = 0,$$

$$\frac{1}{\sin\phi}(u_\phi \sin\phi)_\phi + \frac{1}{\sin^2\phi}u_{\theta\theta} + \mu u = 0,$$

$$R'' + \frac{2}{\rho}R' + \left(\lambda + \frac{2}{\rho} - \frac{\mu}{\rho^2}\right)R = 0.$$

b) Explain why $u = \ell(\ell+1)$, $\ell = 0, 1, 2, \ldots$.

c) Given that $\lambda = -\beta^2$, $\beta > 0$, for physical reasons, follow Exercises 4 and 5, Section 8.1, to show that the only bounded solutions of the R-equation occur when

$$\frac{1}{\beta} + \ell = m = 1, 2, 3, \ldots .$$

d) Show that the product solutions of the problem are

$$e^{-\frac{t}{(m-\ell)^2}} H_n^\ell(\theta, \phi) \left(\frac{\rho}{m-\ell} \right)^\ell e^{-\frac{\rho}{m-\ell}} L_m^{2\ell+1} \left(\frac{\rho}{m-\ell} \right),$$

$$n = 0, 1, 2, \ldots; \ell = 0, 1, 2, \ldots; 2\ell + 1 \leq m = 1, 2, 3, \ldots,$$

where, again, the H_n^ℓ are the spherical harmonics and the L_m^k are the associated Laguerre polynomials.

9.6 The Infinite Wave Equation and Multiple Fourier Transforms

In this final section of the chapter, we'd like to solve the two- and three-dimensional analog of the problem of the infinite string. To that end, we'll need to rely on the two- and three-dimensional versions of the Fourier transform.

To begin, though, let's go back to the infinite string. We have

$$u_{tt} = c^2 u_{xx}, \qquad -\infty < x < \infty, t > 0,$$
$$u(x, 0) = f(x),$$
$$u_t(x, 0) = g(x),$$
$$\lim_{|x| \to \infty} u(x, t) = 0.$$

We may solve this problem as we did in Exercise 5, Section 6.4, by using the Fourier transform

$$U(\alpha, t) = \frac{1}{\sqrt{2\pi}} \int_{-\infty}^{\infty} u(x, t) e^{-i\alpha x} \, dx$$

and inverse transform

$$u(x, t) = \frac{1}{\sqrt{2\pi}} \int_{-\infty}^{\infty} U(\alpha, t) e^{i\alpha x} \, d\alpha.$$

Transforming the PDE and initial condition, we have

$$U_{tt} + c^2\alpha^2 U = 0,$$
$$U(\alpha, 0) = F(\alpha) = \text{Fourier transform of } f(x),$$
$$U_t(\alpha, 0) = G(\alpha) = \text{Fourier transform of } g(x).$$

Solving the "ODE" in t gives us

$$U(\alpha, t) = c_1(\alpha) \cos c\alpha t + c_2(\alpha) \sin c\alpha t,$$

and, applying the initial conditions, we find that the transform of our solution is

$$U(\alpha, t) = F(\alpha) \cos c\alpha t + \frac{G(\alpha)}{c\alpha} \sin c\alpha t,$$

so our solution is the integral

$$u(x, t) = \frac{1}{\sqrt{2\pi}} \int_{-\infty}^{\infty} \left[F(\alpha) \cos c\alpha t + \frac{G(\alpha)}{c\alpha} \sin c\alpha t \right] e^{i\alpha x} \, d\alpha.$$

Not very illuminating as it stands. But we can do better. First, notice that the first term looks like the t-derivative of the second term (with F replacing G, of course). That is,

$$\frac{\partial}{\partial t} \int_{-\infty}^{\infty} \frac{H(\alpha)}{c\alpha} \sin c\alpha \, t e^{i\alpha x} \, d\alpha = \int_{-\infty}^{\infty} H(\alpha) \cos c\alpha \, t e^{i\alpha x} \, d\alpha.^{\ddagger}$$

Thus, we may either simplify the F-part and integrate by t or simplify the G-part and differentiate by t. We choose the latter route, only because it will shed light on our attempt to solve the *three*-dimensional wave problem later in this section.

So we have

$$\int_{-\infty}^{\infty} G(\alpha) e^{i\alpha x} \frac{\sin c\alpha t}{c\alpha} \, d\alpha$$

$$= \frac{1}{2} \int_{-\infty}^{\infty} G(\alpha) e^{i\alpha x} \frac{e^{i\alpha ct} - e^{-i\alpha ct}}{ic\alpha} \, d\alpha$$

$$= \frac{1}{2c} \int_{-\infty}^{\infty} G(\alpha) \int_{-ct}^{ct} e^{i\alpha(x+\xi)} \, d\xi d\alpha \quad (\text{why?})$$

$$= \frac{1}{2c} \int_{-ct}^{ct} \int_{-\infty}^{\infty} G(\alpha) e^{i\alpha(x+\xi)} \, d\alpha d\xi \,^{\S}$$

---

‡So long as we may differentiate under the integral sign.
§So long as we may switch the order of integration.

$$= \frac{1}{2c} \int_{-ct}^{ct} g(x + \xi)d\xi \quad \text{(again, why?)}$$

$$= \frac{1}{2c} \int_{x-ct}^{x+ct} g(\tau)d\tau.$$

This looks familiar, of course! (We could have done this more succinctly, but our approach here is illustrative of what will happen in the three-dimensional case.) Now, replacing G by F, we have

$$\int_{-\infty}^{\infty} F(\alpha) \cos c\alpha t \, e^{-i\alpha x} \, d\alpha = \frac{\partial}{\partial t} \left[\frac{1}{2c} \int_{x-ct}^{x+ct} f(\tau)d\tau \right]$$

$$= \frac{1}{2}[f(x+ct) + f(x - ct)].$$

Next, remembering that $\frac{1}{b-a} \int_a^b h(x)dx$ is the average value of h on $[a, b]$, we see that

$$\frac{1}{2c} \int_{x-ct}^{x+ct} g(\tau)d\tau = t \cdot \bar{g},$$

where $\bar{g}$ is the average value of g on $[x - ct, x + ct]$. Hence, our solution can be written as

$$u(x,t) = \frac{1}{\sqrt{2\pi}} \frac{\partial}{\partial t}(t\bar{f}) + t\bar{g}.$$

Now, let's look again at the wave interpretation of d'Alembert's solution—in this case, though, we choose a point x_0 on the string and see how it is affected by the initial disturbance. First, suppose that the initial shape $f(x)$ is the square wave

$$f(x) = \begin{cases} 1, & \text{if } -1 \le x \le 1, \\ 0, & \text{otherwise}, \end{cases}$$

as in Figure 9.17, and the initial velocity is $g(x) \equiv 0$. Taking x_0 as in that figure, it should be clear that it is affected only by the wave moving to the right,

$$\frac{1}{2}f(x - ct).$$

By watching the wave as it moves right and passes by the point $x = x_0$ (see Figure 9.18), or by looking at the growing interval $(x_0 - ct, x_0 + ct)$ and watching it *hit* the initial disturbance (see Figure 9.19), we find that we have

$$u(x_0, t) = \begin{cases} 0, & \text{if } 0 \le t < \dfrac{x_0 - 1}{c}, \\ \dfrac{1}{2}, & \text{if } \dfrac{x_0 - 1}{c} \le t \le \dfrac{x_0 + 1}{c}, \\ 0, & \text{if } t > \dfrac{x_0 + 1}{c}. \end{cases}$$

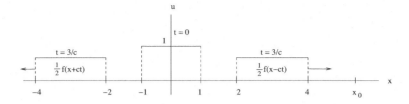

FIGURE 9.17
Half of initial square wave heading toward the point $x = x_0$ at velocity c.

So the wave *hits x_0 sharply* and *leaves just as sharply*. We say that the disturbance has **sharp leading and trailing edges**, and we see that it travels *at* velocity c.

Next we look, instead, at the case where the initial shape is $f(x) \equiv 0$, but the initial velocity is the same square wave

$$g(x) = \begin{cases} 1, & \text{if } -1 \leq x \leq 1, \\ 0, & \text{otherwise.} \end{cases}$$

Now the situation is more complicated and, in particular, it's not clear at all if this disturbance travels in the same way as an initial displacement. However, we *still* may use the second interpretation from above and see what happens as the interval $(x_0 - ct, x_0 + ct)$ grows larger. As we look again at Figure 9.20, we see that the situation is quite different here. To be precise, we have

$$u(x_0, t) = \begin{cases} 0, & \text{if, } 0 \leq t < \dfrac{x_0 - 1}{c}, \\ \dfrac{1}{2c} \int_{x_0 - ct}^{1} d\xi = \dfrac{1 - (x_0 - ct)}{2c}, & \text{if } \dfrac{x_0 - 1}{c} \leq t \leq \dfrac{x_0 + 1}{c}, \\ \dfrac{1}{c}, & \text{if } t > \dfrac{x_0 + 1}{c}. \end{cases}$$

The leading edge is again sharp (although it doesn't hit full-force but, instead, increases gradually from zero). However, *there is no trailing edge in this case*.

More generally, it should be clear that the initial disturbance will eventually disappear at x_0 only if $\int_{-1}^{1} g(x)dx = 0$. The difference, of course, is that in the case $f(x) \neq 0$ and $g(x) = 0$, the solution is affected only by what happens *at* the point $x = x_0 - ct$, while in the case where $g(x) \neq 0$, the solution encompasses the interval $(x_0 - ct, x_0 + ct)$.

In any event, the initial-velocity disturbance in one dimension behaves very differently from the initial-position disturbance, and, in particular, the former does *not* possess a sharp trailing edge. We say that the wave equation, in a given dimension, satisfies **Huygens's Principle** if *every* disturbance travels in a way that its trailing edge is *sharp*. Hence, Huygens's Principle does *not* hold in one dimension.

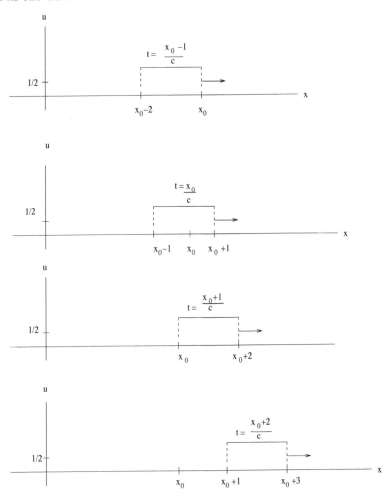

FIGURE 9.18
Half square wave "hitting" and passing the point $x = x_0$.

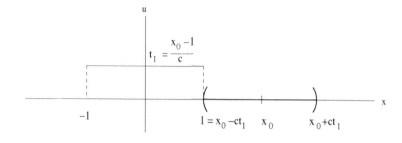

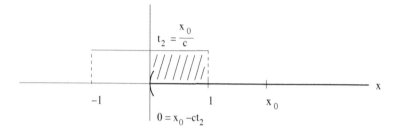

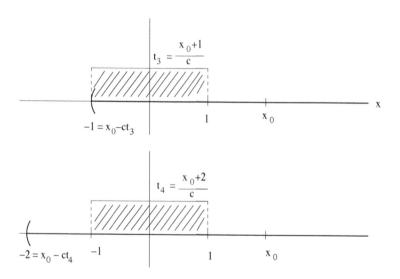

FIGURE 9.19
Interval centered at $x = x_0$ expanding at velocity c.

So what happens in higher dimensions? Our everyday experience with light and sound waves suggests that Huygens's Principle may be true in dimension three, while the ripples from our dropping a pebble in a pool may imply that it is *not* true in two dimensions. Here we must introduce the Fourier transform in two and three (and higher) dimensions. We may do this nonrigorously just by transforming each space variable separately. So, given a function $f(x, y)$ on $-\infty < x < \infty$, $-\infty < y < \infty$, we fix y and look at the x-Fourier transform pair:

$$F_1(\alpha, y) = \frac{1}{\sqrt{2\pi}} \int_{-\infty}^{\infty} f(x, y) e^{i\alpha x} \, dx,$$

$$f(x, y) = \frac{1}{\sqrt{2\pi}} \int_{-\infty}^{\infty} F_1(\alpha, y) e^{-i\alpha x} \, d\alpha.$$

Next, look at the y-Fourier transform of F_1:

$$F(\alpha, \beta) = \frac{1}{\sqrt{2\pi}} \int_{-\infty}^{\infty} F_1(\alpha, y) e^{i\beta y} \, dy,$$

$$= \frac{1}{2\pi} \int_{-\infty}^{\infty} \int_{-\infty}^{\infty} f(x, y) e^{i(\alpha x + \beta y)} \, dx dy,$$

$$F_1(\alpha, y) = \frac{1}{\sqrt{2\pi}} \int_{-\infty}^{\infty} F(\alpha, \beta) e^{-i\beta y} \, dy,$$

from which we get

$$\mathcal{F}[f(x, y)] = F(\alpha, \beta) = \frac{1}{2\pi} \int_{-\infty}^{\infty} \int_{-\infty}^{\infty} f(x, y) e^{i(\alpha x + \beta y)} \, dx dy$$

and

$$\mathcal{F}^{-1}[F(\alpha, \beta)] = f(x, y) = \frac{1}{2\pi} \int_{-\infty}^{\infty} \int_{-\infty}^{\infty} F(\alpha, \beta) e^{-i(\alpha x + \beta y)} \, dx dy.$$

We may, of course, follow the same procedure for dimensions three and higher. We have

$$\mathcal{F}[f(x, y, z)] = F(\alpha, \beta, \gamma)$$

$$= \frac{1}{(2\pi)^{3/2}} \int_{-\infty}^{\infty} \int_{-\infty}^{\infty} \int_{-\infty}^{\infty} f(x, y, z) e^{-i(\alpha x + \beta y + \gamma z)} \, dx dy dz$$

and

$$\mathcal{F}^{-1}[F(\alpha, \beta, \gamma)] = f(x, y, z)$$

$$= \frac{1}{(2\pi)^{3/2}} \int_{-\infty}^{\infty} \int_{-\infty}^{\infty} \int_{-\infty}^{\infty} F(\alpha, \beta, \gamma) e^{i(\alpha x + \beta y + \gamma x)} \, d\alpha d\beta d\gamma$$

and, more generally, for n dimensions,

$$\mathcal{F}[f(\boldsymbol{x})] = F(\boldsymbol{\alpha}) = \frac{1}{(2\pi)^{3/2}} \int_{-\infty}^{\infty} \int_{-\infty}^{\infty} \cdots \int_{-\infty}^{\infty} f(\boldsymbol{x}) e^{-i\boldsymbol{\alpha} \cdot \boldsymbol{x}} \, d\boldsymbol{x}$$

and

$$\mathcal{F}^{-1}[F(\boldsymbol{\alpha})] = f(\boldsymbol{x}) = \frac{1}{(2\pi)^{3/2}} \int_{-\infty}^{\infty} \int_{-\infty}^{\infty} \cdots \int_{-\infty}^{\infty} F(\boldsymbol{\alpha}) e^{i\boldsymbol{\alpha}\cdot\boldsymbol{x}} \, d\boldsymbol{\alpha}.$$

Here, of course, the vectors $\boldsymbol{x}$ and $\boldsymbol{\alpha}$ represent

$$\boldsymbol{x} = (x_1, x_2, \dots, x_n), \boldsymbol{\alpha} = (\alpha_1, \alpha_2, \dots, \alpha_n).$$

It's also straightforward to generalize many of the important properties of Fourier transforms to n dimensions. In particular, we still have the formula for transforms of derivatives,

$$\mathcal{F}\left[\frac{\partial f}{\partial x_j}(\boldsymbol{x})\right] = i\alpha_j F(\boldsymbol{\alpha}),$$

and, as we saw in Exercise 13c of Section 6.4, the inverse transform of a product is a convolution (appropriately defined).

We'll bypass the two-dimensional equation and go right to three dimensions. Then we'll get the 2-D solution almost for free, via the so-called **method of descent** (by which we *descend* from three dimensions to two dimensions, more or less by letting $z = 0$).

Before doing so, we look at an example, where we transform the three-dimensional analog of the square wave defined on the one-dimensional "sphere" $-1 \le x \le 1$. The resulting transform also will help us solve our general three-dimensional wave problem, as it turns out.

Example 1 Find the Fourier transform of the function

$$f(x, y, z) = \begin{cases} 1, & \text{if } x^2 + y^2 + z^2 \le R^2, \\ 0, & \text{otherwise.} \end{cases}$$

We have

$$F(\alpha, \beta, \gamma) = \frac{1}{(2\pi)^{3/2}} \iiint_{x^2+y^2+z^2 \le R^2} e^{-i(\alpha x + \beta y + \gamma z)} \, dx\,dy\,dz.$$

We cannot integrate this in Cartesians coordinates and, of course, the natural inclination is to switch to sphericals. This doesn't work either, as it stands (try it!), but what *does* work is first to rotate the coordinate system so that the vector $\boldsymbol{\alpha} = (\alpha, \beta, \gamma)$ points in the positive z-direction and *then* to use sphericals. As bad as this may sound, it really is quite easy in this particular case, for we wish to define the polar angle ϕ at the point (x, y, z) to be the angle between the vectors $\boldsymbol{\alpha} = (\alpha, \beta, \gamma)$ and $\boldsymbol{x} = (x, y, z)$ and—this is the crux—the *only* function we need to worry about is

$$\alpha x + \beta y + \gamma z = \boldsymbol{\alpha} \cdot \boldsymbol{x}$$
$$= |\boldsymbol{\alpha}||\boldsymbol{x}| \cos \phi$$
$$= |\boldsymbol{\alpha}| \rho \cos \phi.$$

Further—and happily—we have

$$dx dy dz = \rho^2 \sin \phi \, d\rho d\theta d\phi.$$

Thus,

$$F(\alpha, \beta, \gamma) = \frac{1}{(2\pi)^{3/2}} \int_0^R \int_0^{2\pi} \int_0^\pi e^{-i|\boldsymbol{\alpha}|\rho \cos \phi} \rho^2 \sin \phi \, d\phi d\theta d\rho$$

$$= \frac{1}{(2\pi)^{3/2}} \frac{4\pi}{|\boldsymbol{\alpha}|} \int_0^R \rho \sin |\boldsymbol{\alpha}| \rho \, d\rho$$

$$= \sqrt{\frac{2}{\pi}} \frac{1}{|\boldsymbol{\alpha}|^3} [\sin |\boldsymbol{\alpha}| R - |\boldsymbol{\alpha}| R \cos |\boldsymbol{\alpha}| R].^\P$$

By the way, if we were to try to find the *two-dimensional* transform of

$$f(x, y) = \begin{cases} 1, & \text{if } x^2 + y^2 \le R^2, \\ 0, & \text{otherwise}, \end{cases}$$

the same trick will *not* work (try it!). Hence, the method of descent is not just an academic exercise.

Okay, now let's solve

$$u_{tt} = c^2 \nabla^2 u = c^2 (u_{xx} + u_{yy} + u_{zz}),$$
$$-\infty < x < \infty, -\infty < y < \infty, -\infty < z < \infty, t > 0,$$
$$u(x, y, z, 0) = f(x, y, z),$$
$$u_t(x, y, z, 0) = g(x, y, z),$$
$$\lim_{x^2 + y^2 + z^2 \to \infty} u = 0.$$

¶We've been a little sloppy here, as we have avoided the possibility that $|\boldsymbol{\alpha}| = 0$. For the sake of completeness, in this case, we have

$$F(0, 0, 0) = \frac{1}{(2\pi)^{3/2}} \iiint \partial V = \frac{1}{(2\pi)^{3/2}} \frac{4\pi R^3}{3}.$$

Transforming is easy, and we have

$$U_{tt} + c^2(\alpha^2 + \beta^2 + \gamma^2)U = 0,$$
$$U(\alpha, \beta, \gamma, 0) = F(\alpha, \beta, \gamma),$$
$$U_t(\alpha, \beta, \gamma, 0) = G(\alpha, \beta, \gamma).$$

Proceeding as in the one-dimensional case, we get

$$U(\alpha, \beta, \gamma, t) = F(\alpha, \beta, \gamma) \cos c\sqrt{\alpha^2 + \beta^2 + \gamma^2}\, t$$
$$+ \frac{G(\alpha, \beta, \gamma)}{c\sqrt{\alpha^2 + \beta^2 + \gamma^2}} \sin c\sqrt{\alpha^2 + \beta^2 + \gamma^2}\, t.$$

Thus,

$$u(x, y, z, t) = \frac{1}{(2\pi)^{3/2}} \int_{-\infty}^{\infty} \int_{-\infty}^{\infty} \int_{-\infty}^{\infty} \left[F(\alpha, \beta, \gamma) \cos c\sqrt{\alpha^2 + \beta^2 + \gamma^2}\, t \right.$$
$$\left. + \frac{G(\alpha, \beta, \gamma)}{c\sqrt{\alpha^2 + \beta^2 + \gamma^2}} \sin c\sqrt{\alpha^2 + \beta^2 + \gamma^2}\, t \right] e^{i(\alpha x + \beta y + \gamma z)} d\alpha d\beta d\gamma$$
$$= \frac{1}{(2\pi)^{3/2}} \iiint_{\mathbb{R}^3} F(\boldsymbol{\alpha}) e^{i\boldsymbol{\alpha} \cdot \boldsymbol{x}} \cos c|\boldsymbol{\alpha}|t \, dt$$
$$+ \frac{1}{(2\pi)^{3/2}} \iiint_{\mathbb{R}^3} G(\boldsymbol{\alpha}) e^{i\boldsymbol{\alpha} \cdot \boldsymbol{x}} \frac{\sin c|\boldsymbol{\alpha}|t}{c|\boldsymbol{\alpha}|} d\boldsymbol{\alpha}.$$

Transforming back, of course, is the hard part. Our solution looks very much like the one-dimensional solution—as there, the first integral is the time derivative of the second (again, with G replaced by F). So, again, we look at the second integral and, again, see if there's any way that we can do something with the expression

$$\frac{\sin c|\boldsymbol{\alpha}|t}{c|\boldsymbol{\alpha}|}.$$

But we ran into an expression just like this in the second step of evaluating $F(\alpha, \beta, \gamma)$ in Example 1. There we found that

$$\int_0^{2\pi} \int_0^{\pi} e^{-i|\boldsymbol{\alpha}|\rho \cos \phi} \rho^2 \sin \phi \, d\phi d\rho = 4\pi \frac{\rho \sin |\boldsymbol{\alpha}|\rho}{|\boldsymbol{\alpha}|}$$

for any value of ρ so that, letting $\rho = ct$ and rewriting the first integral back there in Cartesians, we have

$$\frac{\sin c|\boldsymbol{\alpha}|t}{c|\boldsymbol{\alpha}|} = \frac{1}{4\pi c^2 t} \iint_{|\boldsymbol{\xi}|=ct} e^{-i\boldsymbol{\alpha} \cdot \boldsymbol{\xi}} \, dS$$
$$= \frac{1}{4\pi c^2 t} \iint_{|\boldsymbol{\xi}|=ct} e^{i\boldsymbol{\alpha} \cdot \boldsymbol{\xi}} \, dS \quad \text{(why?)}.$$

It follows that

$$
\iiint\limits_{\mathbb{R}^3} G(\boldsymbol{\alpha})e^{i\boldsymbol{\alpha}\cdot\boldsymbol{x}}\frac{\sin c|\boldsymbol{\alpha}|t}{c|\boldsymbol{\alpha}|}\,d\boldsymbol{\alpha}
$$

$$
= \frac{1}{4\pi c^2 t}\iiint\limits_{\mathbb{R}^3} G(\boldsymbol{\alpha})\iint\limits_{|\boldsymbol{\xi}|=ct} e^{i\boldsymbol{\alpha}\cdot(\boldsymbol{x}+\boldsymbol{\xi})}\,dS\,d\boldsymbol{\alpha}
$$

$$
= \frac{1}{4\pi c^2 t}\iint\limits_{|\boldsymbol{\xi}|=ct}\left[\iiint\limits_{\mathbb{R}^3} G(\boldsymbol{\alpha})e^{i\boldsymbol{\alpha}\cdot(\boldsymbol{x}+\boldsymbol{\xi})}\,d\boldsymbol{\alpha}\right]dS \;\|
$$

$$
= \frac{1}{4\pi c^2 t}\iint\limits_{|\boldsymbol{\xi}|=ct} g(\boldsymbol{x}+\boldsymbol{\xi})dS.
$$

Now what is this? Well, the integral is performed on the sphere of radius ct, centered at $\boldsymbol{x}$, and, since the area of the sphere is $4\pi k^2 = 4\pi c^2 t^2$, the full expression is just

$$
t\bar{g} = t \cdot (\text{average value of } g \text{ on the sphere of radius } ct, \text{ centered at } \boldsymbol{x}).
$$

Therefore, our solution is

$$
u(x,y,z,t) = \frac{1}{(2\pi)^{3/2}}\left[\frac{\partial}{\partial t}(t\bar{f}) + t\bar{g}\right]
$$

and is known as **Kirchhoff's formula.**** So our solution looks *very* much like the one-dimensional solution. However, there is a significant difference. While the integral in the one-dimensional case encompasses the whole "ball" of radius ct, centered at x, the three-dimensional integration involves only the *boundary* of the ball. To be precise, suppose our initial disturbance is given by

$$
u(x,y,z,0) = 0,
$$

$$
u_t(x,y,z,0) = g(x,y,z) = \begin{cases} 1, & \text{if } x^2 + y^2 + z^2 \leq 1, \\ 0, & \text{otherwise,} \end{cases}
$$

and let's look at a point (x_0, y_0, z_0) which is a distance D from the origin.

‖ Once again, assuming it's okay to switch the order of integration.
** Due to Poisson!

Then,

$$u(x_0, y_0, z_0, t) = \frac{t}{(2\pi)^{3/2}} \iint\limits_{\xi_1^2+\xi_2^2+\xi_3^2=c^2t^2} g(x_0 + \xi_1, y_0 + \xi_2, z_0 + \xi_3)dS$$

$$= \begin{cases} 0, & \text{if } ct < D - 1, \\ \text{nonzero}, & \text{if } D - 1 \le ct \le D + 1, \\ 0, & \text{if } ct > D + 1. \end{cases}$$

See Figure 9.20. In other words, there is no disturbance at (x_0, y_0, z_0) until the leading edge of the initial disturbance "reaches" this point (at $t = \frac{D}{c}$), and none after the disturbance "passes by" (at $t = \frac{D+2}{c}$). So Huygens's Principle *does* hold in three dimensions, as our experience has suggested. (It should be clear that the same thing happens if $f \ne 0$.) This is why electromagnetic disturbances travel *at* the speed of light.

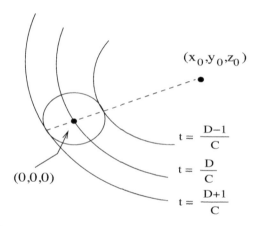

FIGURE 9.20
Huygens's Principle holds in three dimensions: spheres traveling outward from (x_0, y_0, z_0) intersecting the "initial disturbance ball." (Compare with Figure 9.21.)

Now, we use the **method of descent** to compute the solution in two dimensions. Basically, what will happen here is that the region of integration will be projected onto the x-y plane so that, instead of "expanding bubbles," our disturbance will behave like "expanding disks." Thus, as in the one-dimensional case, it looks as though Huygens's principle will *not* hold.

So suppose we have the two-dimensional wave problem

$$u_{tt} = c^2(u_{xx} + u_{yy}), \quad -\infty < x < \infty, -\infty < y < \infty, t > 0,$$
$$u(x, y, 0) = f(x, y),$$
$$u_t(x, y, 0) = g(x, y).$$

The solution of this problem should be identical to that of the *three-dimensional* problem

$$w_{tt} = c^2 \nabla^2 u,$$
$$w(x, y, z, 0) = f(x, y),$$
$$w_t(x, y, z, 0) = g(x, y)$$

(why?), which we've found already. To make matters simpler, let's consider the case where $f(x, y) \equiv 0$. Then the solution will be

$$w(x, y, z) = u(x, y) = t\bar{g}$$
$$= \frac{1}{4\pi c^2 t} \iint\limits_{\xi_1^2 + \xi_2^2 + \xi_3^2 = c^2 t^2} g(x + \xi_1, y + \xi_2) dS.$$

From symmetry, this last integral is twice the integral over the top hemisphere

$$\xi_3 = \sqrt{c^2 t^2 - \xi_1^2 - \xi_2^2}$$

and, since we have solved *explicitly* for the surface in terms of ξ_1 and ξ_2, we may write dS in terms of these variables as

$$dS = \sqrt{1 + \left(\frac{d\xi_3}{d\xi_1}\right)^2 + \left(\frac{d\xi_3}{d\xi_2}\right)^2} d\xi_1 d\xi_2$$
$$= \frac{ct}{\sqrt{c^2 t^2 - \xi_1^2 - \xi_2^2}} d\xi_1 d\xi_2.$$

So our solution is

$$u(x, y) = \frac{1}{2\pi c} \iint\limits_{\xi_1^2 + \xi_2^2 \le c^2 t^2} \frac{g(x + \xi_1, y + \xi_2)}{\sqrt{c^2 t^2 - \xi_1^2 - \xi_2^2}} d\xi_1 d\xi_2,$$

which *now* is an integral on the whole disk of radius ct, centered at (x, y). Again, to be more specific, let

$$g(x, y) = \begin{cases} 1, & \text{if } x^2 + y^2 \le 1, \\ \\ 0, & \text{otherwise.} \end{cases}$$

Then, in Figure 9.21, the disturbance reaches (x_0, y_0) when the disk of radius ct about (x_0, y_0) hits the unit disk and *does not subside thereafter.* Hence, Huygens's Principle does *not* hold for waves in two dimensions.

Finally, we may *descend* one more dimension and recover d'Alembert's solution to the one-dimensional wave equation. In the process of using the method of descent, then, we see that the one- and two-dimensional wave equations also may be interpreted as describing various wave phenomena in *three* dimensions. For obvious reasons, three-dimensional solutions of the one-dimensional wave equation are called **plane waves**; those of the two-dimensional equation are called **cylindrical waves**; and those of the three-dimensional equation are called **spherical waves**.

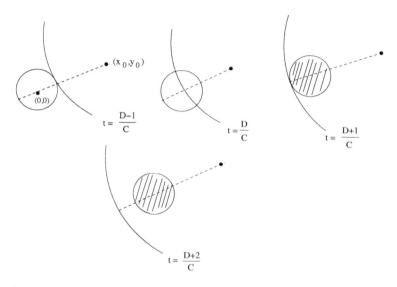

FIGURE 9.21
Huygens's Principle does not hold in two dimensions: disks traveling outward from (x_0, y_0), intersecting the "initial disturbance disk."
(Compare with Figure 9.20.)

Now, what about characteristics in two and three dimensions? Remember that, in one space dimensions, the characteristics were the lines

$$x + ct = \text{constant}, \ x - ct = \text{constant}$$

in the x-t plane. Using these lines, and given any point (x_0, t_0), we defined the *domain of dependence* and the *domain of influence* as in Figure 5.9. The domain of dependence represented, essentially, the *history* of all disturbances that *reach* x_0 at time t_0, while the domain of influence in the *future*, in relation to (x_0, t_0), that is, consists of all points in *space-time*, with $t > t_0$, that eventually are affected by the disturbance at x_0 at time t_0.

It's not difficult to see that, in the two-dimensional case, they are the cones

$$(x - x_0)^2 + (y - y_0)^2 = c^2(t - t_0)^2.$$

As we see in Figure 9.22, the bottom half of the cone (and its interior) represents the *past*, or domain of dependence, of the point (x_0, y_0, t_0), while the top half is the *future*, or region of influence.

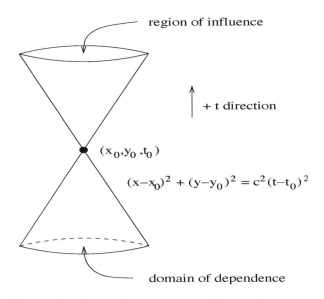

FIGURE 9.22
Characteristic cone for the two-dimensional wave equation.

It follows that, for the three-dimensional case, we have the *four*-dimensional cones

$$(x - x_0)^2 + (y - y_0)^2 + (z - z_0)^2 = c^2(t - t_0)^2,$$

with analogous domains of dependence and influence (except for one *very important* difference—what?). In particular, when studying electromagnetic radiation, the characteristics cones are referred to as **light cones** (the same light cones that we see in the *special theory of relativity*).

Exercises 9.6

1. Show that

$$\frac{\partial}{\partial t} \int_{x-ct}^{x+ct} f(\tau)d\tau = c[f(x+ct) + f(x-ct)].$$

2. Go in the "other direction" in the solution of the one-dimensional wave equation, that is, first compute

$$\int_{-\infty}^{\infty} F(\alpha) \cos c\alpha t \, e^{-i\alpha x} \, dx$$

and then integrate to get the other half of the solution.

3. **Three-dimensional heat equation:** Given a function $f(x, y, z, t)$, define

$$F_1(\alpha, y, z, t) = \frac{1}{\sqrt{2\pi}} \int_{-\infty}^{\infty} e^{-i\alpha x} f(x, y, z, t) dx,$$

$$F_2(\alpha, \beta, z, t) = \frac{1}{\sqrt{2\pi}} \int_{-\infty}^{\infty} e^{-i\beta y} F_1(\alpha, y, z, t) dy,$$

$$F_3(\alpha, \beta, \gamma, t) = \frac{1}{\sqrt{2\pi}} \int_{-\infty}^{\infty} e^{-i\gamma z} F_2(\alpha, \beta, z, t) dz.$$

a) What is $F_3(\alpha, \beta, \gamma, t)$, in terms of f?

b) Solve the heat equation in three space dimensions

$$u_t = \nabla^2 u, \qquad -\infty < x < \infty, -\infty < y < \infty, -\infty < z < \infty,$$
$$u(x, y, z, 0) = f(x, y, z).$$

(Hint: Use the one-dimensional solution derived in Section 6.4.)

c) Now solve the heat equation in half-space

$$u_t = \nabla^2 u, \qquad -\infty < x < \infty, -\infty < y < \infty, 0 < z < \infty,$$
$$u(x, y, z, 0) = f(x, y, z),$$

where the boundary plane is held at temperature zero degrees.

d) Do the same as in part (c), but with the *flux* along the boundary plane equal to zero.

4. a) We *may*, via the method of descent, recover d'Alembert's solution of the *one*-dimensional wave equation directly from the solution in three dimensions. Specifically, from the solution of the three-dimensional problem

$$u_{tt} = c^2 \nabla^2 u, \qquad -\infty < x < \infty, -\infty < y < \infty,$$
$$-\infty < z < \infty, t > 0,$$
$$u(x, y, z, 0) = f(z),$$
$$u_t(x, y, z, 0) = g(z),$$

use spherical coordinates to descend to d'Alembert's solution

$$u = \frac{1}{2}[f(z + ct) + f(z - ct)] + \frac{1}{2c} \int_{z-ct}^{z+ct} g(\xi) d\xi.$$

Notice that this solution is the solution of the one-dimensional wave equation for a string placed along the z-axis and, simultaneously, gives the *plane wave* solution of the original problem.

(Why did we use z and not x or y? How would you have proceeded if we were given, instead, initial conditions dependent upon x only?)

b) Instead, arrive at d'Alembert's solution by letting

$$u(x, y, t) = f(x),$$
$$u_t(x, y, t) = g(x),$$

and descend from the *two-dimensional* solution by integrating out the variable ξ_2.

5. Solve the three-dimensional wave problem

$$u_{tt} = c^2 \nabla^2 u,$$
$$u(x, y, z, 0) = f(\rho),$$
$$u_t(x, y, z, 0) = g(\rho),$$

where, of course, ρ is the radial spherical coordinate. (Hint: Let $u(\rho, t) = \frac{v(\rho,t)}{\rho}$.)

6. **The Dirac delta function in higher dimensions:** The delta function in higher dimensions is just

$$\delta(x, y) = \delta(x)\delta(y) \text{ in two dimensions}$$

and

$$\delta(x, y, z) = \delta(x)\delta(y)\delta(z) \text{ in three dimensions.}$$

Consequently, everything that we did in one dimension can be applied to these cases, as well.

So, compute

$$\int_{-\infty}^{\infty} \int_{-\infty}^{\infty} \delta(x - x_0, y - y_0) dx dy$$

and

$$\int_{-\infty}^{\infty} \int_{-\infty}^{\infty} \int_{-\infty}^{\infty} \delta(x - x_0, y - y_0, z - z_0) dx dy dz.$$

Also, show that, in two dimensions, $\mathcal{F}[\delta(x, y)] = \frac{1}{2\pi}$, while, in three dimensions, we have $\mathcal{F}[\delta(x, y, z)] = \frac{1}{(2\pi)^{3/2}}$. More generally, show that

$$\mathcal{F}[\delta(x - x_0, y - y_0, z - z_0)] = \mathcal{F}[\delta(\boldsymbol{x} - \boldsymbol{x_0})]$$
$$= \frac{e^{-i\boldsymbol{\alpha} \cdot \boldsymbol{x_0}}}{(2\pi)^{3/2}}.$$

(Distributions are defined in higher dimensions as they are in one dimension. So, test functions in three dimensions will be infinitely differentiable in all independent variables and will be identically zero outside a closed and bounded—i.e., compact—subset of the domain. We say that the function has *compact support*.)

7. a) We've shown that the **logarithmic potential**

$$u(x,y) = -\frac{1}{\sqrt{2\pi}} \ln \sqrt{x^2 + y^2}$$

is harmonic in any region not including the origin. It turns out that u satisfies Poisson's equation

$$u_{xx} + u_{yy} = -\delta(x,y), \qquad -\infty < x < \infty, -\infty < y < \infty.$$

What, then, is $\mathcal{F}[\ln \sqrt{x^2 + y^2}]$?

b) It can be shown that $u(x,y) = \frac{i}{4} H_0^{(1)}(k\sqrt{x^2 + y^2})$ satisfies the Helmholtz equation

$$u_{xx} + u_{yy} + k^2 u = -\delta(x,y), \qquad -\infty < x < \infty, -\infty < y < \infty.$$

Here, $H_0^{(1)}$ is the Hankel function of the first kind, of order 0, and $k > 0$. What is $\mathcal{F}[H_0^{(1)}(k\sqrt{x^2 + y^2})]$?

c) Use the result of part (b) to find the formal solution of the PDE

$$u_{xx} + u_{yy} + u_x = -\delta(x,y), \qquad -\infty < x < \infty, -\infty < y < \infty.$$

9.7   Postlude: Eigenvalues and Eigenfunctions of the Laplace Operator; Green's Identities for the Laplacian

In Section 9.2, we found that the eigenvalues and eigenfunctions of the Laplace operator, for the Dirichlet problem on a rectangle, are

$$\lambda_{n,m} = \frac{n^2\pi^2}{a^2} + \frac{m^2\pi^2}{b^2}, \quad \Phi_{n,m}(x,y) = \sin\frac{n\pi x}{a} \sin\frac{m\pi y}{b},$$
$$n = 1, 2, 3, \ldots; m = 1, 2, 3, \ldots .$$

In Section 9.4, we did the same for a disk of radius a and found that

$$\lambda_{n,m} = \frac{x_{n,m}^2}{a^2}, \quad \Phi_{n,m}(r,\theta) = J_n\left(\frac{x_{n,m}r}{a}\right) \begin{cases} \cos n\theta, \\ \sin n\theta, \end{cases}$$

where $x_{n,m}$ is the mth positive root of J_n.

In comparing the situation to that of Chapter 8 for Sturm–Liouville problem, we see that

1) The eigenvalues are real.

2) There are infinitely many eigenvalues. (They are "doubly-infinite" in this case.)

3) An eigenvalue need not have multiplicity one.

4) Eigenfunctions corresponding to different eigenvalues are simply orthogonal on the given domain. (See Exercise 6b.)

5) The eigenfunctions form a complete set in the space of piecewise smooth functions on the given domain. (See Exercise 6a.)

How about on more general domains? It turns out that (1), (2), (4) and (5) remain true for any reasonable bounded domain D. We'll state (2) and (5) without proof; we *will* prove (1) and (4) in Exercise 6. As for (3), it's often possible to find eigenvalues with multiplicity two or greater, sometimes because of the symmetry exhibited in Section 9.4, sometimes just by accident.

It should be no surprise that we need to formulate Green's identities for the Laplacian. So in two dimensions, we'd like to see what we can say about

$$\iint_D u\nabla^2 v \, dA.$$

Although we still can use integration by parts, we have at our disposal Green's Theorem

$$\iint_D \left(\frac{\partial Q}{\partial x} - \frac{\partial P}{\partial y}\right) dA = \oint_{C=\partial D} P \, dx + Q \, dy.$$

Since we need to integrate $uv_{xx} + uv_{yy}$, let's set $Q = uv_x$ and $P = -uv_y$ and see what happens. We have

$$\iint_D \left[\frac{\partial}{\partial x}(uv_x) + \frac{\partial}{\partial y}(uv_y)\right] dA$$

$$= \iint_D u\nabla^2 v \, dA + \iint_D (u_x v_x + u_y v_y) dA$$

$$= \oint_C u(-v_y \, dx + v_x \, dy)$$

or **Green's first identity for the Laplacian**,

$$\iint_D u\nabla^2 v \, dA = \oint_C u\frac{\partial v}{\partial n} ds - \iint_D \nabla u \cdot \nabla v \, dA \text{ }^{\dagger\dagger}$$

[††] These identities often are referred to as "multidimensional integration by parts" formulas.

(make sure you know why we may rewrite the line integral as we have). From there it's trivial to get **Green's second identity for the Laplacian**,

$$\iint_D (u\nabla^2 v - v\nabla^2 u)dA = \oint_C \left(u\frac{\partial v}{\partial n} - v\frac{\partial u}{\partial n}\right)ds.$$

The three-dimensional versions are essentially the same:

$$\iiint_D u\nabla^2 v \, dV = \oiint_{\partial D=S} u\frac{\partial v}{\partial n}dS - \iiint_D \nabla u \cdot \nabla v \, dV$$

and

$$\iiint_D (u\nabla^2 v - v\nabla^2 u)dV = \oiint_S \left(u\frac{\partial V}{\partial n} - v\frac{\partial u}{\partial n}\right)dS.$$

(We prove these in Exercise 5.) Here, of course, we have volume and surface integrals. (Green's identities will loom large in the next chapter, in relation to Green's functions.) Of course, we need all functions, domains and boundaries to be well-enough behaved.

Now, following Section 7.2, suppose that u and v satisfy the condition

$$au + b\frac{\partial u}{\partial n} = 0$$

along the boundary of D, where a and b are constants. Then it's easy to show that

$$u\frac{\partial v}{\partial n} - v\frac{\partial u}{\partial n} = 0 \quad \text{along} \quad \partial D.$$

Therefore, if u and v are well-enough behaved on D, we have, in two dimensions,

$$\iint_D (u\nabla^2 v - v\nabla^2 u)dA = 0$$

or, in three dimensions,

$$\iiint_D (u\nabla^2 v - v\nabla^2 u)dV = 0.$$

As you may have guessed, this is how we define self-adjointness for boundary-value problems in higher dimensions. In fact, we may generalize the idea of Sturm–Liouville equations to dimensions greater than one—see Exercise 8.

So we have the following theorem.

Theorem 9.1 *The Laplace boundary-value problem*

$$\nabla^2 u = 0 \quad on \quad D,$$

$$au + b\frac{\partial u}{\partial n} = 0 \quad on \quad \partial D,$$

where D is a bounded, simply-connected domain, and a and b are constants with at least one nonzero, is **self-adjoint**. (As with one-dimensional problems, we may extend the idea of self-adjointness to problems on unbounded domains, etc.)

Now it's easy to prove properties (1) and (4), which we do in Exercise 6. So we have the following theorem.

Theorem 9.2 *The Laplace boundary-value problem given in Theorem 9.1 possesses an infinite, unbounded sequence of eigenvalues, each of which is real. The eigenfunctions are complete (in the L^2 sense), and eigenfunctions corresponding to different eigenvalues are simply orthogonal on D.*

In physically realistic circumstances it turns out that the eigenvalues are nonnegative and that there is a least eigenvalue (the *fundamental frequency*, again), as earlier and as seen in this section's examples. This can be shown via two- and three-dimensional versions of the *Rayleigh quotient*, as we do in Exercise 7.

Exercises 9.7

1. Find the eigenvalues and eigenfunctions for the two-dimensional Laplace operator on the rectangle $0 < x < a$, $0 < y < b$, subject to the given boundary conditions.

   a) $u_x(0, y) = u_x(a, y) = u_y(x, 0) = u_y(x, b) = 0$

   b) $u(0, y) = u(a, y) = u_y(x, 0) = u_y(x, b) = 0$

   c) $u(0, y) = u_x(a, y) = u(x, 0) = u_y(x, b) = 0$

   d) $u(0, y) = u(a, y) = u(x, 0) - u_y(x, 0) = u(x, b) + u_y(x, b) = 0$

2. Do the same as Exercise 1, but on the disk $0 < r < a$, with boundary condition

   a) $u(a, \theta) = 0$

   b) $u_r(a, \theta) = 0$

3. Find the eigenvalues and eigenfunctions for the three-dimensional problem

$$\nabla^2 u + \lambda u = 0, \qquad 0 < x < \pi, 0 < y < \pi, 0 < z < \pi,$$
$$u(0, y, z) = u(\pi, y, z) = u(x, 0, z) = u(x, \pi z)$$
$$= u(x, y, 0) = u(x, y, \pi) = 0.$$

4. Find the eigenvalues and eigenfunctions for the three-dimensional problem

$$\nabla^2 u + \lambda u = 0, \qquad 0 < \rho < 1,$$
$$u(1, \theta, \phi) = 0.$$

(Hint: Refer to the solution of the problem for diffusion of heat in a ball, at the end of Section 9.5.) What is the multiplicity of each eigenvalue?

5.   a) Using the *Divergence Theorem*

$$\iiint\limits_D \nabla \cdot \boldsymbol{F} \, dV = \oiint\limits_{\partial D} \boldsymbol{F} \cdot \boldsymbol{n} \, dS$$

with $\boldsymbol{F} = u\nabla v$, prove the three-dimensional version of Green's first and second identities.

   b) Show that *Green's Theorem* implies the two-dimensional version of the Divergence Theorem

$$\iint\limits_D \nabla \cdot \boldsymbol{F} \, dA = \oint\limits_{\partial D} \boldsymbol{F} \cdot \boldsymbol{n} \, dS.$$

Thus, we can prove both the two- *and* three-dimensional Green's first identities in one fell swoop.

6. Given the eigenvalue problem

$$\nabla^2 u + \lambda u = 0 \quad \text{on} \quad D$$

subject to the general Robin boundary condition

$$au + b\frac{\partial u}{\partial n} = 0 \quad \text{on} \quad \partial D,$$

show that

   a) If $\lambda_1 \neq \lambda_2$ are eigenvalues with eigenfunctions u_1 and u_2, respectively, then u_1 and u_2 are orthogonal on D.

   b) All eigenvalues of the problem are real (see Theorem 8.4, Section 8.2).

7. In Exercise 26, Section 1.7, we introduced the *Rayleigh quotient* for the eigenvalue problem

$$y'' + \lambda y = 0, \qquad 0 < x < L,$$

with a Dirichlet or Neumann condition at each end; in Exercise 8, Section 8.2, we generalized this idea to regular and periodic Sturm–Liouville problems. Here, we develop the Rayleigh quotient for the Laplace operator.

a) Consider the eigenvalue problem

$$\nabla^2 u + \lambda u = 0 \quad \text{on} \quad D$$

subject either to the Dirichlet condition along ∂D or the Neumann condition along ∂D. Show that if λ_n is an eigenvalue with eigenfunction u_n, then

$$\lambda_n = \frac{\iint\limits_D \|\nabla u\|^2 dA}{\iint\limits_D u^2 \, dA}$$

(in two dimensions, or the same with triple integrals in three dimensions). Thus, we must have $\lambda_n \geq 0$. When will 0 be an eigenvalue?

b) More generally, consider the problem

$$\nabla^2 u + \lambda u = 0 \quad \text{on} \quad D,$$

$$au + b\frac{\partial u}{\partial n} = 0 \quad \text{on} \quad \partial D,$$

where a and b are constant. Show that if $ab \leq 0$, then all of the eigenvalues are nonnegative.

8. **Sturm–Liouville problems in higher dimensions:** An elliptic operator of the form

$$L[u] = \nabla \cdot [r(\boldsymbol{x})\nabla u(\boldsymbol{x})] + q(\boldsymbol{x})u(\boldsymbol{x}),$$

where $\boldsymbol{x} = (x, y)$ or (x, y, z) (or $(x_1, x_2, \ldots, x_n)$), is called a Sturm–Liouville operator, and an eigenvalue problem of the form

$$L[u] + \lambda w(\boldsymbol{x})u = 0 \quad \text{on} \quad D,$$

subject to a Dirichlet, Neumann or Robin condition along ∂D, is called a Sturm–Liouville problem.

a) Show that

$$\iint\limits_D (uL[v] - vL[u])dA = 0$$

for all well-enough behaved functions $u(x, y)$ and $v(x, y)$ which satisfy the boundary conditions.

b) If $r > 0$ and $w > 0$ and r, q and w are continuous on $D \cup \partial D$, show that eigenfunctions corresponding to different eigenvalues are orthogonal with respect to the weight function $w(x, y)$ on D.

c) Under these same assumptions, show that all eigenvalues are real.

Prelude to Chapter 10

Here we look at nonhomogeneous PDEs and, specifically, at the very important *method of Green's functions* for solving them. We've already met a Green's function back in Section 6.4, namely, the *heat kernel*; remember that the solution there was a convolution of that function with the initial temperature. Of course, there we used it to solve the *homogeneous* heat equation.

We begin by looking at Green's functions for ODEs and, fairly quickly, we move from a classical to a distributional setting, as the introduction of the Dirac delta function makes life much easier. Historically, however, Green's functions were introduced in 1828 by the self-taught British mathematician George Green (1793–1841), in the context of solving the Poisson Dirichlet problem. (Of course, distributions didn't show up for another century!) In his self-published *An Essay on the Application of Mathematical Analysis to the Theories of Electricity and Magnetism*, Green proved the Divergence Theorem (which, in two dimensions, is just Green's Theorem) and then used it to prove what came to be called Green's identities, from which his results followed. To this day, it is in the study of elliptic equations that Green's function is most crucial. However, we also derive Green's function for both the heat and wave equations, where we also get to see in greater detail the significance of *Duhamel's Principle* (which we met briefly in Section 6.1).

Incidentally, Green's work remained pretty much unknown until rediscovered by William Thomson, Lord Kelvin (1824–1907), in 1846.

10

Nonhomogeneous Problems and Green's Functions

10.1 Green's Functions for ODEs

Suppose that we have the ODE

$$Ly = -f(x)$$

and rewrite it using the sifting property of the Dirac delta function as

$$Ly = -f(x) = -\int_{-\infty}^{\infty} \delta(\xi - x)f(\xi)d\xi$$

$$\approx -\sum_n \delta(\xi_n - x)f(\xi_n)\Delta\xi_n.$$

Then it *seems* that we need only solve

$$LG_{\xi_n}(x) = -\delta(\xi_n - x) = -\delta(x - \xi_n),$$

then multiply by the "constants" $f(\xi_n)\Delta\xi_n$ and use superposition. So, we would have the solution

$$y \approx \sum_n G_{\xi_n}(x)f(\xi_n)\Delta\xi_n \longrightarrow \int_{-\infty}^{\infty} G_\xi(x)f(\xi)d\xi.$$

Indeed, *formally* we have

$$Ly = L\left[\int_{-\infty}^{\infty} G_\xi(x)f(\xi)d\xi\right] = \int_{-\infty}^{\infty} L[G_\xi(x)]f(\xi)d\xi \ ^*$$

$$= -\int_{-\infty}^{\infty} \delta(\xi - x)f(\xi)d\xi = -f(x).$$

---

*If we can bring L inside the integral, of course. It turns out that we always can, in the setting of distributions.

Further, if $G_\xi(x)$ satisfies the boundary conditions, then so will the solution. (For example, suppose the boundary conditions are $y(0) = y(L) = 0$, so that, for each ξ, $G_\xi(0) = G_\xi(L) = 0$. Then

$$y(0) = \int_{-\infty}^{\infty} G_\xi(0) f(\xi) d\xi = 0$$

and similarly for $y(L)$.)

Instead of writing $G_\xi(x)$, we say that $G(x, \xi)$ is a **Green's function** for the given boundary-value problem. The beauty of Green's function is that it does not depend on the nonhomogeneous right side and, therefore, once we have found it, we automatically have the solution for any well-enough behaved $f(x)$. (In fact, it turns out that one also can arrange to have the Green's function reflect a whole class of boundary conditions. So, for example, we may set $u(0) = \alpha$ and $u(L) = \beta$ and find the corresponding Green's function for arbitrary α and β.)

Of course, George Green lived long before any talk of the Dirac delta function and the like, so it seems that we should be able to "do" Green's functions classically; we introduce them classically in this section. However, there is such a close connection between Green's functions and the delta function that a discussion of one without the other would be misleading.

Let's begin with an example that will lead us to the salient features of Green's functions.

Example 1 Solve the BVP

$$y'' + k^2 y = -f(x),^\dagger \qquad 0 < x < L, \quad k > 0,$$

$$y(0) = y(L) = 0.$$

Two important homogeneous solutions are, of course, $y_1 = \sin kx$ and $y_2 = \cos kx$, and variation of parameters leads us to the particular solution

$$y_p = u_1 y_1 + u_2 y_2,$$

where

$$u_1' = -\frac{1}{k} f(x) \cos kx$$

and

$$u_2' = \frac{1}{k} f(x) \sin kx$$

---

†Here, again, we use $-f$ instead of f. In this case, $f > 0$ will then represent a *source* and $f < 0$ will represent a *sink*.

(see Exercise 8a). Thus,

$$y_p = -\frac{1}{k} \sin kx \int_{c_1}^x f(\xi) \cos k\xi \, d\xi$$
$$+ \frac{1}{k} \cos kx \int_{c_2}^x f(\xi) \sin k\xi \, d\xi$$

for *any* choice of the constants $0 \leq c_1, c_2 \leq L$. Since we're looking to have our Green's function satisfy the boundary conditions, let's choose y_p so that it does the same (in which case y_p will be *the* solution—why?). So,

$$y_p(0) = 0 = \frac{1}{k} \int_{c_2}^0 f(\xi) \sin k\xi \, d\xi$$

while

$$y_p(L) = 0 = -\frac{1}{k} \sin kL \int_{c_1}^L f(\xi) \cos k\xi \, d\xi$$
$$+ \frac{1}{k} \cos kL \int_{c_2}^L f(\xi) \sin k\xi \, d\xi.$$

So we take $c_2 = 0$, and the second equation becomes

$$\sin kL \int_{c_1}^L f(\xi) \cos k\xi \, d\xi = \cos kL \int_0^L f(\xi) \sin k\xi \, d\xi.$$

First, we note that if $\sin kL = 0$, then this equation can only hold for certain functions f; in other words, if k^2 is any eigenvalue of $y'' + \lambda y = 0$, then the boundary-value problem has either/or

1) No solution (if $\int_0^L f(\xi) \sin k\xi \, d\xi \neq 0$)
2) Infinitely many solutions (if $\int_0^L f(\xi) \sin k\xi \, d\xi = 0$).

Now, supposing $\sin kL \neq 0$, the second equation says that

$$\int_{c_1}^L f(\xi) \cos k\xi \, d\xi = \frac{\cos kL}{\sin kL} \int_0^L f(\xi) \sin k\xi \, d\xi,$$

so that

$$\int_{c_1}^x f(\xi) \cos k\xi \, d\xi = \int_{c_1}^L f(\xi) \cos k\xi \, d\xi$$
$$- \int_x^L f(\xi) \cos k\xi \, d\xi$$
$$= \frac{\cos kL}{\sin kL} \int_0^L f(\xi) \sin k\xi \, d\xi$$
$$- \int_x^L f(\xi) \cos k\xi \, d\xi.$$

Finally, we write

$$
y_p = -\frac{\sin kx}{k}\left[\frac{\cos kL}{\sin kL}\int_0^L f(\xi)\sin k\xi\ d\xi - \int_x^L f(\xi)\cos k\xi\ d\xi\right]
$$
$$
+ \frac{\cos kx}{k}\int_0^x f(\xi)\sin k\xi\ d\xi.
$$

Since this representation is different for $\xi < x$ and $\xi > x$, we break up the first integral at x and rewrite

$$
y_p = \int_0^x f(x)\left[\frac{\cos kx\sin k\xi}{k} - \frac{\sin kx\cos kL\sin k\xi}{k\sin kL}\right]d\xi
$$
$$
+ \int_x^L f(\xi)\left[\frac{\sin kx\cos k\xi}{k} - \frac{\sin kx\cos kL\sin k\xi}{k\sin kL}\right]d\xi
$$
$$
= \int_0^L f(\xi)G(x;\xi)d\xi,
$$

where

$$
G(x;\xi) = \begin{cases} \dfrac{\cos kx\sin k\xi}{k} - \dfrac{\sin kx\cos kL\sin k\xi}{k\sin kL}, & \text{if } 0 \le \xi \le x \\[2mm] \dfrac{\sin kx\cos k\xi}{k} - \dfrac{\sin kx\cos kL\sin k\xi}{k\sin kL}, & \text{if } x \le \xi \le L.^{\ddagger} \end{cases}
$$

Notice that G satisfies the boundary conditions; however, this is obvious if we rewrite G as

$$
G(x;\xi) = \begin{cases} \dfrac{1}{k\sin kL}\sin k\xi\sin k(L-x), & \text{if } 0 \le \xi \le x \\[2mm] \dfrac{1}{k\sin kL}\sin kx\sin k(L-\xi), & \text{if } x \le \xi \le L^{\S} \end{cases}
$$

(see Exercise 8b).

So let's list the important properties of this particular Green's function (which, as it turns out, all Green's functions will possess, although the discontinuity in the derivative may behave differently).

1. **Symmetry:** $G(x;\xi) = G(\xi;x)$ for all x,ξ in $[0,L]$.

2. **Continuity:** G is continuous on $[0,L]$ and, specifically, at the point $\xi = x$.

3. **Derivative discontinuous:** The derivative

$$
G_x(x;\xi) = \begin{cases} -\dfrac{1}{\sin kL}\sin x\xi\cos k(L-x), & \text{if } 0 < \xi < x, \\[2mm] \dfrac{1}{\sin kL}\cos kx\sin k(L-\xi), & \text{if } x < \xi < L, \end{cases}
$$

has a jump discontinuity at $x = \xi$. In particular, for this Green's function it is

$$G_x(\xi+;\xi) - G_x(\xi-,\xi)$$

$$= -\frac{1}{\sin kL} \sin k\xi \cos k(L - \xi) - \frac{1}{\sin kL} \cos k\xi \sin k(L - \xi)$$

$$= -1 \quad \text{(why?)}.$$

4. **Satisfies homogeneous equation:** G has a continuous second derivative on $0 < x < \xi$ and $\xi < x < L$ and satisfies the associated homogeneous equation there. (But not at $x = \xi$, of course.)

5. **Satisfies boundary conditions:** G satisfies both boundary conditions, for each value of ξ.

Let's note here that we may rewrite Property 3 as

$$\lim_{\epsilon \to 0} \int_{\xi-\epsilon}^{\xi+\epsilon} (y'' + k^2 y)dx = -1$$

(see Exercise 6) in anticipation of the two- and three-dimensional versions of this property.

Guided by the above example, let's construct the Green's function for the general nonhomogeneous *regular Sturm–Liouville problem*

$$(ry')' + (q + \lambda w)y = -f(x), \qquad a < x < b,$$
$$a_1 y(a) + a_2 y'(a) = b_1 y(b) + b_2 y'(b) = 0.$$

Here, as before, w, q, r and r' are continuous on $a \le x \le b$ and $w(x) > 0$, $r(x) > 0$ on $a \le x \le b$. Also, f is continuous on $a \le x \le b$. (The result can be extended to many of the important singular problems, as well.)

To begin, and guided by Example 1, we assume that λ is not an eigenvalue of the associated homogeneous problem, so that the latter has a unique solution. (We deal with $\lambda =$ an eigenvalue later.) Further, as we'd like Green's function to satisfy the boundary conditions, and we notice from Example 1 that $\sin k(L - x)$ satisfies the boundary condition at $x = L$, and $\sin kx$ that at $x = 0$, we try to do the same here. So, given any two linearly independent solutions z_1, z_2 of the homogeneous problem, it turns out that we may always find constants c_1, c_2, c_3, c_4 such that

$$y_1 = c_1 z_1 + c_2 z_2, \qquad y_2 = c_3 z_1 + c_4 z_2$$

are linearly independent and satisfy the boundary condition at $x = a$ and $x = b$, respectively. (Why? By the way, can y_1, for example, satisfy *both* boundary conditions? Why or why not?)

Now perform variation of parameters, as in the example. We get

$$y_p = u_1 y_1 + u_2 y_2,$$

where

$$u_1' = \frac{f(x)y_2(x)}{r(x)W(x)}, \quad u_2' = -\frac{f(x)y_1(x)}{r(x)W(x)},$$

and W is the Wronskian

$$W(x) = y_1(x)y_2'(x) - y_2(x)y_1'(x)$$

(see Exercise 9a).[¶] (Oh, and how do we know that W is never zero?) Then,

$$u_1 = \int_{c_1}^x \frac{y_2(\xi)}{r(\xi)W(\xi)} f(\xi)d\xi,$$

$$u_2 = -\int_{c_2}^x \frac{y_1(\xi)}{r(\xi)W(\xi)} f(\xi)d\xi$$

and

$$y_p = y_1(x)\int_{c_1}^x \frac{y_2(\xi)}{r(\xi)W(\xi)} f(\xi)d\xi$$

$$- y_2(x)\int_{c_2}^x \frac{y_1(\xi)}{r(\xi)W(\xi)} f(\xi)d\xi.$$

As in the example, we'd like y_p to be our solution, so we'd like c_1 and c_2 to be such that y_p satisfies the boundary conditions. First, our life is made much easier by the fact that $r(x)W(x)$ is constant on $a \le x \le b$.[‖] Further, led by the example, our selection of y_1 and y_2 suggests that we let $c_1 = b$ and $c_2 = a$. Then,

$$y_p = -y_1(x)\int_x^b \frac{y_2(\xi)}{rW} d\xi - y_2(x)\int_a^x \frac{y_1(\xi)}{rW} d\xi,$$

so

$$a_1 y_p(a) + a_2 y_p'(a) = -[a_1 y_1(a) + a_2 y_1'(a)]\int_a^b \frac{y_2(\xi)}{rW} d\xi$$

$$- [a_1 y_2(a) + a_2 y_2'(a)]\int_a^a \frac{y_1(\xi)}{rW} d\xi = 0$$

(see Exercise 9b). Similarly, at $x = b$.

[¶]Why are we not concerned with the possibility that $rW = 0$ at some point?
[‖]See Exercise 3, Section 8.1.

So we have

$$G(x;\xi) = \begin{cases} -\dfrac{y_1(\xi)y_2(x)}{r(\xi)W(\xi)}, & \text{if } a \le \xi \le x \text{ (or } \xi \le x \le b), \\[2ex] -\dfrac{y_1(x)y_2(\xi)}{r(\xi)W(\xi)}, & \text{if } x \le \xi \le b \text{ (or } a \le x \le \xi). \end{cases}$$

It's easy to see that G satisfies Properties (1), (2), (4) and (5), above. What about the jump in the derivative? We have

$$G_x(\xi+,\xi) - G_x(\xi-,\xi) = \frac{y_1'(\xi)y_2(\xi) - y_1(\xi)y_2'(\xi)}{r(\xi)W(\xi)}$$

$$= -\frac{1}{r(\xi)}.$$

We put everything together in a theorem.

Theorem 10.1 *Given the regular Sturm–Liouville problem*

$$(ry')' + (q + \lambda w)y = -f(x), \qquad a < x < b,$$
$$a_1 y(a) + a_2 y'(a) = b_1 y(b) + b_2 y'(b) = 0,$$

suppose that λ is not an eigenvalue and suppose that y_1 and y_2 are solutions of the associated homogeneous equation, satisfying

$$a_1 y_1(a) + a_2 y_1'(a) = b_1 y_2(b) + b_2 y_2'(b) = 0.$$

Then the solution of the problem is given by

$$y = \int_a^b G(x,\xi)f(\xi)d\xi,$$

where **Green's function** *G is given by*

$$G(x;\xi) = \begin{cases} -\dfrac{y_1(\xi)y_2(x)}{r(\xi)W(\xi)}, & \text{if } a \le \xi \le x \text{ (or } \xi \le x \le b), \\[2ex] -\dfrac{y_1(x)y_2(\xi)}{r(\xi)W(\xi)}, & \text{if } x \le \xi \le b \text{ (or } a \le x \le \xi). \end{cases}$$

Further, G satisfies the following properties:

1. *$G(x;\xi) = G(\xi;x)$ for all x, ξ in $[a,b]$. (It turns out that this symmetry is a result of the problem's being self-adjoint. It is often called the property of* **reciprocity** *or* **Maxwell's reciprocity.*****)*

**After James Clark Maxwell.

2. $G(x; \xi)$ is continuous on $a \le x \le b$.

3. $G_x(x; \xi)$ has a jump discontinuity at $x = \xi$ given by

$$G_x(\xi+; \xi) - G_x(\xi-; \xi) = -\frac{1}{r(\xi)}.^{\dagger\dagger}$$

4. G_x and G_{xx} are continuous on $a < x < \xi$ and $\xi < x < b$ and satisfy the associated homogeneous ODE on these intervals.

5. G satisfies the boundary conditions.

The above construction guarantees the existence of Green's function, while uniqueness is left to Exercise 10.

Theorem 10.2 *For the Sturm–Liouville problem of Theorem 10.1, Green's function exists and is unique.*

So we may compute Green's functions either directly, using variation of parameters, or by constructing them via the properties given in Theorem 10.1. Let's look at an example.

Example 2 Find Green's function for the BVP

$$y'' = -f,$$
$$y(0) = y(L) = 0$$

in three different ways.

First, since $\lambda = 0$ is not an eigenvalue of the problem

$$y'' + \lambda y = 0, \qquad 0 < x < L,$$
$$y(0) = y(L) = 0,$$

we are guaranteed that G exists. We begin by finding the solutions $z_1 = 1$ and $z_2 = x$ of the associated homogeneous equation. Now, $y_1 = z_2 = x$ satisfies the boundary condition at $x = 0$, while $y_2 = z_2 - Lz_1 = x - L$ satisfies the other; further, y_1 and y_2 are linearly independent. So Green's function will be

$$G(x; \xi) = \begin{cases} -\dfrac{y_1(\xi)y_2(x)}{rW}, & \text{if } 0 \le \xi \le x, \\[2mm] -\dfrac{y_1(x)y_2(\xi)}{rW}, & \text{if } x \le \xi \le L, \end{cases}$$

$$= \begin{cases} -\dfrac{\xi(x - L)}{L}, & \text{if } 0 \le \xi \le x, \\[2mm] -\dfrac{x(\xi - L)}{L}, & \text{if } x \le \xi \le L, \end{cases}$$

---

†† Again, see Exercise 6.

or, if we'd like,

$$G(x;\xi) = \begin{cases} \dfrac{x(L-\xi)}{L}, & \text{if } 0 \le x \le \xi, \\[2ex] \dfrac{\xi(L-x)}{L}, & \text{if } \xi \le x \le L. \end{cases}$$

Now, instead, let's go *backwards* by using the properties of Green's functions. So, again, $y_1 = x$ and $y_2 = x - L$, and we want the Green's function to be of the form

$$G(x;\xi) = \begin{cases} A(\xi)(x-L), & \text{if } 0 \le \xi \le x, \\[2ex] B(\xi)x, & \text{if } x \le \xi \le L. \end{cases}$$

Now, we know that G is continuous at $x = \xi$, while G_x has a jump of -1 there. So,

$$A(\xi)(\xi - L) = B(\xi)\xi$$

and

$$A(\xi) - B(\xi) = -1.$$

Solving gives us $A(\xi) = -\frac{\xi}{L}$ and $B(\xi) = \frac{L-\xi}{L}$, so that

$$G(x;\xi) = \begin{cases} \dfrac{\xi(L-x)}{L}, & \text{if } 0 \le \xi \le x, \\[2ex] \dfrac{x(L-\xi)}{L}, & \text{if } x \le \xi \le L. \end{cases}$$

Finally, a third method (at least for this example) is just to integrate:

$$y'' = -f(x)$$

$$\Rightarrow y' = -\int_0^x f(\xi)d\xi + c_1$$

$$\Rightarrow y = -\int_0^x \int_0^z f(\xi)d\xi dz + c_1 x + c_2$$

$$= \int_0^x \xi f(\xi)d\xi - x\int_0^x f(\xi)d\xi + c_1 x + c_2$$

(see Exercise 5). Then,

$$y(0) = 0 = c_2$$

and

$$y(L) = 0 = \int_0^L \xi f(\xi)d\xi - L\int_0^L f(\xi)d\xi + c_1 L.$$

So

$$c_1 = \int_0^L f(\xi)d\xi - \frac{1}{L}\int_0^L \xi f(\xi)d\xi,$$

and our solution is

$$y = \int_0^x \xi f(\xi)d\xi - x\int_0^x f(\xi)d\xi + x\left[\int_0^L f(\xi)d\xi - \frac{1}{L}\int_0^L \xi f(\xi)d\xi\right]$$

$$= \int_0^x \frac{\xi(L-x)}{L}f(\xi)d\xi + \int_x^L \frac{x(L-\xi)}{L}f(\xi)d\xi$$

$$= \int_0^L G(x;\xi)f(\xi)d\xi,$$

where G is as given above.

Figure 10.1 shows the graph of $y = G(x;\xi)$ for a fixed value of ξ. Obviously, G is continuous, but not differentiable at the point $x = \xi$. In Figure 10.2, we see the graphs of $y = G(x;\xi_1)$ and $y = G(x;\xi_2)$ and an illustration of the reciprocity property, that $G(\xi_2;\xi_1) = G(\xi_1;\xi_2)$.

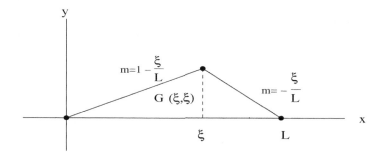

FIGURE 10.1
Graph of $y = G(x;\xi)$ for a fixed value of ξ.

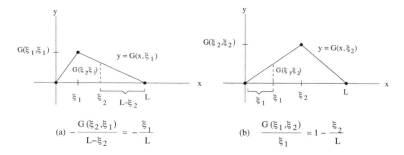

FIGURE 10.2
An illustration of the reciprocity property of Green's function:
$G(\xi_2;\xi_1) = G(\xi_1;\xi_2)$.

FOURIER SERIES REPRESENTATION OF GREEN'S FUNCTION

There is yet another way to compute Green's function. Consider, for example, the nonhomogeneous heat equation

$$u_t = u_{xx} + f(x), \qquad 0 < x < \pi,$$
$$u(x,0) = g(x),$$
$$u(0,t) = u(\pi,t) = 0.$$

We saw that we could solve it by, first, finding the form of the associated homogeneous solution and then looking for a nonhomogeneous solution of similar form. Here, we have

$$u_h = \sum_{n=1}^{\infty} b_n e^{-n^2 t} \sin nx$$

and

$$u_p = \sum_{n=1}^{\infty} c_n(t) \sin nx,$$

respectively, with the functions $c_n(t)$ to be determined by substituting into the PDE and expanding f in its own sine series.

Let's see how to apply this idea in the following example.

Example 3 Find the Green's function for the BVP

$$y'' = -f, \qquad 0 < x < \pi,$$
$$y(0) = y(\pi) = 0.$$

This problem looks *somewhat* like the problem above, so let's try letting

$$y = \sum_{n=1}^{\infty} b_n \sin nx, \quad f = \sum_{n=1}^{\infty} f_n \sin nx,$$

where, of course, the f_n are known and the b_n are to be determined. Also, our choice of a sine series guarantees that the solution will satisfy the boundary conditions.

Substituting into the ODE gives us

$$-\sum_{n=1}^{\infty} n^2 b_n \sin nx = -\sum_{n=1}^{\infty} f_n \sin nx,$$

so

$$b_n = \frac{1}{n^2} f_n, \qquad n = 1, 2, 3, \ldots,$$

$$= \frac{1}{n^2} \frac{2}{\pi} \int_0^\pi f(\xi) \sin n\xi \, d\xi.$$

Therefore, our solution is just

$$y = \sum_{n=1}^\infty \frac{1}{n^2} f_n \sin nx$$

$$= \int_0^\pi \left(\frac{2}{\pi} \sum_{n=1}^\infty \frac{\sin nx \sin n\xi}{n^2} \right) f(\xi) d\xi$$

(so long as we may interchange Σ and $\int$), and it looks like we have the Green's function:

$$G(x; \xi) = \frac{2}{\pi} \sum_{n=1}^\infty \frac{\sin nx \sin n\xi}{n^2}.$$

Then, from Example 2, we must have

$$\frac{2}{\pi} \sum_{n=1}^\infty \frac{\sin nx \sin n\xi}{n^2} = \begin{cases} \dfrac{\xi(\pi - x)}{\pi}, & \text{if } 0 \le \xi \le x, \\ \dfrac{x(\pi - x)}{\pi}, & \text{if } x \le \xi \le x \end{cases}$$

(why?).

Now, what actually happened? In other words, in a more complicated problem, how will we know which functions to use in the Fourier expansion?

Looking more closely, we see that the functions $\sin nx$ are the eigenfunctions of

$$y'' + \lambda y = 0, \qquad 0 < x < \pi,$$
$$y(0) = y(\pi) = 0,$$

while the problem we're actually solving is

$$y'' + 0y = -f,$$
$$y(0) = y(\pi) = 0.$$

And, of course, zero is *not* an eigenvalue of the eigenvalue problem (which is why we were guaranteed a Green's function, of course).

This suggests that, in general, if we're solving the BVP

$$(ry')' + (g + \lambda_0 w)y = L[y] + \lambda_0 wy = -f, \qquad a < x < b,$$

with boundary conditions at $x = a$ and $x = b$, and with λ_0 *not* an eigenvalue of the associated homogeneous problem, we expand y and f in series of eigenfunctions of the problem

$$L[y] + \lambda w y = 0, \qquad a < x < b,$$

with the same boundary conditions.

Suppose, then, that the eigenvalues and eigenfunctions of the latter are, respectively, λ_n and ϕ_n, $n = 1, 2, 3, \dots$. We let

$$y = \sum_{n=1}^{\infty} b_n \phi_n(x)$$

and get

$$
\begin{aligned}
L[y] + \lambda_0 w y &= L\left[\sum_{n=1}^{\infty} b_n \phi_n(x)\right] + \lambda_0 w \sum_{n=1}^{\infty} b_n \phi_n(x) \\
&= \sum_{n=1}^{\infty} b_n L \phi_n(x) + \lambda_0 w \sum_{n=1}^{\infty} b_n \phi_n(x) \\
&= w \sum_{n=1}^{\infty} (\lambda_n - \lambda_0) b_n \phi_n(x),
\end{aligned}
$$

so that the ODE gives us

$$w \sum_{n=1}^{\infty} (\lambda_n - \lambda_0) b_n \phi_n(x) = -f(x).$$

Finally, multiplying both sides by $\phi_m(x)$ and integrating leads to

$$(\lambda_m - \lambda_0) b_m \|\phi_m\|^2 = -\int_a^b f(\xi) \phi_m(\xi) d\xi.$$

Putting everything together, we have

$$
\begin{aligned}
y = \sum_{n=1}^{\infty} b_n \phi_n(x) &= \sum_{n=1}^{\infty} \frac{\int_a^b f(\xi) \phi_n(\xi) d\xi}{(\lambda_0 - \lambda_n) \|\phi_n\|^2} \phi_n(x) \\
&= \int_a^b \left[\sum_{n=1}^{\infty} \frac{\phi_n(x) \phi_n(\xi)}{(\lambda_0 - \lambda_n) \|\phi_n\|^2}\right] f(\xi) d\xi,
\end{aligned}
$$

and Green's function is

$$G(x; \xi) = \sum_{n=1}^{\infty} \frac{\phi_n(x) \phi_n(\xi)}{(\lambda_0 - \lambda_n) \|\phi_n\|^2}.$$

MODIFIED GREEN'S FUNCTION

Now, what about the case where λ *is* an eigenvalue? Going back to Example 1, we saw that if $\sin kL = 0$, in which case $\sin kx$ is an eigenfunction with eigenvalue k^2, then we lose uniqueness. Further, if

$$\int_0^L f(x) \sin kx \, dx = \langle f, \sin kx \rangle \neq 0,$$

we lose existence, as well. The situation is similar—indeed, almost identical—to the case where the linear operators are matrix multiplications of vectors. In that case, if λ is not an eigenvalue of A, then the equation

$$(A - \lambda I)\boldsymbol{x} = \boldsymbol{x}_0$$

has a unique solution $\boldsymbol{x} = (A - \lambda I)^{-1}\boldsymbol{x}_0$. (In fact, although we can't really talk about the inverse of an operator L, it *looks* like we have

$$L^{-1}f = \int_a^b G(x;\xi)f(\xi)d\xi.)$$

If λ *is* an eigenvalue, with eigenvector $\boldsymbol{v}$, then we have

1) No solution if $\langle \boldsymbol{v}, \boldsymbol{x}_0 \rangle \neq 0$

2) Infinitely many solutions if $\langle \boldsymbol{v}, \boldsymbol{x}_0 \rangle = 0$, that is, if $\boldsymbol{v}$ and $\boldsymbol{x}_0$ are perpendicular

(with the obvious generalization for multiplicity greater than one).

The situation for the regular Sturm–Liouville problem (and for matrix operators, as well, with adjustment in terminology) is stated in the following theorem.

Theorem 10.3 (Fredholm Alternative Theorem) *Given the regular Sturm–Liouville problem above, we have the following possibilities.*

a) *Either λ is not an eigenvalue of the associated eigenvalue problem, and the nonhomogeneous problem has a unique solution,*

b) *or λ is an eigenvalue of the associated eigenvalue problem with corresponding eigenfunction ϕ_λ, and*

    i) *either $\int_a^b f(x)\phi_\lambda(x)w(x)dx = \langle f, \phi_\lambda \rangle \neq 0$, in which case the nonhomogeneous problem has no solution.*

    ii) *or $\int_a^b f(x)\phi_\lambda(x)w(x)dx = \langle f, \phi_\lambda \rangle = 0$, in which case the nonhomogeneous problem has infinitely many solutions.*

Can we still talk about Green's function for case (b)? Yes—if (ii) holds, we may guarantee uniqueness by requiring that our solution and the eigenfunction be orthogonal, that is, that $\langle y, \phi_{\lambda_0} \rangle = 0$. If (i) holds—no solution—then there is no Green's function. However, we may solve the problem which is "closest" to this problem by replacing $-f$ with *the component of $-f$ which is orthogonal to ϕ_{λ_0}*.

Remember that we talked about the component of a vector $\boldsymbol{v}$ in the direction of a vector $\boldsymbol{w}$ (see Figure 10.3). This component is just $\boldsymbol{v} \cdot \hat{\boldsymbol{w}}$, where $\hat{\boldsymbol{w}} = \frac{\boldsymbol{w}}{\|\boldsymbol{w}\|}$ is the unit vector in the direction of $\boldsymbol{w}$. Then, $\boldsymbol{v}_p = \boldsymbol{v} - (\boldsymbol{v} \cdot \hat{\boldsymbol{w}})\hat{\boldsymbol{w}}$ is the component of $\boldsymbol{v}$ which is perpendicular to $\boldsymbol{w}$, and it's not hard to show that, of all vectors perpendicular to $\boldsymbol{w}$, this is the vector *nearest* $\boldsymbol{v}$. (To be precise, if $\boldsymbol{w}_p$ represents all vectors perpendicular to $\boldsymbol{w}$, then $\|\boldsymbol{v} - \boldsymbol{w}_p\|$ is minimized by taking $\boldsymbol{w}_p = \boldsymbol{v}_p$.)

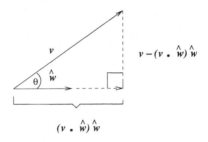

FIGURE 10.3
The components of $\boldsymbol{v}$ with respect to $\hat{\boldsymbol{w}}$.

Of course, we may do the same for functions, using inner product with respect to w instead of dot product. So, if λ_0 is an eigenvalue with eigenfunction ϕ_0, the function we're after is

$$g = -f - \langle -f, \hat{\phi}_0 \rangle \hat{\phi}_0, \text{ where } \hat{\phi}_0 = \frac{\phi_0}{\|\phi_0\|}$$

(see Exercise 11).

Then, the problem

$$(ry')' + (q + \lambda_0 w)y = g$$

satisfies (ii), which we *can* solve. The resulting Green's function is called the **generalized Green's function**, $G_g(x; \xi)$, of the original problem; if we force uniqueness by requiring that $\langle y, \phi_0 \rangle = 0$, we then have the **modified Green's function**, $G_m(x; \xi)$.

Let's conclude with an example.

Example 4 Compute the modified Green's function for the problem

$$y'' = -f, \qquad 0 < x < 1,$$
$$y(0) = y(1) - y'(1) = 0.$$

We solve the associated homogeneous problem and find that $y = cx$ is a solution for any constant c. Remember, we're looking at the problem $y''+0y = 0$, and 0 *is* an eigenvalue. Since the weight function is $w(x) = 1$, we have no solution unless $\int_0^1 xf(x)dx = 0$.

First, we need the eigenfunction ϕ with $\|\phi\| = \left(\int_0^1 \phi^2(x)dx\right)^{1/2} = 1$, so we take $\phi(x) = \sqrt{3}\,x$. Then, the component of f in the direction of ϕ is

$$\langle f, \phi\rangle\phi(x) = \sqrt{3}\,x \int_0^1 \sqrt{3}\,xf(x)dx,$$

so we solve the new problem

$$y'' = -f + 3x \int_0^1 \xi f(\xi)d\xi.$$

Then, as before,

$$y' = -\int_0^x f(\xi)d\xi + \frac{3x^2}{2}\int_0^1 f(\xi)d\xi + c_1$$

$$y = \int_0^x \xi f(\xi)d\xi - x\int_0^x f(\xi)d\xi + \frac{x^3}{2}\int_0^1 \xi f(\xi)d\xi$$
$$+ c_1 x + c_2.$$

The first boundary condition implies that $c_2 = 0$, while the second boundary condition is satisfied identically (see Exercise 7). This is to be expected, as we know that the problem has infinitely many solutions.

At this point, we have the generalized Green's function

$$G_g(x;\xi) = \begin{cases} \xi - x + \dfrac{x^3\xi}{2}, & \text{if } 0 \le \xi \le x, \\ \dfrac{x^3\xi}{2}, & \text{if } x \le \xi \le 1 \end{cases}$$

and

$$y = \int_0^1 G_g(x;\xi)f(\xi)d\xi + cx$$

for any choice of the constant c. Note that G_g is *not* symmetric. In order to find the modified Green's function, we look for the value of c_1 that makes

$\langle \phi, y \rangle = 0$. So, we need

$$0 = \langle x, y \rangle = \int_0^1 x \int_0^x \xi f(\xi) d\xi dx - \int_0^1 x^2 \int_0^x f(\xi) d\xi dx$$

$$+ \int_0^1 \frac{x^4}{2} dx \int_0^1 \xi f(\xi) d\xi + \frac{c_1}{3}$$

$$= \int_0^1 \xi f(\xi) \int_\xi^1 x \, dx d\xi - \int_0^1 f(\xi) \int_\xi^1 x^2 \, dx d\xi$$

$$+ \frac{1}{10} \int_0^1 \xi f(\xi) d\xi + \frac{c_1}{3}$$

$$= \frac{3}{5} \int_0^1 \xi f(\xi) - \frac{1}{3} \int_0^1 f(\xi) d\xi - \frac{1}{6} \int_0^1 \xi^3 f(\xi) d\xi + \frac{c_1}{3}$$

and

$$c_1 = \int_0^1 \left(\frac{\xi^3}{2} - \frac{9\xi}{5} + 1 \right) f(\xi) d\xi.$$

Finally, we put everything together as

$$y = \int_0^x \left[\xi - x + \frac{x^3 \xi}{2} + x \left(\frac{\xi^3}{2} - \frac{9\xi}{5} + 1 \right) \right] f(\xi) d\xi$$

$$+ \int_x^1 \left[\frac{x^3 \xi}{2} + x \left(\frac{\xi^3}{2} - \frac{9\xi}{5} + 1 \right) \right] f(\xi) d\xi,$$

so

$$G_m(x; \xi) = \begin{cases} \dfrac{x^3 \xi}{2} + \dfrac{x \xi^3}{2} - \dfrac{9x\xi}{5} + x, & \text{if } 0 \le x \le \xi, \\[2mm] \dfrac{x^3 \xi}{2} + \dfrac{x \xi^3}{2} - \dfrac{9x\xi}{5} + \xi, & \text{if } \xi \le x \le 1, \end{cases}$$

which is symmetric (though, in general, it need not be) and also continuous at $x = \xi$.[‡‡]

Exercises 10.1

1. Compute Green's function or the generalized Green's function for the BVP

   a) $y'' = -f, y(0) = y'(1) = 0$
   b) $y'' = -f, y(0) + y'(0) = y(L) = 0, L \ne 1$ (see part (f), also)
   c) $y'' = -f, y'(0) = y'(1) = 0$
   d) $y'' + k^2 y = -f, y(0) = y'(L) = 0, k \ne \frac{2n+1}{2L}\pi$ for integral n

---

[‡‡]To see what G_q and G_m do for us, see Exercise 10 in Section 10.2.

e) $y'' - y = 0, y(0) = y(1) = 0$

f) $y'' = -f, y(0) + y'(0) = y(1) = 0$

2. Use the Green's functions derived in the various examples, or in Exercise 1, to solve the nonhomogeneous BVP. (Make sure your answer matches the one you get using traditional methods.)

a) $y'' = -e^{rx}, 0 < x < 1, r \neq 0$ constant,
   $y(0) = y(1) = 0$

b) $y'' + y = 1, 0 < x < \frac{\pi}{2}$,
   $y(0) = y\left(\frac{\pi}{2}\right) = 0$

c) $y'' = x, 0 < x < 1$,
   $y(0) = y'(1) = 0$

3. Use the properties of Green's function to construct Green's function for

a) The singular Cauchy–Euler problem

$$(xy')' - \frac{1}{x}y = -f, \qquad 0 < x < 2,$$

$$y(0) < \infty, \quad y(2) = 0.$$

b) The Bessel's BVP

$$(xy')' - \frac{n^2}{x}y + k^2 xy = -f, \qquad 0 < x < 1,$$

$$y(0) < \infty, \quad y(1) = 0.$$

Here, $n \geq 0$ is an integer, while $k \neq 0$ is real and *not* a root of $J_n(x)$ (why this restriction?).
Hint: Remember that rW is constant. You may use the fact that, in this case, it turns out that

$$rW = xW[J_n, Y_n] = \frac{2}{\pi}.$$

4. Find a Fourier series representation for Green's function for the BVP

a) From Exercise 1a

b) From Example 1 ($k \neq$ an integer)

5. Show that

$$\int_0^x \int_0^z f(\xi)d\xi \; dz = \int_0^x (x - \xi)f(\xi)d\xi.$$

6. a) Verify that if y is continuous at $x = \xi$, then

$$\lim_{\epsilon \to 0} \int_{\xi-\epsilon}^{\xi+\epsilon} (y'' + k^2 y)dx = y'(\xi+) - y'(\xi-).$$

b) More generally, verify that if r, q and y are continuous at $x = \xi$, then

$$\lim_{\epsilon \to 0} \int_{\xi-\epsilon}^{\xi+\epsilon} [(r(x)y'(x))' + q(x)y(x)]dx = r(\xi)[y'(\xi+) - y'(\xi-)].$$

7. Show that the function

$$y = \int_0^x \xi f(\xi)d\xi - x \int_0^x f(\xi)d\xi + \frac{x^3}{2} \int_0^1 \xi f(\xi)d\xi + cx$$

satisfies $y(1) - y'(1) = 0$ for any well-enough behaved function f and any constant c.

8. In Example 1, justify

a) The variation of parameters computations

b) The simplifications leading to the final answer

9. Fill in the details in the computations leading up to Theorem 10.1

a) In the variation of parameters computation

b) In showing that y_p satisfies the boundary conditions

10. Prove that Green's function for the BVP of Theorem 10.1 is unique.

11. a) Verify that, for any well-enough behaved functions f and h, the function

$$g = f - \left\langle f, \frac{h}{\|h\|} \right\rangle \frac{h}{\|h\|}$$

is orthogonal to h.

b) More generally, extend this procedure and create an orthogonal sequence of functions

$$g_1, g_2, g_2, \cdots$$

from a sequence of functions

$$f_1, f_2, f_3, \cdots .$$

This is the **Gram–Schmidt orthogonalization** procedure.

c) If the functions f_n form a linearly independent set, turn the sequence g_n into an orthonormal sequence. Why must we specify linear independence here?

12. Compute both $\int_0^1 G_g(x; \xi)f(\xi)d\xi)$ and $\int_0^1 G_m(x; \xi)f(\xi)d\xi$ for the problem in Example 4, with $f(x) = 2 - 3x$. How do these compare to the problem's actual solution?

13. We may, instead, write our nonhomogeneous BVPs in the form

$$(ry')' + qy = -f, \qquad a < x < b,$$
$$a_1 y(a) + a_2 y'(a) = b_1 y(b) + b_2 y'(b) = 0.$$

a) Then, we'll have a unique solution unless the associated homogeneous problem has a nontrivial solution (as usual). How do we say this using eigenvalue language?

b) Show that, in the case that there is a unique solution, we may solve this problem via generalized Fourier series by expanding y and f in terms of eigenfunctions ϕ_n of the problem

$$(ry')' + (q + \lambda w)y = 0,$$
$$a_1 y(a) + a_2 y'(a) = b_1 y(b) + b_2 y'(b) = 0,$$

for *any* weight function w. Write down Green's function in this case.

c) When will this method not work, and why? (In practice, this generalization turns out not to be very useful as, for most choices of w, the eigenfunctions will be difficult to compute.)

10.2 Green's Function and the Dirac Delta Function

It was suggested at the start of this chapter that Green's function is the solution of the BVP

$$Ly = -\delta(x - \xi), \qquad a < x < b,$$
$$a_1 y(a) + a_2 y'(a) = b_1 y(b) + b_2 y'(b) = 0.$$

Of course, this is a *distributional* equation and care must be taken when we "mix" classical and distributional settings. Keeping this in mind, let's begin by showing that the above statement is true for the problem in Example 1 of the previous section.

Example 1 We computed Green's function for the BVP

$$y'' + k^2 y = -f, \qquad 0 < x < L,$$
$$y(0) = y(L) = 0,$$

where k^2 is not an eigenvalue of the associated homogeneous problem, to be

$$G(x; \xi) = \begin{cases} \dfrac{1}{k \sin kL} \sin k\xi \sin k(L - x), & 0 \le \xi \le x, \\[2mm] \dfrac{1}{k \sin kL} \sin kx \sin k(L - \xi), & x \le \xi \le L. \end{cases}$$

We'd like to show that G is the (distributional) solution of the problem

$$y'' + k^2 y = -\delta(x - \xi), \qquad 0 < x < L,$$
$$y(0) = y(L) = 0.$$

Since G was constructed to satisfy the boundary conditions, we need only show that it satisfies the ODE.

First, we write G using Heaviside functions:

$$G(x;\xi) = \frac{1}{k \sin kL}[H(x - \xi) \sin k\xi \sin k(L - x) + H(\xi - x) \sin kx \sin k(L - \xi)].$$

Then,

$$G_x(x;\xi) = \frac{1}{k \sin kL} \delta(x - \xi)[\sin k\xi \sin k(L - x) - \sin kx \sin k(L - \xi)]$$
$$+ \frac{1}{\sin kL}[-H(x - \xi) \sin k\xi \cos k(L - x) + H(\xi - x) \cos kx \sin k(L - \xi)].$$

Now, the first term is just the zero-function (why?). Then,

$$G_{xx}(x;\xi) = -\frac{1}{\sin kL} \delta(x - \xi)[\sin k\xi \cos k(L - x) + \cos kx \sin k(L - \xi)]$$
$$- k^2 G(x;\xi)$$
$$= -\delta(x - \xi) - k^2 G(x;\xi) \quad \text{(why?)}.$$

Therefore, we have

$$G_{xx} + k^2 G = -\delta(x - \xi).$$

We say that G is the **response** *to a* **unit impulse** or a **unit concentrated source** at $x = \xi$. For example, $G(t;t_0)$ which satisfies

$$G_{tt} + k^2 G = -\delta(t - t_0)$$

is the response of a spring-mass-dashpot system to a unit impulse (a *hammer blow*) at time $t = t_0$.

More generally, we can show that Green's function is always related to δ in this way.

Theorem 10.4 *Green's function given in Theorem 10.1 is the unique distributional solution of the BVP*

$$(ry')' + (q + \lambda_0 w)y = -\delta(x - \xi), \qquad a < x < b,$$
$$a_1 y(a) + a_2 y'(a) = b_1 y(b) + b_2 y'(b) = 0.$$

PROOF   See Exercise 2.   ∎

From earlier, we know that the solution to the Sturm–Liouville problem in Theorem 3.1 is

$$y = \int_a^b G(x;\xi)f(\xi)d\xi,$$

but it's instructive to apply *Green's formula* or *Green's second identity* in this situation. Let's restate Green's first and second identities (which we proved in Exercise 6, Section 8.1). So, given the Sturm–Liouville operator

$$L[y] = (ry')' + qy,$$

regular on $a \leq x \leq b$, we have *Green's first identity*,

$$\int_a^b uL[v]dx = r(x)u(x)v'(x)|_a^b - \int_a^b r(x)u'(x)v'(x)dx + \int_a^b q(x)u(x)v(x)dx,$$

and *Green's second identity*,

$$\int_a^b (uL[v] - vL[u])dx = r(x)[u(x)v'(x) - v(x)u'(x)]|_a^b,$$

for all well-enough behaved u and v. (As we saw, in many cases we can extend these identities to singular operators.)

We now use Green's second identity with $u = y$, the solution, and $v = G(x;\xi)$. So,

$$\int_a^b (yLG - GLy)dx = r(x)[yG_x - Gy']|_{x=a}^{x=b}.$$

We showed back in Section 7.2 that if y_1 and y_2 both satisfy $a_1y(c)+a_2y'(c) = 0$, then $y_1(c)y_2'(c)-y_2(c)y_1'(c) = 0$. Therefore, as y and G satisfy the boundary conditions, our right side is zero.

Now we have

$$0 = -\int_a^b (yLG - GLy)dx$$

$$= -\int_a^b y(x)\delta(x-\xi)dx + \int_a^b G(x;\xi)f(x)dx \ *$$

$$= y(\xi) + \int_a^b G(\xi;x)f(x)dx$$

or

$$y(\xi) = \int_a^b G(\xi;x)f(x)dx,$$

---

*Again, it is not obvious that all of the operations involved—integration by parts, etc.—are valid in the realm of distributions. It turns out that Green's identity *does* still hold in this setting.

that is,

$$y(x) = \int_a^b G(x;\xi) f(\xi) d\xi.$$

We may continue and show that $G(x;\xi)$ defined this way must satisfy the classical properties given in Theorem 10.1. We do so for some of these properties in Exercise 3. In particular, the *symmetry* property $G(x;\xi) = G(\xi;x)$ has a nice physical interpretation in this setting—it says that the response at x due to an impulse at ξ is the same as the response at ξ due to an impulse at x. This *reciprocal* behavior between the points x and ξ is the reason that this symmetry is called the property of *reciprocity*.

A few examples are in order.

Example 2 Repeat Example 2 of the previous section, but this time solve the problem

$$y'' = -\delta(x - \xi), \qquad 0 < x < L,$$
$$y(0) = y(L) = 0.$$

Since $\frac{d}{dx} H(x - \xi) = \delta(x - \xi)$, we have

$$y'' = -\delta(x - \xi) \Rightarrow y' = -H(x - \xi) + c_1$$

$$\Rightarrow y = -\int_0^x H(z - \xi) dz + c_1 x + c_2$$

$$= \begin{cases} c_1 x + c_2, & \text{if } 0 \le x \le \xi, \\ -x + \xi + c_1 x + c_2, & \text{if } \xi \le x \le L. \end{cases}$$

The boundary conditions give us

$$y(0) = 0 = c_2$$
$$y(L) = 0 = -L + \xi + c_1 L + c_2$$

so

$$c_1 = \frac{L - \xi}{L}, \qquad c_2 = 0,$$

and our solution is

$$y = G(x;\xi) = \begin{cases} \dfrac{x(L - \xi)}{L}, & \text{if } 0 \le x \le \xi, \\ \dfrac{\xi(L - x)}{L}, & \text{if } \xi \le x \le L. \end{cases}$$

By the way, we have a nice physical interpretation of this problem, too. The nonhomogeneous heat equation

$$u_t = u_{xx} + \delta(x - \xi)$$

models the temperature distribution in a rod with a unit, time-independent point source of heat at the point $x = \xi$. The steady state version of this problem is, of course,

$$u_{xx} = -\delta(x - \xi),$$

so that $G(x; \xi)$ represents its steady state temperature distribution.

Going a bit further, let's integrate the original ODE over an ϵ-interval centered at $x = \xi$:

$$\int_{\xi-\epsilon}^{\xi+\epsilon} y'' \, dx = -\int_{\xi-\epsilon}^{\xi+\epsilon} \delta(x - \xi) dx$$

or

$$G_x(\xi + \epsilon; \xi) - G_x(\xi - \epsilon, \xi) = -1.$$

Of course, letting $\epsilon \to 0$ gives us Property 3 of Green's function, but here the equation is still true for *any* (small enough) $\epsilon > 0$. What this says, physically, is that the flux across the boundary is equal to the heat generated within the interval; in other words, this is just a statement of the *conservation of heat energy*.

This idea will play a significant role in the computation of Green's function in higher dimensions.

Example 3 Solve the BVP

$$y'' - k^2 y = -\delta(x - \xi), \qquad 0 < x < L, k > 0,$$
$$y(0) = y(L) = 0.$$

First, as solutions of the associated homogeneous equation, we may use $y_1 = e^x$ and $y_2 = e^{-x}$ or $y_1 = \cosh x$ and $y_2 = \sinh x$. However, led by our experiences in Section 10.1, we can make our lives easier by taking y_1 which satisfies the first boundary condition, and y_2 which satisfies the other boundary condition. Taking $y_1 = \sinh x$, it should be clear that $y_2 = \sinh k(L - x)$ will work, *as long as they're linearly independent*. But

$$W[\sinh kx, \sinh k(L - x)] = \begin{vmatrix} \sinh kx & \sinh k(L - x) \\ k \cosh kx & -k \cosh k(L - x) \end{vmatrix}$$

$$= -k[\sinh kx \cosh k(L - x) + \cosh kx \sinh k(L - x)]$$
$$= -k \sinh(kx + k(L - x))$$
$$= -k \sinh kL \neq 0.$$

Then, using variation of parameters, we have $y_p = u_1 y_1 + u_2 y_2$, where

$$u_1' = -\frac{\delta(x - \xi) \sinh k(L - x)}{\sinh kL} \quad \text{and} \quad u_2' = \frac{\delta(x - \xi) \sinh x}{\sinh kL}.$$

We may integrate from zero to x, for example, to get u_1 and u_2, but, again guided by what we've seen, let's take

$$u_1 = \frac{1}{\sinh kL} \int_x^L \delta(z - \xi) \sinh k(L - z) dz$$

$$= \begin{cases} \sinh k(L - \xi) / \sinh kL, & \text{if } 0 \le x \le \xi, \\ 0, & \text{if } \xi < x \le L \end{cases} \quad \text{(why?)}$$

and

$$u_2 = \frac{1}{\sinh kL} \int_0^x \delta(z - \xi) \sinh kz \; dz$$

$$= \begin{cases} 0, & \text{if } 0 \le x < \xi, \\ \dfrac{\sinh k\xi}{\sinh kL}, & \text{if } \xi \le x \le L. \end{cases}$$

The general solution becomes

$$y = G(x; \xi) = \begin{cases} c_1 \sinh kx + c_2 \sinh k(L - x) \\ \quad + \dfrac{\sinh k(L - \xi) \sinh kx}{\sinh kL}, & \text{if } 0 \le x \le \xi, \\ c_1 \sinh kx + c_2 \sinh k(L - x) \\ \quad + \dfrac{\sinh k\xi \sinh k(L - x)}{\sinh kL}, & \text{if } \xi \le x \le L. \end{cases}$$

Finally, from the boundary conditions, $c_1 = c_2 = 0$ (as expected), and our solution is

$$G(x; \xi) = \begin{cases} \dfrac{\sinh kx \sinh k(L - \xi)}{\sinh kL}, & \text{if } 0 \le x \le \xi, \\ \dfrac{\sinh k\xi \sinh k(L - x)}{\sinh kL}, & \text{if } \xi \le x \le L. \end{cases}$$

(By the way, make sure it's clear to you that $G(x; \xi) = G(\xi; x)$ here.)
We may use Fourier series to find Green's function in this setting, as well.

Example 4 Use Fourier series to solve

$$y'' = -\delta(x - \xi), \qquad 0 < x < \pi,$$
$$y(0) = y(\pi) = 0.$$

As before, we let

$$y = \sum_{n=1}^{\infty} b_n \sin nx.$$

Then, taking y'' leads to

$$\sum_{n=1}^{\infty} n^2 b_n \sin nx = \delta(x - \xi),$$

and we multiply both sides by $\sin mx$ and integrate to get

$$\sum_{n=1}^{\infty} n^2 b_n \int_0^\pi \sin nx \sin mx \; dx = \int_0^\pi \delta(x - \xi) \sin mx \; dx$$

or

$$b_n = \frac{2}{\pi n^2} \sin n\xi.$$

So our solution is

$$y(x; \xi) = \frac{2}{\pi} \sum_{n=1}^{\infty} \frac{\sin nx \sin n\xi}{n^2},$$

as expected. In fact, we also have shown that, in a distributional setting, the Fourier sine series for the delta function is

$$\delta(x - \xi) \sim \sum_{n=1}^{\infty} \sin n\xi \sin nx.$$

We also may deal with the modified Green's function using the delta function.

Example 5 We repeat Example 4 of the previous section but this time by solving the problem

$$y'' = -\delta(x - \xi), \qquad 0 < x < 1,$$
$$y(0) = y(1) - y'(1) = 0.$$

As earlier, $\phi(x) = \sqrt{3}\, x$ in the nontrivial solution with norm equal to 1. Then, the projection of $\delta(x - \xi)$ onto ϕ is

$$\langle \delta(x - \xi), \phi \rangle = 3x \int_0^1 x\delta(x - \xi)dx = 3x\xi,$$

so we solve

$$y'' = -\delta(x - \xi) + 3x\xi,$$
$$y(0) = y(1) - y'(1) = 0.$$

Integrating gives us

$$y' = -H(x - \xi) + \frac{3x^2\xi}{2} + c_1$$

$$y' = \begin{cases} \dfrac{x^2\xi}{2} + c_1 x + c_2, & \text{if } 0 \leq x \leq \xi, \\[2mm] -x + \xi + \dfrac{x^2\xi}{2} + c_1 x + c_2, & \text{if } \xi \leq x \leq 1. \end{cases}$$

The first boundary condition gives $c_2 = 0$, while the second boundary condition is satisfied identically (show this yourself!). Finally, we select c_1 so that $\langle x, y \rangle = 0$:

$$0 = \langle x, y \rangle = \int_0^1 x \left(\frac{x^3 \xi}{2} + c_1 x \right) dx$$

$$+ \int_\xi^1 x(-x + \xi) dx$$

or

$$c_1 = \frac{\xi^3}{2} - \frac{9\xi}{5} + 1.$$

Substituting this into our expression for y gives the same result as before.

An advantage of the delta function approach is that it gives us an easy way to deal with nonhomogeneous boundary conditions, which we look at in Exercise 5.

THE FUNDAMENTAL SOLUTION OR FREE-SPACE GREEN'S FUNCTION

We may still talk about Green's function for ODEs on $-\infty < x < \infty$. Of course, we won't have the usual boundary conditions but, as before, we'll have certain integrability conditions.

As we're on $-\infty < x < \infty$, we may use the Fourier transform and, in fact, it gives us an elegant and easy way to deal with the problem. Let's begin with an example.

Example 6 We wish to solve the ODE

$$y'' + k^2 y = -f, \qquad -\infty < x < \infty,$$

in terms of the solution of

$$z'' + k^2 z = -\delta(x - \xi), \qquad -\infty < x < \infty.$$

Let's transform both equations and see what we can do. The latter transforms to

$$Z(\alpha; \xi) = \frac{1}{\sqrt{2\pi}} e^{-i\alpha\xi} \frac{1}{\alpha^2 - k^2},$$

and the former transforms to

$$Y(\alpha) = F(\alpha) \frac{1}{\alpha^2 - k^2}.$$

It's not difficult to find $\mathcal{F}^{-1}[Z(\alpha)]$, and we can write

$$Y(\alpha) = \sqrt{2\pi} \, e^{i\alpha\xi} F(\alpha) Z(\alpha),$$

so our solution will be a convolution. However, the $e^{\pm i\alpha\xi}$ terms seem to get in the way (but see Exercise 10). We *really* need only solve

$$z'' + k^2 z = -\delta(x), \qquad -\infty < x < \infty.$$

Then,

$$Z(\alpha) = Z(\alpha; 0) = \frac{1}{\sqrt{2\pi}} \frac{1}{\alpha^2 - k^2}$$

and

$$Y(\alpha) = \sqrt{2\pi}\, F(\alpha) Z(\alpha),$$

from which our solution is just

$$y(x) = f(x) * z(x).$$

Before continuing the example, let's state an official definition and a theorem.

Definition 10.1 *Given the linear ordinary differential operator $L[y]$, its* **fundamental solution** *or* **free-space Green's function** *is the function $z(x; \xi)$ which satisfies*

$$L[z(x; \xi)] = -\delta(x - \xi), \qquad -\infty < x < \infty.$$

The following theorem will suffice for our purposes.

Theorem 10.5 *The solution of the linear ODE*

$$L[y] = -f, \qquad -\infty < x < \infty$$

with constant coefficients is

$$y(x) = f(x) * z(x; 0),$$

where z is its fundamental solution.

(The proof is an obvious generalization of what we've done in Example 6.)

Example 6 (cont.) So we continue with Example 6. We need only find $\mathcal{F}^{-1}[Z(\alpha)]$. We have

$$\frac{1}{\alpha^2 - k^2} = \frac{1}{2k}\left(\frac{1}{\alpha - k} - \frac{1}{\alpha + k}\right),$$

so we'll need the following, from Chapter 6:

$$\mathcal{F}^{-1}[G(\alpha - c)] = e^{icx}\mathcal{F}^{-1}[G(\alpha)] \quad \text{(Exercise 9d, Section 6.3),}$$

$$\mathcal{F}[\text{sgn } x] = -i\sqrt{\frac{2}{\pi}}\frac{1}{\alpha} \qquad \text{(Section 6.5),}$$

where, again,

$$\operatorname{sgn} x = H(x) - H(-x) = \begin{cases} 1, & \text{if } x > 0, \\ -1, & \text{if } x < 0. \end{cases}$$

Then,

$$\mathcal{F}^{-1}\left[\frac{1}{\alpha - k}\right] = e^{ikx}\mathcal{F}^{-1}\left[\frac{1}{\alpha}\right] = i\sqrt{\frac{\pi}{2}}\, e^{ikx} \operatorname{sgn} x$$

and

$$\mathcal{F}^{-1}\left[\frac{1}{\alpha + k}\right] = i\sqrt{\frac{\pi}{2}}\, e^{-ikx} \operatorname{sgn} x.$$

Finally, then,

$$\mathcal{F}^{-1}\left[\frac{1}{\sqrt{2\pi}}\frac{1}{\alpha^2 - k^2}\right] = \frac{i}{4k}(e^{ikx} - e^{-ikx}) \operatorname{sgn} x$$

$$= -\frac{1}{2k}\sin kx \operatorname{sgn} x = z(x;0),$$

and our solution is

$$y(x) = f(x) * \left(-\frac{1}{2k}\sin kx \operatorname{sgn} x\right)$$

$$= -\frac{1}{2k}\int_{-\infty}^{\infty}\sin k(x-y)\operatorname{sgn}(x-y)f(y)dy$$

$$= \frac{1}{2k}\left[\int_{x}^{\infty}\sin k(x-y)f(y)dy - \int_{-\infty}^{x}\sin k(x-y)f(y)dy\right] \text{ (why?)}$$

$$\left(\text{or } = \frac{1}{2k}\left[\int_{0}^{\infty}\sin ky f(x-y)dy - \int_{-\infty}^{0}\sin ky f(x-y)dy\right]\right).$$

By the way, it's easy to show that the more general fundamental solution is

$$z(x;\xi) = -\frac{1}{2k}\sin k(x-\xi)\operatorname{sgn}(x-\xi),$$

and the solution to our problem is then seen to be what we expect it to be:

$$y = \int_{-\infty}^{\infty} z(x;\xi)f(\xi)d\xi.$$

In fact, in general, if

$$L[z(x;\xi)] = -\delta(x-\xi),$$

then the change of variable $x \to x + \xi$ tells us that

$$L[z(x + \xi; \xi)] = -\delta(x).$$

So, from uniqueness, we have the **translation property** of the fundamental solution

$$z(x - \xi; 0) = z(x; \xi).$$

(See Exercise 8.) Then,

$$f(x) * z(x; 0) = \int_{-\infty}^{\infty} z(x - \xi; 0) f(\xi) d\xi$$

$$= \int_{-\infty}^{\infty} z(x; \xi) f(\xi) d\xi.$$

Theorem 10.6 *The solution of the problem*

$$L[y] = -f, \qquad -\infty < x < \infty$$

is

$$y = \int_{-\infty}^{\infty} z(x; \xi) d(\xi) d\xi,$$

where $z(x; \xi)$ is the fundamental solution.

We may look at modified Green's functions in this distributional setting, as well. Instead, we choose to introduce another Green's-like function, the *Neumann function*. We do so by example.

Example 7 Consider again the (Neumann) problem

$$y'' = -f, \qquad 0 < x < 1,$$
$$y'(0) = y'(1) = 0.$$

If we integrate the ODE from $x = 0$ to $x = 1$, we have

$$\int_0^1 f(x) dx = y'(0) - y'(1) = 0,$$

a compatibility condition which is really just the condition that $\langle 1, f \rangle = 0$, from earlier.

Therefore, the problem

$$G'' = -\delta(x - \xi), \qquad 0 < x < 1,$$
$$G'(0) = G'(1) = 0$$

will *not* have a solution (why not?). However, the condition

$$-1 = -\int_0^1 \delta(x - \xi)dx = G'(1) - G'(0)$$

suggests that, instead of changing the nonhomogeneous part of the ODE (as we did in the case of the generalized Green's function), we might try changing the boundary conditions. So, let's look at the problem

$$H'' = -\delta(x - \xi), \qquad 0 < x < 1,$$
$$H'(0) = a, \quad H'(1) = b,$$

and see what happens. We integrate the ODE to arrive at

$$H(x; \xi) = \begin{cases} c_1 x + c_2, & \text{if } 0 \le x < \xi, \\ \xi - x + c_1 x + c_2, & \text{if } \xi < x \le 1, \end{cases}$$

and, keeping in mind the compatibility equation $a - b = 1$, we have

$$c_1 = a, \quad c_2 \text{ arbitrary.}$$

Really, this just says that we have a solution as long as $a - b = 1$.

What does this do for us? As usual, we apply Green's second identity and, this time, we get

$$\int_0^1 (yH'' - Hy'')dx = (yH' - Hy')|_0^1$$

or

$$y = \int_0^1 H(x; \xi)f(\xi)d\xi + (a - 1)y(1) - ay(0).$$

In other words, we've found the solution of the original problem, as we may treat a as arbitrary.

The function H is called the **Neumann function** for the original problem. This same approach will be used to deal with the Neumann problem in higher dimensions.

Physically, we may look at the problem

$$H'' = -\delta(x - \xi), \qquad 0 < x < 1,$$
$$H'(0) = a, \quad H'(1) = a - 1$$

as a steady state heat problem with constant unit heat source at $x = \xi$, and constant outward flux of heat equal to $-(a - 1) + a = 1$.

Exercises 10.2

1. Redo Exercise 1 of Section 10.1 but this time use the delta function approach.

2. Suppose that y_1 and y_2 are linearly independent solutions of the homogeneous problem

$$(ry')' + (q + \lambda_0 w)y = 0, \qquad a < x < b,$$
$$a_1 y(a) + a_2 y'(a) = b_1 y(b) + b_2 y'(b) = 0$$

(where λ_0 is *not* an eigenvalue). As in Example 1, rewrite the Green's function

$$G(x; \xi) = \begin{cases} -\dfrac{y_1(\xi)y_2(x)}{r(\xi)W(\xi)}, & \text{if } a \le \xi \le x, \\[2mm] -\dfrac{y_1(x)y_2(\xi)}{r(\xi)W(\xi)}, & \text{if } x \le \xi \le b \end{cases}$$

using Heaviside functions, and show that it satisfies the ODE

$$(ry')' + (q + \lambda_0 w)y = -\delta(x - \xi), \qquad a < x < b.$$

3. Here we go *backwards* and show that if we *define* Green's function as the solution of the delta function BVP in this section, then it must satisfy the five properties given in Theorem 10.1.

   a) Use $u = G(x; \xi_1)$ and $v = G(x; \xi_2)$ in Green's second identity to show that G is symmetric, that is, that $G(x; \xi) = G(\xi; x)$ for any numbers x, ξ in the interval in question.

   b) In order to show that

   $$G_x(\xi+; \xi) - G_x(\xi-; \xi) = -\frac{1}{r(\xi)},$$

   first let $u = 1$ and $v = G(x; \xi)$ in Green's second identity and then, having chosen the limits of integration appropriately, treat the result classically.

4. Use the Fourier sine series representation of $\delta(x-\xi)$ and take the Fourier transform of both sides to show that

$$e^{-i\alpha\xi} = i\pi \sum_{n=1}^{\infty} \sin n\xi [\delta(\alpha + n) - \delta(\alpha - n)], \quad \xi \ne 0.$$

5. **Nonhomogeneous boundary conditions**

a) Suppose that the BVP

$$y'' = -f, \qquad a < x < b,$$
$$y(a) = y(b) = 0$$

has Green's function $G(x; \xi)$. Use Green's second identity, with $u = y$, the solution, and $v = G(x; \xi)$, to show that the solution of the BVP

$$y'' = -f, \qquad a < x < b,$$
$$y(a) = \alpha, \quad y(b) = \beta,$$

is

$$y = \int_a^b G(x; \xi) f(\xi) d\xi - \beta G_\xi(x; b) + \alpha G_\xi(x; a).$$

b) Generalize the idea in part (a) to solve

$$L[y] = (ry')' + (q + \lambda_0 w)y = -f, \qquad a < x < b,$$
$$y(a) = \alpha, \quad y(b) = \beta.$$

(Here, λ_0 is not an eigenvalue of the associated homogeneous problem.)

c) Generalize part (b) still further, to the cases where the left end boundary condition is $y(a) = \alpha$ or $y'(a) = \alpha$, while the right end boundary condition is $y(b) = \beta$ or $y'(b) = \beta$.

d) Finally, generalize to the case where the boundary conditions are

$$a_1 y(a) + a_2 y'(a) = \alpha,$$
$$b_1 y(b) + b_2 y'(b) = \beta,$$

where a_1, a_2, b_1 and b_2 are nonzero and $a_1 b_2 - a_2 b_1 \neq 0$.

6. a) Show that the BVP

$$y^{(4)} = 0, \qquad 0 < x < 1,$$
$$y(0) = y'(0) = y(1) = y'(1)$$

has a unique solution.

b) Find Green's function for the BVP

$$y^{(4)} = -f, \qquad 0 < x < 1,$$
$$y(0) = y'(0) = y(1) = y'(1) = 0.$$

7. **Fundamental solution:** Compute the *general* fundamental solution

  a) $y'' - k^2 y = -f, \ \lim_{x \to \pm\infty} y(x) = 0, k > 0$

  b) $y'' = -f, \ \lim_{x \to \pm\infty} y(x) = 0$

  c) $y^{(4)} - k^4 y = -f, \ \lim_{x \to \pm\infty} y(x) = 0, k > 0$

8. Verify, formally, the more general *translation* property for the fundamental solution $F(x; \xi)$, that

$$F(x + a; \xi - a) = F(x; \xi)$$

   for all x, ξ and a.

9.   a) Prove that if g is well-enough behaved, then

$$\frac{d}{dx} \int_a^x g(x, y) dy = g(x, x) + \int_a^x g_x(x, y) dy.$$

   b) Use part (a) to show that the solution given in Example 6 does
      satisfy the ODE, formally.

   c) Do the same for the problem in Exercise 7a.

   d) Do the same for the problem in Exercise 7b.

10. Referring to Example 4 in Section 10.1, suppose we have the problem

$$y'' = -f, \qquad 0 < x < 1,$$
$$y(0) = y(1) - y'(1) = 0,$$

   and suppose that $\langle x, f \rangle = 0$.

   a) Show that

$$\int_0^1 G_g(x; \xi) f(\xi) d\xi$$

      gives the same function for any choice of the generalized Green's
      function G_g.

   b) Show that all solutions of the above problem are of the form

$$y = \int_0^1 G_g(x; \xi) f(\xi) d\xi + C.$$

   c) Show that if there is a solution which is orthogonal to x, then that
      solution is given by

$$y = \int_0^1 G_m(x; \xi) f(\xi) d\xi.$$

11. **Neumann function**

a) Let $f(x) = 2x - 1$ in Example 7, and solve the problem directly via integration.

b) Instead, compute the function $y = \int_0^1 H(x; \xi)(2\xi - 1)d\xi$ for H given in Example 7. Compare with the solution of part (a).

c) Do the same as in parts (a) and (b) for the problem

$$y'' = 3x - 2, \qquad 0 < x < 1,$$
$$y(0) = y(1) - y'(1) = 0.$$

12. One *could* proceed to solve the problem

$$L[y] = -\delta(x - \xi), \qquad a < x < b,$$
$$y(a) = y(b) = 0$$

by first finding the fundamental solution

$$L[z(x; \xi)] = -\delta(x - \xi), \qquad -\infty < x < \infty,$$

then letting $y = z + u$ and solving the resulting u-problem. Do this for the problem

$$y'' + k^2 y = -\delta(x - \xi), \qquad 0 < x < L,$$
$$y(0) = y(L) = 0,$$

and make sure that your solution matches that of Example 1 in Section 10.1.

13. **Method of images—problems on the interval $0 \le x < \infty$:** Suppose we'd like to solve the problem

$$y'' + k^2 y = -\delta(x - \xi), \qquad x > 0,$$
$$y(0) = 0.$$

a) Solve, instead, the problem

$$z'' + k^2 z = -\delta(x - \xi) + \delta(x + \xi), \qquad -\infty < x < \infty.$$

b) Show that the solution to part (a) is actually the solution of the original problem. In particular, make sure you know why

$$z'' + k^2 z = -\delta(x - \xi) \quad \text{on} \quad 0 < x < \infty.$$

This is another version of the *method of images*. Note that we have added a *sink* at $x = -\xi$ in order to cancel out the *source* at $x = \xi$.

10.3 Green's Functions for Elliptic PDEs (I): Poisson's Equation in Two Dimensions

Let's begin our study of Green's functions for PDEs by considering the Dirichlet problem

$$\nabla^2 u = u_{xx} + u_{yy} = -f \quad \text{on} \quad D,$$
$$u = g \quad \text{on} \quad C,$$

where D is a simply-connected domain in $\mathbb{R}^2$ bounded by the simple closed curve C. As we've seen before, if v is the solution of

$$\nabla^2 v = -f \quad \text{on} \quad D,$$
$$v = 0 \quad \text{on} \quad C,$$

and w is the solution of

$$\nabla^2 w = 0 \quad \text{on} \quad D,$$
$$w = g \quad \text{on} \quad C,$$

then $u = v + w$ is the solution we're looking for. Further, the w equation is just Laplace's equation, already solved in Chapters 2 and 9.

Therefore, we search for Green's function for the v-equation. We expect—and will show below—that this $G(x, y; \xi, \eta)$ will satisfy

$$\nabla^2 G = -\delta(x - \xi, y - \eta) \quad \text{on} \quad D,$$
$$G = 0 \quad \text{on} \quad C,$$

where the variables are x and y, and where the two-dimensional delta function is defined, as earlier, by

$$\delta(x, y) = \delta(x)\delta(y).$$

The existence of the boundary still complicates matters, but it turns out that we can even get around *this*. Suppose that $F(x, y; \xi, \eta)$ satisfies

$$\nabla^2 F = -\delta(x - \xi, y - \eta), \qquad -\infty < x < \infty, -\infty < y < \infty,$$

that is, suppose that F is the *fundamental solution* or *free-space Green's function* for Poisson's equation. Then, once we've found F, we need only let $G(x, y; \xi, \eta) = F(x, y; \xi, \eta) + Z(x, y; \xi, \eta)$, where Z satisfies Laplace's equation

$$\nabla^2 Z = 0 \quad \text{on} \quad D,$$
$$Z = -F \quad \text{on} \quad C$$

(which should cause no problem, as F is smooth on $(x, y) \neq (\xi, \eta)$). To repeat, once we have found the fundamental solution of the Laplace operator, then we will have reduced any problem involving Poisson's equation to one which entails solving Laplace's equation on the same domain.

Before doing anything else, let's compute the well-known fundamental solution of the Laplace operator, after which we'll show that Green's function actually does what we'd like it to do. So we wish to solve

$$\nabla^2 F(x, y; \xi, \eta) = -\delta(x - \xi, y - \eta), \qquad -\infty < x < \infty, -\infty < y < \infty.$$

It should be clear that F is symmetric, that is, that

$$F(x, y; \xi, \eta) = F(\xi, \eta; x, y)$$

for all ordered pairs (x, y) and (ξ, η) (why?) and that F is radially symmetric with respect to the point (ξ, η).

We may also ask about the *jump* in "the" first derivative of F at (ξ, η) (realizing that, in two dimensions, there are infinitely many directional derivatives). From the definition of $\delta(x)$, it should be clear (see Exercise 1) that

$$\iint_D \delta(x - \xi, y - \eta) dx dy = 1$$

for *any* domain D with $(\xi, \eta) \in D$. Then, taking D_ϵ to be the disk of radius ϵ centered at (ξ, η), we have, upon integrating the PDE,

$$\iint_{D_\epsilon} \nabla^2 F \, dx dy = -\iint_{D_\epsilon} \delta(x - \xi, y - \eta) dx dy.$$

Using Green's Theorem (or Green's second identity for ∇^2—see Exercise 2), this becomes

$$\oint_{C_\epsilon} \frac{\partial F}{\partial n} ds = -1$$

for any $\epsilon > 0$, where C_ϵ is the circle bounding D_ϵ.

So we're ready to compute F. Unfortunately, at this level we must proceed in a somewhat ad hoc manner, mixing the classical with the distributional; so care must be taken.* F must then satisfy the two conditions

$$\nabla^2 F = 0, \quad (x, y) \neq (\xi, \eta)$$

and

$$\oint_{C_\epsilon} \frac{\partial F}{\partial n} ds = -1,$$

---

*For a rigorous derivation, see R.P. Kanwal's *Generalized Functions: Theory and Technique*.

where C_ϵ is as above. We use the radial symmetry of F to change to the polar coordinates

$$x - \xi = r \cos \theta, \quad y - \eta = r \sin \theta,$$

in which case the conditions become

$$\frac{1}{4} \frac{d}{dr} \left(r \frac{dF}{dr} \right) = 0, \quad r \neq 0$$

and

$$\int_0^{2\pi} \frac{dF}{dr} r \, d\theta = 2\pi \epsilon \frac{dF}{dr}(\epsilon) = -1.$$

The r-equation has solution

$$F = c_1 + c_2 \ln r,$$

and the second condition forces $c_2 = -\frac{1}{2\pi}$. Therefore, the fundamental solution for the Laplace operator is

$$F = C - \frac{1}{2\pi} \ln r$$

for any choice of the constant C. *Usually,* we set $C = 0$ for convenience (although sometimes we need $C \neq 0$, as in Example 4 at the end of this section). So we have the following theorem.

Theorem 10.7 *The fundamental solution of Poisson's equation*

$$\nabla^2 F = -\delta(x - \xi, y - \eta), \quad -\infty < x < \infty, -\infty < y < \infty,$$

is the **logarithmic potential**

$$F = -\frac{1}{2\pi} \ln \sqrt{(x - \xi)^2 + (y - \eta)^2} \ .$$

Note that this 2-D Green's function is *not* continuous at $(x, y) = (\xi, \eta)$. Physically, F represents the *three*-dimensional electric potential due to a unit line source of charge perpendicular to the x-y plane, through the point (ξ, η). That is, it is the energy required to move a point charge from infinity to the point (x, y), under the influence of this field. The graph of F, with $(\xi, \eta) = (0, 0)$, can be seen in Figure 10.4. (Using a finer mesh will emphasize the behavior near the singularity—try it!)

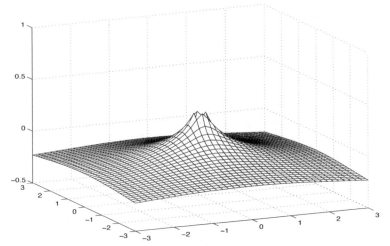

FIGURE 10.4
MATLAB view of the graph of the *logarithmic potential* $F = -\frac{1}{2\pi}\ln r$.

Now we solve the problem

$$\nabla^2 G(x, y; \xi, \eta) = -\delta(x - \xi, y - \eta) \quad \text{on} \quad D,$$
$$G = 0 \quad \text{on} \quad C = \partial D,$$

as mentioned above, by letting $G = F + Z$, where $Z(x, y; \xi, \eta)$ satisfies

$$\nabla^2 Z = 0 \quad \text{on} \quad D,$$
$$Z = \frac{1}{2\pi} \ln \sqrt{(x - \xi)^2 + (y - \eta)^2} \quad \text{on} \quad C.$$

We note that this is a *classical* equation, since the log has no problems along C. Thus, we expect G to have the same kind of singular behavior as F at the point (ξ, η).

As an illustration, let's do an example.

Example 1 Find Green's function for the Dirichlet problem on a disk. Specifically, solve

$$\nabla^2 G = -\delta(x - \xi, y - \eta) \quad \text{on} \quad D,$$
$$G = 0 \quad \text{on} \quad C,$$

where D is the disk $0 \le r < R$.

The solution is

$$G = -\frac{1}{2\pi} \ln \sqrt{(x - \xi)^2 + (y - \eta)^2} + Z,$$

where Z satisfies

$$\nabla^2 Z = 0 \quad \text{on} \quad D,$$

$$Z = \frac{1}{2\pi} \ln \sqrt{(x-\xi)^2 + (y-\eta)^2} \quad \text{on} \quad C.$$

We may find Z via Poisson's integral formula (Section 9.3), but we'll need to get $(x-\xi)^2 + (y-\eta)^2$ as a function of θ. Letting the polar coordinates of (ξ, η) be (r_0, θ_0), we use the law of cosines as we did back in Figure 9.10 to get

$$(x-\xi)^2 + (y-\eta)^2 = r_0^2 + R^2 - 2r_0 R \cos(\theta - \theta_0).$$

Thus, we have

$$Z(r, \theta; r_0, \theta_0) = \frac{1}{4\pi^2} \int_0^{2\pi} \frac{(R^2 - r^2) \ln \sqrt{r_0^2 + R^2 - 2r_0 R \cos(\phi - \theta_0)}}{r^2 + R^2 - 2rR \cos(\theta - \phi)} d\phi$$

and

$$G(r, \theta; , r_0, \theta_0) = -\frac{1}{2\pi} \ln \sqrt{r^2 + R^2 - 2rR \cos(\theta - \theta_0)} + z(r, \theta; r_0, \theta_0)$$

(where we have used the law of cosines again to get the expression under the radical).

Okay, this is not very illuminating. As it turns out, however, there's a more elegant way of solving these problems when the geometry is simple: an extension of the *method of images*. But we're getting ahead of ourselves. We should, at this point, derive the relevant properties of Green's function. In particular, we need to show that Green's function does what we want it to do, that is, that the solution of

$$\nabla^2 u = -f \quad \text{on} \quad D,$$

$$u = 0 \quad \text{on} \quad C = \partial D$$

is really given by

$$u = \iint_D G(x, y; \xi, \eta) f(\xi, \eta) d\xi d\eta.$$

It turns out that we can do even better. At this point, we need to remind ourselves of *Green's second identity* for the Laplacian,

$$\iint_D (U\nabla^2 V - V\nabla^2 U) dA = \oint \left(U \frac{\partial V}{\partial n} - V \frac{\partial U}{\partial n} \right) ds$$

in two dimensions (and, equivalently, in three dimensions). Not surprisingly, it turns out that these Green's identities hold in a distributional setting, as well.

Now, suppose we're given Poisson's equation

$$\nabla^2 u = -f \quad \text{on} \quad D$$

subject to the *nonhomogeneous* boundary condition

$$u = g \quad \text{on} \quad C = \partial D.$$

Let's proceed as we did in the proof of Theorem 10.4 and see what happens. So we let

$$U = u, \quad V = G$$

in Green's identity and get

$$\iint_D (u\nabla^2 G - G\nabla^2 u)dxdy = \oint_C \left(u\frac{\partial G}{\partial n} - G\frac{\partial u}{\partial n} \right) ds.$$

We know that $\nabla^2 G = -\delta(x - \xi, y - \eta)$ and $\nabla^2 u(x,y) = -f(x,y)$ and also that $u = g$ and $G = 0$ along C, so we have

$$\iint_D \delta(x - \xi, y - \eta)u(x,y)dxdy = \iint_D G(x,y;\xi,\eta)f(x,y)dxdy$$
$$- \oint_C g\frac{\partial G}{\partial n}ds.$$

Interchanging (x,y) and (ξ,η) and using the properties of the delta function and Green's function, we have proved the following theorem.

Theorem 10.8 *The solution of the BVP*

$$\nabla^2 u = -f \quad on \quad D,$$
$$u = g \quad on \quad C = \partial D$$

is given by

$$u(x,y) = \iint_D G(x,y;\xi,\eta)f(\xi,\eta)d\xi d\eta - \oint_C g(\xi,\eta)\frac{\partial G}{\partial n}(x,y;\xi,\eta)ds_0,^\dagger$$

where G is the solution of

$$\nabla^2 G = -\delta(x - \xi, y - \eta) \quad on \quad D,$$
$$G = 0 \quad on \quad C.$$

---

†The ds_0 represents integration with respect to ξ and η.

We're also in a position to derive the important properties of G. First, it is easy to show that G is symmetric (see Exercise 2c). Also, as F is not continuous at (ξ, η), then neither is G. Finally, it's easy to show that $\frac{\partial G}{\partial n}$ behaves exactly as $\frac{\partial F}{\partial n}$ at (ξ, η), that is,

$$\lim_{\epsilon \to 0^+} \oint_{C_\epsilon} \frac{\partial G}{\partial n} ds = \lim_{\epsilon \to 0^+} \iint_{D_\epsilon} \nabla^2 G(x, y; \xi, \eta) dx dy = -1,$$

where D_ϵ and C_ϵ are as before (with the added stipulation that $D_\epsilon \subseteq D$).

Putting everything together, we have the following theorem.

Theorem 10.9 *Suppose that $G(x, y; \xi, \eta)$ is the Green's function of the Laplace operator on D. Then,*

1) G is symmetric, that is,

$$G(x, y; \xi, \eta) = G(\xi, \eta; x, y)$$

for any points (x, y) and (ξ, η) in D.

2) G satisfies

$$\lim_{\epsilon \to 0^+} \oint_{C_\epsilon} \frac{\partial G}{\partial n} ds = -1,$$

where C_ϵ is the circle of radius ϵ, centered at (ξ, η).

Example 2 For example, let's use Green's function to solve the Poisson Dirichlet problem on a square,

$$\nabla^2 u = -f, \qquad 0 < x < \pi, 0 < y < \pi,$$
$$u = g \quad \text{on the boundary.}$$

First, we must solve

$$\nabla^2 G = -\delta(x - \xi, y - \eta), \qquad 0 < x < \pi, 0 < y < \pi,$$
$$G = 0 \quad \text{on the boundary.}$$

Example 3 in Section 10.1 suggests we try using a double Fourier sine series. So we have

$$\delta(x - \xi, y - \eta) = \sum_{n=1}^{\infty} \sum_{m=1}^{\infty} b_{n,m} \sin nx \sin my,$$

where

$$b_{n,m} = \frac{4}{\pi^2} \int_0^\pi \int_0^\pi \delta(x - \xi, y - \eta) \sin nx \sin my \, dx dy$$
$$= \frac{4}{\pi^2} \sin n\xi \sin m\eta = b_{n,m}(\xi, \eta).$$

Letting

$$G = \sum_{n=1}^{\infty} \sum_{m=1}^{\infty} c_{n,m} \sin nx \sin my, \text{ where } c_{n,m} = c_{n,m}(\xi, \eta),$$

we have

$$\nabla^2 G = -\sum_{n=1}^{\infty} \sum_{m=1}^{\infty} c_{n,m}(n^2 + m^2) \sin nx \sin my,$$

so that substituting into the PDE and comparing coefficients gives us

$$G(x, y; \xi, \eta) = \frac{4}{\pi^2} \sum_{n=1}^{\infty} \sum_{m=1}^{\infty} \frac{\sin nx \sin my \sin n\xi \sin m\eta}{n^2 + m^2}.$$

(Obviously, G is symmetric.) Then the solution is

$$u(x, y) = \iint_D G(x, y; \xi, \eta) f(\xi, \eta) d\xi d\eta - \oint g(\xi, \eta) \frac{\partial G}{\partial n}(x, y; \xi, \eta) ds$$

$$= \frac{4}{\pi^2} \sum_{n=1}^{\infty} \sum_{m=1}^{\infty} \frac{\sin nx \sin my}{n^2 + m^2} \int_0^\pi \int_0^\pi f(\xi, \eta) \sin n\xi \sin m\eta \, d\xi d\eta$$

$$- \int_0^\pi g(\xi, 0) \left(-\frac{\partial G}{\partial \eta}(x, y; \xi, 0) \right) d\xi$$

$$- \int_0^\pi g(\pi, \eta) \frac{\partial G}{\partial \xi}(x, y; \pi, \eta) d\eta$$

$$- \int_\pi^0 g(\xi, \pi) \frac{\partial G}{\partial \eta}(x, y; \xi, \pi) d\xi$$

$$- \int_\pi^0 g(0, \eta) \left(-\frac{\partial G}{\partial \xi}(x, y; 0, \eta) \right) d\eta$$

$$= \frac{4}{\pi^2} \sum_{n=1}^{\infty} \sum_{m=1}^{\infty} \frac{\sin nx \sin my}{n^2 + m^2} \left[\int_0^\pi \int_0^\pi f(\xi, \eta) \sin n\xi \sin m\eta \, d\xi d\eta \right.$$

$$+ m \int_0^\pi g(\xi, 0) \sin n\xi \, d\xi - n(-1)^n \int_0^\pi g(\pi, \eta) \sin m\eta \, d\eta$$

$$\left. + m(-1)^m \int_0^\pi g(\xi, \pi) \sin n\xi \, d\xi - \int_0^\pi g(0, \eta) \sin m\eta \, d\eta \right]$$

(see Exercise 3).

THE METHOD OF IMAGES

As mentioned, if the geometry is simple enough, we may use a variant of the *method of images* to find Green's function in terms of the fundamental solution. Our procedure is very similar to that in Exercise 13 of the previous section. Let's look at some examples.

Example 3 Let's find Green's function for the Laplacian on the upper half-plane, that is, let's solve

$$\nabla^2 G = -\delta(x - \xi, y - \eta), \qquad -\infty < x < \infty, y > 0,$$
$$G(x, 0; \xi, \eta) = 0.$$

First, we know that the free-space Green's function $F(x, y; \xi, \eta)$ satisfies the PDE, but not the boundary condition. The trick is to *change* the problem *without actually changing it* (whatever that means!). Let's solve, instead, a problem of the form

$$\nabla^2 u = -\delta(x - \xi, y - \eta) + \delta(x - \xi', y - \eta'),$$

where (ξ', η') is chosen judiciously so that the solution satisfies $u = 0$ on the boundary. Here, it shouldn't be surprising that $(\xi', \eta') = (\xi, -\eta)$, the mirror image of (ξ, η) through the boundary curve. Physically, if we start with a positive electric charge at (ξ, η), then we're just placing a negative charge at (ξ', η'). See Figure 10.5.

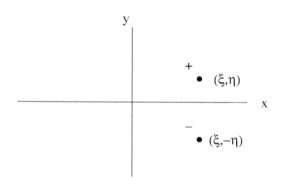

FIGURE 10.5
The image point $(\xi, -\eta)$ for a point (ξ, η) in the upper half-plane.

So, we solve

$$\nabla^2 u = -\delta(x - \xi, y - \eta) + \delta(x - \xi, y + \eta),$$

with the result being

$$u = F(x, y; \xi, \eta) - F(x, y; \xi, -\eta)$$
$$= -\frac{1}{2\pi} \ln \sqrt{(x - \xi)^2 + (y - \eta)^2} + \frac{1}{2\pi} \ln \sqrt{(x - \xi)^2 + (y + \eta)^2},$$

and—presto!—we have $u|_{y=0} = 0$.

But wait a minute; Green's function is supposed to satisfy the *original* PDE

$$\nabla^2 G = -\delta(x - \xi, y - \eta).$$

In fact, it does, on the domain $-\infty < x < \infty$, $y > 0$, since

$$\nabla^2 [\ln \sqrt{(x - \xi)^2 + (y + \eta)^2}] = 0$$

there (why?). So we *have* found Green's function for the problem,

$$
\begin{aligned}
G(x, y; \xi, \eta) &= \frac{1}{2\pi} \left[\ln \sqrt{(x - \xi)^2 + (y + \eta)^2} - \ln \sqrt{(x - \xi)^2 + (y - \eta)^2} \right] \\
&= \frac{1}{4\pi} \ln \frac{(x - \xi)^2 + (y + \eta)^2}{(x - \xi)^2 + (y - \eta)^2}.
\end{aligned}
$$

Note: One may use Fourier transforms to show that the solution of $\nabla^2 u = -f$ on the plane or half-plane is, again, given by

$$u = \iint_D G(x, y; \xi, \eta) f(\xi, \eta) d\xi d\eta,$$

where D is the region in question. If D is the plane, then, of course, $G = F$. Alternatively, one may use Green's second identity by taking a finite bounding curve and then letting it "become infinite."

Example 4 Let's use the same approach to find Green's function for the *interior Dirichlet problem* (on a disk). Specifically, we'd like to solve

$$
\begin{aligned}
\nabla^2 G &= -\delta(x - \xi, y - \eta) \quad \text{on} \quad x^2 + y^2 < R^2, \\
G &= 0 \quad \text{on} \quad x^2 + y^2 = R^2.
\end{aligned}
$$

Again, for (ξ, η) inside the circle, we look for a point (ξ', η') *outside* the circle such that the solution of

$$\nabla^2 u = -\delta(x - \xi, y - \eta) + \delta(x - \xi', y - \eta')$$

is zero on C. First, it should be clear that, if there *is* such a point, it should be of the form $(\xi', \eta') = (\alpha\xi, \alpha\eta)$ for some constant α. (See Figure 10.6.) Then, our solution is

$$
\begin{aligned}
u &= -\frac{1}{2\pi} \ln \sqrt{(x - \xi)^2 + (y - \eta)^2} + \frac{1}{2\pi} \ln \sqrt{(x - \alpha\xi)^2 + (y - \alpha\eta)^2}, \\
&= -\frac{1}{2\pi} \ln r_1 + \frac{1}{2\pi} \ln r'_1, \\
&= \frac{1}{4\pi} \ln \frac{(r'_1)^2}{r_1^2}.
\end{aligned}
$$

For this to be zero along the boundary, we need

$$\frac{r'_1}{r_1} = 1$$

for every point on the boundary, which is impossible. What can we do? Remember that we set the arbitrary constant to zero when deriving the logarithmic potential. So our solution really can be written

$$u = \frac{1}{4\pi} \ln \frac{(r_1')^2}{r_1^2} + c$$

for any constant c, and we need only choose things so that r'/r is constant along the boundary.

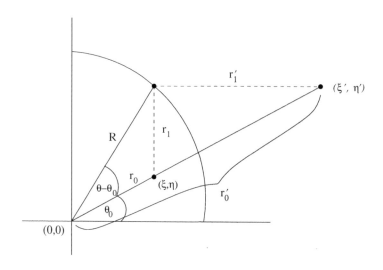

FIGURE 10.6
Finding the image point (ξ', η') for a point (ξ, η) inside the circle $x^2 + y^2 = R^2$.

So for any point (R, θ) on the boundary we have, via the *law of cosines*,

$$r_1^2 = R^2 + r_0^2 - 2Rr_0 \cos(\theta - \theta_0)$$

and

$$(r_1')^2 = R^2 + \alpha^2 \rho_0^2 - 2R\alpha r_0 \cos(\theta - \theta_0).$$

We wish to choose α so that

$$(r_1')^2 = kr_1^2$$

for some constant k, that is, so that

$$R^2 + \alpha^2 r_0^2 - 2R\alpha r_0 \cos\theta = k(R^2 + r_0^2 - 2Rr_0 \cos\theta)$$

for all values of θ. Letting $\theta = \frac{\pi}{2} + \phi$ tells us that

$$R^2 + \alpha^2 r_0^2 = kR^2 + k\rho_0^2,$$

while $\theta = \phi$ then implies that

$$2R\alpha r_0 = 2kRr_0$$

or

$$k = \alpha.$$

Finally,

$$R^2 + \alpha^2 r_0^2 = \alpha R^2 + \alpha r_0^2$$

or

$$\alpha = \frac{R^2}{r_0^2}.$$

Our solution is then

$$u = \frac{1}{4\pi} \ln \frac{(r_1')^2}{r_1^2} - \frac{1}{4\pi} \ln k = \frac{1}{4\pi} \ln \frac{(r_1')^2 r_0^2}{r_1^2 R^2}$$

$$= \frac{1}{4\pi} \ln \left[\frac{\left(x - \frac{R^2}{r_0^2}\xi\right)^2 + \left(y - \frac{R^2}{r_0^2}\eta\right)^2}{(x - \xi)^2 + (y - \eta)^2} \frac{r_0^2}{R^2} \right]$$

and, in polar coordinates (where, of course, (x, y) now represents *any* point in the disk),

$$u = \frac{1}{4\pi} \ln \left[\frac{r^2 + \alpha^2 r_0^2 - 2r\alpha r_0 \cos(\theta - \theta_0)}{r^2 + r_0^2 - 2rr_0 \cos(\theta - \theta_0)} \frac{r_0^2}{R^2} \right]$$

$$= \frac{1}{4\pi} \ln \frac{r^2 r_0^2 + R^4 - 2rr_0 R^2 \cos(\theta - \theta_0)}{R^2 [r^2 + r_0^2 - 2rr_0 \cos(\theta - \theta_0)]}.$$

In retrospect, what we've really done is to choose r_0' so that $r_0 r_0' = r_0 \alpha r_0 = R^2$, or

$$\frac{r_0}{R} = \frac{R}{r_0'}.$$

This tells us that the triangle with vertices $(0,0)$, (ξ, η), (x, y) is similar to the triangle with $(0,0)$, (x, y), (ξ', η'), from which it follows that

$$\frac{r_1'}{R} = \frac{r_1}{r_0}$$

and, taking logs,

$$\ln r_1 = \ln \frac{r_1' r_0}{R}.$$

This means that, in order to cancel out the effect of the potential

$$-\frac{1}{2\pi}\ln r,$$

at the point (x, y), we need only add the potential

$$\frac{1}{2\pi}\ln\frac{r_1' r_0}{R}.$$

Thus, our Green's function is

$$G = \frac{1}{2\pi}\ln\frac{r_1' r_0}{r_1 R},$$

which is what we found above.

Recapping, **Green's function for the Dirichlet problem on a disk of radius R is**

$$G(r, \theta; r_0, \theta_0) = \frac{1}{4\pi}\frac{r^2 r_0^2 + R^4 - 2R^2 r r_0 \cos(\theta - \theta_0)}{R^2[r^2 + r_0^2 - 2r r_0 \cos(\theta - \theta_0)]},$$

where (r_0, θ_0) are the polar coordinates of the point (ξ, η). Note that

$$G(r, \theta; r_0, \theta_0) = G(r_0, \theta_0; r, \theta)$$

for all points in the disk.

It follows that the solution of

$$\nabla^2 u = -f \quad \text{on} \quad x^2 + y^2 < R^2,$$
$$u = g \quad \text{on} \quad x^2 + y^2 = R^2$$

is given by

$$u(x, y) = \iint\limits_{x^2+y^2<R^2} G(x, y; \xi, \eta) f(\xi, \eta) d\xi d\eta$$

$$- \oint\limits_{\substack{C \\ x^2+y^2=R^2}} g(\xi, \eta)\frac{\partial T}{\partial n} ds.$$

Now, what happens if $f \equiv 0$? Changing to polars in the second integral, we have

$$u = -\oint\limits_{C} g(R, \phi)\frac{\partial}{\partial \rho}G(r, \theta; \rho, \phi)ds$$

$$= -\int_0^{2\pi} g(R, \phi)\frac{\partial G}{\partial \rho}\bigg|_{\rho=R} R\, d\phi.$$

(The polar coordinates of (ξ, η) are, here, (ρ, ϕ).)

Now,

$$\frac{\partial G}{\partial \rho} = \frac{1}{4\pi} \frac{2\rho r^2 - 2R^2 r \cos(\theta - \phi)}{\rho^2 r^2 + R^4 - 2R^2 \rho r \cos(\theta - \phi)} - \frac{1}{4\pi} \frac{2\rho - 2r \cos(\theta - \phi)}{r^2 + \rho^2 - 2\rho r \cos(\theta - \phi)}$$

and, at $\rho = R$, we have

$$\frac{\partial G}{\partial \rho}\Big|_{\rho=R} = \frac{1}{2\pi} \frac{r^2 - Rr \cos(\theta - \phi)}{Rr^2 + R^3 - 2R^2 r \cos(\theta - \phi)} - \frac{1}{2\pi} \frac{R - r \cos(\theta - \phi)}{r^2 + R^2 - 2Rr \cos(\theta - \phi)}$$

$$= \frac{1}{2\pi R} \frac{r^2 - R^2}{r^2 + R^2 - 2Rr \cos(\theta - \phi)}.$$

It follows that the solution is

$$u = \frac{1}{2\pi} \int_0^\pi g(R, \phi) \frac{R^2 - r^2}{r^2 + R^2 - 2Rr \cos(\theta - \phi)} d\phi,$$

which should look familiar.

Exercises 10.3

1. Use the fact that

$$\int_I \delta(x - \xi) dx = 1 \quad \text{and} \quad \int_{I'} \delta(x - \xi) dx = 0$$

for any open intervals I and I' with $\xi \in I$ and $\xi \notin I'$ to show that

$$\iint_D \delta(x - \xi, y - \eta) dx dy = 1$$

for any domain D with $(\xi, \eta) \in D$.

2. a) Use Green's Theorem to show that

$$\iint_D \nabla^2 u \, dx dy = \oint_C \frac{\partial u}{\partial n} ds,$$

where D is any domain with boundary C.

b) Do the same, but use *Green's second identity*.

c) Show that Green's function for the Laplacian, on any bounded domain D, is symmetric, that is, show that

$$G(x, y; \xi, \eta) = G(\xi, \eta; x, y)$$

for all pairs (x, y) and (ξ, η).

d) Establish the translation property of the fundamental solution for the Laplacian,

$$F(x, y; \xi, \eta) = F(x - \xi, y; 0, \eta) = F(x, y - \eta; \xi, 0)$$
$$= F(x - \xi, y - \eta; 0, 0)$$

for all (x, y) and (ξ, η); more generally, show that

$$F(x, y; \xi, \eta) = F(x + a, y + b; \xi - a, \eta - b)$$

for any $(x, y), (\xi, \eta), a$ and b.

3. Fill in the details in the solution of Example 2.

4. a) Find a double Fourier series representation of Green's function for the problem

$$\nabla^2 u = -f, \qquad 0 < x < \pi, 0 < y < \pi,$$
$$u(x, 0) = u(x, \pi) = u_x(0, y) = u_x(\pi, y) = 0.$$

b) Modifying the solution given in Theorem 10.8, solve the problem

$$\nabla^2 u = -f, \qquad 0 < x < \pi, 0 < y < \pi,$$
$$u(x, 0) = h_1(x), u(x, \pi) = h_2(x),$$
$$u_x(0, y) = h_3(y), u_x(\pi, y) = h_4(y).$$

c) One cannot find a double Fourier series representation for Green's function for the Neumann problem on the same rectangle. Why is this the case?

5. Often, the double Fourier series in these problems don't possess the nicest convergence properties, whereas single Fourier series behave much better.

a) Solve for Green's function for the Dirichlet problem on the square $0 < x < \pi$, $0 < y < \pi$ by, instead, using a single Fourier sine series

$$G(x, y; \xi, \eta) = \sum_{n=1}^{\infty} b_n(y; \xi, \eta) \sin nx$$

and using Green's function derived in Example 3 of Section 10.2.

b) Proceed similarly, but for the infinite strip $-\infty < x < \infty, 0 < y < \pi$.

6. a) Show that

$$\delta(x - \xi, y - \eta) = \frac{1}{r}\delta(r - r_0, \theta - \theta_0),$$

where (r, θ) and (r_0, θ_0) are the polar coordinates of (x, y) and (ξ, η), respectively.

b) More generally, explain heuristically why, if x and y are transformed to the new coordinates x' and y', then

$$\delta(x - \xi, y - \eta) = \frac{1}{|J|} \delta(x' - \xi', y' - \eta'),$$

where J is the Jacobian of the transformation.

7. Find Green's function for the Laplace operator on the unit disk by, instead, expanding u and $\delta(x - \xi, y - \eta)$ in double Fourier series of the form

$$\sum_{n=0}^{\infty} \sum_{m=1}^{\infty} [A_{n,m}(r_0, \theta_0) J_n(x_{n,m}r) \cos n\theta + B_{n,m}(r_0, \theta_0) J_n(x_{n,m}r) \sin n\theta],$$

where, of course, the J_n-part comes from the Fourier–Bessel expansion. Here, (r_0, θ_0) is the polar representation of (ξ, η), and $x_{n,m}$ is the mth positive zero of J_n.

8. Use the method of images to find Green's function for the *exterior* Dirichlet problem

$$\nabla^2 u = -f, \qquad r > R, -\infty < \theta < \infty,$$
$$u(R, \theta) = 0.$$

9.   a) Use the method of images with *three* image points to find Green's function for the Dirichlet problem on the first quadrant.

  b) Do the same, but with boundary conditions

$$u(x, 0) = 0, \qquad x > 0,$$
$$u_x(0, y) = 0, \qquad y > 0.$$

10. Repeat Exercise 5b but, this time, use the method of images (with *infinitely many* image points).

11. Find Green's function for the Dirichlet problem on the interior of a semicircle, that is, on the domain $0 < r < R, 0 < \theta < \pi$.

12. Compute

$$\mathcal{F}^{-1} \left[\frac{1}{\alpha^2 + \beta^2} \right].$$

13. **Justification of fundamental solution:** Using Green's second identity on the disk of radius R centered at (ξ, η), and then letting $R \to \infty$, show that the solution of

$$\nabla^2 u = -f, \qquad -\infty < x < \infty, -\infty < y < \infty$$

is, in fact,

$$u(x,y) = \int_{-\infty}^{\infty}\int_{-\infty}^{\infty} F(x,y;\xi,\eta)f(\xi,\eta)d\xi d\eta,$$

provided we have

$$\lim_{R\to\infty}[u(R,\theta) - Ru_r(R,\theta)\ln R] = 0.$$

14. **Poisson kernel for the upper half-plane, revisited**

   a) Calculate $\frac{\partial G}{\partial n}\big|_{\eta=0}$ for the Green's function in Example 3, and show that the result is the Poisson kernel from Exercise 8, Section 6.4.

   b) Supposing that $u \to 0$ quickly enough as $x^2 + y^2 \to \infty$, reproduce the solution of that same exercise, using the method of this section.

   c) Show that the Poisson kernel itself is the solution of the problem

$$u_{xx} + y_{yy} = 0, \qquad -\infty < x < \infty, y > 0,$$
$$u(x,0) = \delta(x - \xi).$$

10.4 Green's Functions for Elliptic PDEs (II): Poisson's Equation in Three Dimensions; the Helmholtz Equation

POISSON'S EQUATION IN THREE DIMENSIONS

Not surprisingly, the solution of the 3-D Poisson equation proceeds very much like the 2-D case. We need Green's second identity in three dimensions,

$$\iiint_D (u\nabla^2 v - v\nabla^2 u)dxdydz = \oiint_S (u\nabla v - v\nabla u)\cdot\hat{n}\,dS$$
$$= \oiint_S \left(u\frac{\partial v}{\partial n} - v\frac{\partial u}{\partial n}\right)dS,$$

which we prove in Exercise 8, using the Divergence Theorem. Here, of course, D is now a 3-D domain, with boundary *surface* S.

As before, we begin by finding the fundamental solution of the Laplace operator in three dimensions, that is, the function $F(x,y,z)$ satisfying

$$\nabla^2 F = -\delta(x - \xi, y - \eta, z - \zeta), \qquad -\infty < x < \infty,$$
$$-\infty < y < \infty, -\infty < z < \infty.$$

The 3-D Dirac delta function $\delta(x - \xi, y - \eta, z - \zeta) = \delta(x - \xi)\delta(y - \eta)\delta(z - \zeta)$ possesses all of the obvious properties.

Again as before, F is determined by the two properties

$$\nabla^2 F = 0, \quad (x, y, z) \neq (\xi, \eta, \zeta)$$

and

$$\oiint_{S_\epsilon} \frac{\partial F}{\partial n} dS = -1,$$

where S_ϵ is now the *sphere* of radius ϵ centered at (ξ, η, ζ). We switch to spherical coordinates and use the radial symmetry of F to write the first condition as

$$\frac{1}{\rho^2} \frac{d}{d\rho}\left(\rho^2 \frac{\partial F}{\partial \rho}\right) = 0,$$

which has general solution

$$F = c_1 + \frac{c_2}{\rho}.$$

Then,

$$\oiint_{S_\epsilon} \frac{\partial F}{\partial n} dS = \int_0^\pi \int_0^{2\pi} \frac{\partial F}{\partial \rho} \rho^2 \sin\phi \, d\theta d\phi$$

$$= -4\pi c_2 = -1,$$

so the fundamental solution is the well-known **Newtonian potential**

$$F = \frac{1}{4\pi\rho} = \frac{1}{4\pi\sqrt{(x - \xi)^2 + (y - \eta)^2 + (z - \zeta)^2}}.$$

Physically, F represents the gravitational (Newtonian) potential at (x, y, z) due to a unit point mass, or the electric potential due to a unit charge, placed at (ξ, η, ζ).

As with the 2-D problem, we now find Green's function for

$$\nabla^2 G = -\delta(x - \xi, y - \eta, z - \zeta) \quad \text{on} \quad D,$$
$$G = 0 \quad C = \partial D$$

by letting $G = F + z$ with z satisfying

$$\nabla^2 z = 0 \quad \text{on} \quad D,$$
$$z = -F = -\frac{1}{4\pi\rho} \quad \text{on} \quad C.$$

Alternatively, we may apply the method of images when the geometry is appropriate. And we may show that the theorems of the previous section hold for the 3-D Green's function, as well (with appropriate dimensional changes: line integrals to surface integrals, area integrals to volume integrals, etc.).

Before moving on, let's find Green's function for the Laplace operator on a ball.

Example 1 Solve

$$\nabla^2 G = -\delta(x - \xi, y - \eta, z - \zeta), \qquad x^2 + y^2 + z^2 < R^2,$$
$$G = 0 \quad \text{on the sphere} \quad x^2 + y^2 + z^2 = R^2.$$

We use the method of images, often referred to as the **method of electrostatic images,** for the Laplacian in three dimensions. Letting $\boldsymbol{x} = (x, y, z)$ and $\boldsymbol{x_0} = (\xi, \eta, \zeta)$, we solve

$$\nabla^2 G = -\delta(\boldsymbol{x} - \boldsymbol{x_0}).$$

It should be clear that we proceed exactly as we did in the 2-D case, and we may use the same similar-triangle argument. Thus, we again need

$$\rho_1 = \frac{\rho_1' \rho_0}{R},$$

so the Newtonian potential $\frac{1}{4\pi \rho_1}$ is cancelled on the boundary by adding the potential

$$-\frac{1}{4\pi \rho_1} = -\frac{R}{4\pi \rho_0 \rho_1'}.$$

Physically, we've added a negative change of magnitude $\frac{R}{\rho_0}$ at the image point $\boldsymbol{x_0'}$.

Our Green's function then is

$$G(\boldsymbol{x}, \boldsymbol{x_0}) = \frac{1}{4\pi} \left(\frac{1}{\rho_1} - \frac{R}{\rho_0 \rho_1'} \right)$$
$$= \frac{1}{4\pi} \left(\frac{1}{|\boldsymbol{x} - \boldsymbol{x_0}|} - \frac{R}{\rho_0 |\boldsymbol{x} - \frac{R^2}{\rho_0^2} \boldsymbol{x_0}|} \right)$$
$$= \frac{1}{4\pi} \left(\frac{1}{|\boldsymbol{x} - \boldsymbol{x_0}|} - \frac{1}{|\frac{\rho_0}{R} \boldsymbol{x} - \frac{R}{\rho_0} \boldsymbol{x_0}|} \right).$$

Of course, this means that the solution of the problem

$$\nabla^2 u = -f, \text{ on } x^2 + y^2 + z^2 < R^2,$$
$$u = g \quad \text{on} \quad x^2 + y^2 + z^2 = R^2$$

is given by

$$u(x, y, z) = \iiint_D G(x, y, z; \xi, \eta, \zeta) f(\xi, \eta, \zeta) d\xi d\eta d\zeta$$

$$-\oiint_S g(\xi,\eta,\zeta)\frac{\partial G}{\partial n}(x,y,z;\xi,\eta,\zeta)dS_0,^{\ddagger}$$

where S is the sphere of radius R, D is its interior and $\frac{\partial G}{\partial n} = \frac{\partial G}{\partial \rho_0}$ (where ρ_0 is the spherical radial coordinate, if we write $G = G(\rho,\theta,\phi;\rho_0,\theta_0,\phi_0)$). And, as we did at the end of the previous section, we may recover the 3-D Poisson's integral formula if $f \equiv 0$ (see Exercise 4).

HELMHOLTZ EQUATION

Now we consider that other important elliptic problem, the Helmholtz equation, which, as mentioned earlier, arises from separating variables in the heat and wave equations. So we need to see what Green's second identity becomes for the Helmholtz operator. Writing

$$L[u] = \nabla^2 u + k^2 u,$$

we have

$$uL[v] - vL[u] = u\nabla^2 v - v\nabla^2 u,$$

so Green's second identity is, again,

$$\iint_D (uL[v] - vL[u])dA = \oint_{\partial D} (u\nabla v - v\nabla u)\cdot\hat{n}\,ds$$

$$= \oint_{\partial D} \left(u\frac{\partial v}{\partial n} - v\frac{\partial u}{\partial n}\right)ds$$

(and, equivalently, in three dimensions). As a result, the Helmholtz Green's functions will satisfy the same properties as those of the corresponding Laplace Green's functions. In particular,

1. They will be symmetric.

2. They will exhibit the same discontinuity in $\frac{\partial G}{\partial n}$ at the source point.

We begin by computing the fundamental solution for the 2-D Helmholtz equation. So, we want F that satisfies

$$\nabla^2 F + k^2 F = -\delta(x - \xi, y - \eta).$$

Again, radial symmetry makes life easier. As with Poisson's equation, F must satisfy

$$\nabla^2 F + k^2 F = 0, \quad (x,y) \neq (\xi,\eta)$$

---

‡Again, dS_0 represents integration with respect to the variables ξ, η and ζ.

and (see Exercise 9)

$$\oint_{C_\epsilon} \frac{\partial F}{\partial n} ds = -1.$$

The PDE gives us Bessel's equation of order 0,

$$r^2 F_{rr} + r F_r + k^2 r^2 F = 0,$$

with general solution

$$F = c_1 J_0(kr) + c_2 Y_0(kr).$$

The second condition becomes

$$\epsilon \frac{dF}{dr}(\epsilon) = -\frac{1}{2\pi}.$$

From Section 7.5, we may write F as

$$F = c_1 \left(1 - \frac{1}{2}k^2 r^2 + \cdots\right) + c_2 \frac{2}{\pi} \left[\left(1 - \frac{1}{2}k^2 r^2 + \cdots\right)(\gamma - \ln 2)\right.$$
$$\left. + \left(1 - \frac{1}{2}k^2 r^2 + \cdots\right)(\ln k + \ln r) - \left(\phi(1) - \frac{\phi(2)}{2}k^2 r^2 + \cdots\right)\right],$$

so that

$$\frac{dF}{dr} = \frac{2c_2}{\pi r} + \frac{2c_2}{\pi} r \ln r \cdot (\text{terms involving nonnegative powers of } r)$$
$$+ r \, (\text{terms involving nonnegative powers of } r).$$

So the only way that

$$\epsilon \frac{dF}{dr}(\epsilon) = -\frac{1}{2\pi}$$

for all $\epsilon > 0$ is if $\frac{2c_2}{\pi} = -\frac{1}{2\pi}$ or $c_2 = -\frac{1}{4}$. (Or take $\lim\limits_{\epsilon \to 0^+} \epsilon \frac{dF}{dn}(\epsilon)$ and go from there. To be precise, though, we've only shown that *if* there is a fundamental solution, *then* we must have $c_2 = -\frac{1}{4}$.[§]) So we take our fundamental solution to be

$$F = -\frac{1}{4} Y_0(kr)$$
$$= -\frac{1}{4} Y_0(k\sqrt{(x - \xi)^2 + (y - \eta)^2}).$$

We plot the graph of F in Figure 10.7 (using $k = 1$).

What about the 3-D Helmholtz equation? We have

$$\nabla^2 u + k^2 u = -\delta(x - \xi, y - \eta, z - \zeta) = -\delta(\boldsymbol{x} - \boldsymbol{x}_0)$$

[§]Again, see Kanwal's *Generalized Functions: Theory and Technique.*

and, as in the 2-D case, we'll have radial symmetry. So $F = F(\rho)$ will satisfy the two conditions

$$\nabla^2 F + k^2 F = F_{\rho\rho} + \frac{2}{\rho}F_\rho + k^2 F = 0, \qquad \rho \neq 0,$$

and

$$\lim_{\epsilon \to 0^+} \oiint_{S_\epsilon} \frac{\partial F}{\partial n} dS = \lim_{\epsilon \to 0^+} \int_0^\pi \int_0^{2\pi} \frac{\partial F}{\partial \rho} \rho^2 \sin\phi \, d\theta d\phi \bigg|_{\rho=\epsilon}$$

$$= 4\pi \lim_{\epsilon \to 0^+} \epsilon^2 \frac{\partial F}{\partial \rho}(\epsilon) = -1.$$

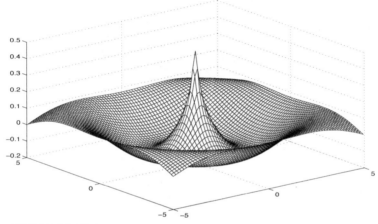

FIGURE 10.7(a)

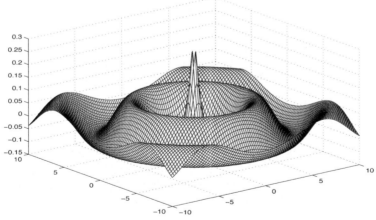

FIGURE 10.7(b)

Two different **MATLAB** views of the graph of the fundamental solution $F = -\frac{1}{4}Y_0(r)$ of the 2-D Helmholtz equation.

The ODE can be solved by rewriting it as

$$(\rho F)'' + k^2(\rho F) = 0,$$

which gives us

$$F = \frac{c_1 \cos k\rho + c_2 \sin k\rho}{\rho}.$$

Then,

$$\lim_{\epsilon \to 0^+} \epsilon^2 \frac{\partial F}{\partial \rho}(\epsilon) = \lim_{\epsilon \to 0^+} [(kc_2\epsilon - c_1) \cos k\epsilon - (kc_1\epsilon + c_2) \sin k\epsilon]$$

$$= -c_1$$

and, letting $c_2 = 0$ (for the sake of convenience), we've found that

$$F = \frac{\cos k\rho}{4\pi\rho} = \frac{\cos k|\boldsymbol{x} - \boldsymbol{x}_0|}{4\pi|\boldsymbol{x} - \boldsymbol{x}_0|}.$$

Note that, as $k \to 0$ in both Helmholtz equations, the 2-D solution approaches $-\frac{1}{2\pi} \ln r$, while the 3-D solution tends to $\frac{1}{4\pi\rho}$, as we might expect.

Now we're in a position to solve the nonhomogeneous Helmholtz equation, of course.

NEUMANN PROBLEMS

What about the Neumann problem? Let's look at

$$\nabla^2 u = -f \quad \text{on} \quad D,$$

$$\frac{\partial u}{\partial n} = g \quad \text{on} \quad C = \gamma D.$$

Integrating the PDE and using Green's Theorem gives us

$$\oint_C g \, ds = -\iint_D f \, dA.$$

Of course, this is just a more general version of the *compatibility condition* we've seen before (in Sections 4.4 and 9.3). Therefore, if f and g do *not* satisfy the condition, then there can be no solution. However, if the condition *is* satisfied, it turns out that we do have a solution (and, hence, an infinite number of solutions—why?). What's happening is that $\lambda = 0$ is an eigenvalue of the system

$$\nabla^2 u + \lambda u = 0 \quad \text{on} \quad D,$$

$$\frac{\partial u}{\partial n} = 0 \quad \text{on} \quad C,$$

or, equivalently, that the associated homogeneous problem has a nontrivial solution. So how does Green's function play into this? Well, we need to solve

$$\nabla^2 u = -\delta(x - \xi, y - \eta) \quad \text{on} \quad D,$$

$$\frac{\partial u}{\partial n} = 0 \quad \text{on} \quad C.$$

But the compatibility condition is not satisfied here because

$$\iint_D \delta(x - \xi, y - \eta)dA = 1 \neq 0.$$

However, we know that the original system *always* has a solution when the compatibility condition is satisfied. So we proceed as we did at the end of Section 10.2—we introduce the idea of the *Neumann function* for this problem. We look at the problem

$$\nabla^2 H = -\delta(x - \xi, y - \eta) \quad \text{on} \quad D,$$

$$\frac{\partial H}{\partial n} = c_1 \quad \text{on} \quad \partial D,$$

where c_1 is a constant to be determined. The compatibility condition becomes

$$-1 = -\iint_D \delta(x - \xi, y - \eta)dA = c_1 \oint_{\partial D} ds,$$

which is satisfied by

$$c_1 = -\frac{1}{\oint_{\partial D} ds}.$$

Note that the denominator is just the arc length of ∂D.

Solving for H, for this value of c_1, and using Green's second identity, we're led to

$$u = \iint_D H(x, y; \xi, \eta)f(\xi, \eta)d\xi d\eta + c_2$$

for any constant c_2.

Of course, for the Helmholtz problem

$$\nabla^2 u + k^2 u = 0 \quad \text{on} \quad D, \qquad k \neq 0,$$

$$\frac{\partial u}{\partial n} = 0 \quad \text{on} \quad \partial D,$$

there are no problems unless k^2 is an eigenvalue of the Laplace operator, with Neumann condition, on D. (See Exercise 5.)

Exercises 10.4

1. Following Example 2 in the previous section, use Fourier series to compute the solution of

$$\nabla^2 u = -f, \qquad 0 < x < \pi, 0 < y < \pi, 0 < z < \pi,$$
$$u = 0 \quad \text{on the boundary.}$$

2. a) Compute Green's function for the exterior Dirichlet problem

$$\nabla^2 u = -f, \quad \rho > R.$$

   b) Find the electric potential due to a point charge q outside a conducting sphere held at constant potential V.

3. Compute Green's function for the Laplace operator

   a) On upper half-space, $z > 0$

   b) On the first octant, $x > 0$, $y > 0$, $z > 0$

4. Use this Green's function approach to derive **Poisson's integral formula**

$$u(\rho, \theta, \phi) = \frac{a(a^2 - \rho^2)}{4\pi}.$$

$$\int_0^{2\pi} \int_0^\pi \frac{g(\alpha, \beta) \sin\alpha}{\{a^2 + \rho^2 - 2a\rho[\cos\theta\cos\alpha + \sin\theta\sin\alpha\cos(\beta - \phi)]\}} d\alpha d\beta$$

   for the solution of the Laplace BVP

$$\nabla^2 u = 0, \qquad 0 < \rho < a,$$
$$u(a, \theta, \phi) = g(\theta, \phi).$$

5. a) Without doing too much work, solve the Helmholtz problem

$$u_{xx} + u_{yy} + 5u = \sin 4x \sin 6y, \qquad 0 < x < \pi, 0 < y < \pi,$$
$$u(x, 0) = u(x, \pi) = u(0, y) = u(\pi, y) = 0.$$

   b) Try to do the same for

$$u_{xx} + u_{yy} + 5u = \sin x \sin 2y.$$

   What's going on?

   c) What restriction(s) on f is (are) necessary in order that

$$u_{xx} + u_{yy} + 5u = -f, \qquad 0 < x < \pi, 0 < y < \pi,$$

   subject to the same boundary conditions, has a solution? What will the solution be?

6. Compute Green's function for the Helmholtz operator $\nabla^2 u + k^2 u$ on the given domain, with Dirichlet boundary condition.

   a) The upper half-plane, $y > 0$
   b) Upper half-space, $z > 0$
   c) The disk, $0 < r < a$
   d) The wedge, $r > 0$, $0 < \theta < \alpha$
   e) The infinite 2-D strip, $0 < x < 1$, $-\infty < y < \infty$

7. a) Show that Green's function for the 3-D Laplace operator, on any bounded domain D, is symmetric, that is, that

$$G(x, y, z; \xi, \eta, \zeta) = G(\xi, \eta, \zeta; x, y, z)$$

   for all ordered triples (x, y, z) and (ξ, η, ζ).

   b) Do the same for the 2-D Helmholtz operator.

   c) What is the general translation property for the fundamental solution of each of these operators?

8. Use the Divergence Theorem to prove *Green's second identity* in three dimensions.

9. a) Use Green's second identity to prove that

$$\oint_{C_\epsilon} \frac{\partial F}{\partial n} ds = -1,$$

   where F is the fundamental solution of the 2-D Helmholtz operator and C_ϵ is the ϵ-circle about the singularity (ξ, η).

   b) Proceed similarly for the 3-D Helmholtz operator.

10.5   Green's Functions for Equations of Evolution

HEAT EQUATION

Let's begin by just solving the heat/diffusion problem

$$F_t - k^2 F_{xx} = \delta(x - \xi)\delta(t - \tau), \qquad -\infty < x < \infty, t > 0,$$
$$F(x, 0) = 0.$$

The solution $F(x, t; \xi, \tau)$ will be called the fundamental solution of the heat equation, of course. Although we couldn't use Fourier transforms for Laplace's equation (we had no idea how to invert the transforms and, in the exercises, we used the solutions to *find* these inverse transforms), it turns out that we may use them here. That being the case, why not just solve the general problem

$$u_t - k^2 \mu_{xx} = g(x,t), \qquad -\infty < x < \infty, t > 0,$$
$$u(x,0) = 0,$$

and then just let $g(x,t) = \delta(x - \xi, t - \tau)\ldots$?

So we transform and get

$$U_t + k^2\alpha^2 U = G(\alpha,t),$$
$$U(\alpha,0) = 0,$$

where $\mathcal{F}[u(x,t)] = U(\alpha,t)$ and $\mathcal{F}[g(x,t)] = G(\alpha,t)$. This is the nonhomogeneous version of the first-order ODE from Section 6.4. Using the integrating factor $e^{k^2\alpha^2 t}$, and applying the initial condition, we have

$$U(\alpha,t) = \int_0^t e^{k^2\alpha^2(s-t)} G(\alpha,s)ds$$
$$= \frac{1}{\sqrt{2\pi}} \int_0^t e^{k^2\alpha^2(s-t)} \int_{-\infty}^\infty e^{-i\alpha y} g(y,s)dy ds \quad \text{(why?)}.$$

The solution, then, is

$$u(x,t) = \mathcal{F}^{-1}[U(\alpha,t)] = \frac{1}{\sqrt{2\pi}} \int_{-\infty}^\infty e^{i\alpha x} U(\alpha,t)d\alpha$$
$$= \frac{1}{2\pi} \int_0^t \int_{-\infty}^\infty g(\xi,\tau) \left[\int_{-\infty}^\infty e^{k^2\alpha^2(\tau-t)+i\alpha(x-\xi)}d\alpha \right] d\xi d\tau,$$

where we have done some order-of-integration switching. Then, back in Section 6.4, we showed that

$$\mathcal{F}^{-1}[e^{-k^2\alpha^2 t}] = \frac{1}{\sqrt{2\pi}} \int_{-\infty}^\infty e^{-k^2\alpha^2 t + i\alpha x} \, d\alpha$$
$$= \frac{1}{k\sqrt{2t}} e^{-\frac{x^2}{4k^2 t}},$$

the *heat kernel*; so, using this to rewrite the α-integral above, we have the solution

$$u(x,t) = \frac{1}{2k\sqrt{\pi}} \int_0^t \int_{-\infty}^\infty g(\xi,\tau) \cdot \frac{1}{\sqrt{t-\tau}} e^{-\frac{(x-\xi)^2}{4k^2(t-\tau)}} d\xi d\tau.$$

Now, what about the original problem? Again, we take $g(x,t) = \delta(x - \xi)\delta(t - \tau)$, with the result that

$$F(x,t;\xi,\tau) = \frac{1}{2k\sqrt{\pi}} \int_0^t \int_{-\infty}^{\infty} \frac{\delta(y - \xi)\delta(s - \tau)}{\sqrt{t - s}} e^{-\frac{(x-y)^2}{4k^2(t-s)}} \, dy \, ds$$

$$= \frac{1}{2k\sqrt{\pi(t - \tau)}} e^{-\frac{(x-\xi)^2}{4k^2(t-\tau)}} H(t - \tau).$$

We have killed two birds with one stone, showing that

a) $F(x,\xi;t,\tau)$ is the solution of the original δ-function problem.

b) The solution of the more general nonhomogeneous problem is

$$u(x,t) = \int_0^t \int_{-\infty}^{\infty} g(\xi,\tau)F(x,t;\xi,\tau)d\xi d\tau.$$

So we're justified in calling the heat kernel F the **fundamental solution** of the one-dimensional (1-D) heat equation. Note the presence of $H(t-\tau)$ in the solution—this is due, of course, to the fact that F is the response to a *heat impulse* at the point $x - \xi$ and *at time* $t = \tau$. Thus, we expect no response before time τ.¶

Now, remember that the solution of the homogeneous heat problem

$$u_t = k^2 u_{xx}, \qquad -\infty < x < \infty, t > 0,$$
$$u(x,0) = \phi(x)$$

is

$$u(x,t) = \frac{1}{2k\sqrt{\pi t}} \int_{-\infty}^{\infty} e^{-\frac{(x-y)^2}{4k^2 t}} \phi(y)dy.$$

Therefore, we may write the solution of the general heat problem

$$u_t = k^2 u_{xx} + g(x,t), \qquad -\infty < x < \infty, t > 0,$$
$$u(x,0) = \phi(x)$$

as

$$u(x,t) = \int_0^t \int_{-\infty}^{\infty} F(x,\xi;t,\tau)g(\xi,\tau)d\xi d\tau$$

$$+ \int_{-\infty}^{\infty} F(x,\xi;t,0)\phi(\xi)d\xi.$$

¶It's standard practice to write $\int_0^t \ldots d\tau$ instead of $\int_0^\infty \ldots d\tau$, even though the $H(t - \tau)$ term is then redundant. The formal notation shows more clearly the causal nature of the effect of g. Because time behaves differently from the space variables in these problems, Green's function often is called the **causal Green's function**. To be consistent, then, we should talk about the *causal fundamental solution*.

The obvious relationship between the homogeneous and nonhomogeneous solutions is another version of *Duhamel's Principle* (see Section 6.1), which we discuss further at the end of this section and in the exercises.

The result is easily generalized to higher dimensions, as

$$\mathcal{F}^{-1}[e^{-k^2(\alpha^2+\beta^2+\gamma^2)t}] = \frac{1}{(2\pi)^{3/2}} \int_{-\infty}^{\infty} e^{-k^2\alpha^2 t + i\alpha x} \, d\alpha \int_{-\infty}^{\infty} e^{-k^2\beta^2 t + i\beta y} \, d\beta$$
$$\cdot \int_{-\infty}^{\infty} e^{-k^2\gamma^2 t + i\gamma z} \, d\gamma.$$

Therefore, the solution of

$$u_t - k^2 \nabla^2 u = \delta(x - \xi)\delta(y - \eta)\delta(t - \tau),$$
$$-\infty < x < \infty, -\infty < y < \infty, t > 0,$$
$$u(\boldsymbol{x}, 0) = 0$$

is

$$F(x, y, t; \xi, \eta, \tau) = \frac{1}{4k^2\pi(t-\tau)} e^{-\frac{(x-\xi)^2+(y-\eta)^2}{4k^2(t-\tau)}}, \quad t > \tau,$$
$$= \frac{1}{4k^2\pi(t-\tau)} e^{-\frac{(x-\xi)^2+(y-\eta)^2}{4k^2(t-\tau)}} H(t - \tau),$$

while the solution of the general nonhomogeneous problem

$$u_t - k^2 \nabla^2 u = g(x, y, t),$$
$$u(x, y, 0) = 0$$

is

$$u(x, y, t) = \int_0^t \int_{-\infty}^{\infty} \int_{-\infty}^{\infty} g(\xi, \eta, \tau) F(x, y, t; \xi, \eta, \tau) d\xi d\eta d\tau.$$

(Similarly for three dimensions. See Exercise 1.)

WAVE EQUATION

Now let's do the same for the wave equation,

$$u_{tt} - c^2 u_{xx} = g(x, t), \qquad -\infty < x < \infty, t > 0,$$
$$u(x, 0) = u_t(x, 0) = 0.$$

Transforming gives us

$$U_{tt} + c^2\alpha^2 U = G(\alpha, t),$$
$$U(\alpha, 0) = U_t(\alpha, 0) = 0.$$

Using variation of parameters, we have (see Exercise 2)

$$U(\alpha, t) = \int_0^t G(\alpha, s) \frac{\sin c\alpha(t - s)}{c\alpha} ds.$$

Then,

$$u(x, t) = \frac{1}{\sqrt{2\pi}} \int_{-\infty}^{\infty} \int_0^t G(\alpha, s) \frac{\sin c\alpha(t - s)}{c\alpha} e^{i\alpha x} \, ds d\alpha$$

$$= \frac{1}{2\sqrt{2\pi} \, c} \int_0^t \int_{-\infty}^{\infty} G(\alpha, s) \int_{x-c(t-s)}^{x+c(t-s)} e^{i\alpha z} \, dz d\alpha ds.$$

(Why? This should look familiar—see Section 9.6.)

$$= \frac{1}{2\sqrt{2\pi} \, c} \int_0^t \int_{x-c(t-s)}^{x+c(t-s)} \int_{-\infty}^{\infty} G(\alpha, s) e^{i\alpha z} \, d\alpha dz ds$$

$$= \frac{1}{2c} \int_0^t \int_{x-c(t-s)}^{x+c(t-s)} g(z, s) dz ds$$

(where, again, we have taken liberties with changing the order of integration). Note that the region of integration, in z-s space-time, is the domain of dependence, or *past history*, of the point (x, t); see Figure 10.8 and compare to Exercise 14, Section 5.3.

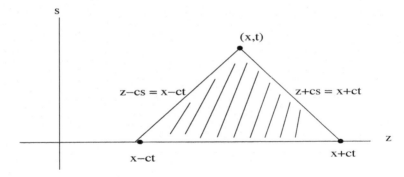

FIGURE 10.8
Domain of dependence or *past history* of the point (x, t) in z-s space-time.

We may write Green's function here using Heaviside functions. Now, we need

$$x - c(t - s) \le z \le x + c(t - s)$$

and

$$t - s \ge 0.$$

It follows that Green's function is

$$F_1(x,t;\xi,\tau) = \frac{1}{2c}H[c(t-\tau) - |x-\xi|].$$

(Here, the subscript 1 represents the spatial dimension of the problem. See the comments following the derivation of the 3-D solution.) F_1 will, of course, satisfy the problem when $g(x,t) = \delta(x-\xi)\delta(t-\tau)$. (See Exercise 5.)

Also, note the similarity between our solution and that part of d'Alembert's solution due to the $u_t(x,0)$ boundary condition—this, again, is Duhamel's Principle in action! (See Exercise 6.)

It follows that the solution of the general problem

$$u_{tt} = c^2 u_{xx} + g(x,t), \qquad -\infty < x < \infty, t > 0,$$
$$u(x,0) = \phi(x),$$
$$u_t(x,0) = \psi(x)$$

is just the above solution, added to d'Alembert's solution:

$$u(x,t) = \int_0^t \int_{-\infty}^\infty F_1(x,t;\xi,\tau)g(\xi,\tau)d\xi d\tau$$
$$+ \frac{1}{2}[\phi(x+ct) + \phi(x-ct)] + \frac{1}{2c}\int_{x-ct}^{x+ct} \psi(\xi)d\xi.$$

As with the heat equation, the last term can be rewritten using F. Further, as before, the middle term looks like the derivative of the last term. Thus, we have

$$[\phi(x+ct) + \phi(x-ct)] = \frac{1}{2c}\frac{\partial}{\partial t}\int_{x-ct}^{x+ct} \phi(\xi)d\xi$$
$$= \frac{\partial}{\partial t}\int_{-\infty}^\infty F_1(x,t;\xi,0)\phi(\xi)d\xi$$

and our solution can be written

$$u(x,t) = \int_0^t \int_{-\infty}^\infty F_1(x,t;\xi,\tau)g(\xi,\tau)d\xi d\tau$$
$$+ \frac{\partial}{\partial t}\int_{-\infty}^\infty F_1(x,t;\xi,0)\phi(\xi)d\xi + \int_{-\infty}^\infty F_1(x,t;\xi,0)\varphi(\xi)d\xi.$$

Once again, Duhamel's Principle is at work.

As for higher dimensions, consider the 3-D problem

$$u_{tt} = c^2\nabla^2 u, \qquad -\infty < x < \infty, -\infty < y < \infty, -\infty < z < \infty,$$
$$u(x,y,z,0) = u_t(x,y,z,0) = 0.$$

Proceeding as in the 1-D case, we arrive at

$$U(\boldsymbol{\alpha}, t) = \int_0^t G(\boldsymbol{\alpha}, s) \frac{\sin c|\boldsymbol{\alpha}|(t - s)}{c|\boldsymbol{\alpha}|} ds,$$

where $\boldsymbol{\alpha} = (\alpha, \beta, \gamma)$ and, below, $\boldsymbol{x} = (x, y, z)$. So

$$u(\boldsymbol{x}, t) = \frac{1}{(2\pi)^{3/2}} \int_0^t \int_{-\infty}^{\infty} \int_{-\infty}^{\infty} \int_{-\infty}^{\infty} G(\boldsymbol{\alpha}, x) \frac{\sin c|\boldsymbol{\alpha}|(t - s)}{c|\boldsymbol{\alpha}|} e^{i\boldsymbol{\alpha} \cdot \boldsymbol{x}} d\boldsymbol{\alpha} ds.$$

But, just as in the 1-D case, we may rewrite the sine-exponential combination as an integral. In fact, we already did this in Section 9.6. Here, we get

$$\frac{\sin c|\boldsymbol{\alpha}|(t - s)}{c|\boldsymbol{\alpha}|} e^{i\boldsymbol{\alpha} \cdot \boldsymbol{x}} = \frac{1}{4\pi c^2 (t - s)} \iint_{|\boldsymbol{z}| = c(t-s)} e^{i\boldsymbol{\alpha} \cdot (\boldsymbol{x} + \boldsymbol{z})} dS.$$

Again, we switch the order of integration and write

$$u(\boldsymbol{x}, t) = \frac{1}{(2\pi)^{3/2}} \frac{1}{4\pi c^2} \int_0^t \frac{1}{t - s} \iint_{|\boldsymbol{z}| = c(t-s)} \left[\int_{-\infty}^{\infty} \int_{-\infty}^{\infty} \int_{-\infty}^{\infty} G(\boldsymbol{\alpha}, s) e^{i\boldsymbol{\alpha} \cdot (\boldsymbol{x} + \boldsymbol{z})} d\boldsymbol{\alpha} \right] dS ds.$$

$$= \frac{1}{4\pi c^2} \int_0^t \frac{1}{t - s} \iint_{|\boldsymbol{z}| = c(t-s)} g(\boldsymbol{x} + \boldsymbol{z}, s) dS ds$$

$$= \frac{1}{4\pi c^2} \int_0^t \iint_{|\boldsymbol{z} - \boldsymbol{x}| = c(t-s)} \frac{g(\boldsymbol{z}, s)}{t - s} dS ds.$$

(Here, $\boldsymbol{z}$ in the dS integration variable.) Note that here, again, the region of integration is the *domain of dependence* of the point $(\boldsymbol{x}, t) = (x, y, z, t)$ (in x-y-z-t space-time, of course).

At this point, we *may* write Green's function as

$$F_3(\boldsymbol{x}, \boldsymbol{\xi}; t, \tau) = \frac{\delta[c(t - \tau) - |\boldsymbol{\xi} - \boldsymbol{x}|]}{4\pi c^2 (t - \tau)} \quad \text{(why?)}.$$

However, it's common practice, instead, to use spatial variables wherever possible. So on the sphere, we have $t - s = \frac{|\boldsymbol{\xi} - \boldsymbol{x}|}{c}$; thus,

$$F_3(\boldsymbol{x}, \boldsymbol{\xi}; t, \tau) = \frac{\delta[c(t - \xi) - |\boldsymbol{\xi} - \boldsymbol{x}|]}{4\pi c |\boldsymbol{\xi} - \boldsymbol{x}|}.$$

In Exercise 9 we show that the 2-D fundamental solution is

$$F_2(\boldsymbol{x}, \boldsymbol{\xi}; t, \tau) = \frac{H[c(t - \tau) - |\boldsymbol{x} - \boldsymbol{\xi}|]}{2\pi c \sqrt{c^2 (t - \tau)^2 - |\boldsymbol{x} - \boldsymbol{\xi}|^2}}.$$

Note that, in contrast with the heat equation, *the fundamental wave solutions depend qualitatively on the space dimension.*

Now, we may write u as an iterated integral, using the spherical coordinates θ and ϕ, with the result being

$$u(\boldsymbol{x}, t) = \frac{1}{4\pi c^2} \int_0^t \int_0^\pi \int_0^{2\pi} \frac{q(z, s)}{t - s} \cdot c^2 (t - s)^2 \sin\phi \; d\theta d\phi dt.$$

This is *almost* a volume integral over the ball of radius ct, and we can get there by substituting $\rho = c(t - s)$, $d\rho = -c \, ds$, to arrive at

$$u(\boldsymbol{x}, t) = \frac{1}{4\pi c^2} \iiint_{|\boldsymbol{z}-\boldsymbol{x}| \leq ct} \frac{g\left(z, t - \frac{|\boldsymbol{z}-\boldsymbol{x}|}{c}\right)}{|\boldsymbol{z} - \boldsymbol{x}|} dV.$$

In this form, we see that the function

$$\frac{1}{4\pi c^2 \rho} = \frac{1}{4\pi c^2 |\boldsymbol{z} - \boldsymbol{x}|}$$

behaves much like the *Newtonian potential* for the 3-D Poisson equation. Here, though, we also have

$$t - \frac{|\boldsymbol{z} - \boldsymbol{x}|}{c}$$

instead of just t. We say that time is *retarded* by the amount $\frac{|\boldsymbol{z}-\boldsymbol{x}|}{c}$, and we call this last form of the solution the **retarded potential** representation of u.

Finally, in order to relate our solution u to the homogeneous solution from Section 9.6, note that

$$\frac{1}{4\pi c^2 (t - s)^2} \iint_{|\boldsymbol{z}-\boldsymbol{x}|=c(t-s)} g(z, s) dS = \bar{g}_{c(t-s)}$$

is the average value of g on the sphere $|\boldsymbol{z} - \boldsymbol{x}| = c(t - s)$, so the solution of the general problem

$$u_{tt} = c^2 \nabla^2 u, \qquad -\infty < x < \infty, -\infty < y < \infty, -\infty < z < \infty, t > 0,$$
$$u(\boldsymbol{x}, 0) = \phi(\boldsymbol{x}),$$
$$u_t(\boldsymbol{x}, 0) = \psi(\boldsymbol{x})$$

can be written as

$$u(\boldsymbol{x}, t) = \int_0^t (t - s) \bar{g}_{c(t-s)} \; ds + \frac{1}{(2\pi)^{3/2}} \left[\frac{\partial}{\partial t} (t\bar{\phi}_{ct}) + t\bar{\psi}_{ct} \right].$$

From this representation, it is again clear that *Huygens's Principle* holds for these problems, as well.

In order to find Green's function and solve problems on proper subsets of the domain in question, we may again use the method of images, series of orthogonal functions, and the like. We do so in the exercises.

DUHAMEL'S PRINCIPLE

To arrive at an "official" statement of Duhamel's Principle in this setting, it's helpful to start with the special case where the forcing (source) function depends only on x. We compare the solutions of the problems

$$u_t = k^2 u_{xx} + g(x), \qquad v_t = k^2 v_{xx}, \qquad -\infty < x < \infty, t > 0,$$
$$u(x,0) = 0, \qquad\qquad v(x,0) = g(x).$$

We find that the Fourier transforms of the solutions are

$$V(\alpha,t) = G(\alpha)e^{-k^2\alpha^2 t}, \quad U(\alpha,t) = \frac{G(\alpha)}{k^2\alpha^2}[1 - e^{-k^2\alpha^2 t}].$$

Therefore, we have

$$U(\alpha,t) = \int_0^t V(\alpha,\tau)d\tau$$

and, transforming back,

$$u(x,t) = \int_0^t v(x,\tau)d\tau.$$

Of course, things are not so obvious when the forcing/source function involves time, as well. So, given

$$u_t = k^2 u_{xx} + g(x,t), \qquad -\infty < x < \infty, t > 0,$$
$$u(x,0) = 0,$$

and solving by Fourier transforms, we have

$$U(\alpha,t) = \int_0^t e^{-k^2\alpha^2(t-\tau)}G(\alpha,\tau)d\tau,$$

where $G(\alpha,t) = \mathcal{F}[g(x,t)]$ (in just the space variable, of course). This suggests that we look at the homogeneous problem with initial temperature $g(x,\tau)$, for values of τ in $0 \le \tau \le t$. So, we let $v_\tau(x,t)$ be the solution of the problem

$$v_t = k^2 v_{xx}, \qquad -\infty < x < \infty, t > \tau,$$
$$v(x,\tau) = g(x,\tau),$$

and we see that the Fourier transform of the solution is, indeed,

$$V_\tau(\alpha,t) = e^{-k^2\alpha^2(t-\tau)}G(\alpha,\tau).$$

It follows that

$$u(x,t) = \int_0^t v_\tau(x,t)d\tau.^{\|}$$

$^{\|}$It's common practice to write $v(x,t;\tau)$ instead of $v_\tau(x,t)$.

We may do the same for higher dimensions, and we may proceed similarly for the wave equation—we do so in the exercises.

One may ask about *reciprocity* or *symmetry* for Green's functions for the heat and wave equations. This turns out to be a delicate affair because of the presence of the time variable and the fact that, in that variable, these problems are *initial*-value problems.

Exercises 10.5

1. Write down the fundamental solution for the 3-D heat problem, and use it to solve the problem

$$u_{tt} = k^2 \nabla^2 u + g(x, y, z, t), \qquad -\infty < x < \infty, -\infty < y < \infty,$$
$$-\infty < z < \infty, t > 0,$$
$$u(x, y, z, 0) = \phi(x, y, z).$$

2. Fill in the details of the derivation of the variation of parameters solution of the 1-D wave problem.

3. Use the method of images to find Green's function for the heat/diffusion problem

   a) $\quad u_t = k^2 u_{xx} + g(x, t), \qquad 0 < x < \infty, t > 0,$
   $$u(x, 0) = u(0, t) = 0$$

   b) $\quad u_t = k^2 u_{xx} + g(x, t), \qquad 0 < x < \infty, t > 0,$
   $$u(x, 0) = u_x(0, t) = 0$$

   c) $\quad u_t = k^2 \nabla^2 u + g(x, y, t), \qquad x > 0, y > 0, t > 0,$
   $$u(x, y, 0) = u(x, 0, t) = u(0, y, t) = 0$$

4. Find Green's function for the problem

$$u_t = k^2 u_{xx} + g(x, t), \qquad 0 < x < \pi, t > 0,$$
$$u(x, 0) = 0,$$
$$u(0, t) = u(\pi, t),$$

   in two different ways:

   a) Using the method of images
   b) By expanding $g(x, t) = \delta(x - \xi)\delta(t - \tau)$ in a Fourier sine series in x

5. Show that the solution of the wave problem

$$u_{tt} = c^2 u_{xx} + \delta(x - \xi)\delta(t - \tau),$$
$$u(x, 0) = 0$$

   is, indeed, $u = F_1(x, t; \xi, \tau) = \frac{1}{2c} H[c(t - \tau) - |x - \xi|].$

6. Write down and justify Duhamel's Principle for the wave problems

$$u_{tt} = c^2 u_{xx} + g(x,t), \qquad -\infty < x < \infty, t > 0,$$
$$u(x,0) = 0,$$
$$u_t(x,0) = 0$$

and

$$v_{tt} = c^2 v_{xx}, \qquad -\infty < x < \infty, t > \tau,$$
$$v(x,\tau) = 0,$$
$$v_t(x,\tau) = g(x,\tau).$$

7. **Source functions and convolutions:** The **source function** for an evolution equation is the solution of the homogeneous problem on all of space, subject to an initial condition $\delta(x)$. So, for example, the source function for the 1-D heat operator is the solution of

$$u_t = k^2 u_{xx}, \qquad -\infty < x < \infty, t > 0,$$
$$u(x,0) = \delta(x),$$

and we found that the solution is the heat kernel

$$S(x,t) = \frac{1}{2k\sqrt{\pi t}} e^{-\frac{x^2}{4k^2 t}}.$$

a) Show that the solution of the heat problem

$$u_t = k^2 u_{xx} + g(x,t), \qquad -\infty < x < \infty, t > 0,$$
$$u(x,0) = \phi(x)$$

can be written as

$$u(x,t) = \int_0^t S(x,t-\tau) * g(x,\tau)d\tau$$
$$+ S(x,t) * \phi(x),$$

where the convolution is in the space variable(s) only.

b) Show that the solution of the wave problem

$$u_{tt} = c^2 u_{xx} + g(x,t), \qquad -\infty < x < \infty, t > 0,$$
$$u(x,0) = \phi(x),$$
$$u_t(x,0) = \psi(x)$$

can be written as

$$u(x,t) = \int_0^t S(x,t-\tau) * g(x,\tau)d\tau$$
$$+ \frac{\partial}{\partial t} S(x,t) * \phi(x) + S(x,t) * \psi(x),$$

where the source function S, called the **Riemann function**, is the solution of the problem

$$u_{tt} = c^2 u_{xx}, \qquad -\infty < x < \infty, t > 0,$$
$$u(x,0) = 0,$$
$$u_t(x,0) = \delta(x).$$

8. **Adjoints and self-adjoint problems:** We may extend the idea of adjoint to PDEs, as well. For the linear, second-order operator

$$L[u] = a_{11}u_{xx} + 2a_{12}u_{xy} + a_{22}u_{yy} + b_1 u_x + b_2 u_y + cu,$$

we define its adjoint to be the operator

$$L^*[u] = a_{11}u_{xx} + 2a_{12}u_{xy} + a_{22}u_{yy} - b_1 u_x - b_2 u_y + cu.$$

Note that, as before, self-adjointness is "ruined" by the presence of a first derivative.

a) Show that, for any such operator, we have a version of Lagrange's identity of the form

$$uL[v] - vL^*[u] = \nabla \cdot \boldsymbol{w},$$

where $\boldsymbol{w}(x,y) = (w_1(x,y), w_2(x,y))$.

b) Show that Green's formula can be written as

$$\iint_D (uL[v] - vL^*[u]) = \oint_{\partial D} \boldsymbol{w} \cdot \boldsymbol{n} \ ds,$$

where D is any bounded domain.

c) Note that we already used the self-adjointness of the Laplace operator in Sections 10.3 and 10.4. Are the heat and wave operators self-adjoint?

9. Find the solution of the 2-D wave problem

$$u_{tt} = c^2\nabla^2 u + \delta(x - \xi)\delta(y - \eta)\delta(t - \tau), \qquad -\infty < x < \infty,$$
$$-\infty < y < \infty, t > 0,$$
$$u(x,y,0) = 0,$$
$$u_t(x,y,0) = 0$$

by treating it as a *3-D* problem, with a *vertical line source*.

Prelude to Chapter 11

In actual practice, most PDEs cannot be solved exactly, and we must resort to approximate methods of solution, that is, to *numerical methods*. In fact, in most cases where we can find an exact solution, it is an infinite series, which must be approximated anyway in any practical applications.

While many numerical methods arrived recently, due to the advent of the computer (and its predecessor, the differential analyzer), some are quite old. In fact, the *finite difference method*, which employs various difference quotients to approximate derivatives, was used to solve problems in astronomy and cartography even before the derivative had been invented/discovered! However, theoretical work on finite difference schemes began in earnest only at the turn of the 20th century, and, as it is with so much of modern mathematics, the names of those involved are too numerous to list (although of particular note is Richard Courant (1888–1972). We should also mention Phyllis Nicolson, who helped to devise the Crank–Nicolson numerical scheme.).

The *finite element method* is like the finite difference method insofar as it involves breaking up the domain of the problem into discrete pieces, but the similarity ends there, as we'll see. This method dates to the years 1915–1925 and, particularly, to the work of Boris Grigorievich Galerkin (1871–1945) and, again, Richard Courant.

The last numerical methods that we look at are the *spectral methods*, so-called because they entail approximating solutions with truncated series of eigenfunctions—indeed, with truncated (generalized) Fourier series. Important contributions in this area were made by Galerkin and by Cornelius Lanczos (1893–1974), among others.

11

Numerical Methods

11.1 Finite Difference Approximations for ODEs

We begin by looking at finite difference approximations for *ordinary* differential equations, as they involve most of the pertinent ideas from PDE methods, yet are easier to implement. The first question we ask is the somewhat obvious question of how to approximate the first derivative. But remember the definition of f' using the *difference quotient*,

$$f'(x) = \lim_{h \to 0} \frac{f(x+h) - f(x)}{h}.$$

Thus, we *should* have

$$f'(x) \approx \frac{f(x+h) - f(x)}{h},$$

with the approximation getting better as h gets smaller.

How will this work? Let's get right to an example.

Example 1 Approximate the solution of the initial-value problem

$$\frac{dy}{dt} + y = 3, \qquad t > 0,$$
$$y(0) = 5.$$

Of course, the exact solution is $y = 2 + 3e^{-t}$, with which we'll compare our approximation.

So let's break the t-axis into intervals of constant length h and form the *grid points* as in Figure 11.1, just as we did when developing the Riemann integral back in calculus. Thus, we have

$$t_0 = 0, t_1 = t_0 + h, t_2 = t_1 + h = 2h, \ldots, t_i = ih, \qquad i = 0, 1, 2, \ldots .$$

FIGURE 11.1
Grid for Example 1.

Now we replace $\frac{dy}{dt}$ by the difference quotient, and the ODE becomes

$$\frac{y(t+h) - y(t)}{h} + y(t) \approx 3$$

or

$$y(t+h) \approx h[3 - y(t)] + y(t).$$

We're now in a position to move *step-wise* from t_0 to t_1, t_1 to t_2, t_2 to t_3, etc. We have

$$
\begin{aligned}
y(t_1) &= y(h) \\
&\approx h[3 - y(t_0)] + y(t_0) \\
&= h[3 - y(0)] + y(0) \\
&= 5 - h, \\
y(t_2) &= y(2h) \\
&\approx h[3 - y(t_1)] + y(t_1) \\
&= 5 - 4h + 2h^2, \\
y(t_3) &\approx \ldots \approx 5 - 6h + 6h^2 - 2h^3, \\
&\;\;\vdots
\end{aligned}
$$

In any single step, from $y(t_i)$ to $y(t_{i+1})$, an obvious source of error is our approximation of the derivative. We should be able to reduce this error by taking smaller values of h. A more mundane, but equally important, error source is the fact that these computations will, in general, be performed on a real computer, which always rounds to a certain number of decimal places.

Each of these errors will accumulate and, since reducing the size of h will increase the number of steps necessary to reach a given time, taking smaller values of h will *increase* the latter error. Thus, we have two competing types of errors:

1. **Truncation or discretization error:*** due to approximation of derivative, reduced by choosing lesser values of h

2. **Roundoff error:** due to the computer's rounding at each calculation, reduced by choosing greater values of h

---

* "Discretization" because we're changing a continuous problem into a discrete problem; "truncation" because, as we'll see below, we can look at this process as a truncation of the Taylor series for y. Also, we actually have various types of truncation errors, each a result of the process of discretization. We'll say more at the end of this section.

Finally, an additional concern is a very practical one, that of computation time. Here, as with roundoff error, decreasing h means adding more steps to get to a given value of t and, thus, more run-time for the computer program.

Example 1 (cont.) Continuing our first example, it's customary to write

$$y(t_i) \approx y_i, \qquad i = 0, 1, \ldots,$$

so that $y(t_i)$ will be the exact solution at t_i, while y_i is the approximate solution there. In this case, the finite difference approximation is written

$$y_{i+1} = h(3 - y_i) + y_i, \qquad i = 0, 1, 2, \ldots .$$

In Table 11.1, we compare the "exact" solution on $0 \le t \le 1$ with the approximate solution, for various values of h. In this case, $i = 0, 1, \ldots, n$, while $h = \frac{1}{n}$.

t	$n = 10$	$n = 100$	$n = 1000$	$3 + 2e^{-t}$
.1	4.8	4.8088	4.8096	4.8047
.2	4.62	4.6358	4.6373	4.6375
.3	4.458	4.4794	4.4814	4.4816
.4	4.3122	4.3379	4.3404	4.3406
.5	4.1810	4.2100	4.2128	3.2131
.6	4.0629	4.0943	4.0973	3.0976
.7	3.9566	3.9897	3.9928	3.9932
.8	3.8609	3.8950	3.8983	3.8987
.9	3.7748	3.8095	3.8128	3.8131
1.0	3.6974	3.7321	3.7354	3.7358

TABLE 11.1
Results of Example 1 on $0 \le t \le 1$. Note that the approximate values seem to converge to the "exact" values (calculated by computer) as h becomes smaller.

Notice that we start at the initial time and *march* to the right, one step at a time. This type of method is called an **explicit one-step method**—the value of y_{i+1} is given *explicitly*, and it depends only on the values t_i and y_i. (More generally, we have *explicit k-step methods*, where y_{i+k} depends on t_j and y_j for $j = i, i + 1, \ldots, i + k - 1$.) By the way, the particular explicit one-step method used here is called **Euler's method**.

On the other hand, **implicit methods** will have y_{i+1} on *both* sides of the equation, and it often will be difficult or impossible to solve for y_{i+1}. In this case, y_{i+1} usually is approximated again. See Exercise 12.

We need to be careful in the h-large-vs-h-small struggle, as the following example shows.

Example 2 We use the method of Example 1, but for the IVP

$$y' = -2y, \quad t > 0,$$
$$y(0) = y_0,$$

the exact solution of which is $y = y_0 e^{-2t}$. The approximation gives us

$$\frac{y_{i+1} - y_i}{h} = -2y_i \quad \text{or} \quad y_{i+1} = (1 - 2h)y_i$$

and, in this simple case, it's easy to show that

$$y_{i+1} = (1 - 2h)^{i+1} y_0$$

(see Exercise 7). Now, as i increases, we see that the behavior depends very much on the value of h. Specifically,

$$\lim_{i \to \infty} (1 - 2h)^{i+1} = \begin{cases} 0, & \text{if } 0 < h \le \dfrac{1}{2}, \\ \pm\infty, & \text{if } h > \dfrac{1}{2}; \end{cases}$$

so, if our time step is too large, our approximate solution oscillates with increasing magnitude and, if $h = \frac{1}{2}$, we get $y_i = 0$ for $i = 1, 2, 3, \dots$. Thus, unless $h < \frac{1}{2}$, our approximation is worthless. This type of behavior often is a signal that we should be using another difference approximation for the problem.

Now, remember that PDEs often are initial-*boundary*-value problems, so we'll need to look at ODE BVPs, the more interesting of which seem to involve y''. Thus, we must recast our approximation of f' in a more general setting that will allow us to approximate higher order derivatives. For this, we turn to the Taylor series.

Remember that, if f is well-enough behaved, we can write

$$f(x + h) = \sum_{n=0}^{\infty} \frac{f^{(n)}(x)}{n!} h^n$$

$$= f(x) + hf'(x) + \frac{h^2}{2!} f''(x) + \frac{h^3}{3!} f'''(x) + \dots ,$$

the series being the **Taylor series** for f, about the point x. Of course, when this series converges, the terms $\to 0$ as $n \to \infty$, so we may approximate

$f(x + h)$ to any desired degree of accuracy just by computing enough terms on the right. In the process, we *ignore* the remaining terms—we *truncate* the series. We can be more precise, though, by introducing notation that will allow us to keep track of the effect of truncating the series.

Definition 11.1 *Given functions $f(h)$ and $g(h)$, if there exist constants $M > 0$ and $\delta > 0$ such that*

$$|f(h)| \leq M|g(h)| \quad for \quad |h| < \delta,$$

then we write

$$f(h) = O(g(h)) \quad as \quad h \to 0$$

and we say "f is big-oh of g as h approaches 0"[†] (and, of course, $g(h) \to 0 \Rightarrow f(h) \Rightarrow 0$). Note that an equivalent statement is that

$$\lim_{h \to 0} \frac{f(h)}{g(h)} = constant.$$

Now, for fixed x, since the terms in the Taylor series $\to 0$, we can show (with a little work) that, for example,

$$\frac{h^2}{2!}f''(x) + \frac{h^3}{3!}f'''(x) + \ldots = h^2 \left[\frac{1}{2}f''(x) + \frac{h}{3!}f'''(x) + \ldots \right]$$
$$= O(h^2)$$

and we write

$$f(x + h) = f(x) + hf'(x) + O(h^2).$$

Similarly,

$$f(x + h) = f(x) + hf'(x) + \frac{h^2}{2!}f''(x) + \frac{h^3}{3!}f'''(x) + O(h^4),$$

and we say that the truncation is $O(h^4)$.

So, let's get back to approximating derivatives. We see that we can repeat our original approximation to f' by writing

$$\frac{f(x + h) - f(x)}{h} = f'(x) + \frac{O(h^2)}{h}$$

or

$$f'(x) = \frac{f(x + h) - f(x)}{h} + O(h) \quad \text{(why } O(h)\text{?).}$$

We call this the **forward difference approximation** to f'. Of course, we can replace h by $-h$, and get

$$\frac{f(x) - f(x - h)}{h} = f'(x) + O(h),$$

[†]We may generalize the definition to include h approaching any value h_0, including $\pm\infty$.

the **backward difference approximation**. However, we can do better. Expanding $f(x + h)$ and $f(x - h)$ in their Taylor series, and subtracting, we get (see Exercise 8a)

$$f'(x) = \frac{f(x + h) - f(x - h)}{2h} + O(h^2),$$

the **central difference approximation**.

How about f''? We now take the Taylor series out to the f'' term, for both $f(x + h)$ and $f(x - h)$, resulting in the **central (second) difference approximation**

$$f''(x) = \frac{f(x + h) - 2f(x) + f(x - h)}{h^2} + O(h^2)$$

(see Exercise 8b). Of course, there are other f'' approximations, as well (like the forward and backward ones for f').

Let's look at an example involving a BVP.

Example 3 Approximate the solution of the BVP

$$y'' + xy = 0, \qquad 0 < x < 1,$$
$$y(0) = 0, y(1) = 1.$$

We don't know how to solve this problem (well, we'll say a bit more, below), so a numerical approximation seems to be the best we can do.

Here, $x_0 = 0, x_n = 1, y_0 = 0, y_n = 1$ and $x_i = ih = \frac{i}{n}, i = 0, 1, \ldots, n$. We choose to use the central difference approximation

$$\frac{y_{i+1} - 2y_i + y_{i-1}}{h^2} + x_i y_i = 0, \qquad i = 1, \ldots, n - 1.$$

Although we can't march forward as we did before, we *do* have $n - 1$ equations in $n - 1$ unknowns, so we may solve *simultaneously*. In matrix form, for $n = 6$, we have $h = \frac{1}{6}$, $x_i = \frac{i}{6}$, and our system looks like

$$\underbrace{\begin{bmatrix} h^2 x_1 - 2 & 1 & 0 & 0 & 0 \\ 1 & h^2 x_2 - 2 & 1 & 0 & 0 \\ 0 & 1 & h^2 x_3 - 2 & 1 & 0 \\ 0 & 0 & 1 & h^2 x_4 - 2 & 1 \\ 0 & 0 & 0 & 1 & h^2 x_5 - 2 \end{bmatrix}}_{A} \underbrace{\begin{bmatrix} y_1 \\ y_2 \\ y_3 \\ y_4 \\ y_5 \end{bmatrix}}_{Y} = \underbrace{\begin{bmatrix} 0 \\ 0 \\ 0 \\ 0 \\ -1 \end{bmatrix}}_{B}.$$

We solve this system, instead, for $n = 10, 100$ and 1000 (that is, for $h = .1, .01$ and $.001$), and we give the results in Table 11.2. There we see that the approximation seems to have converged to five decimal places. So, even for problems that we cannot solve exactly, the seeming convergence of the numerical results

1) Suggests that the problem does, indeed, possess a solution
2) Provides us with a useful, practical representation of the solution.

x	$n = 10$	$n = 100$	$n = 1000$
.1	.013382	.013467	.013467
.2	.046762	.046931	.046932
.3	.100124	.100371	.100374
.4	.173396	.173712	.173716
.5	.266390	.266759	.266763
.6	.378719	.379113	.379117
.7	.509683	.510067	.510071
.8	.658151	.658473	.658476
.9	.822406	.822605	.822607

TABLE 11.2
Results of Example 3, which seem to converge to an accuracy of five decimal places.[‡]

To be certain of what's going on, we could bring existence-uniqueness theory to bear on the problem, and we also have theoretical ways to decide if the numerical solution does, indeed, converge. In practice, for many problems it suffices to perform a seat-of-the-pants numerical approximation, as we often have experimental data with which to compare the results.

Finally, it turns out that we actually *can* find an explicit solution to this particular problem. This solution, however, involves certain Fourier-like integrals called Airy functions, and the numerical solution may turn out to be more useful to us.

Before moving on, let's go back to the matrix in Example 3 and note two important properties.

1) There are three nonzero *bands* running diagonally from top left to bottom right. (In this case, the bands are the main diagonal and those

---

[‡]MATLAB code for the approximation is given with the rest of the MATLAB programs in Appendix E.

diagonals on either side of it—we say the matrix is *tridiagonal.*) We say that the matrix is **banded**. The advantage of having a banded matrix will be obvious to you, once you try to program these problems.

2) "Most" of the entries of the matrix are zero and, as n increases, the ratio of the number of nonzero entries to the total number of entries decreases (in fact, it $\to 0$ as $n \to \infty$!). We say that the matrix is **sparse**. Not surprisingly, computers can handle much larger sparse matrices than those which are nonsparse, allowing us to achieve greater accuracy by choosing greater values of n.

Example 4 Approximate the solution of the BVP

$$y'' + xy = 2, \qquad 0 < x < 1,$$
$$y'(0) = 3, y(1) = 1.$$

This is the problem from Example 3, except for the Neumann condition at the left end. How do we deal with this condition? We approximate it! We *could* write

$$\frac{y_1 - y_0}{h} = 3$$

using the standard forward difference approximation. Then, we'll have the $n - 1$ equations as we had in Example 3, along with this n^{th} equation. But now, of course, we don't have a value for y_0, so we have the n unknowns $y_0, y_1, \ldots, y_{n-1}$, and all is well.

However, remember that the forward difference first derivative approximation is $O(h)$, while the central difference approximation that we're using for the *second* derivative is $O(h^2)$. In order to be consistent, we may want to use an $O(h^2)$ approximation for the left end boundary condition, and the central difference

$$\frac{y(x + h) - y(x - h)}{2h}$$

is the only one available. But if $x = 0$, how do we deal with $y(x - h)$? We introduce the new point

$$x_{-1} = -h, \quad y_{i-1} = y(x_{-1}),$$

which, by virtue of its being outside the domain of the problem, is called a **ghost point**. Then, the new boundary condition

$$\frac{y_1 - y_{-1}}{2h} = 3,$$

along with the additional ODE equation

$$\frac{y_1 - 2y_0 + y_{-1}}{h^2} + x_0 y_0 = 2,$$

gives us $n+1$ equations in the $n+1$ unknowns $y_{-1}, y_0, \ldots, y_{n-1}$. See Exercise 6.

Although we're not in a position to have a detailed discussion on error analysis, we should at least be aware of the concerns. We do so in a specific setting, realizing that the ideas are easily generalized.

So, suppose we're approximating the solution of the ODE

$$y' = f(t, y)$$

with Euler's (explicit) method,

$$\frac{y_{i+1} - y_i}{h} = f(t_i, y_i).$$

The **local truncation error** T_i at each step is what we get when we replace y_i by the exact solution, $y(t_i)$, in the difference approximation:

$$T_i(h) = \frac{y(t_{i+1}) - y(t_i)}{h} - f(t_i, y(t_i)), \text{ for each } i.$$

The **global truncation error** E_i results from the *accumulation* of local truncation error and is the difference between the exact solution of the ODE and the exact solution of the difference approximation:

$$E_i(h) = y(t_i) - y_i, \text{ for each } i.^{\S}$$

Of course, each of these depends on (1) the ODE, (2) the size of h and (3) the difference approximation (the last of which depends on (4) the truncation originally performed on the Taylor series).

Of course, we would like to have

$$T_i \to 0 \quad \text{and} \quad E_i \to 0 \quad \text{as} \quad h \to 0.$$

We say the approximation is

$$\textbf{convergent if} \quad E_i \to 0 \quad \text{as} \quad h \to 0,$$

and

$$\textbf{consistent if} \quad T_i \to 0 \quad \text{as} \quad h \to 0.$$

(Consistency generally is fairly easy to establish. If we *derive* a difference scheme from Taylor series or from other consistent schemes, then the result should be consistent.)

$\S$Be careful–some books refer to T_i as the *truncation error* and E_i as the *discretization error*. (And you may even see the Taylor remainder for y referred to as the *local truncation error!*)

Now, it turns out that a numerical approximation can be consistent without being convergent, and convergence is difficult to prove directly. So we need to introduce a third idea, that of stability. Without going into too much detail, an approximation is **stable** if the total error (including the machine's roundoff error), which propagates as t increases, remains small. More specifically, it is stable if $T_i \to 0$ implies that the total error $\to 0$. As with consistency, stability is easier to establish than convergence, and our life is made easier by the very important **Lax¶ Equivalence Theorem** which states, in essence, that *if an approximation is consistent, then it is convergent if and only if it is stable.*

By the way, when dealing with PDEs, we'll have discretizations in *each* independent variable and, thus, we'll be looking at errors as $h_1, h_2 \to 0$, for example. A given approximation *may* have, say, $E \to 0$ no matter how $h_1, h_2 \to 0$; however, it may also be the case that $E \to 0$ only when $h_1, h_2 \to 0$ in a certain manner (e.g., maybe we need $h_2 \to 0$ faster than h_1). In the latter case, the approximation is said to be *conditionally* convergent (and, similarly, conditionally consistent or stable).

Exercises 11.1

1. Proceed as in Example 1 and use Euler's method, with $n = 4$, to approximate the solution on $0 \le t \le 1$. Compare with the exact (or "exact," if a calculator is needed) solution.

   a) $y' = 4t + 2, y(0) = 2$

   b) $y' + y = t, y(0) = 3$

   c) $y' = ty^2, y(0) = 1$

2. **MATLAB:** Repeat each problem from Exercise 1, using $n = 10$ and $n = 100$ subdivisions, and compare with the exact solution at $t_i = \frac{i}{10}$, $i = 1, \ldots, 10$.

3. Given the BVP

$$y'' = 6, \qquad 0 < x < 1,$$
$$y(0) = 2, y(1) = 5,$$

   a) Follow Example 3 and find the approximate solution for $n = 4$ subdivisions.

   b) **MATLAB:** Repeat part (a) for $n = 10$ and $n = 100$.

   c) Compare all three approximations with the exact solution. What's going on? Use Taylor series to explain what's happening.

¶Peter Lax, 1954.

4. **MATLAB:** Proceed as in Example 3 and approximate the solution of the BVP using $n = 10$ and $n = 100$ subdivisions. In each case, compare with the exact solution

   a) $y'' = \sin \pi x, 0 \leq x \leq 1, y(0) = y(1) = 0$

   b) $y'' - y = 2x, 0 \leq x \leq 3, y(0) = 4, y(3) = -2$

   c) $y'' = \begin{cases} 0, & \text{if } 0 \leq x < 1, \\ x, & \text{if } 1 \leq x \leq 2, \end{cases}$, $y(0) = 1, y(2) = 5$

5. **MATLAB:** Perform the approximation for the problem in Example 4, for $n = 10$ and $n = 100$ subintervals.

6. Proceed as in Example 4 to set up the linear equations for the approximation of the solution of the given problem, using the central difference approximation, with *ghost points* where necessary. In each case, how many equations are there, and what are the unknowns?

   a) $y'' + x^2 y = x - 1, 2y(0) - 3y'(0) = 0, y(4) = 0$

   b) $y'' - y = x, y(0) - y'(0) = 2, y(1) + 4y'(1) = 0$

7. Use mathematical induction to show that if

   $$y_{i+1} = (1 - 2h)y_i, \qquad i = 0, 1, 2, \ldots,$$

   then

   $$y_{i+1} = (1 - 2h)^{i+1} y_0, \qquad i = 0, 1, 2, \ldots.$$

8. a) Derive the central difference approximation by showing that

   $$f'(x) = \frac{f(x + h) - f(x - h)}{2h} + O(h^2).$$

   b) Derive the formula

   $$f''(x) = \frac{f(x + h) - 2f(x) + f(x - h)}{h^2} + O(h^2).$$

   c) Repeat part (b) .but, instead, by using the central difference approximation twice.

9. a) Use Taylor series to decide if the forward, central and backward difference formulas are the *only* approximate formulas for $f'(x)$ involving a linear combination of $f(x)$, $f(x + h)$ and/or $f(x - h)$.

b) Do the same as in part (a), but for the formula

$$f''(x) \approx \frac{f(x+h) - 2f(x) + f(x-h)}{h^2}.$$

10. a) Find the most accurate approximation of $f'(x)$ of the form

$$af(x+2h) + bf(x+h) + cf(x) + df(x-h) + ef(x-2h).$$

b) Find the most accurate approximation of $f''(x)$ of the same form.

11. a) Show that if $f_1(h) = O(g_1(h))$ and $f_2(h) = O(g_2(h))$ as $h \to 0$, then $f_1(h) + f_2(h) = O[g_1(h) + g_2(h)]$ as $h \to 0$.

b) Show that if $f_1(h, k) = O(g_1(h))$ as $h \to 0$ and $f_2(h, k) = O(g_2(k))$ as $k \to 0$, then

$$f_1 + f_2 = O(g_1 + g_2) \quad \text{as} \quad h, k \to 0.$$

12. a) Given the ODE $y' = f(x, y)$, use the fact that

$$y(x_{i+1}) - y(x_i) = \int_{x_i}^{x_{i+1}} y'(t)dt,$$

along with the trapezoidal rule approximation, to derive the implicit method

$$y_{i+1} = y_i + \frac{h}{2}[f(x_i, y_i) + f(x_{i+1}, y_{i+1})].$$

b) Show that a forward difference approximation turns the above method into **Heun's method** or the **improved Euler method**

$$y_{i+1} = y_i + \frac{h}{2}[f(x_i, y_i) + f(x_{i+1}, y_i + hf(x_i, y_i))].$$

c) Generalize part (a) and derive the approximation

$$y_{i+1} = y_i + h[af(x_i, y_i) + (1-a)f(x_{i+1}, y_{i+1})]$$

for the above ODE, where a is any constant such that $0 \le a \le 1$. The term in the brackets is a *weighted average* of $f(x_i, y_i)$ and $f(x_{i+1}, y_{i+1})$. What difference method is used in the case $a = 0$?

d) Generalize part (b) and derive the approximation

$$y_{i+1} = y_i + h[(1-\alpha)f(x_i, y_i) + \alpha f(x_i + \alpha h, y_i + \alpha hf(x_i, y_i)],$$

where α is any constant such that $0 \le \alpha \le 1$.

e) Combine parts (a) and (d), with $\alpha = \frac{1}{2}$, to derive the **modified Euler method**

$$y_{i+1} = y_i + hf\left(x_i + \frac{1}{2}h, y_i + \frac{1}{2}hf(x_i, y_i)\right).$$

Each of these methods is a special case of, or related to, the so-called **Runge–Kutta methods**.

13. **MATLAB:** Approximate the solution of Example 1 using Heun's method and the modified Euler method, for $n = 10$ and $n = 100$ subdivisions, on $0 \le t \le 1$. Compare the results to those in Table 11.1.

11.2 Finite Difference Approximations for PDEs

We may approximate partial derivatives just as we did ordinary derivatives. So, for example, given $u = u(x, t)$, we have

$$u_x(x, t) = \frac{u(x + h, t) - u(x, t)}{h} + O(h)$$

$$= \frac{u(x + h, t) - u(x - h, t)}{2h} + O(h^2)$$

$$= \frac{u(x, t) - u(x - h, t)}{h} + O(h)$$

and

$$u_{xx}(x, t) = \frac{u(x + h, t) - 2u(x, t) + u(x - h, t)}{h^2} + O(h^4)$$

(and similarly for u_t and u_{tt}). Here, however, we also have mixed partials, such as u_{xt}, which we leave to Exercise 8b.

The shorthand we use is similar, as well. If we have a rectangular domain we break it up into a grid, with each x-interval of length Δx and each t-interval of length Δt. Without loss of generality (WLOG), let's take the domain to be the rectangle $0 \le x \le L$, $0 \le t$ (for the heat and wave equations). The x-grid points will be

$$x_0 = 0, x_1 = \Delta x, \ldots, x_i = i\Delta x, \ldots, x_n = n\Delta x = L,$$

while, for the t-coordinates, we have

$$t_0 = 0, t_1 = \Delta t, \ldots, t_j = j\Delta t, \ldots \ .$$

Finally, we write

$$u(x_i, t_j) = u_{i,j}.$$

THE HEAT EQUATION

Explicit scheme

We'd like to begin by discretizing the heat equation

$$u_t = k^2 u_{xx}.$$

The standard explicit scheme entails using the forward difference approximation for u_t and the central difference approximation for u_{xx}, so that we have

$$\frac{u_{i,j+1} - u_{i,j}}{\Delta t} = \frac{u_{i+1,j} - 2u_{i,j} + u_{i-1,j}}{(\Delta x)^2} + T_{i,j}(\Delta x, \Delta t),$$

where the local truncation error $T_{i,j}(\Delta x, \Delta t) = O[\Delta t + (\Delta x)^2]$ (see Exercise 7b). Thus, the scheme is *consistent*, the definition of which is that

$$\lim_{\Delta x \to 0, \Delta t \to 0} T_{i,j}(\Delta x, \Delta t) = 0$$

(in the two-variable setting). This method is **explicit (in time)**, since $u_{i,j+1}$ is given in terms of the values of u at the previous time step. Thus, beginning with the initial condition at $j = 0$, we may march step by step in the t-direction. (Thus, we talk about the time variable in these kinds of equations, when distinguishing between explicit and implicit.)

Solving for $u_{i,j+1}$, we have

$$u_{i,j+1} = \frac{k^2 \Delta t}{(\Delta x)^2} u_{i+1,j} + \left(1 - \frac{2k^2 \Delta t}{(\Delta x)^2}\right) u_{i,j} + \frac{k^2 \Delta t}{(\Delta x)^2} u_{i-1,j}.$$

We say that the quantities $u_{i,j+1}, u_{i+1,j}, u_{i,j}, u_{i,j+1}$ form a **computation molecule** for the given scheme—see Figure 11.2.

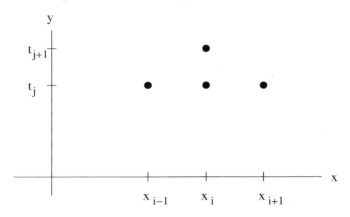

FIGURE 11.2
Computation molecule for the explicit scheme for the heat equation.

In actual practice, we must choose values for Δx and Δt. As before, smaller values of Δx and Δt should reduce the local truncation error, while larger values of Δt will reduce both the number of times that roundoff error is accumulated *and* the run-time of the computations. However, here there is another issue, one that did not show up when dealing with ODEs: we now have to deal with the ratio $\frac{\Delta t}{(\Delta x)^2}$. One can imagine that, if $(\Delta x)^2$ is much smaller than Δt, the coefficients above may be "too big" and may lead to troublesome behavior. Indeed, this turns out to be the case, and we look more closely in Exercise 1c. It can be shown that we must take

$$\frac{k^2 \Delta t}{(\Delta x)^2} \leq \frac{1}{2}$$

in order for the scheme to be *stable*.* (Thus, the scheme is conditionally stable.) Note that this choice guarantees that the coefficients are nonnegative. Note also that the number $1/2$ does not depend on the units or scale that we use, as the number $\frac{k^2(\Delta t)}{(\Delta x)^2}$ is dimensionless (why?). This is a severe restriction. For example, letting $k^2 = 1$, even if we choose Δx to be fairly large, say, $\Delta x = .1$, we're required to take $\Delta t \leq \frac{1}{2}(.1)^2$; if $\Delta x = .01$, instead, the Δt is miniscule, with $\Delta t \leq \frac{1}{2}(.01)^2$. The number of time steps required to march even a short distance in the t-direction makes this scheme quite impractical.

At any rate, once we choose this ratio, the rest of the approximation is fairly easy. The initial condition

$$u(x, 0) = f(x)$$

tells us, of course, that

$$u(x_i, 0) = u_{i,0} = f(x_i), \qquad i = 0, 1, \ldots, n,$$

while the Dirichlet boundary conditions

$$u(0, t) = u(a, t) = 0$$

give us the equations

$$u(0, t_j) = u_{0,j} = u(L, t_j) = u_{n,j} = 0, \qquad j = 0, 1, 2, \ldots .$$

If we have a Neumann or Robin boundary condition, we proceed as in Example 4 of the previous section and use a forward, central or backward difference to approximate the derivative in the boundary condition. In fact, since the local truncation error for the explicit scheme is $O[\Delta t + (\Delta x)^2]$ (see Exercise 7b), it's customary to choose the boundary approximation to be consistent with this error. Thus, we again use the central difference, as its

---

*See, e.g., G.D. Smith's *Numerical Solution of Partial Differential Equations*.

truncation error is $O[(\Delta x)^2]$. Of course, this necessitates the introduction of the *ghost points*

$$u(-\Delta x, t_j) = u_{-1,j} \quad \text{or} \quad u(L + \Delta x, t_j) = u_{n+1,j}, \qquad j = 0, 1, 2, \dots,$$

allowing us to write

$$\frac{u_{1,j} - u_{-1,j}}{2\Delta x} \approx u_x(0, t_j) \text{ or } \frac{u_{n+1} - u_{n-1}}{2\Delta x} \approx u_x(L, t_j).$$

We work out the details in Exercise 2, for example.

The result, for any time $T = m\Delta t$, is a system consisting of the $(n-1)m$ equations

$$u_{i,j+1} = \epsilon u_{i+1,j} + (1 - 2\epsilon)u_{i,j} + \epsilon u_{i-1,j}, \qquad i = 1, \dots, n-1; j = 0, \dots, m-1$$

in the $(n-1)m$ unknowns

$$u_{i,j+1}, \qquad i = 1, \dots, n-1; j = 0, \dots, m-1.$$

Here, $\epsilon = \frac{k^2 \Delta t}{(\Delta x)^2}$. Of course, we need not solve the equations simultaneously, as we can *march* in the time variable by using the results in the jth row to compute those in the $(j+1)^{\text{st}}$.

By the way, if we choose $\epsilon = \frac{1}{2}$, we get the very simple **Bender–Schmidt** explicit scheme

$$u_{i,j+1} = \frac{1}{2}(u_{i+1,j} + u_{i-1,j}).$$

And, of course, there are *numerous* other explicit schemes for the heat equation, a few of which we'll meet in the exercises.

Implicit scheme: Crank–Nicolson

We would rather have schemes that are stable for any choice of $\epsilon = \frac{k^2 \Delta t}{(\Delta x)^2}$. To this end, we proceed as in Exercise 12c from the previous section and look at consistent implicit schemes of the form

$$\frac{u_{i,j+1} - u_{i,j}}{\Delta t} = k^2 \left[a \frac{u_{i+1,j} - 2u_{i,j} + u_{i-1,j}}{(\Delta x)^2} \right.$$
$$\left. + (1-a) \frac{u_{i+1,j+1} - 2u_{i,j+1} + u_{i-1,j+1}}{(\Delta x)^2} \right],$$

where $0 < a < 1$. Thus, the right side is essentially a weighted average of the approximations for $u_{xx}(x_i, t_j)$ and $u_{xx}(x_i, t_{j+1})$. It turns out that, for $0 < a \le \frac{1}{2}$, the scheme is (*unconditionally*) stable. The *computation molecule* can be seen in Figure 11.3.

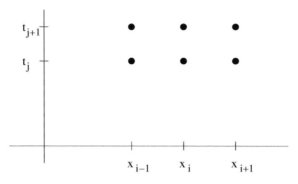

FIGURE 11.3
Computation molecule **for the implicit schemes (including *Crank–Nicolson*) for the heat equation.**

For various reasons, the most popular such scheme is that corresponding to $a = \frac{1}{2}$, the **Crank–Nicolson** scheme

$$(1+\epsilon)u_{i,j+1} = (1-\epsilon)u_{i,j} + \frac{\epsilon}{2}(u_{i+1,j+1} + u_{i-1,j+1} + u_{i+1,j} + u_{i-1,j}),$$

where $\epsilon = \frac{k^2 t \Delta t}{(\Delta x)^2}$.

Let's illustrate with an example.

Example 1 We'll use Crank–Nicolson on the problem

$$u_t = u_{xx}, \qquad 0 < x < 1, t > 0,$$
$$u(x,0) = \sin \pi x,$$
$$u(0,t) = u(1,t) = 0,$$

with $\Delta x = .2$ and $\Delta t = .08$, and we'll compare the solution at time $t = .16$ to the values of $u(x,.16)$ at the x-grid points.

The numerical scheme becomes

$$3u_{i,j+1} = -u_{i,j} + u_{i+1,j+1} + u_{i-1,j+1} + u_{i+1,j} + u_{i-1,j}$$

for $i = 1, 2, 3, 4$ and $j = 0, 1$. The resulting linear system is

$$
\begin{bmatrix}
3 & -1 & 0 & 0 & 0 & 0 & 0 & 0 \\
-1 & 3 & -1 & 0 & 0 & 0 & 0 & 0 \\
0 & -1 & 3 & -1 & 0 & 0 & 0 & 0 \\
0 & 0 & -1 & 3 & 0 & 0 & 0 & 0 \\
1 & -1 & 0 & 0 & 3 & -1 & 0 & 0 \\
-1 & 1 & -1 & 0 & -1 & 3 & -1 & 0 \\
0 & -1 & 1 & -1 & 0 & -1 & 3 & -1 \\
0 & 0 & -1 & 1 & 0 & 0 & -1 & 3
\end{bmatrix}
\begin{bmatrix}
u_{11} \\
u_{21} \\
u_{31} \\
u_{41} \\
u_{12} \\
u_{22} \\
u_{32} \\
u_{42}
\end{bmatrix}
=
\begin{bmatrix}
f_2 - f_1 \\
f_3 - f_2 + f_1 \\
f_4 - f_3 + f_2 \\
-f_4 + f_3 \\
0 \\
0 \\
0 \\
0
\end{bmatrix},
$$

where $f_i = \sin \pi x_i = \sin .2\pi i$, and we solve it as we did in Section 11.1. The results are given in Table 11.3.[†]

x_i	$u_{i,2}$	$u(x_i, .16)$
.2	.118	.121
.4	.190	.196
.6	.190	.196
.8	.118	.121

TABLE 11.3
Crank–Nicolson approximation for the problem in Example 1.

THE WAVE EQUATION

Explicit scheme

For the wave equation

$$u_{tt} = c^2 u_{xx},$$

it seems reasonable that we approximate each second derivative via the central difference, so our scheme will be

$$\frac{u_{i,j+1} - 2u_{i,j} + u_{i,j-1}}{(\Delta t)^2} = c^2 \frac{u_{i+1,j} - 2u_{i,j} + u_{i-1,j}}{(\Delta x)^2} + T_{ij}(\Delta x, \Delta t)$$

or

$$u_{i,j+1} = \epsilon u_{i+1,j} + 2(1 - \epsilon)u_{i,j} + \epsilon u_{i-1,j} - u_{i,j-1},$$

where, here, $\epsilon = \left(\frac{c\Delta t}{\Delta x}\right)^2$. One might suspect that we'll need to have $1 - \epsilon \geq 0$ in order to have stability and, indeed, this turns out to be the case.[‡] However, the restriction on the mesh size is *much* less severe here as, taking $c^2 = 1$, we need only take $\Delta t \leq \Delta x$.

By the way, notice what the condition $\epsilon \leq 1$ says graphically. First, if $\epsilon = 1$, then $c\Delta t = \Delta x$ and the characteristics through the point (x_i, t_{j+1})

[†]Of course, if we were to perform this approximation in earnest, we would use a much smaller mesh size and solve it for a much greater value of t. The problem here is that the matrix is not quite as *nice* as the matrices in the previous section. While it *is* banded and (with small enough mesh size) sparse (why?), life would be much easier if it were, say, tridiagonal. In fact, there are tricks that one can use to turn more complicated linear systems into such simpler systems.

Also, here, again, we may *march* in t, although it requires solving simultaneously for all quantities in row $j + 1$, in terms of those in row j.

[‡]This is the well-known **Courant–Friedrichs–Lewy condition.**

pass directly through the points (x_{i-1}, t_j) and (x_{i+1}, t_j). If $\epsilon < 1$, then the characteristics pass between these points. We say that the *analytical* domain of dependence is a subset of the *numerical* domain of dependence. See Figure 11.4. This says that disturbances must propagate "through the numerical scheme" at a speed $\geq c$ in order for the scheme to be stable.

The initial condition

$$u(x, 0) = f(x)$$

gives us, as in the case of the heat equation,

$$u_{i,0} = f(x_i), \qquad i = 0, 1, \dots, n.$$

However, the second initial condition, of course, involves the time derivative,

$$u_t(x, 0) = g(x).$$

Since the explicit scheme is $O[(\Delta t)^2 + (\Delta x)^2]$ (see Exercise 8d), we again choose the central difference approximation for u_t. Thus, we introduce the *ghost points*

$$u(x_i, -\Delta t) = u_{i,-1}, \qquad i = 0, 1, \dots, n.$$

There are, of course, implicit schemes for the wave equation. However, since the stability condition for the explicit scheme is not strict and allows us to be flexible in our choice of grid, it is not so crucial to find alternatives to this scheme.

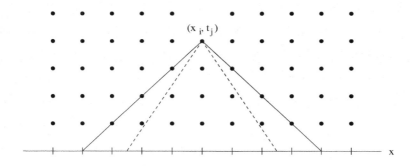

(x_i, t_j)

FIGURE 11.4
The solid lines are called the *numerical characteristics* and bound the *numerical domain of dependence*. Here, we have $\epsilon < 1$, i.e., $c\Delta t < \Delta x$, so the analytic characteristics, represented by the dashed lines, are steeper than the numerical characteristics. Thus, for $\epsilon < 1$, the actual domain of dependence is a proper subset of the numerical domain of dependence.

LAPLACE'S EQUATION

Given Laplace's equation

$$u_{xx} + u_{yy} = 0$$

on the rectangle $0 \leq x \leq a, 0 \leq y \leq b$, we set up the grid as we usually do (of course, both variables are bounded here). So we have

$$x_i = i\Delta x, \qquad i = 0, 1, \ldots, n, \left(\text{so } \Delta x = \frac{a}{n}\right),$$

$$y_j = j\Delta y, \qquad j = 0, 1, \ldots, m, \left(\text{so } \Delta y = \frac{b}{m}\right),$$

and we see no compelling reason *not* to use a central difference in each variable. Then, our scheme is

$$\frac{u_{i+1,j} - 2u_{i,j} + u_{i-1,j}}{(\Delta x)^2} + \frac{u_{i,j+1} - 2u_{i,j} + u_{i,j-1}}{(\Delta y)^2} = 0$$

or

$$u_{i,j} = \frac{1}{2[(\Delta x)^2 + (\Delta y)^2]}[(\Delta x)^2(u_{i,j+1} + u_{i,j-1}) + (\Delta y)^2(u_{i+1,j} + u_{i-1,j})].$$

Notice that the right side is a *weighted average* of the four quantities there (why?) and, if we choose $\Delta x = \Delta y$, we have

$$u_{i,j} = \frac{u_{i,j+1} + u_{i,j-1} + u_{i+1,j} + u_{i-1,j}}{4},$$

the *average* of the four neighboring points (see the computation molecule in Figure 11.5). This should come as no surprise, given the *mean value property* for harmonic functions (Section 9.3). It's not hard to show (try it!) that the maximum (minimum) value of $u_{i,j}$ must occur on the rectangle's boundary.

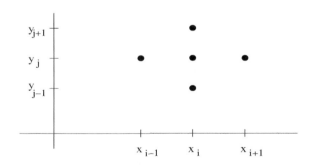

FIGURE 11.5
Computation molecule for the given scheme for Laplace's equation.

Notice that we do not distinguish between explicit and implicit here, because of the bounded domain (so neither variable plays a role similar to t in the heat and wave equations).

As for the boundary, Dirichlet conditions again pose no problem, while for Neumann or Robin conditions, we again need to introduce *ghost points* in order to use the central difference approximation for the first derivative.

Now, for the sake of simplicity, let's look at the Dirichlet problem, with $\Delta x = \Delta y$. In this case, the quantities $u_{0,j}$, $u_{n,j}$, $u_{i,0}$ and $u_{i,n}$ are known for all values of i and j, and we thus have the $(n-1)^2$ equations

$$u_{i,j} = \frac{u_{i,j+1} + u_{i,j-1} + u_{i+1,j} + u_{i-1,j}}{4}, \quad i = 1, 2, \ldots, n-1; j = 1, 2, \ldots, n-1,$$

in the $(n-1)^2$ unknowns

$$u_{i,j}, \quad i = 1, \ldots, n-1; j = 1, \ldots, n-1.$$

As there is no way to *march*, we must solve them as a simultaneous system. So here we have no option but to solve the linear system

$$AY = B.$$

Of course, as we've said all along, with a small enough grid size the matrix A will be quite large.

In order to expedite the approximation, certain **iterative methods** have been developed. These actually involve approximating the solution of the approximate scheme. However, it can be shown that the iterative solution does converge to the solution of the difference scheme in each case.

Jacobi iteration

This is the simplest of the iterative methods. Basically, we begin with an *initial guess* for the solution at each point,

$$u_{i,j}^{(0)}, \quad i = 1, \ldots, n-1; j = 1, \ldots, n-1.$$

Next, we *update* this guess by letting

$$u_{i,j}^{(1)} = \frac{u_{i,j+1}^{(0)} + u_{i,j-1}^{(0)} + u_{i+1,j}^{(0)} + u_{i-1,j}^{(0)}}{4}.$$

We continue this process as long as we'd like, stopping when the

$$\max_{i,j}[u_{i,j}^{(N+1)} - u_{i,j}^{(N)}]$$

is small enough.

Gauss–Seidel iteration

When we implement Jacobi iteration, we have to start somewhere and move to other points in an orderly fashion. Standard practice is to start at $u_{1,1}$ at the bottom left, then proceed along the first row to $u_{n,1}$, then start the second row at $u_{1,2}$ on the left, move along that row, etc. So what happens when we get to $u_{3,2}$, say? We have

$$u_{3,2}^{(1)} = \frac{u_{3,3}^{(0)} + u_{3,1}^{(0)} + u_{4,2}^{(0)} + u_{2,2}^{(0)}}{4}.$$

However, at this point, we've already updated $u_{3,1}^{(0)}$ to $u_{3,1}^{(1)}$ and $u_{2,2}^{(0)}$ to $u_{2,2}^{(1)}$. It *may* make computational sense to use these updated values in our calculation for $u_{3,2}^{(1)}$. Indeed, it turns out that proceeding in this manner *does* speed up the convergence—we call this method **Gauss–Seidel iteration**. In general, it says to take

$$u_{i,j}^{(N+1)} = \frac{u_{i,j+1}^{(N)} + u_{i,j-1}^{(N+1)} + u_{i+1,j}^{(N)} + u_{i-1,j}^{(N+1)}}{4}.$$

Successive overrelation iteration

S-O-R iteration is a generalization of Gauss–Seidel in that, instead of looking only at

$$u_{i,j}^{(N+1)} = u_{i,j}^{(N)} + \frac{1}{4}[u_{i,j+1}^{(N)} + u_{i,j-1}^{(N+1)} + u_{i+1,j}^{(N)} + u_{i-1,j}^{(N+1)} - u_{i,j}^{(N)}]$$

(i.e., Gauss–Seidel), it considers looking at

$$u_{i,j}^{(N+1)} = u_{i,j}^{(N)} + \omega[u_{i,j+1}^{(N)} + u_{i,j-1}^{(N+1)} + u_{i+1,j}^{(N)} + u_{i-1,j}^{(N+1)} - u_{i,j}^{(N)}]$$

for various values of the **relaxation parameter** ω, in order to find ω which gives the fastest rate of convergence.

FIRST-ORDER EQUATIONS

Let's look at the simple first-order PDE

$$u_t + cu_x = 0.$$

We have the option of approximating each of u_t and u_x by a forward, central or backward difference scheme. If the central difference is used in both directions, we have the well-known **leap frog method** (Why the name? See Figure 11.6.). We investigate these schemes in the exercises.

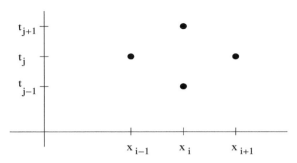

FIGURE 11.6
Computation molecule for the leap frog scheme.

Lax–Wendroff method

Of course, the ultimate goal, beyond the scope of this book, is to solve numerically equations that are difficult or impossible to solve analytically. So, for example, it would be nice to derive a method for the above equation that can be generalized to the setting of the ubiquitous *conservation law equation*

$$u_t + f(u)u_x = 0.$$

To this end, the **Lax–Wendroff method** was developed. This method entails starting off with the t-Taylor series

$$u(x, t + \Delta t) = u(x, t) = \Delta t u_t(x, t) + \frac{(\Delta t)^2}{2!} u_{tt}(x, t) + \frac{(\Delta t)^3}{3!} u_{ttt}(x, t) + \dots$$

and then using the PDE to replace u_t, u_{tt}, etc., by u_x, u_{xx}, etc. How is this done? We know that

$$u_t = -cu_x.$$

Then,

$$u_{tt} = -cu_{xt} = -c(u_t)_x = -c(-cu_x)_x = c^2 u_{xx}$$

and, in general,

$$\frac{\partial^n u}{\partial t^n} = (-1)^n c^n \frac{\partial^n u}{\partial x^n}.$$

Thus,

$$u(x, t + \Delta t) = u(x, t) - c\Delta t u_x(x, t) + \frac{(c\Delta t)^2}{2!} u_{xx}(x, t) + O[(c\Delta t)^3]$$

and, replacing u_x and u_{xx} with central differences, we have the Lax–Wendroff method:

$$u_{i,j+1} = u_{i,j} - \frac{c\Delta t}{2\Delta x}(u_{i+1,j} - u_{i-1,j})$$

$$+ \frac{1}{2}\left(\frac{c\Delta t}{\Delta x}\right)^2 (u_{i+1,j} - 2u_{i,j} + u_{i-1,j})$$

or

$$u_{i,j+1} = (1 - \epsilon^2)u_{i,j} + \frac{\epsilon}{2}(1 + \epsilon)u_{i-1,j} + \frac{\epsilon}{2}(\epsilon - 1)u_{i+1,j},$$

where $\epsilon = \frac{c\Delta t}{\Delta x}$. It can be shown that the local truncation error is $O[(\Delta t)^2 + (\Delta x)^2]$ and that the scheme is stable for $0 < \epsilon \leq 1$.

Exercises 11.2

1. Consider, again, the problem from Example 1,

$$u_t = u_{xx}, \qquad 0 < x < 1, t > 0,$$
$$u(x, 0) = \sin \pi x,$$
$$u(0, t) = u(1, t) = 0.$$

a) Apply the explicit scheme

$$u_{i,j+1} = \epsilon u_{i+1,j} + (1 - 2\epsilon)u_{i,j} + \epsilon u_{i-1,j}, \epsilon = \frac{\Delta t}{(\Delta x)^2}$$

to the problem, using $\Delta x = .2$ and $\Delta t = .02$; compute by hand the approximate solution at time $t = .04$, by *marching* from each time step to the next.

b) **MATLAB:** Extend part (a) and write a program which approximates the solution for any $t_n = .02n$. Compute the solution for $t = .16$ and compare the results to those in Table 11.3.

c) Use the same scheme, but with $\Delta t = .08$, and compute the approximate solution at $t = .16$. Compare to the results above and in Table 11.3. What's going on?

2. Here we apply the same explicit scheme to the heat problem with Neumann boundary condition

$$u_t = u_{xx}, \qquad 0 < x < 5, t > 0,$$
$$u(x, 0) = \cos \frac{2\pi x}{5},$$
$$u_x(0, t) = u_x(5, t) = 0.$$

a) Use $\Delta x = 1$ and $\Delta t = \frac{1}{3}$, and *march* to the solution at $t = \frac{2}{3}$. Approximate the u_x terms using a scheme which has truncation error consistent with that of the PDE scheme. Compare with the exact solution.

b) **MATLAB:** Extend part (a) to approximate the solution at time $t = 10$. Again, compare with the exact solution.

3. Given the wave equation problem

$$u_{tt} = u_{xx}, \qquad 0 < x < 1, t > 0,$$
$$u(x, 0) = \sin \pi x,$$
$$u_t(x, 0) = 0,$$
$$u(0, t) = u(1, t) = 0,$$

use the explicit scheme

$$u_{i,j+1} = \epsilon u_{i+1,j} + 2(1 - \epsilon)u_{i,j} + \epsilon u_{i-1,j} - u_{i,j-1}, \epsilon = \left(\frac{\Delta t}{\Delta x}\right)^2$$

to approximate its solution.

a) Use $\Delta x = .2$ and $\Delta t = .1$, and compute by hand the solution at time $t = .2$. Compare with the exact solution.

b) **MATLAB:** Extend part (a) and write a program to approximate the solution at any time $t_n = n(.1)$. Compare the approximate solution with the exact solution for time $t = 1$.

c) Repeat part (a), but for the initial conditions

$$u(x, 0) = 0,$$
$$u_t(x, 0) = \sin \pi x.$$

Make sure that the scheme you use for u_t is consistent with that used for the PDE, as far as truncation error is concerned.

d) **MATLAB:** Repeat part (b), but for the problem in part (c).

4. In matrix form, write down the equations for the approximation of the potential problem

$$\nabla^2 u = u_{xx} + u_{yy} = 0, \qquad 0 < x < 1, 0 < y < 1,$$
$$u(0, y) = u(1, y) = u(x, 1) = 0,$$
$$u_y(x, 0) = f(x),$$

using the scheme

$$4u_{i,j} = u_{i,j+1} + u_{i,j-1} + u_{i+1,j} + u_{i-1,j}$$

with $\Delta x = \Delta y = .25$. (Hint: Your coefficient matrix should be 15×15.)

5. Given the first-order problem

$$u_t + u_x = 0, \qquad x > 0, t > 0,$$
$$u(x, 0) = f(x),$$
$$u(0, t) = g(t),$$

use the Lax–Wendroff scheme

$$2u_{i,j+1} = 2(1 - \epsilon^2)u_{i,j} + \epsilon(1 + \epsilon)u_{i-1,j} + \epsilon(\epsilon - 1)u_{i+1,j},$$

$$\epsilon = \frac{\Delta t}{\Delta x},$$

to approximate its solution on $0 \le x \le 1$. (Hint: You will need to use an initial x-interval which is bigger than $0 \le x \le 1$. If we'd like to compute the solution after n time steps, we need to consider the initial x-interval $0 \le x \le 1 + n\Delta x$.)

   a) Use $f(x) = \sin \pi x, g(t) = -\sin \pi t$, $\Delta x = .2$ and $\Delta t = .1$, and compute the approximate solution at time $t = .2$. Compare your results with the exact solution.

   b) **MATLAB:** Extend part (a) and approximate the solution at time $t = 1$. Again, compare with the exact solution.

   c) **MATLAB:** Instead, use $f(x) = 1 - x$ and $g(t) = 1 + t$ and repeat part (b). What's going on?

   d) **MATLAB:** Repeat part (c), but using $\Delta x = .1$ and $\Delta t = .2$. What's going on *here*?

6. **Burger's equation:** Apply the Lax–Wendroff scheme to the Burger's equation problem

$$u_t + uu_x = 0, \qquad x > 0, t > 0,$$
$$u(x, 0) = 3x,$$
$$u(0, t) = 0,$$

on $0 \le x \le 1$. (Note that the exact solution is $u = \frac{3x}{1+3t}$.) Use $\Delta x = .2$ and $\Delta t = .1$, and compute the solution at $t = .2$. Compare with the exact solution

7.   a) Write out the approximating equation for the *leap frog method*.

   b) Show that its local truncation error is $T_{i,j}(\Delta x, \Delta t) = O[(\Delta t)^2 + (\Delta x)^2]$.

   c) How would you go about implementing the leap frog method? Specifically, supposing $\Delta x = .2$ on $0 \le x \le 1$ and $\Delta t = .1$, how would you find the values of the approximating solution at a later time, say, $t = .3$?

8. Verify the local truncation error for the given scheme.

   a) For $u_t + cu_x = 0$, the scheme using explicit forward differences in both x and t; $T_{i,j}(\Delta x, \Delta t) = O[\Delta t + (\Delta x)]$

b) The explicit scheme for the heat equation (Exercise 1); $T_{i,j}(\Delta x, \Delta t)$
   $= O[\Delta t + (\Delta x)^2]$

c) *Crank–Nicolson*; $T_{i,j}(\Delta x, \Delta t) = O[(\Delta t)^2 + (\Delta x)^2]$

d) The explicit scheme for the wave equation (Exercise 3); $T_{i,j}(\Delta x, \Delta t)$
   $= O[(\Delta t)^2 + (\Delta x)^2]$

e) The scheme for Laplace's equation, with $\Delta x = \Delta y$ (Exercise 4);
   $T_{i,j}(\Delta x, \Delta y) = O[(\Delta x)^2]$

9. a) Derive the double Taylor series for a function $f(x, y)$ formally by,
    first, expanding $f(x + \Delta x, y + \Delta y)$ in a Taylor series in the x-
    coordinate and then expanding each term in a Taylor series in the
    y-coordinate. Include terms out to the fourth partial derivative.

    b) Derive the formula

$$u_{xy}(x, y) = \frac{1}{4(\Delta x)(\Delta y)} [u(x + \Delta x, y + \Delta y) - u(x + \Delta x, y - \Delta y)$$
$$- u(x - \Delta x, y + \Delta y) + u(x - \Delta x, y - \Delta y)]$$
$$+ O[(\Delta x)^2 + (\Delta Y)^2].$$

11.3   Spectral Methods and the Finite Element Method

Although finite difference approximations are, in some sense, the "easiest,"
most straightforward numerical methods for ODEs and PDEs, they are by
no means the only ones. In this section, we introduce two classes of methods
which, in the right circumstances, are quite powerful.

SPECTRAL METHODS

To make a long story short, spectral methods essentially involve plugging
in a truncated Fourier series for the unknown function and determining the
coefficients so that we have a solution. This is really the same as the method
we used in Section 4.4 to solve nonhomogeneous PDEs, except that we now
cannot solve the problem exactly and must use a computer to calculate the
approximate coefficients.

   But let's return to the nonhomogeneous heat equation to remember what
we did.

Example 1 Given the heat problem

$$u_t = u_{xx} + F(x), \qquad 0 < x < \pi, t > 0,$$
$$u(x,0) = f(x),$$
$$u(0,t) = u(\pi,t) = 0,$$

we know that the functions $\{\sin nx\}_{n=1}^{\infty}$ *span* the interval $0 \leq x \leq \pi$. Thus, we look for a solution of the form

$$U = \sum_{n=1}^{\infty} b_n(t) \sin nx.$$

Then, we "plug" U into the PDE and compare coefficients, the latter being possible because the functions $\{\sin nx\}_{n=1}^{\infty}$ are (simply) *orthogonal* on $0 \leq x \leq \pi$. Finally, we solve the resulting ODEs and use the initial condition to determine the arbitrary constants.

This is an example of a **Fourier sine spectral method**. It works because

1) The functions $\{\sin nx\}_{n=1}^{\infty}$ form a *basis* for the set of functions on $0 \leq x \leq \pi$.

2) Each function $\sin nx$ already satisfies the boundary conditions.

One can imagine a situation where we can't determine analytically the coefficients $b_n(t)$ and will need to approximate them numerically. Or, it *may* happen that the only way to get the coefficients b_n is by solving a linear system of equations. In either case, the best we can do is to compute

$$u_N(x,t) = \sum_{n=1}^{N} b_n(t) \sin nx \approx u(x,t).$$

Note the two sources of error: the truncation of the series *and* the approximation involved in computing the coefficients.

In general, any spectral method requires a set of basis functions. However, it may be difficult to find basis functions which also satisfy the boundary conditions. As a result, we give a special name to spectral methods that employ basis functions which satisfy the boundary conditions—they are called **Galerkin* methods,** and the method used in Example 1 is called a **Fourier–Galerkin** (or **Fourier-sine-Galerkin**) spectral method.

---

*Galerkin methods refer to numerous numerical methods that involve the use of basis functions which also satisfy other useful properties. Most commonly, the "useful property" is that they satisfy the boundary conditions. If we can*not* find such basis functions, then we may use the **tau** or **collocation** spectral methods, and we meet the former method below.

In general, the Galerkin spectral methods produce, in theory, exact solutions, with numerical approximation being necessary for the same reason that we estimate Fourier series to N terms. However, it may not always be feasible to require that the basis functions also satisfy the boundary conditions, as in the following example.

Example 2 Let's consider the ODE boundary-value problem

$$y'' = -f, \qquad 0 < x < \pi,$$
$$y(0) - y'(0) = y(\pi) + 2y'(\pi) = 0.$$

A straightforward way for finding basis functions which satisfy the boundary condition is, as always, to solve the eigenvalue problem

$$y'' + \lambda y = 0, \qquad 0 < x < \pi,$$
$$y(0) - y'(0) = y(\pi) + 2y'(\pi) = 0.$$

However, we cannot solve for the eigenvalues explicitly (try it), and so the Galerkin method is not feasible.

We still know, though, that the functions $\{\sin nx\}_{n=1}^{\infty}$ form a basis for functions on $0 \le x \le \pi$, so we let

$$y \approx y_N = \sum_{n=1}^{N} c_n \sin nx$$

in the ODE. We have

$$-\sum_{n=1}^{N} n^2 c_n \sin nx \approx -\sum_{n=1}^{\infty} f_n \sin nx, \; f_n = \frac{2}{\pi} \int_0^{\pi} f(x) \sin nx \; dx,$$

and we know from Chapter 3 that the best we can do is to take

$$n^2 c_n = f_n, \qquad n = 1, 2, \dots, N.$$

However, y_N does not satisfy the boundary conditions. The way around this is, instead, to require that

$$c_n = \frac{f_n}{n^2}, \qquad n = 1, 2, \dots, N - 2,$$

and determine c_{N-1} and c_N by *requiring* that y_N satisfy the boundary conditions. Thus, we replace the $n = N - 1, N$ equations with the two equations

$$y(0) - y'(0) = 0 = -\sum_{n=1}^{N} nc_n$$

and

$$y(\pi) + 2y'(\pi) = 0 = 2\sum_{n=1}^{N}(-1)^n n c_n.$$

This method was invented by Cornelius Lanczos in 1938 and is called the **(Lanczos) tau method**.[†] Thus, the method of the example is referred to as the **Fourier-tau method** (or **Fourier-sine-tau method**).

As we've seen, the trigonometric functions are certainly not the only sets of basis functions. For example, we also have the orthogonal polynomials from Chapter 7.

Example 3 Here we approximate the solution of the convection problem

$$u_t + u_x = t, \qquad -1 < x < 1, t > 0,$$
$$u(x,0) = f(x),$$
$$u(-1,t) = 0.$$

As we're on the interval $-1 \le x \le 1$, we have both kinds of Chebyshev polynomials, as well as the Legendre polynomials, at our disposal. Here we use the Legendres—remember that they are the polynomials

$$P_0(x) = 1, P_1(x) = x, P_2(x) = \frac{3}{2}x^2 - \frac{1}{2}, P_3(x) = \frac{5}{2}x^3 - \frac{3}{2}x, \dots$$

and, in general,

$$nP_n(x) = (2n-1)xP_{n-1}(x) - (n-1)P_{n-2}(x).$$

We also know that each P_n satisfies

$$P_n(\pm 1) = (\pm 1)^n.$$

Finally, we'll need the facts, which we prove in Exercise 5, that if

$$f(x) = \sum_{n=0}^{N} a_n P_n(x),$$

then

$$f'(x) = \sum_{n=0}^{N-1}(2n+1)\sum_{\substack{p=n+1 \\ p+n \text{ odd}}}^{N} a_p P_n(x).$$

[†] "Tau" only because Lanczos used the letter τ to represent the error.

Now, we may use either the *Galerkin* or the *tau* method at this point. As the functions $P_n(x)$ don't satisfy the boundary conditions, we choose the latter (although the Galerkin method is easy here, as we'll see in Exercise 2), so we set

$$u_N(x,t) = \sum_{n=0}^{N} b_n(t) P_n(x)$$

and, substituting $u = u_N$ into the PDE, we have

$$\sum_{n=0}^{N} b'_n(t) P_n(x) + \sum_{n=0}^{N-1} (2n+1) \sum_{\substack{p=n+1 \\ p+n \text{ odd}}}^{N} b_p(t) P_n(x) = t P_0(x).$$

Thus, we have the N equations, in $N+1$ unknowns,

$$b'_0 + \sum_{\substack{p=1 \\ p \text{ odd}}}^{N} b_p = t,$$

$$b'_n + (2n+1) \sum_{\substack{p=n+1 \\ p+n \text{ odd}}}^{N} b_p = 0, \qquad n = 1, \ldots, N-1,$$

along with the boundary condition

$$u_N(-1,t) = 0 = \sum_{n=0}^{N} b_n(t)(-1)^n.$$

Spectral methods also turn out to be extremely useful in solving eigenvalue problems. See Exercise 3.

FINITE ELEMENT METHOD

A *very* powerful and popular method for the numerical solution of PDEs is the **finite element method**, about which much has been written. We don't pretend to cover the method here with any sophistication; rather, we scratch the surface, with the intent of giving a very basic idea of how it works.

The finite element approach is somewhat more difficult to implement than finite difference or spectral methods. However, this disadvantage is more than compensated for by the fact that this method is much more broadly applicable. It *does* share similarities with both—with difference methods, in that the first step is to break the domain into subdivisions, and with spectral methods, as the solution again is approximated by a finite sum of function. However, the similarities end there because

1. The subdivisions need not be rectangular (allowing the method to be applied to domains of more-or-less arbitrary shape).

2. The approximating sum is *not* smooth, but is a continuous, piecewise polynomial function. (Note that, in the spectral methods that we used, the basis functions were infinitely differentiable).

The second point means that the approximating sum can*not* be "plugged into" the differential equation. In fact, if the sum is piecewise *linear* (which often is the case), then the second derivative will be either zero or nonexistent at each point. Thus, a reformulation of the problem is necessary; it is recast in its so-called **weak formulation**. But let's introduce these ideas in an example.

Example 4 We'll illustrate the *finite element method* as applied to the simple BVP

$$y'' = -f(x), \qquad 0 < x < L,$$
$$y(0) = y(L) = 0.$$

We begin by breaking the domain into n equal subdivisions $x_{i-1} \le x \le x_i$ of length $h = \frac{L}{n}$, and we'd like to approximate the solution by a function $v(x)$ that is continuous on $0 \le x \le L$ and linear on each subdivision. Thus, v will be piecewise linear.

The standard way to do this is to create a set of functions v_i, $i = 0, 1, \ldots, n$, which span the space of functions which are continuous and linear on the subintervals. Although not obvious, this can be accomplished by taking the v_i to be the **tent functions** or **hat functions**

$$v_0(x) = \begin{cases} 1 - \dfrac{1}{h}(x - x_1), & \text{if } x_0 = 0 \le x \le x_1, \\ 0, & \text{otherwise,} \end{cases}$$

$$v_n(x) = \begin{cases} \dfrac{1}{h}(x - x_{n-1}), & \text{if } x_{n-1} \le x \le x_n = L, \\ 0, & \text{otherwise,} \end{cases}$$

$$v_i(x) = \begin{cases} \dfrac{1}{h}(x - x_{i-1}), & \text{if } x_{i-1} \le x \le x_i, \\ 1 - \dfrac{1}{h}(x - x_i), & \text{if } x_i \le x \le x_{i+1}, \\ 0, & \text{otherwise,} \end{cases}$$

as in Figure 11.7. (See Exercise 8.) However, in this problem, the boundary conditions $y(0) = y(L) = 0$ allow us to neglect the functions v_1 and v_n (thus, $v_1, \ldots, v_{n-1}$ spans the subspace consisting of the above-mentioned functions that *also* satisfy these boundary conditions). Thus, we form the **finite element Galerkin approximation**

$$v(x) = \sum_{i=1}^{n-1} c_i v_i(x).$$

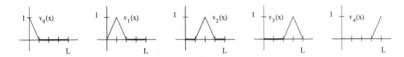

FIGURE 11.7
The *tent functions* v_0 through v_n for $n = 4$.

Now, how do we reformulate the ODE? Basically, we use that *smoothing operator*, the integral. Instead of searching for functions which satisfy

$$y'' = f(x),$$

we begin by multiplying both sides by any piecewise smooth function ϕ that satisfies the boundary conditions, that is, for which

$$\phi(0) = \phi(L) = 0.$$

Then we have

$$\int_0^L y''(x)\phi(x)dx = -\int_0^L f(x)\phi(x)dx$$

and, upon integrating by parts,

$$\langle y', \phi' \rangle = \int_0^L y'(x)\phi'(x)dx = \int_0^L f(x)\phi(x)dx = \langle f, \phi \rangle.$$

Thus, a weak solution of the problem is any function that satisfies this last equation, for all such ϕ, and which satisfies the boundary conditions.

What of an approximate solution, $v(x)$? We require only that it satisfy

$$\langle v', v_i' \rangle = \langle f, v_i \rangle, \qquad i = 1, 2, \ldots, n-1.$$

Thus, our finite element approximation to the solution is the function $v(x)$, with the $n-1$ constants $c_1, \ldots, c_{n-1}$ satisfying the $n-1$ linear equations

$$\sum_{i=1}^{n-1} c_i \langle v_i', v_j' \rangle = \langle f, v_j \rangle, \qquad j = 1, 2, \ldots, n-1.$$

We rewrite this system as

$$A\boldsymbol{c} = \boldsymbol{b},$$

where

$$
\boldsymbol{c} = \begin{bmatrix} c_1 \\ c_2 \\ \vdots \\ c_{n-1} \end{bmatrix}, \quad \boldsymbol{b} = \begin{bmatrix} \langle f, v_1 \rangle \\ \langle f, v_2 \rangle \\ \vdots \\ \langle f, v_{n-1} \rangle \end{bmatrix}
$$

and the so-called **stiffness**[‡] **matrix** A satisfies

$$
A_{ij} = \langle v_i', v_j' \rangle, \qquad i = 1, \ldots, n-1; j = 1, \ldots, n-1.
$$

One may try, instead, piecewise quadratic or cubic polynomials, for example, and one need not use equal subdivisions (for example, we may need a finer mesh at one end than at the other). But a major concern is that we choose our polynomials and mesh so that the stiffness matrix is **sparse**.

Let's close with a look at one more example, this one in two dimensions.

Example 5 Consider the two-dimensional Dirichlet problem

$$
\nabla^2 u = -f, \text{ on } D,
$$
$$
u = 0 \text{ on } \partial D,
$$

where $D \cup \partial D$ is the convex polygonal region in Figure 11.8a. As in Example 3, we begin by choosing the functions which will appear in our approximating sum. Again, we choose piecewise linear functions (knowing that we also could, instead, use quadratic, cubic, etc. functions).

The standard idea is as follows. We'll *triangulate* $D \cup \partial D$, as in Figure 11.8b. Note that any vertex of a triangle is also only a vertex of other triangles, as well; no vertex lies along the edge of another triangle.

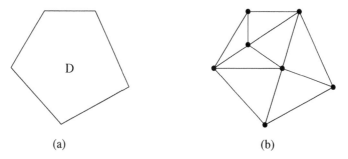

(a)   (b)

FIGURE 11.8
Polygonal domain (a) and one possible triangulation of it (b).

[‡]The name comes from the fact that the finite element method was first used to solve problems in the theory of elasticity.

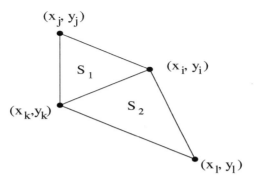

FIGURE 11.9
The determination of $v_i(x, y)$ on S_1 and S_2.

Now we create our (Galerkin) linear approximating functions as follows. Given any interior vertex (x_i, y_i), we define $v_i(x, y)$ so that it is nonzero on all triangles which share (x_i, y_i) as a vertex and zero elsewhere. So, for example, on triangle S_1 in Figure 11.9, we define $v_i(x, y) = a_1 x + b_1 y + c_1$ and require that

$$a_1 x_i + b_1 y_i + c_1 = 1,$$
$$a_1 x_j + b_1 y_j + c_1 = 0,$$
$$a_1 x_k + b_1 y_k + c_1 = 0.$$

Similarly, on triangle S_2, we have $v_i(x, y) = a_2 x + b_2 y + c_2$, where a_2, b_2 and c_2 are determined via $v_i(x_j, y_j) = v_i(x_\ell, y_\ell) = 0$ and $v_i(x_i, y_i) = 1$. Thus, each v_i is a *tent function*, with a height of one at vertex (x_i, y_i) and with sides that are planes which "stop" at a height of zero along each edge opposite the shared vertex (x_i, y_i).

In this particular example in Figure 11.8b, we have only $n = 2$ approximating functions, v_1 and v_2. Then the approximation to our solution is given by

$$v(x, y) = \sum_{i=1}^{n} c_i v_i(x, y).$$

Next, we must give the *weak formulation* of the PDE as, again, the functions v_i are not smooth. As before, we multiply both sides by any $\phi(x, y)$ which satisfies the boundary condition (again, it's a *Galerkin* approximation that we're doing) and integrate over D:

$$\iint_D \phi \nabla^2 u \, dA = - \iint_D \phi f \, dA.$$

Then, Green's first identity for the Laplacian (Section 10.3) is our 2-D integration by parts formula. Thus, we have

$$\iint_D \nabla \phi \cdot \nabla u \, dA = \iint_D \phi f \, dA \quad \text{(why?)}.$$

Finally, as before, this equation suggests that we *require* v and v_i to satisfy

$$\iint\limits_{D} \nabla v \cdot \nabla v_i \, dA = \iint\limits_{D} f v_i \, dA, \qquad i = 1, \ldots, n,$$

or, if you prefer,

$$\langle \nabla v, \nabla v_i \rangle = \langle f, v_i \rangle.$$

As with finite difference approximations, we may use finite elements when the domain has a curvilinear boundary. This, along with the flexibility we have in triangulating the domain and in choosing the degree of the approximating functions, is what makes the finite element method so powerful and so popular.

Exercises 11.3

1. a) Use the functions $\{\sin nx\}_{n=1}^{N}$ and perform the Fourier-sine-Galerkin spectral approximation for the problem

$$2y'' + y = -f(x), \qquad 0 < x < \pi,$$
$$y(0) = y(\pi) = 0.$$

(Write down the equations satisfied by the constants c_n.)

   b) Instead, use the functions $\{\cos nx\}_{n=0}^{N}$ to write down the *tau* equations for the same problem.

   c) Why can't we use the functions $\{\sin nx\}_{n=1}^{N}$ to perform the Fourier-sine-Galerkin method on the system

$$y'' + y = -f(x), \qquad 0 < x < \pi,$$
$$y(0) = y(\pi) = 0?$$

2. Here we repeat Example 3, but using the Galerkin method.

   a) Show that the functions

$$\phi_n(x) = P_n(x) - (-1)^n$$
$$= P_n(x) - (-1)^n P_0(x), \qquad n = 1, 2, \ldots, N,$$

   span the space of "reasonably behaved" functions f on $-1 \le x \le 1$ which also satisfy $f(-1) = f(1) = 0$.

   b) Use these functions to perform the Legendre–Galerkin approximation for the problem given in Example 4. Write down the ODEs satisfied by the functions $b_n(t)$.

3. Use the Legendre-tau or Legendre–Galerkin spectral method to approximate the eigenvalues of the problem

$$y'' + \lambda y = 0, \qquad -1 < x < 1,$$
$$y(-1) = y(1) = 0.$$

Write down the basic equations for the constants c_n and then explain how to compute values for λ.

4. a) Reformulate the BVP

$$y'' + 2y' - 3y = -f(x), \qquad 0 < x < 4,$$
$$y(0) = y(4) = 0,$$

so that one may perform a Legendre spectral method to approximate its solution.

   b) Reformulate the BVP

$$y'' + xy = -f(x), \qquad 1 < x < 2,$$
$$y(1) = y(2) = 0,$$

so that one may perform a Fourier-sine spectral method to approximate its solution.

5. One may use Rodrigues's formula (Section 7.6) to prove that the Legendre polynomials satisfy

$$P_n'(x) - P_{n-2}'(x) = (2n-1)P_{n-1}(x).$$

   a) Use this, along with mathematical induction, to show that

$$P_n'(x) = (2n-1)P_{n-1}(x) + (2n-5)P_{n-3}(x) + \dots$$

$$+ \begin{cases} \dots + 7P_3(x) + 3P_1(x), & \text{if } n \text{ is even,} \\ \dots + 5P_2(x) + P_0(x), & \text{if } n \text{ is odd.} \end{cases}$$

   b) Thus, show that

$$P_n'(x) = \sum_{\substack{k=0 \\ k+n \text{ odd}}}^{n-1} (2k+1)P_k(x).$$

   c) More generally, derive the formula given in the text, that

$$\frac{d}{dx} \sum_{n=0}^{\infty} a_n P_n(x) = \sum_{n=0}^{\infty} (2n+1) \sum_{\substack{p=n+1 \\ p+n \text{ odd}}}^{\infty} a_p P_n(x).$$

(Hint: You'll need to switch the order of summation.)

6. a) Work through the Galerkin finite element method to approximate
    the solution of

    $$y'' = x, \qquad 0 \le x \le 2,$$
    $$y(0) = y(2) = 0,$$

    using only the one tent function $v_1(x) = \begin{cases} x, & \text{if } 0 \le x \le 1, \\ 2-x, & \text{if } 1 \le x \le 2. \end{cases}$

    Compare your result with the exact solution.

    b) Do the same, but use three tent functions (so, one each on $0 \le x \le 1$, $\frac{1}{2} \le x \le \frac{3}{2}$ and $1 \le x \le 2$). Again, compare with the exact solution.

7. a) Work through the Galerkin finite element approximation of the
    solution of the Poisson Dirichlet problem

    $$u_{xx} + u_{yy} = 1, \qquad 0 < x < 2, 0 < y < 2,$$
    $$u(x,0) = u(x,2) = u(0,y) = u(2,y) = 0$$

    using piecewise linear functions and the triangulation of Figure 11.10.

    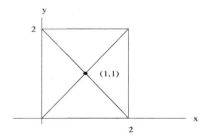

    FIGURE 11.10
    Triangulation of domain for Exercise 7a.

    b) Proceed as in part (a), but for the problem

    $$u_{xx} + u_{yy} = 1, \qquad 0 < x < 4, 0 < y < 2,$$
    $$u(x,0) = u(x,2) = u(0,y) = u(4,y) = 0,$$

    using the triangulation of Figure 11.11.

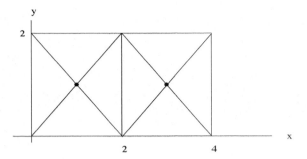

FIGURE 11.11
Triangulation of domain for Exercise 7b.

8.　a) Show that any continuous piecewise linear function

$$f(x) = \begin{cases} ax + b, & 0 \le x \le 1, \\ cx + d, & 1 \le x \le 2, \end{cases}$$

can be written as a linear combination of the tent functions

$$v_0(x) = \begin{cases} 1 - x, & \text{if } 0 \le x \le 1, \\ 0, & \text{if } 1 \le x \le 2, \end{cases}$$

$$v_1(x) = \begin{cases} x, & \text{if } 0 \le x \le 1, \\ 2 - x, & \text{if } 1 \le x \le 2, \end{cases}$$

$$v_2(x) = \begin{cases} 0, & \text{if } 0 \le x \le 1, \\ x - 1, & \text{if } 1 \le x \le 2. \end{cases}$$

　　b) Generalize part (a) to any interval $0 \le x \le N$, where N is a natural number, and where f is continuous and piecewise linear on the intervals $i - 1 \le x \le i$, $i = 1, \ldots, N$.

　　c) Show that any piecewise continuous function

$$f(x) = \begin{cases} ax + b, & 0 \le x \le 1, \\ cx + d, & 1 \le x \le 2, \end{cases}$$

with $f(0) = f(2) = 0$, is just a constant multiple of the tent function $v_1(x)$ from above.

　　d) Generalize part (c) as we did part (a).

A

Uniform Convergence; Differentiation and Integration of Fourier Series

Here we look at *uniform* convergence and its relation to trigonometric Fourier series. Most results are stated without proof. If you're interested in looking deeper, see the references listed at the end of this appendix.

UNIFORM CONVERGENCE AND FOURIER SERIES

Before defining uniform convergence of a sequence of functions, we look at an example.

Example 1 We saw that the sequence of continuous functions

$$f_n(x) = x^n, \qquad n = 0, 1, 2, \ldots, 0 \le x \le 1,$$

converges to the discontinuous function

$$f(x) = \begin{cases} 0, & \text{if } 0 \le x < 1, \\ 1, & \text{if } x = 1. \end{cases}$$

See Figure A.1. In order to prove this, for each x_0 in $[0, 1]$ we must show that, for any $\epsilon > 0$, there exists a natural number N such that

$$n > N \Rightarrow |x_0^n - f(x_0)| < \epsilon.$$

This is easy for $x_0 = 1$, so let's take x_0 in $[0,1)$. Then, given ϵ, we look at

$$|x_0^n - f(x_0)| = x_0^n < \epsilon,$$

and we see that we need only take

$$N > \frac{\ln \epsilon}{\ln x_0}$$

(make sure you work this out for yourself!).

Notice that as $x_0 \to 1$, the denominator $\to 0$; thus, for a given ϵ, as $x_0 \to 1$, $N = N(\epsilon, x_0)$ (since it depends on both ϵ and on x_0) increases without bound.

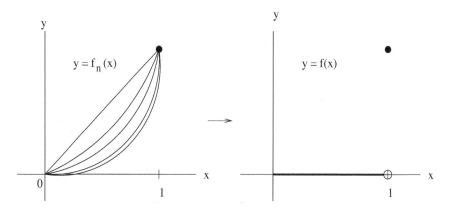

FIGURE A.1
The sequence of continuous functions $f_n(x)$ converges to the discontinuous function $f(x)$.

If we, instead, look at exactly the same problem, but on $0 \le x \le a < 1$, the worst case scenario occurs at $x = a$, where we need

$$N > \frac{\ln \epsilon}{\ln a}.$$

Thus, given ϵ, this same N works for *every* x, that is, $N = N(\epsilon)$ depends only on ϵ.

We say, in the latter case, that the sequence of functions converges **uniformly** to f, while, in the former case, although the sequence converges to f, it does *not* do so uniformly.

(By the way, convince yourself that using open intervals doesn't change anything. In essence, if an interval is not closed, we may close it off without consequence.)

Definition A.1 *Given a sequence $f_n(x)$ on $a \le x \le b$, we say that f_n con-* **verges to f uniformly** *on $a \le x \le b$ if, for every $\epsilon > 0$ there exists an $N = N(\epsilon)$ such that*

$$n > N \Rightarrow |f_n(x) - f(x)| < \epsilon \text{ for all } x \text{ in } [a, b].$$

Graphically speaking, if $f_n \to f$ uniformly, then, given ϵ, for all $n > N$ the function $f_n(x)$ lies inside an "ϵ-band" on either side of $y = f(x)$, as in Figure A.2. Note that *uniform convergence implies pointwise convergence.*

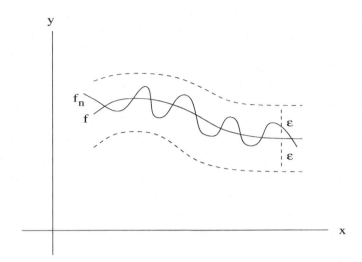

FIGURE A.2
Uniform convergence.

Now, since we define the sum of a series to be the limit of its nth partial sums, we may extend our definition to series of functions.

Definition A.2 *We say that the series* $\sum\limits_{i=1}^{\infty} f_i(x)$ **converges uniformly** *to* $f(x)$ *on* $a \leq x \leq b$ *if the sequence*

$$F_n(x) = \sum_{i=1}^{n} f_i(x)$$

converges uniformly to f *on* $a \leq x \leq b$.

How does all this relate to Fourier series? Remember, these are series of *continuous* functions which, in many cases, converge to *discontinuous* functions—we suspect that, in these cases, the convergence cannot be uniform. In fact, we have the following theorem.

Theorem A.1 *Suppose that* $f_n \to f$ *uniformly on* $a \leq x \leq b$, *and suppose that each* f_n *is continuous. Then,* f *must be continuous.*

PROOF We must show that, given any x_0, for any $\epsilon > 0$ there is a $\delta > 0$ such that
$$|x - x_0| < \delta \Rightarrow |f(x) - f(x_0)| < \epsilon.$$
We *know* that

a) For any $\epsilon_1 > 0$ there is an N such that

$$n > N \Rightarrow |f_n(x) - f(x)| < \epsilon_1 \text{ for all } x \text{ in } a \leq x \leq b.$$

b) Each f_n is continuous at x_0. In particular, if we fix $N_1 > N$ then, for any $\epsilon_2 > 0$, there is a $\delta > 0$ such that

$$|x - x_0| < \delta \Rightarrow |f_{N_1}(x) - f_{N_1}(x_0)| < \epsilon_2.$$

So, the standard trick is to use the triangle inequality:

$$\begin{aligned}|f(x) - f(x_0)| &= |f(x) - f_{N_1}(x) + f_{N_1}(x) - f_{N_1}(x_0) + f_{N_1}(x_0) - f(x_0)| \\ &\leq |f(x) - f_{N_1}(x)| + |f_{N_1}(x) - f_{N_1}(x_0)| + |f_{N_1}(x_0) - f(x_0)| \\ &\leq \epsilon_1 + \epsilon_2 + \epsilon_1 \end{aligned}$$

if $|x - x_0| < \delta$ (using the fact that $N_1 > N$, of course). After "cleaning it up" (e.g., letting $\epsilon_1 = \epsilon_2 = \frac{\epsilon}{3}$; the reader should do all this), we've proven the theorem. ∎

(Note that this theorem actually says that if $a \leq x_0 \leq b$, then

$$\lim_{x \to x_0} [\lim_{n \to \infty} f_n(x)] = \lim_{n \to \infty} [\lim_{x \to x_0} f_n(x)].)^*$$

Thus, if f is not continuous, the various Fourier series for f cannot converge uniformly to f. Even if f *is* continuous, its periodic extension may not be, in which case we *still* cannot have uniform convergence (why?).

So when *will* a Fourier series converge uniformly?

Theorem A.2 *Suppose that f is continuous on $-L \leq x \leq L$, f' is piecewise continuous on $-L < x < L$ and $f(-L) = f(L)$. Then the trigonometric Fourier series of f converges absolutely and uniformly to f on $-L \leq x \leq L$ (and, thus, to its periodic extension on $-\infty < x < \infty$).*

You should decide what additional constraints, if any, are necessary to guarantee uniform convergence of the Fourier cosine and sine series on $0 \leq x \leq L$.

Now, we can relax the conditions of Theorem A.2 if we're willing to avoid any "bad" points of f.

Theorem A.3 *Suppose that f is piecewise smooth on $-L \leq x \leq L$. Then the Fourier series of f converges uniformly to f on any closed interval subset of $[-L, L]$ throughout which f is continuous (with a similar statement concerning f's periodic extension).*

---

*If $x_0 = a$, then we have $\lim\limits_{x \to a^+}$; if $x_0 = b$, $\lim\limits_{x \to b^-}$.

TERM-BY-TERM DIFFERENTIATION AND INTEGRATION OF FOURIER SERIES

We often want to know if we can *interchange the operations of summation and differentiation*, that is, when can we say that

$$\frac{d}{dx} \sum_{n=1}^{\infty} f_n(x) = \sum_{n=1}^{\infty} f_n'(x)?$$

In particular, when may we do this with Fourier series? The following example shows that we must be careful.

Example 2 The Fourier sine series for $f(x) = 1$ on $0 \leq x \leq L$ is

$$F_s(x) = \sum_{k=0}^{\infty} \frac{4}{(2k+1)L} \sin \frac{(2k+1)\pi x}{L}.$$

F_s converges to f on $0 < x < L$ (and uniformly on any closed interval subset of $(0, L)$, as we saw in Theorem A.3). So, taking $L = \pi$, we may write

$$f(x) = \frac{4}{\pi} \sum_{k=0}^{\infty} \frac{\sin(2k+1)x}{2k+1} \quad \text{on} \quad 0 < x < \pi.$$

Then, $f'(x) = 0$ on $0 < x < \pi$ but, if we try to differentiate the series term-by-term, we get

$$\frac{4}{\pi} \sum_{k=0}^{\infty} \cos(2k+1)x,$$

which diverges on $0 < x < \pi$ (why? Actually, except at $x = \frac{\pi}{2}$—why?).

The following theorem tells us when we're able to differentiate a Fourier series *term-by-term*.

Theorem A.4 *Suppose that f is continuous on $-L \leq x \leq L$, f' is piecewise continuous on $-L < x < L$ and $f(-L) = f(L)$. (Then, of course, we know that*

$$f(x) = \frac{a_0}{2} + \sum_{n=1}^{\infty} \left(a_n \cos \frac{n\pi x}{L} + b_n \sin \frac{n\pi x}{L} \right)$$

uniformly on $-L \leq x \leq L$, where the a_n and b_n are the Fourier coefficients.)
If $f''(x_0)$ exists, $-L \leq x_0 \leq L$,[†] then

$$f'(x_0) = \frac{\pi}{L} \sum_{n=1}^{\infty} n \left(-a_n \sin \frac{n\pi x}{L} + b_n \cos \frac{n\pi x}{L} \right).$$

[†]With obvious extension to $-\infty < x < \infty$ via periodic extension.

It's easier to *integrate* Fourier series term-by-term.

Theorem A.5 *Suppose that f is piecewise smooth on $-L \leq x \leq L$; thus, f has convergent Fourier series*

$$f(x) \sim \frac{a_c}{2} + \sum_{n=1}^{\infty} \left(a_n \cos \frac{n\pi x}{L} + b_n \sin \frac{n\pi x}{L} \right).$$

Then,

$$\int_{-L}^{x} f(z) dz = \frac{a_0}{2} \int_{-L}^{x} dz + \sum_{n=1}^{\infty} \left(a_n \int_{-L}^{x} \cos \frac{n\pi z}{L} dz + b_n \int_{-L}^{x} \sin \frac{n\pi z}{L} dz \right).$$

Note that the result is *not* a Fourier series. By the way, it turns out that, even if f is only piecewise continuous, the formal antiderivative of its Fourier series actually converges to the function

$$\frac{F(x+) + F(x-)}{2},$$

where $F(x)$ is the periodic extension of the function $\int_{-L}^{x} f(z) dz$. (Why is this the case?)

References: For further study, see the following references listed after Chapter 11: Churchill and Brown; Kirkwood; Marsden and Hoffman; Myint-U and Debnath; and Pinsky.

B

Other Important Theorems

We begin by giving somewhat informal statements of the three big theorems from vector analysis—Green's, Stokes's and Gauss's, or the Divergence, Theorems. In each case, we assume that all regions are bounded and that all functions, regions and boundaries are smooth enough not to give us any problems. Further, any boundary integrals go "once around" the boundary, with standard positive orientation.

Theorem B.1 (Green's Theorem) *In* $\mathbb{R}^2$,

$$\oint_C p\ dx + q\ dy = \iint_D (q_x - p_y)dA,$$

where $p = p(x, y), q = q(x, y)$ *and* $C = \partial D$.

Theorem B.2 (Stokes's Theorem) *In* $\mathbb{R}^3$,

$$\iint_S \nabla \times \boldsymbol{F} \cdot \hat{\boldsymbol{n}}\ dS = \oint_C F_1\ dx + F_2\ dy + F_3\ dz,$$

where $\boldsymbol{F}(x, y, z) = (F_1(x, y, z), F_2(x, y, z), F_3(x, y, z)), S$ *is a surface with boundary curve* C *and*

$$\nabla \times \boldsymbol{F} = curl\ \boldsymbol{F} = \begin{vmatrix} \boldsymbol{i} & \boldsymbol{j} & \boldsymbol{k} \\ \dfrac{\partial}{\partial x} & \dfrac{\partial}{\partial y} & \dfrac{\partial}{\partial z} \\ F_1 & F_2 & F_3 \end{vmatrix}.$$

Theorem B.3 (Divergence Theorem or Gauss's Theorem or Ostrogradsky's Theorem) *In* $\mathbb{R}^3$,

$$\iiint_D \nabla \cdot \boldsymbol{F}\ dV = \iint_S \boldsymbol{F} \cdot \hat{\boldsymbol{n}}\ dS,$$

where F is as above, D is a three-dimensional region with boundary surface S and

$$\nabla \cdot \boldsymbol{F} = F_{1x} + F_{2y} + F_{3y} = \text{the divergence of } \boldsymbol{F}.$$

It's interesting to note the following:

1. Green's Theorem is a special case of Stokes's Theorem.

2. If, in Green's Theorem, we let $q = F_1$ and $p = -F_2$, and $\boldsymbol{F} = (F_1, F_2)$, then we get

$$\iint\limits_D \nabla \cdot \boldsymbol{F} \, dA = \oint\limits_C \boldsymbol{F} \cdot \hat{\boldsymbol{n}} \, dS,$$

which is just the two-dimensional version of the Divergence Theorem.

3. If, in the Divergence Theorem, we let $\boldsymbol{F} = \nabla \times \boldsymbol{G}$, we get

$$\iiint\limits_D \nabla \cdot \boldsymbol{F} \, dV = 0 \quad \text{(why?)}$$

$$= \iint\limits_S \nabla \times \boldsymbol{G} \cdot \hat{n} \, dS,$$

where S is the closed boundary of D. Stokes's Theorem (for $\boldsymbol{G}$) follows almost immediately (why?).

4. If we take the one-dimensional **Fundamental Theorem of Calculus**

$$\int_a^b f'(x)dx = f(b) - f(a),$$

it can be made to look like the Divergence Theorem, since the integrand is the one-dimensional divergence of f and the right side is an "oriented" sum of the values of f on the boundary.

But we digress. For a detailed treatment, see any standard book on advanced calculus.

Theorems B.4–B.12 involve the interchange of limiting operations, while Theorems B.13 and B.14 provide us with tests for uniform convergence of series.

Theorem B.4 *Suppose that the sequence of continuous functions $f_n(x)$ converges uniformly to $f(x)$ on $a \leq x \leq b$. Then, for each x_0 in the interval,*

$$\lim_{x \to x_0} [\lim_{n \to \infty} f_n(x)] = \lim_{n \to \infty} [\lim_{x \to x_0} f_n(x)].$$

(This is just Theorem A.1.)

Theorem B.5 *Suppose that $f_n \to f$ pointwise on $a < x < b$, that each $f_n'(x)$ is continuous on $a < x < b$ and that $f_n' \to g$ uniformly on $a < x < b$. Then, $f'(x)$ exists and $f'(x) = g(x)$ on $a < x < b$.*

We may extend Theorem B.5 for infinite series of functions. As before, we now look at n^{th} partial sums.

Theorem B.6 *Suppose that $\sum\limits_{n=1}^{\infty} f_n(x) = f(x)$ pointwise on $a < x < b$, each $f_n'(x)$ is continuous on $a < x < b$ and $\sum\limits_{n=1}^{\infty} f_n'(x) = g(x)$ uniformly on $a < x < b$. Then, $f'(x)$ exists and $f'(x) = g(x)$ on $a < x < b$, that is,*

$$\frac{d}{dx} \sum_{n=1}^{\infty} f_n(x) = \sum_{n=1}^{\infty} f_n'(x) \quad on \quad a < x < b.$$

Theorem B.7 *Suppose that each function $f_n(x)$ is integrable on $a \le x \le b$, and suppose that $f_n \to f$ uniformly on $a \le x \le b$. Then f is integrable on the same interval, and*

$$\int_a^b f(x)dx = \lim_{n \to \infty} \int_a^b f_n(x)dx,$$

that is,

$$\int_a^b [\lim_{n \to \infty} f_n(x)]dx = \lim_{n \to \infty} \int_a^b f_n(x)dx.$$

(Compare Theorem B.7 to Theorem A.5. Theorem B.7, in turn, is a stricter version of the well-known and very important **Lebesgue's Dominated Convergence Theorem**.)

We may extend Theorem B.7 to the case where f is an infinite series of functions where, as before, $f_n(x)$ is replaced by the nth partial sum.

Theorem B.8 *Suppose that each function $f_n(x)$ is integrable on $a \le x \le b$, and suppose that $f(x) = \sum\limits_{n=1}^{\infty} f_n(x)$ uniformly on $a \le x \le b$. Then*

$$\int_a^b f(x)dx = \sum_{n=1}^{\infty} \int_a^b f_n(x)dx,$$

that is,

$$\int_a^b \left[\sum_{n=1}^{\infty} f_n(x) \right] dx = \sum_{n=1}^{\infty} \int_a^b f_n(x)dx.$$

How about interchanging differentiation and integration?

Theorem B.9 *Suppose that $f(x, y)$ is a continuous real-valued function on the rectangle $a \leq x \leq b$, $c \leq y \leq d$, and suppose that $\frac{\partial f}{\partial x}$ is also continuous on the same rectangle. Then*

$$\frac{d}{dx}\left[\int_c^d f(x,y)dy\right] = \int_c^d \frac{\partial f}{\partial x}(x,y)dy.$$

(We may replace c by $-\infty$ or d by ∞, with the additional assumptions that both integrals converge, with the second converging uniformly* on $a \leq x \leq b$.)

Theorem B.10 (Leibniz's Rule) *Suppose that $f(x, y)$ and $\frac{\partial f}{\partial x}(x, y)$ are continuous on the rectangle $a \leq x \leq b$, $c \leq y \leq d$, and suppose that $u(x)$ and $v(x)$ are functions from $a \leq x \leq b$ to $c \leq y \leq d$ such that u' and v' are continuous on $a \leq x \leq b$. Then*

$$\frac{d}{dx}\int_{u(x)}^{v(x)} f(x,y)dy = f(x,v(x)) \cdot v'(x) - f(x,u(x)) \cdot u'(x)$$

$$+ \int_{u(x)}^{v(x)} \frac{\partial f}{\partial x}(x,y)dy.$$

Theorem B.11 (Fubini's Theorem) *Suppose that D is the rectangular region $a \leq x \leq b$, $c \leq y \leq d$, and suppose that $f(x, y)$ is continuous on D. Then*

$$\iint_D f \; dA = \int_a^b \left[\int_c^d f(x,y)dy\right] dx = \int_c^d \left[\int_a^b f(x,y)dx\right] dy.$$

Theorem B.12 *Given a function $f(x, y)$ on an open set D, if $\frac{\partial f}{\partial x \partial y}$ and $\frac{\partial f}{\partial y \partial x}$ are continuous on D, then*

$$\frac{\partial f}{\partial x \partial y} = \frac{\partial f}{\partial y \partial x} \quad on \quad D.$$

Theorem B.13 (Weierstrass M-test) *Given the functions $f_n(x)$, $n = 1, 2, 3, \ldots$, on an interval I, suppose there exist constants M_n such that*

$$|f_n(x)| < M_n$$

for all x in I, for each $n = 1, 2, 3, \ldots$, and such that $\sum\limits_{n=1}^{\infty} M_n$ converges. Then $\sum\limits_{n=1}^{\infty} f_n(x)$ converges absolutely and uniformly on I.

---

* $\int_c^{\infty} g(x,y)dy$ converges uniformly to $G(x)$ on $a \leq x \leq b$ if the sequence of functions $g_n(x) = \int_c^n g(x,y)dy$ converges uniformly to $G(x)$ on $a \leq x \leq b$.

Theorem B.14 (Abel's test) *Suppose we have functions $f_n(t)$ and $g_n(t)$, $n = 1, 2, 3, \ldots$, on an interval I, such that (a) $g_{n+1}(t) \leq g_n(t)$ for all t in I, (b) there exists a constant M such that $|g_n(t)| \leq M$ for all t in I and $n = 1, 2, 3, \ldots$ and (c) $\sum_{n=1}^{\infty} f_n(t)$ converges uniformly on I. Then, the series*

$$\sum_{n=1}^{\infty} g_n(t) f_n(t)$$

also converges uniformly on I.

Reference: A nice treatment of Theorems B.4–B.14 can be found in the excellent text by Marsden and Hoffman listed in the References.

C

Existence and Uniqueness Theorems

In this appendix we state and prove theorems concerning the existence and uniqueness of solutions for the one-dimensional (1-D) heat and wave equations, and uniqueness for the two-dimensional (2-D) Poisson's equation, on finite domains.

EXISTENCE—HEAT AND WAVE EQUATIONS

1-D heat equation on a finite interval

Theorem C.1 *Suppose that $f(x)$ is continuous on $0 \leq x \leq \pi$, $f'(x)$ is piecewise continuous on $0 < x < \pi$ and $f(0) = f(\pi) = 0$. Then the heat problem*

$$
\begin{aligned}
u_t &= k^2 u_{xx}, & 0 < x < \pi, t > 0, \\
u(x,0) &= f(x), & 0 \leq x \leq \pi, \\
u(0,t) &= u(\pi,t) = 0, & t \geq 0,
\end{aligned}
$$

has solution

$$
u(x,t) = \sum_{n=1}^{\infty} b_n e^{-n^2 k^2 t} \sin nx,
$$

where the b_n are the Fourier sine coefficients

$$
b_n = \frac{2}{\pi} \int_0^{\pi} f(x) \sin nx \; dx.
$$

PROOF We'll show the following:

(1) $\sum_{n=1}^{\infty} b_n e^{-n^2 k^2 t} \sin nx$ converges uniformly on $0 \leq x \leq \pi$, for any (fixed) $t > 0$.

(2) $\sum_{n=1}^{\infty} b_n e^{-n^2 k^2 t} \sin nx$ converges uniformly on any interval $t_0 \leq t \leq T$, where $0 < t_0 < T$, for any (fixed) x in $0 \leq x \leq \pi$.

Then, (1) will allow us to differentiate by x, term-by-term, and also to show that $u(0, t) = u(\pi, t) = 0$ for $t > 0$, and (2) will allow us to differentiate by t, term-by-term.

So, we prove (1) and (2) as follows. First, we show that the sequence b_n is bounded. Indeed, by Theorem A.2, the Fourier sine series

$$\sum_{n=1}^{\infty} b_n \sin nx$$

converges uniformly to $f(x)$ on $0 \leq x \leq \pi$ and, thus, $b_n \to 0$ as $n \to \infty$. Of course, any convergent sequence is bounded, so there is some constant M with

$$|b_n| \leq M \quad \text{for all} \quad n = 1, 2, 3, \ldots .$$

Then, choosing t_0 and T arbitrarily, with $0 < t_0 < T$, we have

$$|b_n e^{-n^2 k^2 t} \sin nx| \leq M e^{-n^2 k^2 t_0}, \qquad 0 \leq x \leq \pi, \quad t_0 \leq t \leq T.$$

Using the ratio test, we see that $\sum_{n=1}^{\infty} M e^{-n^2 k^2 t_0}$ converges, so the *Weierstrass M-test* (Theorem B.10) tells us that

$$\sum_{n=1}^{\infty} b_n e^{-n^2 k^2 t} \sin nx$$

converges absolutely and uniformly on $0 \leq x \leq \pi$, $t_0 \leq t \leq T$.

Therefore, from Theorem B.3, we may differentiate by x or by t term-by-term. Then, a similar argument shows that we may do the same for u_x. Thus,

$$u_t = -k^2 \sum_{n=1}^{\infty} n^2 b_n e^{-n^2 k^2 t} \sin nx$$

and

$$u_{xx} = -\sum_{n=1}^{\infty} n^2 b_n e^{-n^2 k^2 t} \sin nx,$$

and, since t_0 and T are arbitrary, we have

$$u_t = k^2 u_{xx} \quad \text{on} \quad 0 < x < \pi, \quad t > 0.$$

As for the boundary conditions, since a uniform limit of continuous functions is continuous (Theorem A.1), we may "plug in" $x = 0$ and $x = \pi$ to get

$$u(0, t) = u(\pi, t) = 0 \quad \text{for all} \quad t > 0.$$

Notice that we can*not* do the same for $t = 0$ (why?). But we *do* know, again, that

$$f(x) = \sum_{n=1}^{\infty} b_n \sin nx$$

converges uniformly on $0 \le x \le \pi$; for fixed x_0, then, it converges uniformly as a function of t on any interval $0 \le t \le T$. Then, letting

$$f_n(t) = b_n \sin n x_0, \quad g_n(t) = e^{-n^2 k^2 t},$$

we see that $g_{n+1}(t) \le g_n(t)$ on $0 \le t \le T$, and, thus, from *Abel's test* (Theorem B.11),

$$\sum_{n=0}^{\infty} b_n e^{-n^2 k^2 t} \sin n x$$

does converge uniformly on $0 \le t \le T$ so that we *may* plug in $t = 0$ and get

$$u(x, 0) = \sum_{n=1}^{\infty} b_n \sin n x = f(x).$$

∎

Note that if $f(x)$ is only piecewise smooth, or if $f(0) \ne 0$ or $f(\pi) \ne 0$, then everything *still* works except for the last step, where we'll only have equality *in the mean*. Finally, we may proceed similarly for Neumann or Robin conditions at either end.

1-D Wave equation on a finite interval

Theorem C.2 *Given $f(x)$ and $g(x)$ on $0 \le x \le \pi$, let $F(x)$ and $G(x)$ be their odd periodic extensions. Suppose that F, F', F'', G and G' are continuous (thus, we must have $f(0) = f(\pi) = 0$, etc.). Then, the wave problem*

$$
\begin{aligned}
u_{tt} &= c^2 u_{xx}, & 0 &< x < \pi, t > 0, \\
u(x, 0) &= f(x), & 0 &\le x \le \pi \\
u_t(x, 0) &= g(x), & 0 &\le x \le \pi, \\
u(0, t) &= u(\pi, t) = 0, & t &\ge 0
\end{aligned}
$$

has solution

$$u(x, t) = \sum_{n=1}^{\infty} \sin n x (a_n \cos n c t + b_n \sin n c t),$$

where the a_n and b_n are the Fourier coefficients

$$a_n = \frac{2}{\pi} \int_0^{\pi} f(x) \sin n x \, dx, \quad b_n = \frac{2}{n \pi c} \int_0^{\pi} g(x) \sin n x \, dx.$$

PROOF This proof is easier than the one for the heat equation, as we'll be able to write our solution in terms of the Fourier sine series for f and g. So, we know that

$$f(x) = \sum_{n=1}^{\infty} a_n \sin n x \quad \text{and} \quad g(x) = \sum_{n=1}^{\infty} n c b_n \sin n x$$

absolutely and uniformly on $0 \leq x \leq \pi$ (from Theorem A.2). Thus,

$$F(x) = \sum_{n=1}^{\infty} a_n \sin nx \quad \text{and} \quad G(x) = \sum_{n=1}^{\infty} ncb_n \sin nx$$

absolutely and uniformly on $-\infty < x < \infty$. Now, let's rewrite u as we did in Exercise 8, Section 4.2. Using trigonometric identities, we have

$$\sum_{n=1}^{\infty} a_n \sin nx \cos nct = \frac{1}{2} \left[\sum_{n=1}^{\infty} a_n \sin n(x - ct) + \sum_{n=1}^{\infty} a_n \sin n(x + ct) \right]$$

$$= \frac{1}{2}[F(x + ct) + F(x - ct)].$$

Similarly,

$$\sum_{n=1}^{\infty} b_n \sin nx \sin nct = \frac{1}{2} \left[\sum_{n=1}^{\infty} b_n \cos n(x - ct) - \sum_{n=1}^{\infty} b_n \cos n(x + ct) \right]$$

$$= \frac{1}{2c} \int_{x-ct}^{x+ct} G(z)dz,$$

where the term-by-term integration is okay because of Theorem B.5. It follows that each of these series converges absolutely and uniformly for *any* choice of x and t. Thus, we can do everything we need to do.

So we may differentiate (term-by-term) to get

$$u_x(x, t) = \frac{1}{2}[F'(x + ct) + F'(x - ct)] + \frac{1}{2C}[G(x + ct) - G(x - ct)]$$

and

$$u_t(x, t) = \frac{1}{2}[F'(x + ct) - F'(x - ct)] + \frac{1}{2}[G(x + ct) + G(x - ct)].$$

Each of them is, essentially, an absolutely and uniformly convergent series (why?), so we can do it again:

$$u_{xx}(x, t) = \frac{1}{2}[F''(x + ct) + F''(x - ct)] + \frac{1}{2C}[G(x + ct) - G(x - ct)]$$

and

$$u_{tt}(x, t) = \frac{c^2}{2}[F''(x + ct) + F''(x - ct)] + \frac{c}{2}[G'(x + ct) + G'(x - ct)],$$

from which it follows that

$$u_{tt} + c^2 u_{xx} \quad \text{for all} \quad x, t.$$

The boundary conditions are easy (but you should do them to make sure), as is the initial condition (ditto).   ∎

Of course, our conditions on f and g are fairly restrictive. Again, we may relax them and still have a solution *in the mean-square sense*.

UNIQUENESS—HEAT AND WAVE EQUATIONS

We can prove uniqueness fairly easily for the *non*homogeneous heat and wave equations, with more general boundary conditions.

Theorem C.3 *Suppose that* $u(x,t)$ *is a solution of the heat problem*

$$u_t = k^2 u_{xx} + f(x,t), \qquad\qquad\qquad 0 < x < L, t > 0,$$
$$u(x,0) = g(x), \qquad\qquad\qquad\qquad 0 \le x \le L,$$
$$a_1 u(0,t) + a_2 u_x(c,t) = b_1 u(L,t) + b_2 u_x(L,t) = 0, \qquad t \ge 0,$$
$$a_1, a_2, b_1 \text{ and } b_2 \text{ constant},$$

with u *continuous on* $0 \le x \le L$, $t \ge 0$, *and with* u_t *and* u_{xx} *continuous on* $0 < x < L$, $t > 0$. *Then* u *is the problem's only such solution.*

PROOF   Suppose that u_1 and u_2 are both such solutions, and let $v = u_1 - u_2$. Then v satisfies the homogeneous problem

$$v_t = k^2 v_{xx}, \qquad 0 < x < L, t > 0,$$
$$v(x,0) = 0, \qquad 0 \le x \le L,$$
$$a_1 v(0,t) + a_2 v_x(0,t) = b_1 v(L,t) + b_2 v_x(L,t) = 0, \qquad t \ge 0.$$

Now, v is continuous on $t \ge 0$, so the function

$$I(t) = \int_0^L v^2(x,t)\,dx$$

is a continuous function of t, as well. Further,

$$I(t) \ge 0 \text{ for } t \ge 0 \quad \text{(why?)},$$

and

$$I(0) = \int_0^L v^2(x,0)\,dx = 0.$$

Next, we have, from Theorem B.6, that

$$I'(t) = 2 \int_0^L v(x,t) v_t(x,t)\,dx$$

and, using the PDE and integration by parts, we see that

$$\int_0^L v v_t \, dx = \int_0^L v v_{xx} \, dx$$
$$= (v v_x)\Big|_{x=0}^{x=L} - \int_0^L v_x^2(x,t)\,dx.$$

Now, back in Section 8.2, we showed that the boundary conditions force the boundary term to be zero and, as the last integral is nonnegative, we must have

$$I'(t) \leq 0 \quad \text{for} \quad t > 0.$$

Finally, $I(0) = 0$ and $I'(t) \leq 0$ for $t > 0$ imply that we must have $I(t) \leq 0$ for $t > 0$ (prove this!), from which it follows that we must have

$$I(t) = \int_0^L v^2(x,t)dx = 0 \quad \text{for all} \quad t \geq 0.$$

Thus, we must have

$$v(x,t) = 0 \quad \text{for} \quad 0 \leq x \leq L \quad \text{and} \quad t \geq 0.$$

∎

Theorem C.4 *Suppose that $u(x,t)$ is a solution of the wave problem*

$$
\begin{aligned}
u_t &= c^2 u_{xx} + f(x,t), & 0 < x < L, t > 0, \\
u(x,0) &= g(x), & 0 \leq x \leq L, \\
u_t(x,0) &= h(x), & 0 \leq x \leq L, \\
a_1 u(0,t) + a_2 u_x(0,t) &= b_1 u(L,t) + b_2 u_x(L,t) = 0, & t \geq 0,
\end{aligned}
$$

where a_1, a_2, b_1 and b_2 are constants, u is continuous on $0 \leq x \leq L$, $t \geq 0$ and u_{tt} and u_{xx} are continuous on $0 \leq x \leq L$, $t > 0$. Then u is the problem's **only** *solution.*

PROOF   Our proof is very similar to that of Theorem C.3, except that the function $I(t)$ doesn't pop out of nowhere!

Again, we begin by assuming two solutions u_1 and u_2 and then showing that their difference, $v = u_1 - u_2$, must be the zero-function. As above, v satisfies

$$
\begin{aligned}
v_{tt} &= c^2 v_{xx}, & 0 < x < L, t > 0, \\
v(x,0) &= v_t(x,0) = 0, & 0 \leq x \leq L, \\
a_1 v(0,t) + a_2 v_x(0,t) &= b_1 v(L,t) + b_2 v_x(L,t) = 0, & t \geq 0.
\end{aligned}
$$

As before, v has the same smoothness properties as u_1 and u_2. Thus, the function

$$E(t) = \frac{\rho}{2} \int_0^L (c^2 v_x^2 + v_t^2)dx$$

is continuous on $t \geq 0$. Here, ρ is the constant mass density of the string.

Where did we come up with E? It can be shown that $E(t)$ is the energy (kinetic plus potential) of the string at time t. Further, conservation of energy suggests that E should be constant, and we show this to be the case.

First, let

$$I(t) = \int_0^L (c^2 v_x^2 + v_t^2)dx.$$

Theorem B.6 then tells us that

$$I'(t) = 2\int_0^L (c^2 v_x v_t + v_t v_{tt})dx$$

and, integrating the first term by parts, we have

$$I'(t) = 2[c^2 v_x v_t]_{x=0}^{x=L} + 2\int_0^L v_t(v_{tt} - c^2 v_{xx})dx.$$

With a little work, we can again show that the boundary term disappears (try it) and, using the PDE, the integrand disappears, as well. Thus,

$$I'(t) = 0 \quad \text{and} \quad I(t) = \text{constant}.$$

What happens when $t = 0$? We have

$$I(0) = \int_0^L [c^2 v_x^2(x, 0) + v_t^2(x, 0)]dx.$$

Now, we know that $v_t(x, 0) = 0$ and $v(x, 0) = 0$, with the latter implying that $v_x(x, 0) = 0$. So $I(0) = 0$ and, since I is constant, we must have $I(t) = 0$ for $t \geq 0$.

Finally, this can happen only if $v_x(x, t) = v_t(x, t) = 0$ for all x and t, implying $v(x, t) = \text{constant}$ and, since $v(x, 0) = 0$, it follows that

$$v(x, t) = 0 \quad \text{for} \quad 0 \leq x \leq L \quad \text{and} \quad t \geq 0.$$

∎

THE MAXIMUM PRINCIPLE FOR THE LAPLACIAN AND UNIQUENESS FOR THE POISSON DIRICHLET PROBLEM

Uniqueness for Poisson's equation on a bounded domain follows immediately from the maximum principle that we mentioned in Section 9.3.

Theorem C.5 (Maximum Principle for harmonic functions) *Suppose that $\nabla^2 u = 0$ on a domain D and that u is continuous on $D \cup \partial D$. Then u attains its maximum and minimum values on ∂D.*

Let's get right to the uniqueness theorem and prove Theorem C.5 afterward.

Theorem C.6 *Suppose that $u(x, y)$ is a solution of the Poisson problem*

$$u_{xx} + u_{yy} = f(x, y) \text{ on } D,$$
$$u = 0 \text{ on } \partial D$$

with the properties that u_{xx} and u_{yy} are continuous on the bounded domain D, with u continuous on $D \cup \partial D$. Then u is the problem's only such solution.

PROOF   As usual, suppose there are two such solutions u_1 and u_2, from which we have that $v = u_1 - u_2$ satisfies

$$v_{xx} + v_{yy} = 0 \text{ on } D,$$
$$v = 0 \qquad \text{on } \partial D.$$

Also, u_1, u_2 and v all share the smoothness properties in the statement of the theorem, of course. Then, from Theorem C.5, v attains its maximum and minimum values on ∂D; thus, $v = 0$ on D.   ∎

This result can be generalized to the case where the boundary condition is $au + b\frac{\partial u}{\partial n} = 0$, for constants a and b. If $a \neq 0$, then the solution is unique; if $a = 0$, we know from earlier that the solution is *not* unique, but it turns out that the difference between any two solutions is a constant.*

PROOF of Maximum Principle   Let

$$v(x, y) = u(x, y) + \epsilon(x^2 + y^2), \text{ for } \epsilon > 0.$$

Then

$$\nabla^2 v = u_{xx} + u_{yy} + 4\epsilon = 4\epsilon > 0 \text{ on } D.$$

Thus, v does not attain a maximum in D (since, for a max to occur at (x_0, y_0) in D, we must have $v_{xx}(x_0, y_0) \leq 0$ and $v_{yy}(x_0, y_0) \leq 0$). Thus, the maximum value of v occurs on ∂D; let's say that it is $M_1 = v(x_1, y_1)$.

Now, let M be the maximum value of u on ∂D. We know that

$$M_1 = \max_{\partial D}[u(x, y) + \epsilon(x^2 + y^2)] \leq \max_{\partial D} u(x, y) + \epsilon \max_{\partial D}(x^2 + y^2)$$
$$= M + \epsilon \max_{\partial D}(x^2 + y^2).$$

Then, on $D \cup \partial D$,

$$u(x, y) = v(x, y) - \epsilon(x^2 + y^2)$$
$$\leq v(x, y)$$
$$\leq M + \epsilon \max_{\partial D}(x^2 + y^2)$$

---

*See, e.g., Churchill and Brown's *Fourier Series and Boundary Value Problems*.

and, since ϵ is arbitrary, we must have

$$u(x,y) \leq M \text{ on } D \cup \partial D.$$

In order to prove that the minimum value of u also occurs on ∂D, apply the Maximum Principle to $-u$. ∎

THE HEAT EQUATION REVISITED—THE MAXIMUM PRINCIPLE

Actually, there also is a maximum principle for the heat equation, and we can use *this* principle to prove uniqueness, as well.

Theorem C.7 (Maximum Principle for the heat equation) *Suppose that u satisfies the heat equation*

$$u_t = k^2 u_{xx}, \qquad 0 < x < L, t > 0$$

and, given any $T > 0$, suppose that u is continuous on the closed rectangular region $0 \leq x \leq L$, $0 \leq t \leq T$. Thus, u attains a maximum value on the rectangular region. Then, this maximum occurs on the bottom or the sides of the rectangle, that is, on the set

$$C = \{(x,0): \ 0 \leq x \leq L\} \ \cup \ \{(0,t): \ 0 \leq t \leq T\} \ \cup \ \{(L,t): \ 0 \leq t \leq T\}.$$

(Similarly for the minimum value of u.)

PROOF   First, let

$$D = \{(x,t): \ 0 < x < L, 0 < t < T\}.$$

Now, since C is closed and bounded, we know that u attains a maximum M on C. We wish to show that

$$u(x,t) \leq M \text{ on } D \cup \partial D.$$

We begin by letting

$$v(x,t) = u(x,t) + \epsilon x^2 \text{ on } D \cup \partial D$$

and noting that

$$v_t - k^2 v_{xx} = -2k^2 \epsilon < 0.$$

So what can we say about v? First, suppose that v attains its maximum at a point (x_0, t_0) in D. Then,

$$v_t(x_0, t_0) = 0 \quad \text{and} \quad v_{xx}(x_0, t_0) \leq 0,$$

so that

$$v_t(x_0, t_0) - k^2 v_{xx}(x_0, t_0) \geq 0,$$

a contradiction. Thus, its maximum occurs on the boundary. Can it occur at (x, T)? If so, we must have

$$v_{xx}(x, T) \leq 0 \quad \text{and} \quad v_t(x, T) \geq 0 \quad \text{(why?)}.$$

Again, this leads to a contradiction. So the maximum value M_1 of v occurs on C.

It follows that

$$M_1 = \max_C [u(x, t) + \epsilon x^2] \leq \max_C u(x, t) + \epsilon \max_C x^2$$
$$= M + \epsilon \max_C x^2.$$

Thus, on $D \cup \partial D$,

$$u(x, t) = v(x, t) - \epsilon x^2$$
$$\leq v(x, t)$$
$$\leq M + \epsilon \max_C x^2$$

and, since ϵ is arbitrary, we must have

$$u(x, t) \leq M \text{ on } D \cup \partial D.$$

∎

Uniqueness for the heat problem then follows easily.

Theorem C.8 *Suppose that $u(x, t)$ is a solution of the heat problem*

$$u_t = k^2 u_{xx} + f(x, t), \qquad\qquad 0 < x < L, t > 0,$$
$$u(x, 0) = g(x), \qquad\qquad 0 \leq x \leq L,$$
$$u(0, t) = h_1(t), u(L, t) = h_2(t), \qquad t \geq 0,$$

with u continuous on $0 \leq x \leq L$, $t \geq 0$, and u_t and u_{xx} continuous on $0 < x < L$, $t > 0$. Then u is the problem's only such solution.

By the way, note that the maximum/minimum principle says that, at any point in time $t = T$, the maximum/minimum temperatures up until then must have occurred initially or at an endpoint (in the absence of a source $f(x, t)$, of course).

D

A Menagerie of PDEs

This appendix gathers together all of the PDEs studied in this book, along with many other important equations that are not covered. In each case, ∇ is the *del operator* in the space variables. Therefore, in one dimension, $\nabla^2 u = u_{xx}$; in two dimensions, $\nabla^2 u = u_{xx} + u_{yy}$; and in three dimensions, $\nabla^2 u = u_{xx} + u_{yy} + u_{zz}$ (in Cartesian coordinates, of course).

D.1 The Big Three and Other Important PDEs

We begin by listing the equations most frequently referred to in this book, followed by a larger compilation based on areas of application.

Heat/diffusion equation: $u_t = k^2 \nabla^2 u + f$

Wave equation: $u_{tt} = c^2 \nabla^2 u + f$

> In each of the above, the unknown u and the source term f are functions of the space and time variables. If $f \equiv 0$, we have, of course, the homogeneous versions of these equations.

Poisson's equation: $\nabla^2 u = -f$

> Here, u and f are functions of the space variables. If $f \equiv 0$, we have

Laplace's equation: $\nabla^2 u = 0$

> When separating out the time variable in the heat and wave equations, we encountered the

Helmholtz equation: $\nabla^2 u + k^2 u = 0$

We also have the

Convection or **advection** or **linear transport equation:** $u_t + v \cdot \nabla u = 0$

Here, the velocity v is a function of the independent variables.

A very important first-order equation is the linear

Continuity equation: $\rho_t + \nabla \cdot \Phi = 0$

Here, ρ usually is a concentration or density, while Φ is the flux, and the equation is just a statement of the conservation of energy, charge or the like. Examples include the following.

Heat/diffusion: $(\sigma \rho u)_t + \nabla \cdot \Phi = 0$

(σ = specific heat, ρ = mass density, u = temperature, Φ = flux)

Fluid flow: $\rho_t + \nabla \cdot (\rho v) = 0$

(ρ = density, v = velocity)

Electric current: $\rho_t + \nabla \cdot J = 0$

(ρ = charge density, J = current)

These continuity equations are linear examples of the more general

Conservation law: $\rho_t + \nabla \cdot \Phi(u, x, y, z, t) = q(u, x, y, z, t)$

a well-known example of which is the nonlinear

Burger's equation: $u_t + \frac{\partial}{\partial x}\left(\frac{1}{2}u^2\right) = u_t + u u_x = 0$

(see below, under Fluid Dynamics)

RELATED EQUATIONS

Related to the heat equation is the parabolic

Fokker–Planck equation: $u_t = u_{xx} + x u_x + u$

from statistical mechanics. Two important hyperbolic equations are the

Telegraph equation: $u_{xx} = CL u_{tt} + (RC + GL)u_t = RGu$

(see below, under Electrical Circuits) and the

Hanging chain equation: $u_{tt} = g(x u_{xx} + u_x)$

Here, g is the gravitational acceleration.

Of course, the source term f in the heat equation can depend on u. In this case, the equation may be nonlinear. An example is the following *nonlinear heat equation*:

Fisher's equation: $u_t = u_{xx} + u(1 - u)$

This equation is an example of a so-called *reaction-diffusion* equation which arises in cell biology.

D.2 Schrödinger's Equation

The cornerstone of the *wave* approach to the study of quantum mechanics is *Schrödinger's equation* (sometimes called *Schrödinger's wave equation*), which is a PDE satisfied by the *wave function* ψ.

Schrödinger's equation:

$$i\hbar\psi_t + \frac{\hbar^2}{2m}\nabla^2\psi - V(x, y, z)\psi = 0$$

Here,

$$\psi(x, y, z, t) = \text{wave function of particle}$$
$$= \text{probability that the particle is at location } (x, y, z)$$
$$\text{at time } t$$
$$h = 2\pi\hbar = \text{Planck's constant}$$
$$m = \text{mass of particle}$$
$$V(x, y, z) = \text{quantum mechanical potential at point } (x, y, z)$$

It's customary to separate out time by letting $\psi = e^{\frac{-iEt}{\hbar}}\Psi(x, y, z)$, resulting in the

Time-independent Schrödinger's equation:

$$\nabla^2\Psi + \frac{2m}{\hbar^2}[E - V(x, y, z)]\Psi = 0$$

Here, $\Psi(x, y, z)$ is called the *quantum state* of the particle when the particle has *energy* equal to the eigenvalue E.

When the particle is the lone electron in the hydrogen atom, then the potential is $V = -\frac{e^2}{\rho}$, where e is the electron's charge and ρ its (spherical coordinate) distance from the nucleus.

Schrödinger's equation for the hydrogen atom:

$$i\hbar\psi_t + \frac{\hbar^2}{2m}\nabla^2\psi + \frac{e^2}{\rho}\psi = 0$$

Another version of Schrödinger's equation which often leads to eigenvalues is that for the one-dimensional *harmonic oscillator* or *"particle in a box."* In this case, x is the only space variable and $V(x) = \frac{1}{2}mw^2x^2$, where w is the "classical frequency" of the oscillator. Upon separation of variables, we end up with the following ODE.

Time-independent Schrödinger's equation for linear harmonic oscillator:

$$\psi'' + \alpha(\lambda - x^2)\psi = 0$$

for constants α and (eigenvalues) λ.

There also is a *nonlinear* Schrödinger's equation, which shows up in the study of *solitary waves* or *solitons* (see below).

D.3 Maxwell's Equations

Maxwell's equations are the basic equations of the electromagnetic field. If

$\boldsymbol{E}(x,y,z,t) =$ electric field strength at point (x,y,z), at time t

$\boldsymbol{B}(x,y,z,t) =$ magnetic field strength at point (x,y,z), at time t

$\rho(x,y,z,t) =$ charge density (per unit volume) at point (x,y,z), at time t

$\boldsymbol{J}(x,y,z,t) =$ current density (per unit area) at point (x,y,z), at time t

then Maxwell showed that the vectors $\boldsymbol{E}, \boldsymbol{B}$ and $\boldsymbol{J}$, and the scalar ρ, satisfy the partial differential equations

$$\nabla \cdot \boldsymbol{E} = 4\pi\rho, \qquad\qquad \nabla \cdot \boldsymbol{B} = 0,$$
$$\boldsymbol{E}_t = c\nabla \times \boldsymbol{B} - 4\pi\boldsymbol{J}, \qquad \boldsymbol{B}_t = -c\nabla \times \boldsymbol{E}.$$

Here, c is the speed of light. If the configuration is static, so that $\boldsymbol{E}$ and $\boldsymbol{B}$ do not change over time and there is no current, the last two equations become

$$\nabla \times \boldsymbol{B} = \boldsymbol{0}, \qquad \nabla \times \boldsymbol{E} = \boldsymbol{0}.$$

To be precise, the above are the so-called **vacuum** or **microscopic Maxwell's equations**. These equations work well for a small number of sources of charge and current. They are *not* appropriate when dealing with macroscopic aggregates of matter. For these cases, we have the so-called **macroscopic Maxwell's equations**

$$\nabla \cdot \boldsymbol{D} = 4\pi\rho, \qquad \nabla \cdot \boldsymbol{B} = 0,$$
$$\boldsymbol{D}_t = c\nabla \times \boldsymbol{H} - 4\pi\boldsymbol{J}, \qquad \boldsymbol{B}_t = -c\nabla \times \boldsymbol{E},$$

where $\boldsymbol{E}$ and $\boldsymbol{B}$ here are averages of the microscopic $\boldsymbol{E}$ and $\boldsymbol{B}$, and $\boldsymbol{D}$ and $\boldsymbol{H}$ are vector fields related to $\boldsymbol{E}$ and $\boldsymbol{B}$, respectively.

D.4 Elasticity

Among other things, the theory of elasticity studies the static and vibration behavior of objects like strings, beams and plates—mathematical idealizations of these real physical objects. As we've seen, the motions of strings, membranes and the like are described by the

wave equation: $u_{tt} = c^2\nabla^2 u + f$

> Here, the *load* f is a function of all of the independent variables. If $f \equiv 0$, we have our old homogeneous wave equation, while if there is no time dependence, then $u_{tt} \equiv 0$ and u represents a static shape (e.g., a string hanging under the influence of gravity).

The difference between a *string* and a *beam* is that a beam is *stiff*, a property that can be quantified. When deriving the PDE for the vibrating string, we saw that the only force exerted on a differential element by the rest of the string is the *tension*, which, in a perfectly elastic string, is tangent to the string at each point. Thus, if this tension is instantly removed at a point, the only reaction would be for the string to *unstretch*.

Suppose, instead, that we have a *cantilever beam*, as shown in Figure D.1. If we pull down on the free end and then let go, the beam has the tendency both to unbend and to vibrate in the vertical direction. What's happening is that there are many small forces acting on any cross section of the beam, in many different directions. We may sum all of these and decompose the result into a vertical force, called the *shear*, and a moment, called the bending moment. (See Figure D.2.) In the simplest model, we make a number of assumptions— e.g., that w and w_x are small, that the beam is *thin*—that allow us to say, e.g., that the shear is, indeed, vertical and that it acts on the midsection of the beam.

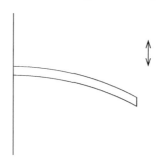

FIGURE D.1
The cantilever beam.

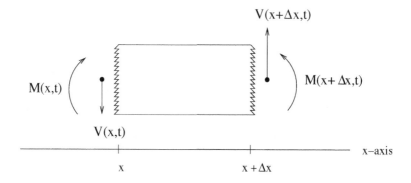

FIGURE D.2
Differential element for the E–B beam, showing the shear force V and the bending moment M. These forces/moments are applied to the element at each end by the rest of the beam.

It's not hard to show that, in the simplest case, we must have

$$M_x(x, t) = V(x, t),$$

where M is the bending moment and V is the shear, and that

$$M = EIw_{xx},$$

where w is the deflection of the beam and E and I are "beam constants," related to its composition and shape—in fact, the product EI is referred to as the beam's *stiffness*.

Finally, we may put everything together by, again, summing forces which act on a differential element. The result is the

Euler–Bernoulli (E–B) beam equation: $w_{tt} + \alpha^4 w_{xxxx} = f(x, t)$

The constant $\alpha^4 = \frac{EI}{\rho}$, where ρ is the density per unit length of the beam, and $f(x,t)$ is the load. One may incorporate more effects into the model, resulting in the (linear) *Rayleigh* and *Timoshenko* *beams*.

There are four sets of naturally occurring boundary conditions for the beam—you should convince yourself that they do make sense, physically.

clamped:	$w(0,t) = w_x(0,t) = 0$
pinned or **simply-supported:**	$w(0,t) = w_{xx}(0,t) = 0$
roller-supported:	$w_x(0,t) = w_{xxx}(0,t) = 0$
free:	$w_{xx}(0,t) = w_{xxx}(0,t) = 0$

The two-dimensional analog of the E–B beam is the

Kirchhoff thin plate equation: $w_{tt} + \alpha^4 \nabla^4 w = f(x,y,t)$

Here, $\nabla^4 = (\nabla^2)^2 = \Delta^2$, and the constant α^4 again is related to the stiffness of the plate. Note that the bending moment and shear boundary conditions for the plate are *not* $w_{xx} = 0$, $w_{xxx} = 0$, but are more complicated.

We note that the *longitudinal* vibration of beams/bars is also governed by the wave equation. Another equation which arises in this context is the

Boussinesq equation: $u_{tt} - c^2 \nabla^2 u - u \nabla^2 u_{tt} = 0$

We meet this equation in the study of fluid dynamics, as well.

For more information on beam and plate equations, see, e.g., Timoshenko and MacCullough's *Elements of Strength of Materials*, along with Timoshenko and Woinowsky–Krieger's *Theory of Plates and Shells*.

D.5 Electric Current in a Wire

The flow of electricity in a wire or cable is modeled by the

Transmission line equations:

$$i_x + Cv_t + Gv = 0$$
$$v_x + Li_t + Ri = 0$$

Here, we have

$$x = \text{position along wire}$$
$$i(x,t) = \text{current (at location } x, \text{ at time } t)$$
$$v(x,t) = \text{potential}$$
$$C = \text{capacitance per unit length}$$
$$R = \text{resistance per unit length}$$
$$L = \text{self-inductance per unit length}$$
$$G = \text{leakage per unit length.}$$

Then, by elimination, we see that i and v both satisfy the

Telegraph or **telephone equation:** $u_{xx} = CLu_{tt} + (RC + GL)u_t + GRu$

D.6   Fluid Dynamics

There are many PDEs which arise in the study of the dynamics of liquids and gases.

Convection or **advection** or **linear transport equation:** $u_t + vu_x = 0$, $v = v(x,t) = \text{velocity}$

Generalized Burger's equation: $u_t + f(u)u_x = \nu u_{xx}$

If $f(u) = u$, we have

Burger's equation (with dissipation): $u_t + uu_x = \nu u_{xx}$

If we set the dissipation coefficient, ν, equal to zero, we get

Burger's equation: $u_t + uu_x = 0$

We also have the following equations.

Tricomi equation: $u_{xx} + xu_{yy} = 0$

Euler–Poisson–Darboux equation: $u_{xx} - a^2 u_{yy} - bu_y = 0$

Boussinesq equation: $u_{tt} - c^2 u_{xx} - \mu u_{xxtt} = \frac{1}{2}(u^2)_{xx}$

Euler's equations of motion for a perfect inviscid fluid:

Euler's equations, in two space dimensions, for the motion of a perfect inviscid (nonviscous) fluid are

$$u_t + uu_x + vu_y + \frac{1}{\rho}p_x = 0,$$

$$v_t + uv_x + vv_y + \frac{1}{\rho}p_y = 0,$$

$$\rho_t + (\rho u)_x + (\rho v)_y = 0,$$

$$\left(\frac{p}{\rho^\gamma}\right)_t + u\left(\frac{p}{\rho^\gamma}\right)_x + v\left(\frac{p}{\rho^\gamma}\right)_y = 0,$$

where

$u(x, y, t) = x$-component of velocity at point (x, y), at time t

$v(x, y, t) = y$-component of velocity

$\rho(x, y, t) = $ density

$p(x, y, t) = $ pressure

$\gamma = $ constant, dependent upon fluid, $\gamma > 1$

The three-dimensional equations are analogous. For gases, γ is close to one, while for liquids, it can be much larger. In the latter case, we often set $\gamma = \infty$, allowing us to neglect the last equation. In this case, it can also be shown that ρ is a constant, in which case we refer to the fluid as *incompressible*. Note that, for an incompressible fluid, the third equation becomes

$$\nabla \cdot (u, v) = 0.$$

We note that the time derivatives in Euler's equations are *not* the standard time derivatives, in the sense that, here, they follow the motion of the fluid elements. Hence, we give them the special name **Eulerian derivatives**. Further, we note that we actually gave a system of only four equations, in five unknowns. The fifth equation is a relationship between density and pressure, called the **equation of state**.

Now, in the case of sound waves, we assume that the vibrations involved are small, from which it follows that the changes in ρ will be small. Under this assumption, we simplify Euler's equations, with the result being the

Linearized equations of acoustics:

$$u_t = c\rho_x, \quad v_t = c\rho_y, \quad \rho_t + \rho_0(u_x + v_y) = 0$$

Here, ρ_0 is the initial density, and c is a constant depending upon the initial density and pressure.

The simplest two-dimensional model for the flow of an incompressible *viscous* fluid is the set of

Navier–Stokes equations:

$$\rho u_t + \rho u u_x + \rho v u_y + p_x = \mu \nabla^2 u$$
$$\rho u_t + \rho u v_x + \rho v v_y + p_y = \mu \nabla^2 v$$
$$\nabla \cdot (u, v) = u_x + v_y = 0$$

We call μ the *coefficient of viscosity*.

D.7 Solitons

A **solitary wave** or **soliton**, famously first observed by J.S. Russell in England in 1834, is a single, lone wave which propagates without changing shape. It is a wave with particle-like behavior in that it is stable, localizable (we can say where it is) and possesses finite energy. Further, a collision of two solitons leads to the creation of new solitons and not to the break-up of the original waves. Because of this wave-particle duality, we should not be *too* surprised to see Schrödinger's equation showing up in this context, as well.

In each case, $u(x, t)$ is the shape of the wave at time t.

KdV (Kortweg–de Vries) equation:

$$u_t + \sigma u_{xxx} + c_0 \left(1 + \frac{3}{2} \frac{u}{h_0} \right) u_x = 0$$

This equation often is seen in its *canonical* form

$$u_t + 6 u u_x + u_{xxx} = 0.$$

Also, h_0 is the constant depth of the water channel. If h_0 is large, we may neglect the term $\frac{u}{h_0}$, leading to the

Linearized KdV equation:

$$u_t + c_0 u_x + \sigma u_{xxx} = 0$$

BBM (Benjamin–Bona–Mahony) equation:

$$u_t + u_x + u u_x - u_{xxt} = 0$$

Nonlinear or cubic Schrödinger's equation (canonical form):

$$i\psi_t + \psi_{xx} + \gamma|\psi|^2\psi = 0$$

(Note that $|\psi|^2\psi \neq \psi^3$ unless ψ is real!)

Sine-Gordon equation: $u_{tt} - u_{xx} + \sin u = 0$

Kadomtsev–Petviashvili equation (two dimensions):

$$(u_t + 6uu_x + u_{xxx})_x + u_{yy} = 0$$

D.8 Financial Mathematics—The Black–Scholes Equation

Prominent these days in the study of financial mathematics is the **Black–Scholes** model for options pricing. If we let

$$t = \text{time}$$
$$s = \text{market value of given asset}$$
$$\sigma = \text{(constant) volatility of the asset}$$
$$r = \text{(constant) interest rate}$$
$$v(s,t) = \text{value of option on the asset}$$

then v satisfies the

Black–Scholes equation: $v_t + \frac{\sigma^2 s^2}{2}v_{ss} + rsv_s - rv = 0$

for $s > 0, 0 \leq t \leq T$. Here, $t = 0$ represents the time when the asset is purchased, while $t = T$ is the time when the asset has reached maturity.

Much of the work done in financial mathematics entails developing numerical methods to deal with Black–Scholes and other equations. However, it *is* possible to reduce Black–Scholes to the heat equation.

D.9. Particle Physics

Klein–Gordon equation:

$$u_{tt} - c^2\nabla^2 u + m^2 u + gu^3 = 0$$

Linearized Klein–Gordon equation:

$$u_{tt} - c^2 \nabla^2 u + m^2 u = 0$$

In each case, m is the mass of the particle, c is the speed of light and g is a constant, as well.

Important, too, in the study of the physics of elementary particles are the **Dirac equation** and the **Yang–Mills equations** (from *gauge theory*).

D.10 Miscellaneous

Here we list a few other interesting PDEs.

Airy's equation: $u_t + u_{xxx} = 0$ (optics)

Eikonal equation: $|\nabla u|^2 = \frac{1}{c^2}$

(geometrical optics; c is a function of the space variables)

Minimal surface equation: $(1 + u_y^2)u_{xx} - 2u_x u_y u_{xy} + (1 + u_x^2)u_{yy} = 0$

(from the calculus of variations and related to the famous **Plateau's problem**)

Hamilton–Jacobi equation: $u_t + Hu = 0$

(from mechanics—H is a differential operator, in the space variables, known as the **Hamiltonian** of the system)

Bratu's equation: $\nabla^2 u + e^u = 0$

(which models spontaneous combustion and is very important in the theoretical study of nonlinear PDEs)

E

MATLAB Code for Figures and Exercises

SECTION 1.7

Figure 1.2

```
hold off;
% The hold off command clears any previous graphs.
% The domain interval and mesh:
x = 0:.01:12;
T = tan(x);
Y = -x;
Z = 0;
% Without the following two lines, the asymptotes will appear.
I = find(abs(T)>16);
T(I)=NaN;
plot(x,T);        % Plots y = tan x.
hold on;          % Keeps all plots on the same graph.
plot(x,Y,'-.');         % Plots y = -x.
plot(x,Z);        % Plots x-axis.
axis([0 12 -12 12]);
```

Figure 1.3

```
function sol = Fig1_3

% The following line makes all of the plots different: solid,
%     dash, dot, etc
set(0,'DefaultAxesColorOrder',[0 0 0],...
      'DefaultAxesLineStyleOrder','-|-.|--|:')

for n = 1:5

% This code generates all five graphs of figure 1.3, but each
%    will be solid
```

```
% In order to differentiate graphs using dashes and/or dots,
%    each graph must be done individually, then overlaid.

     % Here we provide an initial guess for the eigenvalue:

          lambda = (((2*n-1)^2)*(pi)^2)/4;
          solinit = bvpinit(linspace(0,1,10),@guess,lambda);
          sol = bvp4c(@odes,@bcs,solinit);
          xint = linspace(0,1,100);
          Sxint = deval(sol,xint);
          axis([0 1 -.3 .6]);


             plot(xint,Sxint(1,:));
             hold all;

end;


% The subroutine bvpinit needs an initial guess and
%    its derivative:
function v = guess(x)
v = [ sin(5*pi*x); 5*pi*cos(5*pi*x) ];


% The eigenvalue differential equation:
function dydx = odes(x,y,lambda)
dydx = [y(2); -(lambda)*y(1) ];


% The boundary conditions:
function res = bcs(ya,yb,lambda)
res = [ ya(1); yb(1)+yb(2); ya(2)-1];


% We compute the eigenvalues in Table 1.1 individually, rather
%    than in the for loop. (The eigenvalue will appear in the
%    command window.)
```

SECTION 3.2

Figure 3.2

```
x=0:.001:3*pi;
S = sin(2*x) + cos(4*x);
Y = 0.0;
plot(x,S);
hold on;
plot(x,Y);
axis([0 3*pi -3 2]);
```

Figure 3.5

```
x=-pi:.001:pi;
S = sin(2*x).*sin(4*x);
Y = 0;
plot(x,S);
hold on;
plot(x,Y);
axis([-pi pi -1 1]);
```

SECTION 3.4

Figure 3.7

```
x=1:.001:2;
F = x;
Y = 0;
plot(x,F);
hold on;
plot(x,Y);
x = 2:.001:3;
F = 1./(x-2);
plot(x,F);
axis([1 3 0 6]);
```

Figure 3.9

```
x = -2:.001:2;
F = sign(x).*abs(x.^(1/3));

% The function sign(x) is -1 if x < 0 and 1 if x > 0. Writing F
%   in this way allows MATLAB to select the real (and not a
%   complex) cube root.

Y = 0
plot(x,F);
hold on;
plot(x,Y);
axis([-2 2 -2 2]);
```

SECTION 3.5

Figure 3.13

```
t = -3*pi:.001:3*pi;
fs = (1/2) * ones(size(t));
```

```
% Here we use only n = 40.

for n = 1:40
            fs = fs + (2/pi)*sin((2*n-1)*t) / (2*n-1);
end;
plot(t,fs);
```

SECTION 4.1

Figure 4.1

```
x = 0:.001:pi;
u = (0) * ones(size(x));

for t = 0:4
        u = 0;
            for k = 1:10
                u = u + (8/pi) * (exp(-(2*k-1)*(2*k-1)*t)) *
sin((2*k-1)*x)/(2*k-1)/(2*k-1)/(2*k-1);

% This last statement must be entered on one line.

                hold on;
        end;
        plot(x,u);
        hold all;
end;
```

Figure 4.2

```
x = 0:.01:pi;
u = (3) * ones(size(x));

% The following line makes all the plots different: solid,
%    dash, dot, etc.

    set(0,'DefaultAxesColorOrder',[0 0 0],...
       'DefaultAxesLineStyleOrder','-|-.|--|:')
axis([0 3.15 1.5 4.2]);

for t = 0:4
        u = 3;
            for k = 1:50
                u = u - (12/(pi^2)) * exp((-(2*k-1)*(2*k-1)*
(pi^2)*t)/9) * (cos((k*(pi)*x)/3))/
```

```
((2*k-1)^2);
```

```
% Again, the last statement must be entered on one line.
```

```
        hold all;
        end;
      plot(x,u);
      hold on;
end;
```

SECTION 5.1

Figure 5.4

```
x=(-10:.2:10)';
t=(0:.2:5)';
[x,t] = meshgrid(x,t);
R=x-(2./3).*t;
z= sin(R);
mesh (x,t,z,'EdgeColor','black');
```

SECTION 6.1

Figure 6.1

```
x = 0:.01:2;
Y = erf(x);
Z= erfc(x);
plot(x,Y);
hold on;
plot(x,Z,'-.');
axis([0 2 -.1 1.1]);
```

SECTION 6.3

Figure 6.2

```
x = -10:.01:10;
Y = exp(-4.*x.^2);
Z= (1/(2.*sqrt(2))).*exp(-(x.^2)/16);
plot(x,Y);
hold on;
plot(x,Z,':');
axis([-10 10 0 1.2]);
```

SECTION 6.4

Figure 6.4

```
% f = 1/(4*a*sqrt(pi*t)) * e^((-x-y)^2)/4t
% This particular code is for t=5.


f = '(1./(4.*1.*sqrt(pi.*5))).*exp((-(x-y).^2)./(4.*5))';
F = inline(f,'y','x');

a = 1;
for x = -10:.02:10;
          u = quad(F,-a,a,[],[],x);
          plot(x,u);
          axis([-10 10 0 .28]);
          hold on;
end;
hold all;
% Run the program for t=1, t=3 and t=5. As usual, the hold on
%    command will keep each graph. The result will be a grid
%    with three graphs on it, one for each value of t.
```

Figure 6.5

```
% This particular code is for a=1.

f = '(1./(4.*1.*sqrt(pi.*1))).*exp((-(x-y).^2)./(4.*1))';

% a must be changed above. For example, for a=3,
%    f = '(1./(4.*3.*sqrt(pi.*1))).*
%    exp((-(x-y).^2)./(4.*1))';

F = inline(f,'y','x');

a = 1;
for x = -10:.02:10;
          u = quad(F,-a,a,[],[],x);
          plot(x,u);
          axis([-10 10 0 .28]);
          hold on;
end;
```

SECTION 7.2

Figure 7.1a

```
x = -1:.05:1;
P0 = ones(size(x));
P2 = .5*(3*x.^2 - 1);
```

```
P4 = .125*(35*x.^4 - 30*x.^2 + 3);
plot (x,P0,'k');
hold on;
plot (x,P2,'k-.');
plot (x,P4,'k--');
axis([-1 1 -.6 1.2]);
hold off;
```

SECTION 7.4

Figure 7.2

```
x = -4:.001:5;
y = gamma(x);
plot(x,y);
hold on;
axis([-4.4 5 -10 10]);
```

SECTION 7.5

Figure 7.3

```
axis([0 20 -1 1])
x=0:0.01:20;
for k = 0:2
    y= besselj(k,x);
  hold on;
    if k == 0
       plot(x,y,'k');
    end;
    if k == 1
       plot(x,y,'k:')
    end;
    if k == 2
       plot(x,y,'k--')
    end;
end;
```

Figure 7.4

```
axis([0 20 -2 1])
x=0:0.01:20;
for k = 0:2
    y= bessely(k,x);
    hold on;
    if k == 0
```

```
        plot(x,y,'k');
    end;
    if k == 1
        plot(x,y,'k:')
    end;
    if k == 2
        plot(x,y,'k--')
    end;
end;
```

Figure 7.5

```
axis([0 4 0 10])
x=0:.1:4;
for k = 0:2
    y= besseli(k,x);
    hold on;
    if k == 0
        plot(x,y,'k');
    end;
    if k == 1
        plot(x,y,'k:')
    end;
    if k == 2
        plot(x,y,'k--')
    end;
end;
```

Figure 7.6

```
axis([0 2 0 10])
x=0:.01:4;
for k = 0:2
    y= besselk(k,x);
    hold on;
    if k == 0
        plot(x,y,'k');
    end;
    if k == 1
        plot(x,y,'k:')
    end;
    if k == 2
        plot(x,y,'k--')
    end;
end;
```

SECTION 9.2

Figure 9.4

```
[X,Y] = meshgrid([0:.1:2],[0:.1:3]);
colormap([1 1 1]);

subplot(3,3,1)
V = sin((pi.*X)./2).*sin((pi.*Y)/3);
surf(X,Y,V);

subplot(3,3,2)
V = sin((pi.*X)./2).*sin((2.*pi.*Y)/3);
surf(X,Y,V);

subplot(3,3,3)
V = sin((pi.*X)./2).*sin((pi.*Y));
surf(X,Y,V);

subplot(3,3,4)
V = sin((pi.*X)).*sin((pi.*Y)/3);
surf(X,Y,V);

subplot(3,3,5)
V = sin((pi.*X)).*sin((2.*pi.*Y)/3);
surf(X,Y,V);

subplot(3,3,6)
V = sin((pi.*X)).*sin((pi.*Y));
surf(X,Y,V);

subplot(3,3,7)
V = sin((3.*pi.*X)./2).*sin((pi.*Y)/3);
surf(X,Y,V);

subplot(3,3,8)
V = sin((3.*pi.*X)./2).*sin((2.*pi.*Y)/3);
surf(X,Y,V)

subplot(3,3,9)
V = sin((3.*pi.*X)./2).*sin((pi.*Y));
surf(X,Y,V)
```

Figure 9.6

```
[X,Y] = meshgrid([0:.05:1],[0:.05:1]);
```

```
V1 = sin((pi.*X)).*sin((2.*pi.*Y));
V2 = sin((2.*pi.*X)).*sin((pi.*Y));

colormap([1 1 1]);

subplot(1,2,1)
surf(X,Y,V1);
axis normal;
view(-37.5,16);

subplot(1,2,2)
surf(X,Y,V2);
axis normal;
view(-37.5,16);
```

Figure 9.8

```
[X,Y] = meshgrid([0:.05:1],[0:.05:1]);
V1 = sin((pi.*X)).*sin((2.*pi.*Y));
V2 = sin((2.*pi.*X)).*sin((pi.*Y));

colormap([1 1 1]);

subplot(1,2,1)
V = -V1 - V2;
surf(X,Y,V);
axis normal;
view(15,15);

subplot(1,2,2)
V = V1 - V2;
surf(X,Y,V);
axis normal;
view(15,15);
```

SECTION 9.4

Figure 9.11

We would like to thank Udaak Z. George, and her adviser, Dr. Anotida
Madzvamuse, University of Sussex, UK, for pointing out that this figure was
incorrect in the first edition, and for providing us with the correct MATLAB
code.

```
[Th,r] = meshgrid([0:10:360]*pi/180,[0:.1:1]);
```

```
[X,Y] = pol2cart(Th,r);
colormap([1,1,1]);

subplot(3,3,1)
V1 = besselj(0,2.40483.*sqrt(X.^2+Y.^2)).*cos(0*Th);
surf(X,Y,V1);
hold on;
hold off;

subplot(3,3,2)
V2 = besselj(1,3.83171.*sqrt(X.^2+Y.^2)).*cos(1*Th);
surf(X,Y,V2);
hold on;
hold off;

subplot(3,3,3)
V3 = besselj(2,5.13562.*sqrt(X.^2+Y.^2)).*cos(2*Th);
surf(X,Y,V3);

subplot(3,3,4)
V4 = besselj(0,5.52008.*sqrt(X.^2+Y.^2)).*cos(0*Th);
surf(X,Y,V4);

subplot(3,3,5)
V5 = besselj(1,7.01559.*sqrt(X.^2+Y.^2)).*cos(1*Th);
surf(X,Y,V5);

subplot(3,3,6)
V6 = besselj(2,8.41724.*sqrt(X.^2+Y.^2)).*cos(2*Th);
surf(X,Y,V6);

subplot(3,3,7)
V7 = besselj(0,8.65373.*sqrt(X.^2+Y.^2)).*cos(0*Th);
surf(X,Y,V7);

subplot(3,3,8)
V8 = besselj(1,10.17347.*sqrt(X.^2+Y.^2)).*cos(1*Th);
surf(X,Y,V8);

subplot(3,3,9)
V9 = besselj(2,11.61984.*sqrt(X.^2+Y.^2)).*cos(2*Th);
surf(X,Y,V9);
```

SECTION 9.5

Figure 9.16

```
axis([0 10 -.4 1])
x=0:0.01:10;

k = 0;
Y= sqrt(pi./(2.*x)).*besselj(k+.5,x);
hold on;
plot(x,Y,'k-');

k = 1;
Y= sqrt(pi./(2.*x)).*besselj(k+.5,x);
hold on;
plot(x,Y,'k:');

k = 2;
Y= sqrt(pi./(2.*x)).*besselj(k+.5,x);
hold on;
plot(x,Y,'k-.');


k = 3;
Y= sqrt(pi./(2.*x)).*besselj(k+.5,x);
hold on;
plot(x,Y,'k--');
```

SECTION 10.3

Figure 10.4

```
subplot(1,2,1)
[X,Y] = meshgrid([-3:.01:3],[-3:.01:3]);
Z = (-1./(4.*pi)).*log(X.^2 + Y.^2);
surf(X,Y,Z);
axis([-3 3 -3 3 -.5 1]);
view(-24, 19);

subplot(1,2,2)
[X,Y] = meshgrid([-3:.15:3],[-3:.15:3]);
colormap([1,1,1]);
Z = (-1./(4.*pi)).*log(X.^2 + Y.^2);
surf(X,Y,Z);
axis([-3 3 -3 3 -.5 1]);
view(-24, 19);
```

SECTION 10.4

Figure 10.7a

```
[X,Y] = meshgrid([-5:.14:5],[-5:.14:5]);
colormap([1,1,1]);
Z = (-1/4).*bessely(0,sqrt(X.^2+Y.^2));
surf(X,Y,Z);
```

Figure 10.7b

```
[X,Y] = meshgrid([-10:.2:10],[-10:.2:10]);
colormap([1,1,1]);
Z = (-1/4).*bessely(0,sqrt(X.^2+Y.^2));
surf(X,Y,Z);
```

SECTION 11.1

Table 11.2

```
% n = number of rows of A = number of columns of A.

n = 9;

% Read zeros into coefficient matrix A:

A = zeros(n,n);

% Main diagonal of A:

for i = 1:n;
    for j = 1:n;
        if i == j;
            A(i,j) = (i./(n.^3)) - 2;
        end;
    end;
end;

% Diagonals above and below main diagonal:

for k = 1:n;
    for m = 1:n-1;
        if k == m+1
            A(k,m) = 1;
    A(m,k) = 1;
        end;
    end;
```

```
end;

% Read in column vector B, the right side of the matrix
%    equation:

B = zeros(n,1);

for k = 1:n
    if k == n
        B(k,1) = -1;
    end;
end;

% Compute A-inverse:.

C = inv(A);

% Left-multiply B by A-inverse. D is the solution vector.

D = C*B;

fprintf('The solution matrix is: \n');
fprintf('\n');

for a = 1:n
    fprintf('%g\n',D(a,1));
end;

% D will appear in the command window. You can check
%    your matrices in workspace.
%The solution matrix is:

%0.224487
%0.336037
%0.446205
%0.553924
%0.657844
%0.756349
%0.847592
%0.929534
```

F

Answers to Selected Exercises

CHAPTER 1

Section 1.1

5.   d) $a^2 + b^2 = 1$

6.   a) $u(x, y) = f(y)$, where f is an arbitrary function

   b) infinitely many solutions (need $f(0) = 0$)

   c) $u(x, y) = y^2 - \cos y$

   d) no solution

Section 1.2

1. $u(x, y) = 2xy + f(x)$

3. $u(x, y, z) = -\cos x + x \cos y + f(y, z)$

5. $u(x, y) = \frac{x^2 y}{2} - \frac{xy^2}{2} + f(x) + g(y)$

7. $u(x, y) = -\frac{y^2}{8} \sin 2x + y f(x) + g(x) + x h(y) + k(y)$

9. $u(x, y, z) = z f(x, y) + g(x, y) + h(x, z) + k(y, z)$

11. $u(x, y) = e^{4y} f(x)$

13. $u(x, y) = e^{xy^2} f(y)$

15. $u(x, y) = \frac{2}{x} + e^{-xy} f(x), x \neq 0$

17. $u(x, y) = e^{-2x} f(y) + e^x g(y)$

19.   a) $u(x, y) = e^{2x} y^2$

   c) no solution

Section 1.3

   1. yes

   4. no

   5. $\sin 2L \neq 0$

Section 1.4

   1. nonlinear

   5. linear

Section 1.5

   9. $u_p = -\cos x + x \cos y$

   11. $u_p = -3$

Section 1.6

   1. $3X' + \lambda X = 2Y' + \lambda Y = 0$

   3. $X' + \lambda x^2 X = Y' - \lambda y^2 Y = 0$

   5. $X'' + \lambda X = T'' + \lambda T = 0$

   7. $X'' - X' + \lambda X = 0 = 2Y'' + 3Y' - \lambda Y = 0$

   9. $X'' + (\lambda - x^2)X = T' + i\lambda T = 0$

   11. $r^2 R'' + rR' - \lambda R = \Theta'' + \lambda\Theta = 0$

   13. $X^{(4)} + \lambda X = T'' - \lambda T = 0$

   15. $X' + \lambda_1 X = Y' - \lambda_2 Y = Z' + (1 - \lambda_1 + \lambda_2)Z = 0$

   17. $x^2 X' + \lambda_1 X = y^3 Y' + \lambda_2 Y = 4zZ' + (\lambda_1 - \lambda_2)Z = 0$

   19. $X'' + \lambda_1 X = Y'' + (\lambda_2 - \lambda_1)Y = T'' + \lambda_2 T = 0$

   22. $u_\lambda = ce^{-\lambda(\frac{x}{3} + \frac{y}{2})}$

   24. $u_\lambda = ce^{\frac{\lambda}{3}(y^3 - x^3)}$

   26. $\lambda > 0$: $u_\lambda = (c_1 \cos\sqrt{\lambda}\,x + c_2 \sin\sqrt{\lambda}\,x)(c_3 \cos\sqrt{\lambda}\,t$
      $+c_4 \sin\sqrt{\lambda}\,t)$,
     $\lambda = 0$: $u_0 = (c_1 x + c_2)(c_3 t + c_4)$,
     $\lambda < 0$: $u_\lambda = (c_1 \cosh\sqrt{-\lambda}\,x + c_2 \sinh\sqrt{-\lambda}\,x)(c_3 \cosh\sqrt{-\lambda}\,t$
      $+c_4 \sinh\sqrt{-\lambda}\,t)$

28. $\lambda < -\frac{9}{8}$: $u_\lambda = (c_1 \cosh \sqrt{1-\lambda}\,x + c_2 \sinh \sqrt{1-\lambda}\,x)e^{-\frac{3y}{4}}$
$(c_1 \cos \frac{\sqrt{-9-8\lambda}}{4}y - c_2 \sin \frac{\sqrt{-9-8\lambda}}{4}y)$,
$\lambda = -\frac{9}{8}$: $u_{-9/8} = (c_1 \cosh \frac{17x}{8} + c_2 \sinh \frac{17x}{8})e^{-\frac{3y}{4}}(c_3 y + c_4)$,
$-\frac{9}{8} < \lambda < 1$: $u_\lambda = (c_1 \cosh \sqrt{1-\lambda}\,x + c_2 \sinh \sqrt{1-\lambda}\,x)e^{-\frac{3y}{4}}$
$(c_3 e^{\frac{\sqrt{9+8\lambda}\,y}{4}} + c_4 e^{-\frac{\sqrt{9+8\lambda}\,y}{4}})$,
$\lambda = 1$: $u_1 = (c_1 x + c_2)e^{-\frac{3y}{4}}(c_3 e^{\frac{\sqrt{17}\,y}{4}} + c_4 e^{-\frac{\sqrt{17}\,y}{4}})$,
$\lambda > 1$: $u_\lambda = (c_1 \cos \sqrt{\lambda-1}\,x + c_2 \sin \sqrt{\lambda-1}\,x)e^{-\frac{3y}{4}}(c_3 e^{\frac{\sqrt{9+8\lambda}\,y}{4}} +$
$c_4 e^{-\frac{\sqrt{9+8\lambda}\,y}{4}})$

29. $\lambda > 0$: $u_\lambda = (c_1 r^{\sqrt{\lambda}} + c_2 r^{-\sqrt{\lambda}})(c_1 \cos \sqrt{\lambda}\,\theta + c_2 \sin \sqrt{\lambda}\,\theta)$,
$\lambda = 0$: $u_0 = (c_1 + c_2 \ln r)(c_3 \theta + c_4)$,
$\lambda < 0$: $u_\lambda = (c_1 \cos \ln \sqrt{-\lambda}\,r + c_2 \sin \ln \sqrt{-\lambda}\,r)(c_3 \cosh \sqrt{-\lambda}\,\theta +$
$c_4 \sinh \sqrt{-\lambda}\,\theta)$

31. $u_\lambda = ce^{-\lambda_1 x + \lambda_2 y + (\lambda_1 - \lambda_2 - 1)z}$

37. a) $u = c_1(x - y) + c_2$

38. a) $X(0) = 0$, or $T(t) = 0$ for all t

39. a) no

Section 1.7

1. $\lambda_n = \frac{n^2 \pi^2}{25}, y_n = \sin \frac{n\pi x}{5}, n = 1, 2, \ldots$

3. $\lambda_n = \frac{(2n-1)^2}{4}, y_n = \cos \frac{(2n-1)x}{2}, n = 1, 2, \ldots$

5. $\lambda_0 = -1, y_0 = \cosh x + \sinh x$;
$\lambda_n = n^2 \pi^2, y_n = n\pi \cos n\pi x + \sin n\pi x, n = 1, 2, \ldots$

7. $\lambda_n = \frac{n^2 \pi^2}{4} + 1, y_n = x^{-1} \sin \frac{n\pi}{2} \ln x, n = 1, 2, \ldots$

9. $\lambda_{2k-1} = \frac{(2k-1)^2 \pi^2}{4}, y_{2k-1} = \cos \frac{(2k-1)\pi x}{2}, k = 1, 2, \ldots$
$\lambda_{2k} = k^2 \pi^2, y_{2k} = \sin k\pi x, k = 1, 2, \ldots$

11. $\lambda_n = -n^4 \pi^4, y_n = \sin n\pi x, n = 1, 2, \ldots$

13. λ_n: $\sqrt{\lambda_n} = \tan 2\sqrt{\lambda_n}, y_n = \sin \sqrt{\lambda_n}\,x, n = 1, 2, \ldots; \lambda_n > 0$

16. a) $\lambda_n = \frac{n^2 \pi^2}{L^2}, y_n = \sin \frac{n\pi x}{L}, n = 1, 2, \ldots$
b) $\lambda_n = \frac{(2n-1)^2 \pi^2}{4L^2}, y_n = \sin \frac{(2n-1)\pi x}{2L}, n = 1, 2, \ldots$
c) $\lambda_n = \frac{(2n-1)^2 \pi^2}{4L^2}, y_n \cos \frac{(2n-1)\pi x}{2L}, n = 1, 2, \ldots$
d) $\lambda_0 = 0, y_0 = 1; \lambda_n = \frac{n^2 \pi^2}{L^2}, y_n = \cos \frac{n\pi x}{L}, n = 1, 2, \ldots$

18. $u_n(x, t) = \sin n\pi x(c_n \cos n\pi t + d_n \sin n\pi t), n = 1, 2, \ldots$

CHAPTER 2

Section 2.2

1.   a) $\frac{\text{distance}^2}{\text{time}}$

   c) $\frac{\text{calories}}{\text{distance}\cdot\text{time}\cdot\text{degrees}}$

2.   b) $u_t = \frac{.63}{(.215)(2.7)} u_{xx}$

11.   a) $\sigma^2 s^2 S''(s) + 2rsS'(s) + 2(\lambda - r)S(s) = T'(t) - \lambda T(t) = 0$

Section 2.4

1.   a) $u(x,0) = 20, u_x(5,t) = 0, t > 0,$
   $u(x,0) = 50, 0 \le x \le 5,$
   $u(0,t) = 20, u_x(5,t) = 0, t > 0$

3.   a) $u(x,t) = \frac{T_2 - T_1}{L} x + T_1$

4.   a) $u_{tt} = u_{xx}, 0 < x < 8, t > 0,$

$$u(x,0) = \begin{cases} x/2, & \text{if } 0 \le x \le 4, \\ 4 - x/2, & \text{if } 4 \le x \le 8, \end{cases}$$
   $u_t(x,0) = 0,$
   $u(0,t) = u(8,t) = 0$

7.   a) $v(x) = 4x + 10, w(x,0) = f(x) - 4x - 10$

9. Fourier's Law

Section 2.5

8. $u(r) = c_1 + c_2 \ln r$

Section 2.6

3. $u(x,t) = 3 + e^{-4\pi^2 t} \cos 2\pi x$

5. $u(x,t) = \sum_{n=1}^{\infty} c_n e^{-\frac{\alpha^2 n^2 \pi^2 t}{L^2}} \sin \frac{n\pi x}{L}$

7. $u(x,t) = \sum_{n=1}^{\infty} \sin n\pi x (c_n \cos n\pi t + d_n \sin n\pi t)$

9. $u(x,t) = 5 \sin 2x \cos 4t - 7 \sin 4x \cos 8t$

10.   b) $u(x,y) = \sin 3x \cosh 3y + \frac{1}{\sinh 1} \sin x \sinh y - \frac{\cosh 3}{\sinh 3} \sin 3x \sinh 3y$

CHAPTER 3

Section 3.2

1. fundamental period $= 4$

3. not periodic

9. neither

17.   a) $\frac{1024}{9}$

Section 3.3

1. $F(x) = 1 - \frac{4}{\pi} \sum\limits_{k=1}^{\infty} \frac{1}{2k-1} \sin(2k-1)\pi x$

3. $F(x) = \frac{10}{\pi} \sum\limits_{n=1}^{\infty} \frac{(-1)^{n+1}}{n} \sin \frac{n\pi x}{5}$

5. $F(x) = \frac{1}{3} + \frac{4}{\pi^2} \sum\limits_{n=1}^{\infty} \frac{(-1)^n}{n^2} \cos n\pi x$

7. $F(x) = \frac{1}{2} - \frac{4}{\pi^2} \sum\limits_{k=1}^{\infty} \frac{1}{(2k-1)^2} \cos(2k-1)\pi x$

9. $F(x) = \frac{1}{4} - \frac{1}{\pi} \sum\limits_{n=1}^{\infty} \frac{1}{n} \left\{ \sin \frac{n\pi}{2} \cos nx + \left[(-1)^n - \cos \frac{n\pi}{2} \right] \sin nx \right\}$

11. $F(x) = f(x)$

13.   c) neither, neither

Section 3.4

1. continuous and piecewise smooth

3. continuous, not piecewise smooth

5. none

7.   b) continuous, not piecewise smooth

13.

15. false

17. true

Section 3.5

7. a) each $= 5$
    c) $f'_L(0) = f'(0-) = f'(0+) = 0$, $f'_R(0)$ does not exist

9. a) false
    c) false

Section 3.6

1. $F_s(x) = \frac{2}{\pi} \sum\limits_{n=1}^{\infty} \frac{1-\cos\frac{n\pi}{2}}{n} \sin\frac{n\pi x}{4}$, $F_c(x) = \frac{1}{2} + \frac{2}{\pi} \sum\limits_{k=1}^{\infty} \frac{(-1)^{k+1}}{2k-1} \cos\frac{(2k-1)\pi x}{4}$

3. $F_s(x) = \sin x$, $F_c(x) = \frac{2}{\pi} + \frac{4}{\pi} \sum\limits_{k=1}^{\infty} \frac{1}{1-4k^2} \cos 2kx$

9. true

10. false

CHAPTER 4

Section 4.1

1. a) $u(x,t) = \frac{80}{\pi} \sum\limits_{k=1}^{\infty} \frac{1}{2k-1} e^{-2(2k-1)^2 t} \sin(2k-1)x$

    c) $u(x,t) = \frac{40}{\pi} \sum\limits_{n=1}^{\infty} \frac{1}{n} \left(1 - \cos\frac{n\pi}{2}\right) e^{-\frac{n^2\pi^2 t}{2}} \sin\frac{n\pi x}{2}$

2. a) $u(x,t) = \frac{\pi^2}{3} + 4 \sum\limits_{n=1}^{\infty} \frac{(-1)^n}{n^2} e^{-4n^2 t} \cos nx$

3. a) $u(x,t) = \frac{400}{\pi} \sum\limits_{n=1}^{\infty} \frac{1}{2n-1} e^{-\frac{(2n-1)^2 t}{4}} \sin\frac{(2n-1)x}{2}$

5. a) $u(x,t) = \sum\limits_{n=1}^{\infty} b_n e^{-\frac{\alpha^2 n^2 \pi^2 t}{L^2}} \sin\frac{n\pi x}{L}$, $b_n = \frac{2}{L} \int_0^L f(x) \sin\frac{n\pi x}{L} dx$

    c) $u(x,t) = \sum\limits_{n=1}^{\infty} c_n e^{-\frac{\alpha^2 (2n-1)^2 \pi^2 t}{L^2}} \sin\frac{(2n-1)\pi x}{2L}$,
    $c_n = \frac{2}{L} \int_0^L f(x) \sin\frac{(2n-1)\pi x}{2L} dx$

    e) $u(x,t) = \sum\limits_{n=1}^{\infty} c_n e^{-\alpha^2 k_n^2 t} \sin k_n x$, $k_n = n^{\text{th}}$ positive zero of $k+\tan kL$,
    $c_n = \frac{2}{L} \int_0^L f(x) \sin k_n x \, dx$

7. $u(x,t) = \sum\limits_{n=1}^{\infty} b_n e^{-(1+n^2)t} \sin nx$, $b_n = \frac{2}{\pi} \int_0^{\pi} f(x) \sin nx \, dx$

9. b) conservation of heat energy

Section 4.2

1. a) $u(x,t) = 3\sin 2x \cos 2\sqrt{5}\, t + \frac{1}{\sqrt{5}} \sin x \sin \sqrt{5}\, t - \frac{7}{4\sqrt{5}} \sin 4x \sin 4\sqrt{5}\, t$

   c) $u(x,t) = \frac{24}{\sqrt{5}\,\pi^2} \sum\limits_{k=1}^{\infty} \frac{1}{(2k-1)^2} \sin \frac{(2k-1)\pi x}{2} \sin \frac{(2k-1)\pi\sqrt{5}\, t}{2}$

2. b) $u(x,t) = \frac{2}{\pi} + \frac{4}{\pi} \sum\limits_{k=1}^{\infty} \frac{1}{1-4k^2} \cos 2kx \cos 4kt$

3. a) $u(x,t) = \frac{8}{\pi^2} \sum\limits_{n=1}^{\infty} \frac{1}{(2n-1)^2} \sin \frac{(2n-1)\pi x}{2} \sin \frac{(2n-1)\pi t}{2}$

4. a) $u(x,t) = \sum\limits_{n=1}^{\infty} \sin \frac{n\pi x}{L} \left[c_n \cos \frac{n\pi ct}{L} + d_n \sin \frac{n\pi ct}{L} \right]$,

   $c_n = \frac{2}{L} \int_0^L f(x) \sin \frac{n\pi x}{L} dx$, $d_n = \frac{2}{n\pi c} \int_0^L g(x) \sin \frac{n\pi x}{L} dx$

   c) $u(x,t) = \sum\limits_{n=1}^{\infty} \sin \frac{(2n-1)\pi x}{2L} \left[c_n \cos \frac{(2n-1)\pi ct}{2L} + d_n \sin \frac{(2n-1)\pi ct}{L} \right]$, $c_n =$

   $\frac{2}{L} \int_0^L f(x) \sin \frac{(2n-1)\pi x}{2L} dx$, $d_n = \frac{4}{(2n-1)\pi c} \int_0^L g(x) \sin \frac{(2n-1)\pi x}{2L} dx$

5. a) $u(x,t) = \frac{4}{\pi} e^{-2t} \Big[\sin x \cosh \sqrt{3}\, t + \sum\limits_{k=2}^{\infty} \frac{1}{2k-1} \sin(2k-1)x$

   $\cdot \cos \sqrt{4k^2 - 4k - 3}\, t \Big]$

   c) $u(x,t) = e^{-t} \Big[t\sin x + \frac{1}{\sqrt{3}} \sin 2x \sin \sqrt{3}\, t \Big]$

9. b) Each overtone's frequency is an integral multiple of the fundamental frequency, but not every integral multiple of the fundamental gives an overtone.

   c) xylophone, e.g.

Section 4.3

1. $u(x,y) = \frac{40}{\pi} \sum\limits_{k=1}^{\infty} \frac{\sin(2k-1)\pi x \sinh(2k-1)\pi y}{(2k-1)\sinh 2(2k-1)\pi}$

3. $u(x,y) = 2 \sum\limits_{n=1}^{\infty} \frac{(-1)^{n+1}}{n} \sin ny (\cosh nx - \coth n \sinh nx)$

5. $u(x,y) = \frac{4}{\pi} \sum\limits_{n=1}^{\infty} \frac{(-1)^{n+1}}{n} \sin \frac{n\pi y}{2} \left[\cosh \frac{n\pi x}{2} + \frac{1}{\sinh \frac{n\pi}{2}} \left(2 - \cosh \frac{n\pi}{2}\right) \sinh \frac{n\pi x}{2} \right]$

6. $u(x,y) = \frac{3}{\cosh 5} \sin y \sinh x - \frac{5}{4\cosh 20} \sin 4y \sinh 4x$

9. no solution

11.   a) $u(x,y) = \frac{9}{\pi} \sum\limits_{n=1}^{\infty} \sin \frac{n\pi x}{3} \frac{(-1)^{n+1}}{n} \left[\cosh \frac{n\pi y}{3} \right.$

$\left. + \left(2 \operatorname{csch} \frac{2n\pi}{3} - \coth \frac{2n\pi}{3} \right) \sinh \frac{n\pi y}{3} \right]$

$- \frac{72}{\pi^3} \sum\limits_{k=1}^{\infty} \frac{1}{(2k-1)^3} \operatorname{csch} \frac{2(2k-1)\pi}{3} \sinh \frac{(2k-1)\pi y}{3} \sin \frac{(2k-1)\pi x}{3}$

12.   a) $u = u_1 + u_2$, where $u_1 =$ solution of Exercise 1 and $u_2 =$ solution of Exercise 5

13.   a) $u(x,y) = \sum\limits_{n=1}^{\infty} \sin \frac{n\pi x}{a} \left(c_n \cosh \frac{n\pi y}{a} + d_n \sinh \frac{n\pi y}{a} \right)$,

where $c_n = \frac{2}{n\pi} \int_0^a f(x) \cos \frac{n\pi x}{a} dx$, $c_n \cosh \frac{n\pi b}{a} + d_n \sinh \frac{n\pi b}{a}$
$= \frac{2}{n\pi} \int_0^a g(x) \cos \frac{n\pi x}{a} dx$; must have $\int_0^a f(x)dx = \int_0^a g(x)dx = 0$

Section 4.4

1.   b) $w(x,t) = u(x,t) - ax - T$; $w_{tt} = w_{xx}$, $w(x,0) = f(x) - ax - T$,
$w_t(x,0) = g(x)$, $w(0,t) = w_x(L,t) = 0$

5. $u(x,t) = \frac{40}{\pi} \sum\limits_{k=1}^{\infty} \frac{1}{(2k-1)^3} [1 - e^{-(2k-1)^2 t}] \sin(2k-1)x + 3e^{-t} \sin x$
$-4e^{-4t} \sin 2x + 5e^{-9t} \sin 3x$

7. $u(x,t) = \frac{2}{\pi} \sum\limits_{n=1}^{\infty} \frac{(-1)^{n+1}}{n^3} \left[1 + (n^2 - 1)e^{-n^2 t} \right] \sin nx$
$+ \frac{4}{\pi} \sum\limits_{k=1}^{\infty} \frac{1}{(2k-1)} \left[t + \frac{1}{(2k-1)^3} \{ e^{-(2k-1)^2 t} - 1 \} \right] \sin(2k-1)x$

9. $u(x,t) = -15 + 10t$

11. $u(x,t) = \sin x(1 - \cos t) + \sin 3x \cos 3t + \frac{1}{5} \sin 5x \sin 5t$

17. $u(x,y) = 4\pi \sum\limits_{n=1}^{\infty} \frac{(-1)^{n+1}}{n^5} \sin nx [\cosh ny + (\operatorname{csch} n\pi - \coth n\pi) \sinh ny]$
$+ 2\pi^3 \sum\limits_{n=1}^{\infty} \frac{(-1)^{n+1}}{n^3} \operatorname{csch} n\pi \sin nx \sinh ny$
$+ \frac{16}{\pi} \sum\limits_{k=1}^{\infty} \frac{1}{(2k-1)^7} \sin(2k-1)x \{ -\cosh(2k-1)y$
$+ [\coth(2k-1)\pi - \operatorname{csch}(2k-1)\pi] \sinh(2k-1)y \}$
$- 8 \sum\limits_{k=1}^{\infty} \frac{\operatorname{csch}(2k-1)\pi}{(2k-1)^5} \sin(2k-1)x \sinh(2k-1)y$

CHAPTER 5

Section 5.1

1. $u(x,y) = 4 \sin \frac{5}{7}(7x + 5y)$

3. no solution; $y = -2x + 4$ is a characteristic

5. $u(x, y) = (x + y - 3)^2 e^{2(y-3)}$

7. $u(x, y) = 3x^2(y + 1) - \frac{1}{2}x^3 + 4(x - 2y - 2) + \frac{1}{2}(x - 2y - 2)^3$

9. $u(x, y) = y - 1 + [1 + x - y + \sin(y - x)]e^{-x}$

11. speed $= 2/3$

15. $u(x, y) = (x - y)^2/4$

19. can solve $\Leftrightarrow 2A + 3B \neq 0$

Section 5.2

1. $u(x, y) = \frac{(x^2 + y^2)^2}{25}$; characteristics: $x^2 + y^2 = c$

3. $u(x, y) = \sin 3(y - x^3)$; characteristics: $y = x^3 + c$

8. sine wave spreads out as $t \to \infty$

9. $u(x, y, z) = \frac{1}{3}(2x - y)(z - 3x) + \frac{2}{9}(z - 3x)^2$

11. $u(x, y, z) = (x^3 - y)(x + z)$

13. b) $u(6, 8) = \frac{12}{17}, u(0, 5) = 0$

    c) $u(x, t) = \frac{2x}{1 + 2t}$

Section 5.3

3. $u(x, t) = \frac{3}{2}[e^{-(x+t)^2} + e^{-(x-t)^2}]$

11. $u(5, 5) = 1, u(10, 6) = 0$

Section 5.4

1. $u(x, t) = \begin{cases} \frac{3}{2}(e^{x-2t} + e^{-x-2t}), & \text{if } x \geq 2t, \\ \frac{3}{2}(e^{-x-2t} - e^{x-2t}), & \text{if } x < 2t \end{cases}$

11. $u(x, t) = \begin{cases} \frac{1}{2}[f(x + ct) + f(x - ct)] + \frac{1}{2c}\int_{x-ct}^{x+ct} g(z)\,dz, & \text{if } x \geq ct, \\ \frac{1}{2}[f(x + ct) + f(ct - x)] \\ \qquad + \frac{1}{2c}\left[\int_0^{x+ct} g(z)\,dz + \int_0^{ct-x} g(z)\,dz\right], & \text{if } x < ct \end{cases}$

Section 5.5

1. parabolic, $u(x, y) = x\phi(4x - y) + \psi(4x - y)$

3. hyperbolic, $u(x, y) = \phi(x + 2y) + \psi(x - 2y)$

5. elliptic

7. hyperbolic, $u(x, y) = \phi(3x + y) + \psi(2x - y)$

15. hyperbolic for $|y| > \sqrt[4]{4}$, parabolic for $|y| = \sqrt[4]{4}$, elliptic for $|y| < \sqrt[4]{4}$

20. $u(x, y) = \begin{cases} f\left(x - \frac{a}{b}y\right), & \text{if } bx - ay \geq 0, \\ g\left(y - \frac{b}{a}x\right), & \text{if } bx - ay < 0. \end{cases}$

CHAPTER 6

Section 6.1

1.   a) $u(x, t) = 3 + 2H\left(t - \frac{x}{2}\right)$

2.   b) $u(x, t) = 10e^{t-x} + 10\int_0^t (1 + t - \tau + e^{t-\tau})\,\text{erfc}\left(\frac{x}{2\sqrt{\tau}}\right) d\tau$

3.   b) $u(x, t) = \begin{cases} -\frac{gt^2}{2}, & \text{if } x \geq ct, \\ \frac{g}{2c^2}(x^2 - 2cxt), & \text{if } x < ct \end{cases}$

9.   a) $u(x, t) = -\int_0^t \frac{1}{\sqrt{\pi\tau}}e^{-x^2/4\tau}\,d\tau$

Section 6.2

2. $F_c(\alpha) = \frac{2(1-\alpha^2)}{\pi(\alpha^2+1)^2}$

4.   b) $y = \frac{1}{2}(9e^{-\sqrt{3}\,x} - e^{-x})$

5. $u(x, t) = \frac{1}{2\sqrt{\pi t}}\int_0^\infty [e^{-\frac{(x-z)^2}{4t}} + e^{-\frac{(x+z)^2}{4t}}]f(z)\,dz$

Section 6.3

1.   a) $F(\alpha) = \frac{1}{\sqrt{2\pi}\,\alpha}(e^{-ib\alpha} - e^{-ia\alpha}), g(x) = \begin{cases} 1, & \text{if } a < x < b, \\ \frac{1}{2}, & \text{if } x = a \text{ or } x = b, \\ 0, & \text{otherwise} \end{cases}$

   c) $F(\alpha) = \frac{-2\sqrt{2}\,ic\alpha}{(\alpha^2+c^2)^2}, g(x) = f(x)$

e) $F(\alpha) = \begin{cases} \sqrt{\frac{\pi}{2}}, & \text{if } |\alpha| \le 1, \\ 0, & \text{if } \alpha > 1, \end{cases}$ $g(x) = \begin{cases} f(x), & \text{if } x \ne 0, \\ 1, & \text{if } x = 0 \end{cases}$

3.   a) $F(\alpha) = \frac{1}{\sqrt{2\pi}} e^{-im\alpha - \frac{\sigma^2\alpha^2}{2}}$

11.   a) $F(\alpha) = -\frac{i\alpha\sqrt{\pi}}{2\sqrt{2}} e^{-\alpha}$

   d) $F(\alpha) = \frac{1}{2\pi\sqrt{\pi}} \int_{-\infty}^{\infty} \frac{e^{-(\alpha-\beta)^2}}{\beta^2+1} d\beta$

Section 6.4

1.   a) $u(x,t) = \frac{T_1}{2} \operatorname{erfc}\left(\frac{x}{2k\sqrt{t}}\right) + \frac{T_2}{2}\left[1 + \operatorname{erf}\left(\frac{x}{2k\sqrt{t}}\right)\right]$

4.   a) $u(x,t) = \frac{e^{-t}}{2\sqrt{\pi t}} \int_{-\infty}^{\infty} e^{-\frac{(x-\xi)^2}{4t}} f(\xi)d\xi$

5. $u(x,t) = \frac{1}{2\sqrt{\pi t}} \int_{-\infty}^{\infty} e^{-\frac{(x-\xi-t)^2}{4t}} f(\xi)d\xi$

6.   a) $u(x,t) = \frac{1}{2\pi}\Big[f(x) * \int_{-\infty}^{\infty} e^{i\alpha x} \cos\alpha^2 t \, d\alpha + g(x)$

   $* \int_{-\infty}^{\infty} e^{i\alpha x} \frac{\sin\alpha^2 t}{\alpha^2} d\alpha\Big]$

8. $u(x,t) = \frac{1}{2\sqrt{\pi t}} \int_{-\infty}^{\infty} e^{-\frac{(x-\xi)^2}{4t}} f(\xi)d\xi + \frac{2x}{\sqrt{\pi}} \int_0^t \tau^{-3/2} e^{-\frac{x^2}{4\tau}} g(t-\tau)d\tau$

11.   b) $u(x,t) = \frac{1}{\sqrt{2\pi}} \int_{-\infty}^{\infty}\Big[F(\alpha) \cos\sqrt{m^2 + c^2\alpha^2}\, t$

   $+ \frac{G(\alpha)}{\sqrt{m^2+c^2\alpha^2}} \sin\sqrt{m^2 + c^2\alpha^2}\, t\Big]e^{i\alpha x}\, d\alpha$

13.   b) $u(x,y,t) = \frac{1}{4\pi t} \int_0^{\infty} \int_{-\infty}^{\infty} e^{-\frac{(x-\xi)^2}{4t}}[e^{-\frac{(y-\eta)^2}{4t}} + e^{-\frac{(y+\eta)^2}{4t}}]f(\xi,\eta)d\xi d\eta$

Section 6.5

2.   a) T c) F e) F

9.   a) $i\sqrt{2\pi}\, e^{icx} H(x)$

10.   b) $I(t) = 2H(t-3)\sinh\frac{t-3}{2}$

Section 6.6

5. $w(y,\alpha)$ satisfies the heat problem

$$w_\alpha = w_{yy}, \qquad -\infty < y < \infty, \alpha > 0,$$
$$w(y,0) = f(y)$$

6. $f(x) = \sqrt{\frac{2}{\pi}} \frac{1}{ab} \frac{b-a}{x^2+(b-a)^2}$

CHAPTER 7

Section 7.1

   4.   f) $a = \ell, b = -\frac{1}{2}$

Section 7.2

   1.   a) $x = -2$ singular point

   3.   a) $\lambda_n = n^2, n = 0,1,2,\ldots$; $T_0(x) = 1, T_1(x) = x, T_2(x) = 2x^2 - 1, T_3(x) = 4x^3 - 3x$

       b) $\lambda_n = n(n+2), n = 0,1,2,\ldots$; $S_0(x) = 1, S_1(x) = 2x, S_2(x) = 4x^2 - 1, S_3(x) = 8x^3 - 4x$

       c) $\lambda_n = 2n, n = 0,1,2,\ldots$; $H_0(x) = 1, H_1(x) = 2x, H_2(x) = 4x^2 - 2, H_3(x) = 8x^3 - 12x$

Section 7.3

   1.   b) $x = 0$ irregular

   3. $y_2 = x^{-1/2}\left[1 - x - \sum_{n=2}^{\infty} \frac{2^{n-2}}{n(n-1)(2n-3)!}x^n\right]$

   6.   a) $y_1 = x^2 \sum_{n=0}^{\infty} (n+1)x^n, y_2 = y_1 \ln x + x^2 \sum_{n=0}^{\infty} x^n$

   7.   a) $y_1 = x^\pi \sum_{k=0}^{\infty} \frac{\pi(-1)^k}{4^k k!(0+\pi)(1+\pi)\cdots(k+\pi)}x^{2k}$

       c) $y_2 = x^{-3/2}\left[1 + \frac{1}{2}x^2 - \frac{1}{8}x^4 + \cdots\right]$

Section 7.4

   2.   b) $\Gamma\left(\frac{5}{2}\right) = \frac{3\sqrt{\pi}}{4}$

Section 7.5

   1.   a) $y = c_1 J_{\sqrt{5}}(x) + c_2 J_{-\sqrt{5}}(x) = c_3 J_{\sqrt{5}}(x) + c_4 Y_{\sqrt{5}}(x)$

   7.   b) $\lambda_{0,0} = 0, y_{0,0} = 1; \lambda_{n,m} = \frac{z_{n,m}}{L}$, where $z_{n,m}$ is the mth positive zero of $J_n'(x), y_{n,m} = J_n\left(\frac{z_{n,m}}{L}x\right)$

   11.   a) $\frac{dv}{v-bv^2+c} = \frac{dx}{x}$

Section 7.6

1. $P_4(x) = \frac{1}{8}(35x^4 - 30x^2 + 3), P_5(x) = \frac{1}{8}(63x^5 - 70x^3 + 15x);$
   $T_4(x) = 8x^4 - 8x^2 + 1, T_5(x) = 16x^5 - 20x^3 + 6x;$
   $S_4(x) = 16x^4 - 12x^2 + 1, S_5(x) = 32x^5 - 32x^3 + 6x;$
   $H_4(x) = 16x^4 - 48x^2 + 12, H_5(x) = 32x^5 - 160x^3 + 120x;$
   $L_3(x) = 1 - 3x + \frac{3}{2}x^2 - \frac{1}{6}x^3, L_4(x) = 1 - 4x + 3x^2 - \frac{3}{2}x^3 + \frac{1}{24}x^4,$
   $L_5(x) = 1 - 5x + 5x^2 - \frac{5}{3}x^3 + \frac{5}{24}x^4 - \frac{1}{120}x^5$

6.   b) $m = 0, 1, 2, \ldots; n = 0, 2, 4, \ldots$

CHAPTER 8

Section 8.1

1.   a) $(e^{3x}y')' - 2e^{3x}y = 0$

     e) $(e^{-x^2}y')' + 2ne^{-x^2}y = 0$

2.   c) $(\sqrt{1-x^2}y')' + \frac{\lambda}{\sqrt{1-x^2}}y = 0$, singular

Section 8.2

1.   b) $\lambda_n = \frac{n^2\pi^2}{(\ln 2)^2}, y_n = x\sin\frac{n\pi \ln x}{\ln 2}, n = 1, 2, \ldots$

7.   a) no real eigenvalues; does not contradict Theorem 8.1 (of course!)

10.   b) $a_1 a_2 \leq 0, b_1 b_2 \geq 0$

Section 8.3

1.   a) none

2.   b) $w(x) = e^{2x}$

4.   a) no   b) yes

5. $ab = 1$

8.   a) i) yes   ii) no

9.   a) $a_0 y'' - a_1 y' + a_2 y = 0, (a_0 + a_1)y(0) - a_0 y'(0) = 0, (a_1 - a_0)y(1) - a_0 y'(1) = 0$

Section 8.4

2.   a) $\psi_0 = \frac{1}{\sqrt{L}}, \psi_n = \sqrt{\frac{2}{L}}\cos\frac{n\pi x}{L}, n = 1, 2, \ldots$

3. a) $\psi_n = \sqrt{\frac{2}{\pi}} S_n$, $n = 0, 1, 2, \ldots$, where S_n is the nth Chebyshev polynomial of the second kind

10. a) $c_1 = \frac{2}{\pi} \int_0^\pi f(x) \sin x \; dx$ (which should look familiar!)

11. a) $c_0 = \frac{1}{5}, c_1 = 0, c_2 = \frac{4}{7}$

Section 8.5

2. a) $\frac{1}{4} P_0 + \frac{1}{2} P_1 + \frac{5}{4} P_2 + 0 P_3 + \cdots$

5. $\frac{1}{3} L_0 + \frac{2}{9} L_1 + \frac{4}{27} L_2 + \frac{8}{81} L_3 + \cdots$

10. no

12. c) $\nu_n = \frac{k_n}{4\pi} \sqrt{\frac{g}{L}}, k_n = n^{\text{th}}$ positive zero of J_0

CHAPTER 9

Section 9.2

3. a) $u(x,t) = \sum\limits_{n=1}^{\infty} c_{n,0} e^{-\frac{n^2 \pi^2 t}{a^2}} \sin \frac{n\pi x}{a} + \sum\limits_{n=1}^{\infty} \sum\limits_{m=1}^{\infty} c_{n,m} e^{-\pi^2 (\frac{n^2}{a^2} + \frac{m^2}{b^2})t}$
    $\sin \frac{n\pi x}{a} \cos \frac{m\pi y}{b}$

4. a) $u(x,y,t) = \sum\limits_{n=1}^{\infty} \sum\limits_{m=1}^{\infty} \sin \frac{n\pi x}{a} \cos \frac{(2m-1)\pi y}{2b} [c_{n,m} \cos \sqrt{\lambda_{n,m}} \, t$
    $+ d_{n,m} \sin \sqrt{\lambda_{n,m}} \, t]$, where $\lambda_{n,m} = \frac{n^2 \pi^2}{a^2} + \frac{(2m-1)^2 \pi^2}{4b^2}$

5. a) $u(x,y,t) = 2 + 5e^{-2\pi^2 t} \cos \pi x \cos \pi y$

6. a) $F(x,y) = \frac{16}{\pi^2} \sum\limits_{j=1}^{\infty} \sum\limits_{k=1}^{\infty} \frac{1}{(2j-1)(2k-1)} \sin(2j-1)x \sin(2i-1)x$

9. a) $u(x,y,t) = \frac{16}{\pi^2} \sum\limits_{j=1}^{\infty} \sum\limits_{k=1}^{\infty} \frac{1}{(2j-1)(2k-1)} e^{-[(2j-1)^2 + (2k-1)^2]t}$
    $\sin(2j-1)x \sin(2k-1)y$

12. $u(x,y,t) = e^{-(\alpha x + \beta y)} \sum\limits_{n=1}^{\infty} \sum\limits_{m=1}^{\infty} c_{n,m} e^{-\lambda_{n,m} t} \sin \frac{n\pi x}{a} \sin \frac{m\pi y}{b}$,

   where $\lambda_{n,m} = \frac{n^2 \pi^2}{a^2} + \frac{m^2 \pi^2}{b^2} + \alpha^2 + \beta^2 + k$ and

   $c_{n,m} = \frac{4}{ab} \int_0^a \int_0^b f(x,y) e^{\alpha x + \beta y} \sin \frac{n\pi x}{a} \sin \frac{m\pi y}{a} \; dydx$

15. b)

$$
g_2(x,y) = \begin{cases} f(x,y), & \text{if } 0 \le x \le \pi \text{ and } 0 \le y \le \pi, \\ f(-x,y), & \text{if } -\pi \le x < 0 \text{ and } 0 \le y \le \pi, \\ -f(-x,-y), & \text{if } -\pi \le x < 0 \text{ and } -\pi \le y < 0, \\ -f(x,-y), & \text{if } 0 \le x \le \pi \text{ and } -\pi \le y < 0 \end{cases}
$$

16. a) start with $g_1(x,y) = [f(x,y) + f(-x,-y) - f(-x,y) - f(x,-y)]$

18. a) $u(x,y,t) = \frac{1}{13} \sin 2x \sin 3y(1 - e^{-13t}) + e^{-65t} \sin 4x \sin 7y$

Section 9.3

1. a) 0

2. a) $2r^4 \sin\theta(18 \sin\theta + \cos\theta)$

3. c) $u = 4$

4. $u(r,\theta) = \frac{a_0}{2} + \sum\limits_{n=1}^{\infty} \left(\frac{a}{r}\right)^n (a_n \cos n\theta + b_n \sin n\theta)$,
   where $a_n = \frac{1}{\pi} \int_{-\pi}^{\pi} f(\theta) \cos n\theta \, d\theta$, $b_n = \frac{1}{\pi} \int_{-\pi}^{\pi} f(\theta) \sin n\theta \, d\theta$

6. $u(r,\theta) = c + \sum\limits_{n=1}^{\infty} \frac{r^n}{na^{n-1}}(a_n \cos n\theta + b_n \sin n\theta)$,
   where $a_n = \frac{1}{\pi} \int_{-\pi}^{\pi} g(\theta) \cos n\theta \, d\theta$, $b_n = \frac{1}{\pi} \int_{-\pi}^{\pi} g(\theta) \sin n\theta \, d\theta$; must have
   $\int_{-\pi}^{\pi} g(\theta) = 0$

8. a) $u(r,\theta) = \sum\limits_{n=1}^{\infty} b_n \left(\frac{r}{a}\right)^{\frac{n\pi}{\alpha}} \sin \frac{n\pi\theta}{\alpha}$, where $b_n = \frac{2}{\alpha} \int_0^{\alpha} f(\theta) \sin \frac{n\pi\theta}{\alpha} \, d\theta$

17. for $0 < r < a$,

$$
\mathbf{E} = -\sum_{n=1}^{\infty} \frac{nr^{n-1}}{a^n}[(a_n \hat{\imath} + b_n \hat{\jmath}) \cos(n-1)\theta + (b_n \hat{\imath} - a_n \hat{\jmath}) \sin(n-1)\theta],
$$

   where $a_n = \frac{1}{\pi} \int_{-\pi}^{\pi} f(\theta) \cos n\theta \, d\theta$, $b_n = \frac{1}{\pi} \int_{-\pi}^{\pi} f(\theta) \sin n\theta \, d\theta$

19. b) hyperbolas $xy = c$
    h) lines $y = c(x+1)$ and $x = -1$

Section 9.4

1. b) $u(r, \theta, t) = 2 \sum\limits_{n=1}^{\infty} \frac{1}{k_n J_0(k_n)} J_0(k_n r) \sin k_n t$, where $k_n = n^{\text{th}}$ positive zero of J_0

2. a) $u(r, \theta, t) = \sum\limits_{n=1}^{\infty} c_n J_0(k_n r) e^{-\alpha^2 k_n^2 t}$, where $k_n = n^{\text{th}}$ positive zero of J_0 and $c_n = \frac{2}{J_1^2(k_n)} \int_0^1 f(r) J_0(k_n r) dr$

4. a) $u(r, \theta, t) = 5 J_4(x_{4,1}, r) \cos 4\theta \cos c x_{4,1} t - J_2(x_{2,3} r) \sin 2\theta \cos c x_{2,3} t$, where $x_{n,m} = m^{\text{th}}$ positive zero of J_n

6. a) $u(r, \theta, z) = \sum\limits_{n=1}^{\infty} c_n J_0(k_n r) \sinh k_n z$, where $k_n = n^{\text{th}}$ positive zero of J_0 and $c_n = \frac{2 \operatorname{csch} k_n L}{J_1^2(k_n)} \int_0^1 f(r) J_0(k_n r) dr$

7. a) $u(r, \theta, z) = \sum\limits_{n=0}^{\infty} \sum\limits_{m=1}^{\infty} J_n(x_{n,m} r)(a_{n,m} \cos n\theta + b_{n,m} \sin n\theta)$
    $\cdot [\tanh(x_{n,m} L) \cosh x_{n,m} z - \sinh x_{n,m} z]$,
    where $x_{n,m} = m^{\text{th}}$ positive zero of J_n and
    $a_{0,m} = \frac{\coth x_{0,m} L}{\pi J_1^2(x_{0,m})} \int_0^1 \int_{-\pi}^{\pi} r f(r, \theta) J_0(x_{n,m} r) d\theta dr$,
    $a_{n,m} = \frac{2 \coth x_{0,m} L}{\pi J_1^2(x_{n,m})} \int_0^1 \int_{-\pi}^{\pi} r f(r, \theta) J_n(x_{n,m} r) \cos n\theta \, d\theta dr$,
    $b_{n,m} = \frac{2 \coth x_{0,m} L}{\pi J_1^2(x_{n,m})} \int_0^1 \int_{-\pi}^{\pi} r f(r, \theta) J_n(x_{n,m} r) \sin n\theta \, d\theta dr$

Section 9.5

3. a) $u(\rho, \theta, \phi) = 3\rho^5 P_5(\cos \phi) - 7\rho^2 P_2(\cos \phi)$

4. a) $u(\rho, \theta, \phi) = \sum\limits_{n=0}^{\infty} c_n \rho^{-n-1} P_n \cos \phi$, where

$$c_n = \frac{2n+1}{2} \int_0^{\pi} f(\phi) P_n(\cos \phi) \sin \phi \, d\phi$$

8. $u(\rho, \theta, \phi) = \sum\limits_{k=1}^{\infty} \sum\limits_{n=0}^{\infty} \sum\limits_{m=0}^{\infty} H_m^n(\theta, \phi) j_m(y_{m,k} \rho)(c_{n,m,k} \cos k\theta + d_{n,m,k} \sin k\theta)$,
    where $y_{m,k} = k^{\text{th}}$ positive root of $J_{m+\frac{1}{2}}$

10. a) $3xz$ c) 0 f) 0, if $n = 0$; m, if $n \neq 0$

12. c) $j_2(x) = \frac{(3-x^2) \sin x - 3x \cos x}{x^3}$

Section 9.6

3. b) $u(x,y,z,t) = \frac{1}{8(\pi t)^{3/2}} \int_{-\infty}^{\infty} \int_{-\infty}^{\infty} \int_{-\infty}^{\infty} e^{-\frac{(x-\xi)^2+(y-\eta)^2+(z-\zeta)^2}{4t}}$
   $f(\xi,\eta,\zeta)d\xi d\eta d\zeta$

7. b) $\frac{-4i}{\alpha^2+\beta^2-k^2}$

Section 9.7

1. b) $\lambda_{n,m} = \pi^2\left(\frac{n^2}{a^2}+\frac{m^2}{b^2}\right)$, $u_{n,m}(x,y) = \sin\frac{n\pi x}{a}\cos\frac{m\pi y}{b}$;
   $n = 1,2,\ldots; m = 0,1,2,\ldots$

2. a) $\lambda_{n,m} = \frac{x_{n,m}^2}{a^2}$, $u_{n,m}(r,\theta) = J_n\left(\frac{x_{n,m}}{a}r\right)(c_n\cos n\theta + d_n\sin n\theta)$, $n = 0,1,2,\ldots; m = 1,2,\ldots$, where $x_{n,m} = m^{th}$ *nonnegative* zero of J_n

4. $\lambda_{m,k} = y_{m,k}^2$, $u_{m,k,n}(\rho,\theta,\phi) = H_m^n(\theta,\phi)j_m(y_{m,k}\rho)$, $k = 0,1,2,\ldots,m = 0,1,2,\ldots$, for all $n = 0,1,2,\ldots$, where $y_{m,k} = k^{th}$ positive zero of $J_{m+\frac{1}{2}}$. Multiplicity of each $\lambda_{m,k}$ is infinity.

CHAPTER 10

Section 10.1

1. a) $G(x;\xi) = \begin{cases} x, & \text{if } 0 \le x \le \xi, \\ \xi, & \text{if } \xi \le x \le 1 \end{cases}$

   c) $G_g(x;\xi) = \begin{cases} x^2/2, & \text{if } 0 \le x \le \xi, \\ \xi - x + x^2/2, & \text{if } \xi \le x \le 1 \end{cases}$

   d) $G(x;\xi) = \begin{cases} \frac{\sin kx \cos kL-\xi)}{k\cos kL}, & \text{if } 0 \le x \le \xi, \\ \frac{\sin k\xi \cos k(L-x)}{k\cos kL}, & \text{if } \xi \le x \le 1 \end{cases}$

3. b) $G(x;\xi) = \begin{cases} \frac{\pi J_n(kx)[Y_n(k)J_n(k\xi)-J_n(k)Y_n(k\xi)]}{2J_n(k)}, & \text{if } 0 \le x \le \xi, \\ \frac{\pi J_n(k\xi)[Y_n(k)J_n(kx)-J_n(k)Y_n(kx)]}{2J_n(k)}, & \text{if } \xi \le x \le 1 \end{cases}$

4. a) $G(x;\xi) = \frac{8}{\pi^2}\sum_{n=1}^{\infty}\frac{1}{(2n-1)^2}\sin\frac{(2n-1)\pi x}{2}\sin\frac{(2n-1)\pi\xi}{2}$

13. b) $G(x;\xi) = \sum_{n=1}^{\infty}\frac{\phi_n(x)\phi_n(\xi)}{\lambda_n\|\phi_n\|^2}$

Section 10.2

5. b) $y = \int_a^b G(x;\xi)f(\xi)d\xi - \beta r(b)G_\xi(x;b) + \alpha r(a)G_\xi(x;a)$

6. b)

$$G(x;\xi) = \begin{cases} \dfrac{1}{3}x^3\xi^3 - \dfrac{1}{2}x^2\xi^3 - \dfrac{1}{2}x^3\xi^2 \\[2mm] \quad + x^2\xi^2 + \dfrac{1}{6}x^3 - \dfrac{1}{2}x^2\xi, & \text{if } 0 \le x \le \xi, \\[3mm] \dfrac{1}{3}x^3\xi^3 - \dfrac{1}{2}x^3\xi^2 - \dfrac{1}{2}x^2\xi^3 \\[2mm] \quad + x^2\xi^2 + \dfrac{1}{6}\xi^3 - \dfrac{1}{2}x\xi^2, & \text{if } \xi \le x \le 1 \end{cases}$$

7. b) $F(x;\xi) = -\frac{1}{2}\sqrt{\frac{\pi}{2}}\,\operatorname{sgn}(x-\xi) * \operatorname{sgn}(x-\xi)$

11. a) $y = -\frac{x^3}{6} + \frac{x^2}{4} + c$

    b) $y = -\frac{x^3}{6} + \frac{x^2}{4}$

13. a) $z(x;\xi) = \sqrt{2\pi}[\,\operatorname{sgn}(x+\xi)\sin k(x+\xi) - \operatorname{sgn}(x-\xi)\sin k(x-\xi)]$

Section 10.3

4. a) $u(x,y) = \frac{8}{\pi^2}\sum\limits_{m=1}^{\infty} \frac{\sin my}{m^2} \int_0^\pi \int_0^\pi f(\xi,\eta)\sin m\eta\, d\xi d\eta$

    $+\frac{4}{\pi^2}\sum\limits_{m=1}^{\infty}\sum\limits_{n=1}^{\infty} \frac{\cos nx \sin my}{n^2+m^2} \int_0^\pi \int_0^\pi f(\xi,\eta)\sin n\xi \sin m\eta\, d\xi d\eta$

5. a) $G(x,y;\xi,\eta) = \frac{2}{\pi}\sum\limits_{n=1}^{\infty}\sin nx \sin n\xi\, G_n(y;\xi)$, where

$$G_n(y,\xi) = \begin{cases} \dfrac{\sinh ny \sinh n(\pi-\eta)}{\sinh n\pi}, & \text{if } 0 \le y \le \eta, \\[3mm] \dfrac{\sinh n\eta \sinh n(\pi-y)}{\sinh n\pi}, & \text{if } \eta \le y \le \pi \end{cases}$$

8. $G(r,\theta;r_0,\theta_0) = \frac{1}{4\pi}\ln\frac{r^2 r_0^2 + R^4 - 2rr_0 R^2\cos(\theta-\theta_0)}{R^2[r^2+r_0^2-2rr_0\cos(\theta-\theta_0)]}$

9. a) $G(x,y;\xi,\eta) = \frac{1}{4\pi}\ln\frac{[(x-\xi)^2+(y+\eta)^2][(x+\xi)^2+(y-\eta)^2]}{[(x-\xi)^2+(y-\eta)^2][(x+\xi)^2+(y+\eta)^2]}$

Section 10.4

2. b) $u(\boldsymbol{x}) = \frac{1}{4\pi}\left(\frac{1}{|\boldsymbol{x}-\boldsymbol{x_0}|} - \frac{R}{\rho_0|\boldsymbol{x}-\frac{R^2}{\rho_0^2}\boldsymbol{x_0}|}\right) \iint\limits_{|\boldsymbol{x_0}|=R} \frac{V(R^2-\rho_0^2)}{4\pi R|\boldsymbol{x}-\boldsymbol{x_0}|^3}\,dS_0$, where $\boldsymbol{x_0}$ is
    the location of the charge and $\rho_0 = |\boldsymbol{x_0}|$

3. a) $G(\boldsymbol{x};\boldsymbol{x_0}) = \frac{1}{4\pi}\left[\frac{1}{\sqrt{(x-\xi)^2+(y-\eta)^2+(z-\zeta)^2}} - \frac{1}{\sqrt{(x-\xi)^2+(y-\eta)^2+(z+\eta)^2}}\right]$

5. b) $\lambda = 5$ is an eigenvalue of ∇^2, $\sin x \sin 2y$ is a corresponding eigenfunction

Section 10.5

1. $G(\boldsymbol{x},t;\boldsymbol{x_0},\tau) = \frac{1}{8k^3[\pi(t-\tau)]^{3/2}}e^{-\frac{|\boldsymbol{x}-\boldsymbol{x_0}|^2}{4k^2(t-\tau)}}H(t-\tau)$

4. b) $G(x,t;\xi,\tau) = \frac{2}{\pi}H(t-\tau)\sum_{n=1}^{\infty}e^{-n^2k^2(t-\tau)}\sin nx \sin n\xi$

CHAPTER 11

Section 11.1

1. b) $y\left(\frac{1}{4}\right) = 2.36520313$, $y\left(\frac{1}{2}\right) = 1.92612264$, $y\left(\frac{3}{4}\right) = 1.63946621$, $y(1) = 1.47151776$, $y_1 = 2.25$, $y_2 = 1.75$, $y_3 = 1.4375$, $y_4 = 1.265625$

3. a) $y\left(\frac{1}{4}\right) = y_1 = 2.1875$, $y\left(\frac{1}{2}\right) = y_2 = 2.75$, $y\left(\frac{3}{4}\right) = y_3 = 3.6875$

6. a) $2y_0 - 3\frac{y_1-y_{-1}}{h} = 0$, $\frac{y_{i+1}-2y_i+y_{i-1}}{h^2}+x_i^2 y_i = x_i - 1$, $i = 0,1,\ldots,n-1$, $n+1$ equations in $n+1$ unknowns $y_{-1}, y_0, y_1, \ldots, y_{n-1}$

Section 11.2

1. a)

x_i $u_{i,2}$	$e^{-.04\pi^2}\sin(.2)i\pi$
.2 .385	.396
.4 .622	.641
.6 .622	.641
.8 .385	.396

5. a)

x_i $u_{i,2}$	$\sin\pi(x_i - .2)$
.2 .140	0.0
.4 .607	.588
.6 .953	.951
.8 .935	.951
1.0 .560	.588

8.   a) $f(x+\Delta x, y+\Delta y) \sim \sum_{n=0}^{\infty} \sum_{m=0}^{\infty} \frac{(\Delta x)^n}{n!} \frac{(\Delta y)^m}{m!} D_x^n D_y^m f(x,y)$, where $D_x^n = \frac{\partial^n}{\partial x^n}$

Section 11.3

1.   b) $y \sim \frac{a_0}{2} + \sum_{n=1}^{N} a_n \cos nx$, where

$$a_n(2n^2 - 1) = \frac{2}{\pi} \int_0^{\pi} f(x) \cos nx \, dx, \qquad n = 0, 1, \ldots, N-2,$$

$$\sum_{n=0}^{N} a_n = 0,$$

$$\sum_{n=0}^{N} (-1)^n a_n = 0$$

4.   a) $\frac{1}{4} \frac{d^2 y}{dz^2} + \frac{dy}{dz} - 3y = -f\left(\frac{z+1}{2}\right)$,
      $y(-1) = y(1) = 0$

6.   a) $v(x) = \begin{cases} -\frac{1}{2}x, & \text{if } 0 \le x \le 1, \\ \frac{1}{2}x - 1, & \text{if } 1 \le x \le 2 \end{cases}$

References

Partial Differential Equations, Boundary-Value Problems and Fourier Analysis

[1] L.C. Andrews, *Elementary Partial Differential Equations with Boundary Value Problems*, Academic Press, Orlando, 1986.

[2] N.H. Asmar, *Partial Differential Equations and Boundary Value Problems*, 2nd ed., Prentice-Hall, Upper Saddle River, New Jersey, 2004.

[3] D. Betounes, *Partial Differential Equations for Computational Science*, Springer-Verlag, New York, 1998.

[4] D. Bleecker and G. Csordas, *Basic Partial Differential Equations*, International Press, Cambridge, Massachusetts, 1996.

[5] H.S. Carslaw, *Introduction to the Theory of Fourier's Series and Integrals*, 3rd ed., Dover, New York, 1951.

[6] R.V. Churchill and J.W. Brown, *Fourier Series and Boundary Value Problems*, 6th ed., McGraw-Hill, New York, 2000.

[7] C. Constanda, *Solution Techniques for Elementary Partial Differential Equations*, 2nd ed., Chapman & Hall/CRC Press, Boca Raton, Florida, 2010.

[8] J. Cooper, *Introduction to Partial Differential Equations with MATLAB*, Birkhäuser, Boston, 1998.

[9] R. Dennemeyer, *Introduction to Partial Differential Equations and Boundary Value Problems*, McGraw-Hill, New York, 1968.

[10] P. Du Chateau and D. Zachmann, *Applied Partial Differential Equations*, Dover, New York, 2002.

[11] P. Du Chateau and D. Zachmann, *Schaum's Outline of Partial Differential Equations*, McGraw-Hill, New York, 1986.

[12] G.F.D. Duff and D. Naylor, *Differential Equations of Applied Mathematics*, Wiley, New York, 1966.

[13] L.C. Evans, *Partial Differential Equations*, 2ⁿᵈ ed., American Mathematical Society, Providence, Rhode Island, 2010.

[14] S. Farlow, *Partial Differential Equations for Scientists and Engineers*, Dover, New York, 1993.

[15] P. Garabedian, *Partial Differential Equations*, 2ⁿᵈ ed., Chelsea, New York, 1998.

[16] M.S. Gockenbach, *Partial Differential Equations: Analytical and Numerical Methods*, Society for Industrial and Applied Mathematics, Philadelphia, 2002.

[17] K.E. Gustafson, *Introduction to Partial Differential Equations*, Wiley, New York, 1980.

[18] R. Haberman, *Applied Partial Differential Equations with Fourier Series and Boundary Value Problems*, 4ᵗʰ ed., Pearson Prentice-Hall, Upper Saddle River, New Jersey, 2003.

[19] A. Jeffrey, *Applied Partial Differential Equations, an Introduction*, Elsevier Science, Academic Press, Burlington, Massachusetts, 2002.

[20] F. John, *Partial Differential Equations*, 4ᵗʰ ed., Springer-Verlag, New York, 1982.

[21] R.P. Kanwal, *Generalized Functions: Theory and Technique*, 2ⁿᵈ ed., Academic Press, New York, 1998.

[22] M.K. Keane, *A Very Applied First Course in Partial Differential Equations*, Prentice-Hall, Saddle River, New Jersey, 2001.

[23] T.W. Korner, *Fourier Analysis*, Cambridge University Press, Cambridge, England, 1989.

[24] P.K. Kythe, M.R. Schäferkotter and P. Puri, *Partial Differential Equations and Boundary Value Problems with Mathematica*, 2ⁿᵈ ed., Chapman & Hall/CRC Press, Boca Raton, Florida, 2002.

[25] T. Myint-U and L. Debnath, *Partial Differential Equations for Scientists and Engineers*, North-Holland, New York, 1987.

[26] K.A. Nguyen, *Qualitative Theory*, CM Press, Bridgeport, Connecticut, 1982.

[27] M.A. Pinsky, *Partial Differential Equations and Boundary-Value Problems with Applications*, 3ʳᵈ ed., McGraw-Hill, New York, 1998.

[28] D.L. Powers, *Boundary Value Problems*, 3ʳᵈ ed., Harcourt Brace Jovanovich, San Diego, 1987.

[29] M. Renardy and R.C. Rogers, *An Introduction to Partial Differential Equations*, 2ⁿᵈ ed., Springer-Verlag, New York, 2004.

[30] H. Sagan, *Boundary and Eigenvalue Problems in Mathematical Physics*, Dover, New York, 1989.

[31] I.N. Sneddon, *Fourier Transforms*, Dover, New York, 1995.

[32] I.N. Sneddon, *Elements of Partial Differential Equations*, Dover, New York, 2006.

[33] S.L. Sobolev, *Partial Differential Equations of Mathematical Physics*, Dover, New York, 2011.

[34] M. Spiegel, *Schaum's Outline of Fourier Analysis*, McGraw-Hill, New York, 1974.

[35] I. Stakgold, *Green's Functions and Boundary Value Problems*, 2nd ed., Wiley, New York, 1997.

[36] W.A. Strauss, *Partial Differential Equations*, 2nd ed., Wiley, New York, 2007.

[37] G.P. Tolstov, *Fourier Series*, Dover, New York, 1976.

[38] D. Vvevedensky, *Partial Differential Equations with Mathematica*, Addison-Wesley, New York, 1993.

[39] H.F. Weinberger, *A First Course in Partial Differential Equations*, Dover, New York, 1995.

[40] N. Wiener, *The Fourier Integral and Certain of Its Applications*, Dover, New York, 1958.

[41] E.C. Zachmanoglou and D.W. Thoe, *Introduction to Partial Differential Equations with Applications*, Dover, New York, 1989.

[42] A.H. Zemanian, *Distribution Theory and Transform Analysis: An Introduction to Generalized Functions, with Applications*, Dover, New York, 1987.

[43] A. Zygmund, *Trigonometric Series*, 2nd ed., Cambridge University Press, Cambridge, England, 1977.

Special Functions

[1] F. Bowman, *Introduction to Bessel Functions*, Dover, New York, 2010.

[2] H. Hochstadt, *The Functions of Mathematical Physics*, Dover, New York, 2012.

[3] N.N. Lebedev and R.A. Silverman, *Special Functions and their Application*, Dover, New York, 1972.

[4] E.D. Rainville, *Special Functions*, Macmillan, New York, 1960.

Numerical Analysis and Numerical PDEs

[1] W.F. Ames, *Numerical Methods in Partial Differential Equations*, Academic Press, Waltham, Massachusetts, 1992.

[2] R. Burden and J. Faires, *Numerical Analysis*, 9th ed., Brooks/Cole, Pacific Grove, California, 2010.

[3] P. Du Chateau and D. Zachmann, *Schaum's Outline of Partial Differential Equations*, McGraw-Hill, New York, 1986.

[4] F.B. Hildebrand, *Introduction to Numerical Analysis*, McGraw-Hill, New York, 1956.

[5] E. Isaacson and H.B. Keller, *Analysis of Numerical Methods*, Dover, New York, 1994.

[6] T. Myint-U and L. Debnath, *Partial Differential Equations for Scientists and Engineers*, North-Holland, New York, 1987.

[7] M.A. Pinsky, *Partial Differential Equations and Boundary-Value Problems with Applications*, 3rd ed., McGraw-Hill, New York, 1998.

[8] G.D. Smith, *Numerical Solution of Partial Differential Equations*, 3rd ed., Clarendon, Oxford, 1985.

[9] E. Süli and D. Mayers, *An Introduction to Numerical Analysis*, Cambridge University Press, Cambridge, England, 2003.

Mathematical Methods, etc.

[1] M.L. Boas, *Mathematical Methods in the Physical Sciences*, Wiley, New York, 1966.

[2] R.V. Churchill, *Operational Mathematics*, 3rd ed., McGraw-Hill, New York, 1972.

[3] R. Courant and D. Hilbert, *Methods of Mathematical Physics, Volumes 1 and 2*, Interscience, New York, 1962.

[4] J.W. Dettman, *Mathematical Methods in Physics and Engineering*, Dover, New York, 1988.

[5] D.G. Duffy, *Advanced Engineering Mathematics with MATLAB*, 2nd ed., Chapman & Hall/CRC Press, Boca Raton, Florida, 2003.

[6] I. Kreyszig, *Advanced Engineering Mathematics*, 7th ed., Wiley, New York, 1992.

[7] J. Mathews and R.L. Walker, *Mathematical Methods of Physics*, W.A. Benjamin, New York, 1970.

[8] P.M. Morse and H. Feshbach, *Methods of Theoretical Physics, Parts 1 and 2*, McGraw-Hill, New York, 1953.

[9] P.V. O'Neil, *Advanced Engineering Mathematics*, 2nd ed., Wadsworth, Belmont, California, 1987.

[10] G. Strang, *Introduction to Applied Mathematics*, Wellesley-Cambridge Press, Cambridge, Massachusetts, 1986.

[11] P.R. Wallace, *Mathematical Analysis of Physical Problems*, Dover, New York, 1984.

[12] R. Weinstock, *Calculus of Variations*, Dover, New York, 1974.

Physics, etc.

[1] J.R. Den Hartog, *Mechanical Vibrations*, Dover, New York, 1985.

[2] M. Elmore and M.A. Heald, *Physics of Waves*, Dover, New York, 1985.

[3] K.F. Graff, *Wave Motion in Elastic Solids*, Clarendon, Oxford, 1975.

[4] J.D. Jackson, *Classical Electrodynamics*, 3rd ed., Wiley, New York, 1998.

[5] S. Timoshenko and G.H. MacCullough, *Elements of Strength of Materials*, 3rd ed., Van Nostrand, New York, 1949.

[6] S. Timoshenko and J.N. Goodier, *Theory of Elasticity*, McGraw-Hill, New York, 1970.

[7] S. Timoshenko and S. Woinowsky–Kreiger, *Theory of Plates and Shells*, 2nd ed., McGraw-Hill, New York, 1987.

Mathematical Tables, etc.

[1] M. Abramowitz and I.A. Stegun (eds), *Handbook of Mathematical Functions*, Dover, New York, 1974.

[2] *CRC Standard Mathematical Tables and Formulae*, 31st ed., CRC Press, Boca Raton, Florida, 2002.

Journal Articles

[1] F. Cajori, The early history of partial differential equations and of partial differentiation and integration, *American Mathematical Monthly*, **35** (1928), 459–467.

[2] M. Kac, Can one hear the shape of a drum? *American Mathematical Monthly* **73** (1966), 1–23.

Calculus and Ordinary Differential Equations

[1] W.C. Boyce and R.C. DiPrima, *Elementary Differential Equations and Boundary Value Problems*, 7th ed., Wiley, New York, 2000.

[2] M. Braun, *Differential Equations and their Applications: An Introduction to Applied Mathematics*, 3rd ed., Springer-Verlag, New York, 1983.

[3] E.A. Coddington and N. Levinson, *Theory of Ordinary Differential Equations*, McGraw-Hill, New York, 1955.

[4] H.F. Davis and J.D. Snider, *Introduction to Vector Analysis*, 7th ed., WCB/McGraw-Hill, New York, 1995.

[5] E.L. Ince, *Ordinary Differential Equations*, Dover, New York, 1956.

[6] A.L. Rabenstein, *Introduction to Ordinary Differential Equations*, Academic Press, New York, 1972.

[7] E.D. Rainville, *Infinite Series*, Macmillian, New York, 1967.

[8] L.F. Shampine, I. Gladwell, and S. Thompson, *Solving ODEs with MATLAB*, Cambridge University Press, Cambridge, England, 2003.

[9] G.F. Simmons, *Differential Equations, with Applications and Historical Notes*, 2nd ed., McGraw-Hill, New York, 1991.

[10] M. Spivak, *Calculus*, 4th ed., Publish or Perish, Berkeley, California, 2008.

[11] M. Spivak, *Calculus on Manifolds: A Modern Approach to Classical Theorems of Advanced Calculus*, Westview Press, Boulder, Colorado, 1971.

[12] J. Stewart, *Calculus*, 7th ed., Brooks/Cole, Pacific Grove, California, 2011.

Real and Complex Analysis

[1] L.V. Ahlfors, *Complex Analysis*, 3rd ed., McGraw-Hill, New York, 1979.

[2] T. Apostol, *Mathematical Analysis*, 2nd ed., Addison-Wesley, New York, 1974.

[3] D. Bressoud, *A Radical Approach to Real Analysis*, 2nd ed., The Mathematical Association of America, Washington, DC, 2006.

[4] J.W. Brown and R.V. Churchill, *Complex Variables with Applications*, 7th ed., McGraw-Hill, New York, 2003.

[5] J.R. Kirkwood, *An Introduction to Analysis*, 2nd ed., PWS/Thomson, Boston, 1995.

[6] J.E. Marsden and M.J. Hoffman, *Elementary Classical Analysis*, 2nd ed., W.H. Freeman and Company, New York, 1993.

[7] W. Rudin, *Principles of Mathematical Analysis*, 3rd ed., McGraw-Hill, New York, 1976.

History of Mathematics

[1] U. Bottazzini, *The Higher Calculus: A History of Real and Complex Analysis from Euler to Weierstrass*, Springer-Verlag, New York, 1986.

[2] C.B. Boyer and U.C. Merzbach, *A History of Mathematics*, Wiley, New York, 2011.

[3] D. Bressoud, *A Radical Approach to Real Analysis*, The Mathematical Association of America, Washington, DC, 2006.

[4] F. Cajori, *History of Mathematics*, Chelsea, New York, 1980.

[5] H.W. Eves, *An Introduction to the History of Mathematics*, 5th ed., Saunders College Publishing, Philadelphia, 1983.

[6] H.H. Goldstine, *Numerical Analysis from the 16th to the 19th Century*, Springer-Verlag, New York, 1977.

[7] I. Grattan-Guinness (ed), *From the Calculus to Set Theory, 1630–1910*, Gerald Duckworth & Co., London, 1980.

[8] I. Grattan-Guinness (in collaboration with J.R. Ravetz), *Joseph Fourier 1768–1830*, MIT Press, Cambridge, Massachusetts, 1972.

[9] I. Grattan-Guinness, *The Rainbow of Mathematics*, W.W. Norton and Co., New York, 2000.

[10] J. Herivel, *Joseph Fourier: The Man and the Physicist*, Oxford University Press, Oxford, 1975.

[11] M. Kline, *Mathematical Thought from Ancient to Modern Times*, in three volumes, Oxford University Press, Oxford, 1990.

[12] D.E. Smith, *A Source Book in Mathematics*, Dover, New York, 1984.

[13] D.E. Smith, *History of Mathematics*, Dover, New York, 1958.

[14] D.J. Struik, *A Concise History of Mathematics*, 4th ed., Dover, New York, 1987.

Financial Mathematics

[1] M. Baxter and A. Rennie, *Financial Calculus*, Cambridge University Press, Cambridge, England, 1996.

[2] R. Jarrow and S. Turnbull, *Derivative Securities*, South-Western College Publishing, Cincinnati, Ohio, 1999.

MATLAB

[1] D.G. Duffy, *Advanced Engineering Mathematics with MATLAB*, 3rd ed., Chapman & Hall/CRC Press, Boca Raton, Florida, 2010.

[2] D. Etter and D. Kuncicky, *Introduction to MATLAB 6*, 2nd ed., Prentice-Hall, Upper Saddle River, New Jersey, 2003.

[3] L.F. Shampine, I. Gladwell, and S. Thompson, *Solving ODEs with MATLAB*, Cambridge University Press, Cambridge, England, 2003.

[4] T.A. Davis, MATLAB Primer, 8th ed., Chapman & Hall/CRC Press, Boca Raton, Florida, 2010.

Index

Abel's test, 589, 593
acoustics, 184, 207, 609
adjoint, 327, 334–336, 344, 346ff, 536
    boundary conditions, 349–350
advection, 136, 163 (see *convection*)
advection equation, 5, 136, 163 (see
    convection equation)
Airy's equation, 316, 332, 612
Alembert, d', see *d'Alembert*
amplitude, 141
angular frequency, 192
approximation, numerical, 539
associated homogeneous equation, 18,
    39, 469ff, 485ff, 497, 523
associated Laguerre polynomial, 282,
    439
associated Laguerre's equation, 282
associated Legendre's equation, 281–
    283, 291, 349, 429
associated Legendre function, 292, 324,
    365, 429ff
average value, 121, 137, 441, 449, 532

backward difference, 544ff
band-limited, 229, 242
banded matrix, 546, 556
bell-shaped curve (see *normal den-
    sity function*)
Bender–Schmidt explicit scheme, 554
bending moment, 605–607
Benjamin–Bona–Mahony equation, 610
Bernoulli, Daniel (1700-1782), 1, 19,
    41, 77, 103, 161, 275, 371
Bernoulli, James (1655-1705), 1, 102
Bernoulli, John (1667-1748), 1, 41,
    103

Bessel, Friedrich Wilhelm (1784-1824),
    275
Bessel function, 275, 277, 305ff, 324,
    325, 368, 371
    modified, 278, 313ff, 415, 418
    of the first kind, 271, 306ff, 317,
        416ff
        table of zeros, 312
        zeros of, 310–311, 324–325, 368,
            416ff, 431
    of the second kind, 308ff, 317,
        416, 419
    of the third kind, 313, 317, 455
        (see *Hankel* function)
    properties, 317–318
    spherical, 432, 436, 437
Bessel's equation, 275, 277, 286, 292–
    293, 297, 299, 300, 305ff, 317,
    333, 345, 349, 416, 419, 436
    modified, 278, 313ff, 415, 418
    spherical, 431–432
Bessel's inequality, 327, 360, 364
big-oh (see *O*)
biharmonic operator, 67
Black–Scholes equation, 48, 52, 611
boundary condition, 10ff, 24ff, 59ff,
    67ff, 348ff
    adjoint, 349–350
    at a singular point, 278–279, 282,
        346, 349
    at infinity, 224, 345–346
    for Dirichlet problem, 249, 409
    for heat equation, 59ff, 214ff,
        242ff
    for Hermite's equation, 279, 349
    for Laguerre's equation, 282,
        349

for wave equation, 59ff, 181, 249, 439
  Dirichlet, 24, 59ff
  Neumann, 24, 60ff
  nonhomogeneous, 65, 151ff
  periodic, 341
  Robin, 24, 61ff
  separated, 331, 340
boundary-value problem, 12, 26ff, 69, 542
Boussinesq equation, 607–608
Bratu's equation, 612
Burger's equation, 4, 15, 179, 564–565, 602, 608
  generalized (see *generalized Burger's equation*)

canonical form, 202ff (see *standard form*)
cantilever beam, 605–606
Cauchy, Augustin-Louis (1789-1857), 166, 211, 327
Cauchy data, 166
Cauchy principal part (see *Cauchy principal value*)
Cauchy principal value, 266
Cauchy problem, 166
Cauchy–Euler equation, 30, 278, 286, 310, 332, 415, 418, 428
Cauchy–Riemann equations, 70
Cauchy's inequality, 360
causal Green's function, 527
central difference, 544ff
chain (see *hanging chain*)
characteristic, 161, 163, 165ff, 452–453, 556–557
  method of characteristics, 161, 163, 213
characteristic cone, 452
characteristic coordinates, 166–167, 179, 207
characteristic curve (see *characteristic*)
Chebyshev, Pafnuti (1821-1894), 275

Chebyshev polynomial, 89, 277, 283ff, 323, 344, 349, 365, 568
  of the first kind, 89, 284, 291, 323, 344, 346, 357
  properties, 319–320
  of the second kind, 284, 291, 320, 323, 344, 346
  properties, 320–321
Chebyshev's equation
  of the first kind, 319
  of the second kind, 320, 333
Clairaut, Alexis-Claude (1713-1765), 77, 89
coefficient of diffusion, 47
compatibility condition, 150, 152–153, 494, 522–523
complementary error function, 215–216
complete, 124ff, 330, 337ff, 365ff, 430, 433, 456, 458
complete orthogonal set, 124ff, 134, 371
completeness, 124ff, 134–135, 327, 361, 365ff, 430
complex Fourier series (see *Fourier series, complex*)
computation molecule, 552ff
concentrated force (see *impulse* and *Dirac delta function*)
conditionally stable numerical scheme, 553
conformal mapping, 386
conservation law, 49, 179, 561, 602
conservation of energy, 596, 602
conservation of heat energy, 49, 244–245, 488, 632
consistent numerical scheme, 547–548
continuity, equation of (see *equation of continuity*)
continuous spectrum (see *spectrum, continuous*)
convection, 136, 163
convection equation, 4–5, 136, 163ff, 218, 567, 602, 608
convergence, 17, 389, 412, 579ff

in the mean (see *convergence, mean-square*)

mean-square, 121ff, 327, 354ff

of Fourier integrals, 222, 232, 265ff

of Fourier series, 77, 95, 98ff, 104ff, 121ff, 327, 354ff

pointwise, 98ff, 104ff, 327, 354ff, 580, 587

uniform, 98, 269, 579ff

convolution, 216–217, 228, 236ff, 252, 260, 264, 272, 446, 463, 492, 535–536

multiplicative, 272

Courant, Richard (1888-1972), 537

Courant–Friedrichs–Lewy condition, 557

Crank–Nicolson numerical scheme, 537, 554–556, 564

cutoff frequency, 229, 241

cylindrical wave, 373, 452

d'Alembert, Jean le Rond (1717-1783), 1, 19, 41, 180

d'Alembert's solution for the wave equation, 180ff, 373, 441, 452, 454, 529ff

damped wave equation (see *telegraph equation*)

del operator, 601

delta function (see *Dirac delta function*)

descent, method of (see *method of descent*)

dependence, domain of (see *domain of dependence*)

differential element, 44–45, 51–52, 54–55, 164, 375–376, 378, 385, 605–606

diffusion, 44ff, 136, 253, 375ff, 430, 459

diffusion equation, 43ff, 136, 375ff, 430ff (see *heat equation*)

diffusion-convection equation, 136, 248, 264, 377, 383

Dirac delta function, 198, 245–246, 254ff, 463, 466, 484ff

sifting property, 256, 465

three-dimensional, 455, 516–517

two-dimensional, 455, 500

Dirichlet, Peter Gustav Lejeune (1805-1859), 58, 77, 95

Dirichlet condition (see *boundary condition, Dirichlet*)

Dirichlet kernel, 105, 266–267

Dirichlet problem, 69, 398, 463, 500ff, 558, 571ff

exterior, 69, 409, 411, 434–435, 515, 523

in a cylinder, 412, 423–425

interior, 69, 403ff, 502ff

on a ball, 412, 426ff

on a disk, 403ff, 503

on a rectangle, 147ff, 456

on an annulus, 409

Dirichlet's theorem, 95

discrete spectrum (see *spectrum, discrete*)

discretization error, 540, 547

dispersion, 189, 192

dissipative wave equation (see *telegraph equation*)

distribution, 254ff, 262, 463, 484ff

distributional derivative, 262

divergence theorem, 383, 460, 585–586

domain of dependence, 186ff, 452, 529, 556–557

double Fourier series (see *Fourier series, double*)

drumhead (see *vibrating membrane*)

Duhamel, Jean Marie Constant (1797-1872), 217

Duhamel's principle, 214, 217, 219, 463, 528ff, 533ff

eigenfunction, 26ff, 72ff, 122–123, 129ff, 288ff, 310ff, 327ff, 387ff, 456ff, 476ff, 537

eigenfunction expansion (see *expansion in eigenfunctions*)
eigenvalue, 1, 25ff, 68, 72ff, 122–123, 129ff, 288ff, 310ff, 327ff, 387ff, 456ff, 476ff, 566
   nonnegative, 38, 342, 458, 460 (see *Rayleigh quotient*)
eigenvector, 329–330, 334–335, 478
eikonal equation, 4–5, 14, 24, 612
electric current, 47–50, 52–53, 57, 382, 602, 607
electric potential, 68ff, 380–381, 502, 517, 523, 608
electrical network, 263
electromagnetic field, 604
electrostatics, 67, 380, 412, 423
elliptic, 43, 207ff, 211, 250, 460, 463, 500ff, 516ff
equation of continuity, 49–50, 53, 382, 602
equipotential
   curve, 412, 414
   surface, 412
error, 540ff, 547, 566
   discretization, 540, 547
   global truncation, 547
   local truncation, 547ff, 552ff, 564–565
   mean-square, 361
   roundoff, 540–541, 547
error function, 215–216
Euler, Leonhard (1707-1783), 1, 41, 77, 89, 102, 161, 254, 275, 370, 373
Euler–Bernoulli beam, 4–5, 23, 25, 146, 192, 248, 352, 605–607
Euler–Mascheroni constant (see *Euler's constant*)
Euler–Poisson–Darboux equation, 608
Eulerian derivative, 609
Euler's constant, 303
Euler's equations, 609
Euler's formula, 94, 106, 192, 241, 283, 316
Euler's method, 541ff

improved (see *Heun's method*)
   modified, 551
even extension, 118ff, 246
even function, 82ff, 93, 116, 230
   Dirac delta function as, 260
   orthogonal polynomials as, 422
existence
   of Green's function, 472, 478
   of solutions of partial differential equations, 11–12, 177, 591ff
expansion in eigenfunctions, 121ff, 330, 337, 361ff
explicit numerical scheme, 541ff

Fick's law, 47, 377
financial mathematics, 48, 611
finite difference numerical method, 537ff
finite element numerical method, 537, 569ff
finite Fourier transform (see *Fourier transform, finite*)
first-order partial differential equations, 6ff, 161ff, 560ff
Fisher's equation, 603
Fischer, Ernst (1875-1959), 327
fluid, 50, 53, 66, 379, 382, 384–385, 602, 607–610
   incompressible, 382, 384, 609–610
   inviscid, 609
   irrotational, 382, 384
fluid dynamics, 608–610
flux, 46, 49, 59, 153, 376, 406, 454, 488, 495, 602
Fokker–Planck equation, 602
forward difference, 543ff
Fourier, Jean Baptiste Joseph (1768-1830), 41, 77, 127, 211, 327, 373
Fourier coefficient, 90ff, 105, 420, 591, 593
   cosine, 117ff
   generalized, 363ff
   sine, 117ff, 220

Fourier cosine series, 83, 117ff, 127, 130ff, 367, 370, 582
Fourier cosine transform, 220ff
Fourier integral, 211, 232ff, 265ff (see *Fourier transform*)
Fourier integral formula, 266ff
  cosine, 221
  sine, 221
Fourier series, 77ff, 220, 327, 329, 340, 354–355, 386–387
  and completeness, 121ff
  complex, 94–95, 241
  double, 389ff, 400–401, 420, 514–515
    for Green's function, 514–515
  for Green's function, 475ff, 514–515
  generalized, 91, 122, 327, 329, 361ff, 484, 539
  mean-square convergence, 121ff, 327, 354ff
  pure cosine series, 93, 116
  pure sine series, 93, 116, 400
  pointwise convergence, 98ff, 104ff
  term-by-term differentiation (see *term-by-term differentiation*)
  term-by-term integration (see *term-by term integration*)
  triple, 399
  uniform convergence, 579ff
Fourier sine series, 80, 83, 117ff, 156, 220, 229, 241, 363, 370, 389, 583, 591–593
  double, 390, 396ff, 506
  for Dirac delta function, 489–490, 496, 534
  for Green's function, 475ff, 514
Fourier sine transform, 220ff
Fourier transform, 211, 213, 230ff, 265ff, 439ff
  and Duhamel's principle, 532–533
  and fundamental solution, 491–492
  and Green's function, 525–526

finite, 220
generalized, 254, 258ff
in higher dimensions, 252–253, 445ff
inverse transform, 232ff
table of transforms and properties, 274
Fourier transform pair, 233, 445
Fourier's law, 45–46, 376–377
Fourier–Bessel
  coefficients, 318, 368
  series, 318, 368, 417, 420
  transform (see *Hankel transform*)
Fourier–Chebyshev
  coefficients, 320–321
  series, 320–321, 369
Fourier–Hermite
  coefficients, 322, 367
  series, 322, 367
Fourier–Laguerre
  coefficients, 321, 367
  series, 321, 367, 369
Fourier–Legendre
  coefficients, 319, 363, 367
  series, 319, 363, 367
Fredholm alternative theorem, 478
free-space Green's function (see *fundamental solution*)
frequency, 143ff, 270, 371, 391ff, 421, 604
  cutoff, 229, 241
  fundamental (see *fundamental frequency*)
Frobenius, method of (see *method of Frobenius*)
Fubini's theorem, 588
functional, 254–255
fundamental frequency, 143ff, 458
fundamental mode, 143
fundamental period, 81ff
fundamental solution, 243, 491ff
  for one-dimensional heat equation, 244, 525ff (see *heat kernel*)

for one-dimensional wave equation, 528ff

for three-dimensional Laplacian, 516–517

for three-dimensional wave equation, 530ff

for two-dimensional Helmholtz operator, 519ff

for two-dimensional Laplacian, 500–502

translation property of, 494, 497, 513, 525

fundamental theorem of calculus, 158, 183, 594

Galerkin, Boris Gregorievich (1871-1945), 537

Galerkin method, 566ff

Galois, Evariste (1811-1832), 327

gamma function, 300ff

Gauss's theorem (see *divergence theorem*)

Gauss–Seidel iteration, 560

Gauss–Weierstrass kernel (see *Weierstrass kernel*)

Gaussian (see *normal density function*)

general solution, 7ff, 72ff, 130ff, 388ff

generalized Burger's equation, 608

generalized Fourier series (see *Fourier series, generalized*)

generalized Green's function (see *Green's function, generalized*)

ghost point, 548ff

Gibbs, Josiah Willard, 115

Gibbs phenomenon, 103, 114ff, 121, 270

Gram–Schmidt orthogonalization, 483

gravitational potential, 127, 373, 381, 517

gravity, 55, 63, 381

Green, George (1793-1841), 463, 466

Green's first identity, 38
    for Laplacian, 456ff

for Sturm–Liouville operators, 335, 342–343, 486

Green's formula (see *Green's second identity*)

Green's function, 457, 463ff, 478ff
    causal, 527
    for heat equation, 527ff
    for ordinary differential equations, 463ff, 484ff
    for three-dimensional Laplacian, 516ff
    for two-dimensional Helmholtz operator, 519ff
    for two-dimensional Laplacian, 500ff
    for wave equation, 528ff
    Fourier series representation of, 475ff
    generalized, 480–481
    modified, 478ff, 490–491
    Neumann function, 494ff, 523
    reciprocity, 468, 472, 475, 487, 501, 506–507, 534

Green's identities (see *Green's first identity* and *Green's second identity*)

Green's second identity, 38, 536
    for Helmholtz operator, 518–519
    for Laplacian, 456ff, 501ff, 523, 525
    for Sturm–Liouville operators, 335, 348, 350, 486ff

Green's theorem, 152, 191, 200, 383–384, 457, 459, 463, 501, 513, 522, 585–586

Hamilton–Jacobi equation, 612

Hamiltonian operator, 612

hanging chain, 371, 602

Hankel function, 313, 317, 455 (see *Bessel function of the third kind*)

Hankel transform, 271

harmonic function, 67, 406ff, 432–433, 558

maximum principle for, 407, 410, 597–599

mean value property for, 406, 558

minimum principle for, 410

harmonic operator, 67 (see *Laplacian*)

harmonic oscillator, 23, 279, 604

harmonics
  solid, 430
  spherical, 373, 430–433, 437, 439
  zonal, 433

hat function (see *tent function*)

heat content, 45–47, 243–244, 375

heat equation, 61, 66, 463, 599
  one-dimensional, 4, 10–11, 13, 21, 36, 41ff, 58ff, 70ff, 127, 129ff, 204
    and Black–Scholes equation, 48, 52
    and Tychonov's example, 253
    existence of solutions, 591–593
    fundamental solution, 243, 525ff
    Green's function, 525ff
    infinite space, on, 224ff, 242ff, 253, 310, 525ff
    maximum principle for solutions, 599–600
    minimum principle for solutions, 600
    nonhomogeneous, 50–51, 151ff, 401, 475, 488–489, 565, 595
    numerical solution, 552ff
    semi-infinite space, on, 211ff, 247ff
    uniqueness of solution, 253, 595–596
  three-dimensional, 15, 67, 375–377, 382–383
    in spherical coordinates, 430ff
    infinite space, on, 454
  two-dimensional, 18, 23–24, 67, 70, 251–252, 373, 386ff, 402
    in polar coordinates, 422–423

heat kernel, 244, 264, 463, 526–527, 535

heat/diffusion equation, 601 (see *heat equation* and *diffusion equation*)

heat-exchange coefficient, 61

Heaviside, Oliver (1850-1925), 211, 217, 254, 257

Heaviside function, 218, 258ff, 526–527, 529

Helmholtz equation, 71, 207, 395, 410, 430ff, 601
  fundamental solution, 519ff
  Green's function, 455, 516, 519ff

Hermite, Charles (1822-1901), 275

Hermite polynomial, 275, 279, 290–291, 322, 365
  properties, 322

Hermite's equation, 279, 285, 291, 322, 346, 349

Hermitian adjoint matrix, 334–335

Heun's method, 550

Hilbert, David (1863-1943), 327

Hilbert transform, 271

homogeneous, 15ff

Hooke's law, 55
  for fluid, 384

Huygens, Christiaan (1629-1695), 189, 373

Huygens's principle, 190, 373, 442ff, 532

hydrogen atom, 280, 438, 603–604

hyperbolic, 43, 203ff, 250, 602

images, method of (see *method of images*)

implicit numerical scheme, 542ff

impulse, 485, 487, 527 (see *Dirac delta function*)

incompressible, 382, 384, 609–610

indicial equation, 294ff

influence, region of (see *region of influence*)

initial condition, 10–12, 59ff, 71ff, 130ff, 166ff, 181ff, 416ff

initial curve, 166ff

initial disturbance, 185, 194, 441ff

initial-boundary-value problem, 11, 60–
64, 70ff, 127ff, 138ff, 153ff,
542
initial-value problem, 11, 166, 171ff,
181
inner product, 85ff, 254, 479
complex, 94, 334–335
with respect to weight function,
355ff, 479
insulated, 11, 45, 52, 56, 59ff, 134,
136, 248
integral equation, 273
integral transforms, 213ff, 271–273
table, 272
interval of dependence, 185–186
inverse Fourier sine and cosine transforms, 220ff
inverse Fourier transform, 232ff
in higher dimensions, 445–446
inverse Laplace transform, 214–217
inverse Mellin transform, 272
inversion formula (see *the corresponding transform's inverse transform*)
inviscid, 609
irrotational, 382, 384
isotherm, 412

Jacobi iteration, 559
Jacobian, 515
jump discontinuity, 95, 112–113, 469,
472

Kadomtsev–Petviashvili equation, 611
KdV equation (see *Kortweg–de Vries
equation*)
Kelvin, Lord (see *William Thomson*)
kernel, 270–272
Kirchhoff thin plate equation, 607
Kirchhoff's formula, 449
Klein–Gordon equation, 252, 611–612
Kortweg–de Vries equation, 610

L^2, 354ff
L^2 convergence (see *convergence,
mean-square*)

L^2 norm, 354
Lacroix, Francois (1765-1843), 77
Lagrange, Joseph-Louis (1736-1813),
77, 161, 180, 373
Lagrange's identity, 335, 347, 536
Laguerre, Edmond (1834-1886), 275,
349
Laguerre polynomial, 89, 275, 277,
292ff, 321–323, 365
associated, 282, 439
properties, 321–322
Laguerre's equation, 280–282, 286, 289,
292ff, 321, 333, 346, 349
associated, 282
Lamé constants, 380, 385
Lanczos, Cornelius (1893-1974), 537
Laplace, Pierre-Simon de (1749-1827),
41–42, 77, 125, 211, 275, 373
Laplace coefficient, 275 (see *Legendre
polynomial*)
Laplace transform, 211ff, 240, 271
table of transforms and properties, 218
Laplace's equation, 4–5, 18, 23–24,
41, 43, 66ff, 127, 278, 380–
382, 558ff, 601
eigenvalues and eigenfunctions,
395, 456ff
fundamental solution
in three dimensions, 516–517
in two dimensions, 501–502
Green's function
in three dimensions, 516ff
in two dimensions, 500ff
maximum principle, 407, 410, 597–
599
minimum principle, 410, 599
numerical solution, 558ff
on a ball, 426ff
on a cylinder
finite, 423–425
infinite, 411
on a disk, 24, 402ff
on a half-plane, 249–250

on a rectangle, 24, 147ff, 207, 209

on a rectangular solid, 397–398

on a wedge, 410

on an annulus, 409

uniqueness of solution, 597–599

Laplacian, 66, 205, 377, 379

eigenvalues and eigenfunctions, 395, 456ff

fundamental solution

in three dimensions, 515–517

in two dimensions, 500–502

Green's function

in three dimensions, 516ff

in two dimensions, 500ff

Green's identities for, 457

first, 573

second, 504

in polar coordinates, 24, 402–403

in spherical coordinates, 278, 412, 426, 433–434

invariance

rotational, 71

translational, 71

law of cosines, 406–407, 503, 510

Lax equivalence theorem, 548

Lax–Wendroff numerical method, 561–562

Lebesgue's dominated convergence theorem, 587

Legendre, Adrien-Marie (1752-1833), 275, 373

Legendre duplication formula, 304

Legendre function, 292

associated, 292, 323, 365, 429ff

Legendre polynomial, 88, 275, 277, 285, 288ff, 318–319, 323, 349, 357, 361, 363, 365, 427, 568, 575

properties, 318–319

Legendre's equation, 278–279, 281, 284–285, 287ff, 318, 332, 346, 349, 351, 427

associated, 281–283, 292, 349, 429

Leibniz, Gottfried Wilhelm, 103

Leibniz's rule, 588

light cone, 453

linear, 1, 12ff

linear combination, 16ff

linear operator, 13

linearized equations of acoustics, 609

Liouville, Joseph (1809-1882), 327

Liouville normal form, 337

Liouville's theorem, 410

local truncation error (see *error, local truncation*)

logarithmic potential, 70, 456, 502–503, 509, 511

longitudinal vibration, 55–56, 607

magnetostatics, 381

mass density, 46

linear, 55

per unit area, 383

maximum principle

for harmonic functions, 407, 410, 597–599

for heat functions, 599–600

Maxwell, James Clerk (1831-1879), 68

Maxwell's equations, 68, 380–381, 386, 604–605

Maxwell's reciprocity (see *Green's function, reciprocity*)

mean value property for harmonic functions, 406, 558

mean-square, 364–365, 594

error (see *error, mean-square*)

norm, 356

mean-square convergence (see *convergence, mean-square*)

Mellin transform, 271–272

method of characteristics, 161, 163, 213 (see *characteristics*)

method of descent, 446–447, 450ff

method of electrostatic images, 518 (see *method of images*)

method of Frobenius, 292ff, 305, 313, 415

method of images, 195, 200, 226, 247ff, 499, 504, 507ff, 518, 532, 534
minimal surface equation, 612
minimum principle
    for harmonic functions, 410, 599
    for heat functions, 600
mode of vibration, 371, 391–394
    circular membrane, 417, 421ff
    hanging chain, 371
    rectangular membrane, 391ff
    string, 139ff
modified Bessel function, 278, 313–315, 415, 418
modified Bessel's equation, 278, 313ff, 415, 418
modified Green's function, 478ff, 490–491, 494
Monge, Gaspard (1746-1818), 77, 161
multiplicity of eigenvalue, 329, 339–341, 394–395, 478
    for Laplacian, 456, 459
music, 41, 143–144
    Bach family, 102

Navier–Stokes equations, 610
Neumann, Carl Gottfried (1832-1925), 60
Neumann condition (see *boundary condition, Neumann*)
Neumann function, 494ff, 523
Neumann problem, 69, 494–495
    and Green's function in two dimensions, 522ff
    exterior, 69
    interior, 69, 405–406, 409–410
    on a disk, 405ff
    on a rectangle, 148ff, 514
    on a wedge, 410
    Poisson's integral formula for, 411
Newton's law of cooling, 61
Newton's second law of motion, 54, 378, 385
Newtonian gravity, 381
nodal curve (see *nodal line*)
nodal line, 392ff, 417, 422

node, 142–143, 392
nonhomogeneous
    boundary condition, 153ff
        for Laplace's equation, 147ff
    ordinary differential equation, 156
        Green's function for, 165ff
    partial differential equation, 15–16, 155ff, 463
        Green's function for, 500ff
nonlinear partial differential equation, 13ff
    and shocks, 180
    Burger's, 180
    heat (see *Fisher's equation*)
    Schrödinger's, 604, 611
normal density function, 215, 233–234, 238
normal mode of vibration, 141ff

O, 214, 543
odd extension, 118ff, 246
odd function, 82ff, 93, 116, 230
    orthogonal polynomials as, 322
operator, 12ff, 486
    adjoint, 344, 346, 536
    del, 601
    Hamiltonian, 612
    Laplace (see *Laplacian*)
    linear, 13ff
    self-adjoint, 344ff, 452
    Sturm–Liouville, 331ff, 486
order, 4ff
orthogonal, 37, 80, 84ff, 94, 122–124, 134–136, 156, 329, 334–335, 351, 389, 397, 434–435, 437–438,
    and Gram–Schmidt orthogonalization, 483
    pairwise (see *pairwise orthogonal*)
    simply (see *simply orthogonal*)
    with respect to weight function, 338ff, 355ff
orthogonal polynomials, 88, 275ff
    and spectral methods, 567–568

properties, 317ff
orthogonality
  of associated Legendre functions, 434–435
  of eigenfunctions of the Laplacian, 456ff
  of spherical Bessel functions, 437–438
orthonormal, 356ff
Ostrogradsky's theorem (see *divergence theorem*)
overshoot, 111–113 (see *Gibbs phenomenon*)
overtone, 144, 146

pairwise orthogonal, 85, 94, 339
parabolic, 43, 205ff, 250, 602
Parseval, Marc-Antoine (d. 1836), 327
Parseval's equality, 365–366
  for Fourier transform, 229, 241
  for generalized Fourier series, 327, 361ff
  for trigonometric Fourier series, 366
partial differential equation, 3(ff!)
partial fraction expansion, 224
past history, 191, 529 (see *domain of dependence*)
perfect fluid, 382
period, 36, 81ff, 99ff, 118
  fundamental, 81ff
periodic, 36, 81ff, 100ff, 118, 417
  Sturm–Liouville problem, 337, 340ff, 348–349, 365, 405
periodic extension, 100ff, 146, 582–584, 593
piecewise continuous, 96ff, 108ff, 214, 582–584, 591
piecewise smooth, 96ff, 108ff, 220, 222, 230, 232, 262–263, 265, 269, 330, 337, 355, 456, 570, 582, 584, 591, 593
plane wave, 452, 454
plucked string, 64, 142

Poisson, Siméon-Denis (1781-1840), 70, 127, 211, 214, 373, 449, 463
Poisson kernel, 250, 406, 410, 516
Poisson's equation, 15, 70, 127, 157, 158, 373, 380–381, 455, 575, 601
  and Green's function
    in three dimensions, 516ff
    in two dimensions, 500ff
  uniqueness of solution, 597–599
Poisson's integral formula, 373, 402ff, 503, 524
  for exterior Dirichlet problem, 402ff, 411
  for interior Neumann problem, 411
  for upper half-plane, 250
potential
  electric (see *electric potential*)
  gravitational (see *gravitational potential*)
  logarithmic (see *logarithmic potential*)
  retarded (see *retarded potential*)
potential drop, 52
potential energy, 23, 135, 280, 596
potential equation, 41, 43, 66ff, 380 (see *Laplace's equation*)
potential function, 23, 66ff, 135, 280, 426
potential theory, 66
power series, 285ff
probability integral (see *error function*)
product solution, 1, 19ff, 70ff, 373
  for heat equation
    in one dimension, 21, 36, 130ff
    in rectangular coordinates, 24, 388
    in spherical coordinates, 431
  for Laplace's equation
    in polar coordinates, 24, 404
    in rectangular coordinates, 24, 147ff

in spherical coordinates, 428–429
for wave equation
in one dimension, 24, 36, 138ff
in polar coordinates, 416, 419
in rectangular coordinates, 24, 391
pure cosine series, 93, 116
pure sine series, 93, 116, 400

quantum mechanics, 4, 135, 207, 276, 279, 280, 438, 603–604
quasi-linear equation, 179

radial vibrations of a ball, 425
radius of convergence, 286
Rameau, Jean-Philippe (1683-1764), 41
Rayleigh quotient, 38, 458
for Laplace operator, 460
for Sturm–Liouville operator, 342
reciprocity (see *Green's function, reciprocity*)
reduced wave equation (see *Helmholtz equation*)
reflection, 193ff
region of influence, 187, 190, 192–193, 452–453
relativity, special theory of, 453
retarded potential, 532
Riemann, Georg Friedrich Bernhard (1826-1866), 327
Riemann function, 536
Riemann sum, 85, 113, 221
Riemann–Lebesgue Lemma, 108ff, 268, 270
Robin condition (see *boundary condition, Robin*)
roundoff error, 540–541, 547, 552
Runge–Kutta numerical methods, 551

sampling, 229–230, 242
sampling theorem, 230, 242
Schrödinger's equation, 4, 23, 275, 280, 603–604, 610, 611

for harmonic oscillator, 279, 604
for hydrogen atom, 604
for "particle in a box," 135
nonlinear, 604, 611
Schwarz inequality, 359
Schwarz integral formula, 250 (see *Poisson's integral formula for the upper half-plane*)
Schwartz, Laurent (1915-2002), 255
self-adjoint, 327
eigenvalue problem, 345ff
form, 323
general second-order linear operator, 535–536
Laplace's eigenvalue problem as, 458–459
linear operator, 344ff, 452
matrix, 334–335
separable, 19ff
ordinary differential equation, 8, 316
separation of variables, 1, 19ff, 71ff, 275, 277ff
and Sturm–Liouville problems, 330–331
for heat equation
in one dimension, 21, 129ff
in rectangular coordinates, 23, 70, 387
in spherical coordinates, 430
for Helmholtz equation, 395–396
for Laplace's equation
in polar coordinates, 404
in rectangular coordinates, 23, 147ff
in spherical coordinates, 426ff
for wave equation
in one dimension, 23, 138ff
in polar coordinates, 415ff
in rectangular coordinates, 23, 70, 391
sharp leading/trailing edge (see *Huygens's principle*)
shear force, 605–607
shear modulus, 57

shock wave, 180
side condition, 9–11
sifting property of Dirac delta function, 256, 465
sign function (see *signum function*)
signum function, 260
simply orthogonal, 89, 339ff, 357–358, 363, 565
    eigenfunctions of Laplace operator as, 456ff
    spherical harmonics as, 437–438
simply orthonormal, 357–358
sine-Gordon equation, 611
singular point, 278, 285ff
    irregular, 292–293
    regular, 292–293
sink, 50, 375
solitary wave (see *soliton*)
soliton, 604, 610
sound wave, 379, 384, 444, 609–610
source, 600–601, 603
    and Duhamel's principle, 532–533
    concentrated, 485, 488 (see *Dirac delta function*)
    for convection problems, 173
    heat, 50–51, 375, 402, 488, 495
    line, 536
    voltage, 263
source function, 535
sparse matrix, 546, 556, 572
special functions, 275ff
    properties, 317ff
specific heat, 46, 50–51, 62, 375, 382, 602
spectral method, 537, 565ff
spectrum, 143, 146
    continuous, 270
    discrete, 269
spherical Bessel function, 432ff
spherical Bessel's equation, 431–432
spherical coordinates, 426
    heat equation in, 430ff
    Laplacian in, 278–280, 373, 412, 425ff

wave equation in, 282, 425, 454, 531
spherical harmonic (see *harmonics, spherical*)
spherical wave, 373, 452
square wave, 183ff, 441ff
    and Fourier transform, 232, 262, 270
stable numerical scheme, 548
standard form (see *canonical form*)
steady state, 64, 67, 75, 136–137, 158
stiffness matrix, 572
Stokes's theorem, 384, 585–586
Sturm, Charles (1803-1852), 327
Sturm comparison theorem, 37, 311
Sturm–Liouville form, 331–333
Sturm–Liouville operator, 331
    Green's identities for, 335
    in higher dimensions, 461
    self-adjointness of, 345, 461
Sturm–Liouville problem, 327ff
    in higher dimensions, 461
    periodic, 340ff, 348–349, 365, 405
    regular, 331ff, 337ff, 348–349, 365
    and Green's function, 469ff
    singular, 331ff, 345ff
successive overrelaxation iteration, 560
superposition, 16ff
    of wave solutions, 184ff
symmetric (see *Green's function, reciprocity*)

tau method of Lanczos, 568–569
Taylor series, 50, 54, 285, 540, 542ff
    double, 564
Tychonov's example, 253
telegraph equation, 48, 53, 56, 58, 145, 202, 251, 602, 608
telephone equation, 608 (see *telegraph equation*)
temperature function, 10, 44ff, 375ff
temperature gradient, 46, 376
tension, 55, 58, 62, 64, 378, 385
tent function, 570ff
term-by-term

differentiation, 17, 156, 253, 583
integration, 80, 90, 98, 316, 583–584
test function, 255ff
thermal conductivity, 46, 50–52, 62, 375, 382
thermal diffusivity, 47, 51, 62, 242, 377, 430
Thomson, William (1824-1907), 463
torsional vibration, 56
translation property for fundamental solution, 494, 497, 514, 525
transmission line equations, 607–608
transport, 4, 163, 602, 608 (see *convection*)
triangulation, 572–573
Tricomi equation, 207, 608
truncation, 545, 567 (see *error, global truncation* and *error, local truncation*)

uniform convergence, 99, 269, 579ff, 587ff
uninsulated rod, 248
uniqueness
    and Fredholm alternative theorem, 479, 484
    of Fourier series, 91
    of Green's function, 472, 483, 485
    of solution, 11–12, 595ff
        and characteristics, 171, 177
        heat equation, 595–596, 600
        Poisson's equation, 597–599
        wave equation, 596–597
variation of parameters, 466ff
vibrating ball, 282, 425, 435
vibrating beam (see *Euler–Bernoulli beam*)
vibrating chain (see *hanging chain*)
vibrating drumhead (see *vibrating membrane*)
vibrating membrane, 277–278, 373, 378ff
    circular, 275, 373, 414ff
    rectangular, 387ff

vibrating plate, 605, 607
vibrating string, 11, 41, 54ff, 61ff, 77, 138ff, 161, 417, 605
vibration
    longitudinal, 56–57, 607
    torsional, 57
vibration spectrum
    for Euler–Bernoulli beam, 145
    for violin string, 141
violin string, 141–142

wave
    cylindrical, 373, 452
    plane, 452, 454
    spherical, 373, 452
wave equation, 1, 4–5, 18, 23–24, 36, 41, 43, 66, 127, 601, 605
    damped, 59 (see *telegraph equation*)
    dissipative, 59 (see *telegraph equation*)
    nonhomogeneous, 56, 153ff, 161, 191, 596–597
    one-dimensional, 23–24, 36, 54ff, 61ff, 74, 127, 138ff, 452, 454
        existence of solutions, 593–595
        fundamental solution, 528ff
        Green's function, 528ff
        infinite space, on, 180ff, 439ff, 528ff
        numerical solution, 556ff
        semi-infinite, 192ff, 203
        uniqueness of solution, 596–597
    reduced (see *Helmholtz equation*)
    three-dimensional, 282, 373, 379–380
        fundamental solution, 530ff
        Green's function, 530ff
        in spherical coordinates, 282, 425, 454, 531
        infinite space, on, 447ff, 530ff
    two-dimensional, 23–24, 70, 373, 378–379, 383, 386, 390ff
        fundamental solution, 536

in polar coordinates, 414ff
infinite space, on, 450ff, 536
wave function, 135, 280, 438, 603
wave number, 146, 192
wave speed, 56, 63, 172, 189, 379, 449, 556
and dispersion, 189
weak formulation, 570ff
Weber function, 308, 317 (see *Bessel's function of the second kind*)
Weierstrass M-test, 588, 592
Weierstrass kernel, 272
Weierstrass transform, 271–272
weight function, 89, 339ff, 461
well-posed, 11–12, 41, 59, 72
Dirichlet problem on a disk, 405
infinite heat problem not, 253
Neumann problem not, 148ff
Wiener–Hopf equation, 273
Wilbraham, 112
Wilbraham–Gibbs phenomenon, 112 (see *Gibbs phenomenon*)

Yang–Mills equations, 612
Young's modulus, 57

zeros of Bessel functions (see *Bessel function of the first kind, zeros of*)
zonal harmonics, 433